Mathematical Tools for Physics

JAMES NEARING
Physics Department
University of Miami

DOVER PUBLICATIONS, INC.
Mineola, New York

Copyright

Bibliographical Note

This Dover edition, first published in 2010, is the first publication in book form of the electronic version of the work, originally available on the following website: www.physics.miami.edu/nearing/mathmethods.

Library of Congress Cataloging-in-Publication Data

Nearing, James.
Mathematical tools for physics / James Nearing. — Dover ed.
p. cm.
"... is the first publication in book form of the electronic version of the work [copyrighted c2003], originally available on the following website: www.physics.miami.edu/nearing.mathmethods"—
Includes bibliographical references and index.
ISBN-13: 978-0-486-48212-5
ISBN-10: 0-486-48212-X
1. Mathematical physics—Textbooks. I. Title.

QC20.N43 2010
510.2453—dc22

2010010191

Manufactured in the United States by LSC Communications
48212X008 2021
www.doverpublications.com

Preface to the Dover Edition

I wrote this text for a one semester course for physics majors at the sophomore-junior level. Our experience with students taking junior-level courses is that even if they've had the mathematical prerequisites, they usually need more experience to handle the mathematics efficiently and to possess usable intuition about the processes involved. If you've seen infinite series in a calculus course, you may have no idea that they're good for anything. If you've taken a differential equations course, which of the scores of techniques that you've seen are really used a lot? The world is (at least) three dimensional so you clearly need to understand multiple integrals, but will everything be rectangular?

Some of the material is here by student request. When one person asked for help understanding solid angle and scattering in order to analyze some of her experimental data, that became part of chapter 8. When another was doing a Monte Carlo calculation that required him to differentiate the output, that required differentiating noisy data as in chapter 11. When I first learned about matrices, the subject appeared as an unmotivated mess of numbers with arbitrary rules for manipulation. Eventually I learned that matrix theory follows from one equation defining the components of an operator, and that matrix multiplication is nothing but composition of functions as in chapter 7.

This text was originally constructed with hyperlinks, to be read on the screen. If there's a reference to an equation or problem, click on it and it takes you there, and bookmarks do the same. In practice, most students simply print the text to read it as a standard book, and the paperless society remains a fiction. This Dover edition addresses that need by providing a full, bound edition ready to go. The convenience of a bound text still overrides the other advantages of e-books.

Dover Publishers have agreed that the electronic version of this text will remain online at www.physics.miami.edu/nearing/mathmethods/ in order to provide a linked and searchable rendering of the material. The same site will also show errata to the printed text as they are (inevitably) discovered.

James Nearing
University of Miami
Physics Department
2009

Contents

Introduction

How do you learn intuition?

When you've finished a problem and your answer agrees with the back of the book or with your friends or even a teacher, you're not done. The way do get an intuitive understanding of the mathematics and of the physics is to analyze your solution thoroughly. Does it make sense? There are almost always several parameters that enter the problem, so what happens to your solution when you push these parameters to their limits? In a mechanics problem, what if one mass is much larger than another? Does your solution do the right thing? In electromagnetism, if you make a couple of parameters equal to each other does it reduce everything to a simple, special case? When you're doing a surface integral should the answer be positive or negative and does your answer agree?

When you address these questions to every problem you ever solve, you do several things. First, you'll find your own mistakes before someone else does. Second, you acquire an intuition about how the equations ought to behave and how the world that they describe ought to behave. Third, It makes all your later efforts easier because you will then have some clue about why the equations work the way they do. It reifies the algebra.

Does it take extra time? Of course. It will however be some of the most valuable extra time you can spend.

Is it only the students in my classes, or is it a widespread phenomenon that no one is willing to sketch a graph? ("Pulling teeth" is the cliché that comes to mind.) Maybe you've never been taught that there are a few basic methods that work, so look at section 1.8. *And keep referring to it.* This is one of those basic tools that is far more important than you've ever been told. It is astounding how many problems become simpler after you've sketched a graph. Also, until you've sketched some graphs of functions you really don't know how they behave.

When I taught this course I didn't do everything that I'm presenting here. The two chapters, Numerical Analysis and Tensors, were not in my one semester course, and I didn't cover all of the topics along the way. Several more chapters were added after the class was over, so this is now far beyond a one semester text. There is enough here to select from if this is a course text, but if you are reading it on your own then you can move through it as you please, though you will find that the first five chapters are used more in the later parts than are chapters six and seven. Chapters 8, 9, and 13 form a sort of package. I've tried to use examples that are not all repetitions of the ones in traditional physics texts but that do provide practice in the same tools that you need in that context.

For polar and cylindrical coordinate systems it is common to use theta for the polar angle in one and phi for the polar angle in the other. Here I chose to try for consistency and in all three systems (plane polar, cylindrical, spherical) this angle is $\phi = \tan^{-1}(y/x)$. In line integrals it is common to use ds for an element of length, and many authors will use dS for an element of area. I have tried to avoid this confusion by sticking to $d\ell$ and dA respectively (with rare exceptions).

In many of the chapters there are "exercises" that precede the "problems." These are supposed to be simpler and mostly designed to establish some of the definitions that appeared in the text.

I'd like to thank the students who found some, but probably not all, of the mistakes in the text. Also Howard Gordon, who used it in his course and provided me with many suggestions for improvements. Prof. Joseph Tenn of Sonoma State University has given me many very helpful ideas, correcting mistakes, improving notation, and suggesting ways to help the students.

Bibliography

Mathematical Methods for Physics and Engineering by Riley, Hobson, and Bence. Cambridge University Press For the quantity of well-written material here, it is surprisingly inexpensive in paperback.

Mathematical Methods in the Physical Sciences by Boas. John Wiley Publ About the right level and with a very useful selection of topics. If you know everything in here, you'll find all your upper level courses much easier.

Mathematical Methods for Physicists by Arfken and Weber. Academic Press At a more advanced level, but it is sufficiently thorough that will be a valuable reference work later.

Mathematical Methods in Physics by Mathews and Walker. More sophisticated in its approach to the subject, but it has some beautiful insights. It's considered a standard, though now hard to obtain.

Mathematical Methods by Hassani. Springer At the same level as this text with many of the same topics, but said differently. It is always useful to get a second viewpoint because it's commonly the second one that makes sense — in whichever order you read them.

Schaum's Outlines by various. There are many good and inexpensive books in this series: for example, "Complex Variables," "Advanced Calculus," "German Grammar." Amazon lists hundreds.

Visual Complex Analysis by Needham, Oxford University Press The title tells you the emphasis. Here the geometry is paramount, but the traditional material is present too. It's actually fun to read. (Well, I think so anyway.) The Schaum text provides a complementary image of the subject.

Complex Analysis for Mathematics and Engineering by Mathews and Howell. Jones and Bartlett Press Another very good choice for a text on complex variables. Despite the title, mathematicians should find nothing wanting here.

Applied Analysis by Lanczos. Dover Publications This publisher has a large selection of moderately priced, high quality books. More discursive than most books on numerical analysis, and shows great insight into the subject.

Linear Differential Operators by Lanczos. Dover publications As always with this author, useful insights and unusual ways to look at the subject.

Numerical Methods that (usually) **Work** by Acton. Mathematical Association of America Practical tools with more than the usual discussion of what can (and will) go wrong.

Numerical Recipes by Press et al. Cambridge Press The standard current compendium surveying techniques and theory, with programs in one or another language.

A Brief on Tensor Analysis by James Simmonds. Springer This is the only text on tensors that I will recommend. To anyone. Under any circumstances.

Linear Algebra Done Right by Axler. Springer   Don't let the title turn you away. It's pretty good.

Advanced mathematical methods for scientists and engineers by Bender and Orszag. Springer   Material you won't find anywhere else, with clear examples. "... a sleazy approximation that provides good physical insight into what's going on in some system is far more useful than an unintelligible exact result."

Probability Theory: A Concise Course by Rozanov. Dover   Starts at the beginning and goes a long way in 148 pages. Clear and explicit and cheap.

Calculus of Variations by MacCluer. Pearson   Both clear and rigorous, showing how many different types of problems come under this rubric, even "... operations research, a field begun by mathematicians, almost immediately abandoned to other disciplines once the field was determined to be useful and profitable."

Special Functions and Their Applications by Lebedev. Dover   The most important of the special functions developed in order to be useful, not just for sport.

The Penguin Dictionary of Curious and Interesting Geometry by Wells. Penguin   Just for fun. If your heart beats faster at the sight of the Pythagorean Theorem, wait 'til you've seen Morley's Theorem, or Napoleon's, or when you first encounter an unduloid in its native habitat.

Basic Stuff

1.1 Trigonometry

The common trigonometric functions are familiar to you, but do you know some of the tricks to remember (or to derive quickly) the common identities among them? Given the sine of an angle, what is its tangent? Given its tangent, what is its cosine? All of these simple but occasionally useful relations can be derived in about two seconds if you understand the idea behind one picture. Suppose for example that you know the tangent of θ, what is $\sin\theta$? Draw a right triangle and designate the tangent of θ as x, so you can draw a triangle with $\tan\theta = x/1$.

The Pythagorean theorem says that the third side is $\sqrt{1+x^2}$. You now read the sine from the triangle as $x/\sqrt{1+x^2}$, so

$$\sin\theta = \frac{\tan\theta}{\sqrt{1+\tan^2\theta}}$$

Any other such relation is done the same way. You know the cosine, so what's the cotangent? Draw a different triangle where the cosine is $x/1$.

Radians

When you take the sine or cosine of an angle, what units do you use? Degrees? Radians? Cycles? And who invented radians? Why is this the unit you see so often in calculus texts? That there are 360° in a circle is something that you can blame on the Sumerians, but where did this other unit come from?

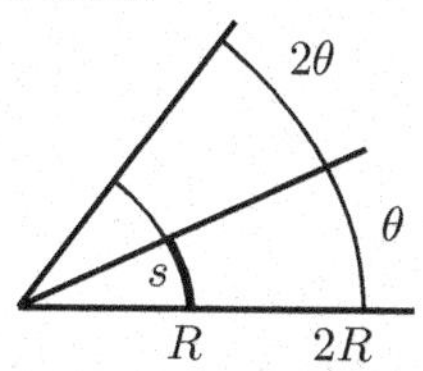

It results from one figure and the relation between the radius of the circle, the angle drawn, and the length of the arc shown. If you remember the equation $s = R\theta$, does that mean that for a full circle $\theta = 360°$ so $s = 360R$? No. For some reason this equation is valid only in radians. The reasoning comes down to a couple of observations. You can see from the drawing that s is proportional to θ — double θ and you double s. The same observation holds about the relation between s and R, a direct proportionality. Put these together in a single equation and you can conclude that

$$s = CR\theta$$

where C is some constant of proportionality. Now what is C?

You know that the whole circumference of the circle is $2\pi R$, so if $\theta = 360°$, then

$$2\pi R = CR\,360°, \qquad \text{and} \qquad C = \frac{\pi}{180}\,\text{degree}^{-1}$$

It has to have these units so that the left side, s, comes out as a length when the degree units cancel. This is an awkward equation to work with, and it becomes *very* awkward when you try to do calculus. An increment of one in$\Delta\ \theta$ is big if you're in radians, and small if you're in degrees, so it should be no surprise that $\Delta \sin\theta/\Delta\theta$ is much smaller in the latter units:

$$\frac{d}{d\theta}\sin\theta = \frac{\pi}{180}\cos\theta \qquad \text{in degrees}$$

This is the reason that the radian was invented. The radian is the unit designed so that the proportionality constant is one.

$$C = 1\,\text{radian}^{-1} \qquad \text{then} \qquad s = \left(1\,\text{radian}^{-1}\right)R\theta$$

In practice, no one ever writes it this way. It's the custom simply to omit the C and to say that $s = R\theta$ with θ restricted to radians — it saves a lot of writing. How big is a radian? A full circle has circumference $2\pi R$, and this equals $R\theta$ when you've taken C to be one. It says that the angle for a full circle has 2π radians. One radian is then $360/2\pi$ degrees, a bit under $60°$. Why do you always use radians in calculus? Only in this unit do you get simple relations for derivatives and integrals of the trigonometric functions.

Hyperbolic Functions

The circular trigonometric functions, the sines, cosines, tangents, and their reciprocals are familiar, but their hyperbolic counterparts are probably less so. They are related to the exponential function as

$$\cosh x = \frac{e^x + e^{-x}}{2}, \qquad \sinh x = \frac{e^x - e^{-x}}{2}, \qquad \tanh x = \frac{\sinh x}{\cosh x} = \frac{e^x - e^{-x}}{e^x + e^{-x}} \tag{1.1}$$

The other three functions are

$$\operatorname{sech} x = \frac{1}{\cosh x}, \qquad \operatorname{csch} x = \frac{1}{\sinh x}, \qquad \coth x = \frac{1}{\tanh x}$$

Drawing these is left to problem 1.4, with a stopover in section 1.8 of this chapter.

Just as with the circular functions there are a bunch of identities relating these functions. For the analog of $\cos^2\theta + \sin^2\theta = 1$ you have

$$\cosh^2\theta - \sinh^2\theta = 1 \tag{1.2}$$

For a proof, simply substitute the definitions of cosh and sinh in terms of exponentials and watch the terms cancel. (See problem 4.23 for a different approach to these functions.) Similarly the other common trig identities have their counterpart here.

$$1 + \tan^2\theta = \sec^2\theta \qquad \text{has the analog} \qquad 1 - \tanh^2\theta = \operatorname{sech}^2\theta \tag{1.3}$$

The reason for this close parallel lies in the complex plane, because $\cos(ix) = \cosh x$ and $\sin(ix) = i\sinh x$. See chapter three.

The inverse hyperbolic functions are easier to evaluate than are the corresponding circular functions. I'll solve for the inverse hyperbolic sine as an example

$$y = \sinh x \qquad \text{means} \qquad x = \sinh^{-1} y, \qquad\qquad y = \frac{e^x - e^{-x}}{2}, \quad \text{solve for } x.$$

Multiply by $2e^x$ to get the quadratic equation

$$2e^x y = e^{2x} - 1 \qquad \text{or} \qquad \left(e^x\right)^2 - 2y\left(e^x\right) - 1 = 0$$

The solutions to this are $e^x = y \pm \sqrt{y^2 + 1}$, and because $\sqrt{y^2 + 1}$ is always greater than $|y|$, you must take the positive sign to get a positive e^x. Take the logarithm of e^x and

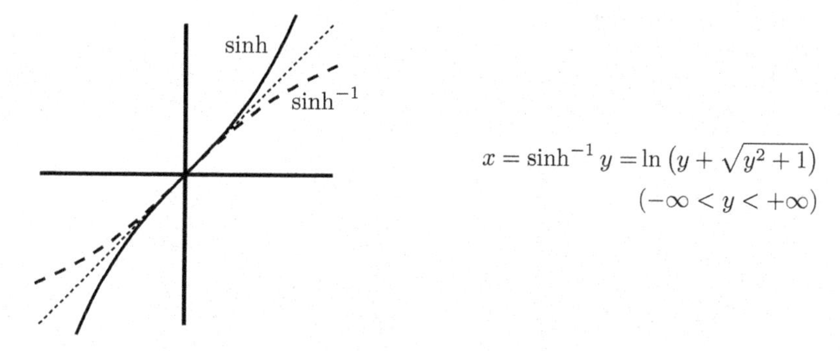

$$x = \sinh^{-1} y = \ln\left(y + \sqrt{y^2 + 1}\right) \qquad (-\infty < y < +\infty)$$

As x goes through the values $-\infty$ to $+\infty$, the values that $\sinh x$ takes on go over the range $-\infty$ to $+\infty$. This implies that the domain of $\sinh^{-1} y$ is $-\infty < y < +\infty$. The graph of an inverse function is the mirror image of the original function in the 45° line $y = x$, so if you have sketched the graphs of the original functions, the corresponding inverse functions are just the reflections in this diagonal line.

The other inverse functions are found similarly; see problem 1.3

$$\begin{aligned} \sinh^{-1} y &= \ln\left(y + \sqrt{y^2 + 1}\right) \\ \cosh^{-1} y &= \ln\left(y \pm \sqrt{y^2 - 1}\right), && y \geq 1 \\ \tanh^{-1} y &= \frac{1}{2} \ln \frac{1 + y}{1 - y}, && |y| < 1 \\ \coth^{-1} y &= \frac{1}{2} \ln \frac{y + 1}{y - 1}, && |y| > 1 \end{aligned} \tag{1.4}$$

The $\cosh^{-1}$ function is commonly written with only the + sign before the square root. What does the other sign do? Draw a graph and find out. Also, what happens if you add the two versions of the $\cosh^{-1}$?

The calculus of these functions parallels that of the circular functions.

$$\frac{d}{dx} \sinh x = \frac{d}{dx} \frac{e^x - e^{-x}}{2} = \frac{e^x + e^{-x}}{2} = \cosh x$$

Similarly the derivative of $\cosh x$ is $\sinh x$. Note the plus sign here, not minus.

Where do hyperbolic functions occur? If you have a mass in equilibrium, the total force on it is zero. If it's in *stable* equilibrium then if you push it a little to one side and release it, the force will push it back to the center. If it is *unstable* then when it's a bit to one side it will be pushed farther away from the equilibrium point. In the first case, it will oscillate about the equilibrium position and for small oscillations the function of time will be a circular trigonometric function — the common sines or cosines of time, $A\cos\omega t$. If the point is unstable, the motion will will be described by hyperbolic functions of time, $\sinh\omega t$ instead of $\sin\omega t$. An ordinary ruler held at one end will swing back and forth, but if you try to balance it at the other end it will fall over. That's the difference between cos and cosh. For a deeper understanding of the relation between the circular and the hyperbolic functions, see section 3.3

1.2 Parametric Differentiation

The integration techniques that appear in introductory calculus courses include a variety of methods of varying usefulness. There's one however that is for some reason not commonly done in calculus courses: *parametric differentiation.* It's best introduced by an example.

$$\int_0^\infty x^n e^{-x}\,dx$$

You could integrate by parts n times and that will work. For example, $n = 2$:

$$= -x^2e^{-x}\Big|_0^\infty + \int_0^\infty 2xe^{-x}\,dx = 0 - 2xe^{-x}\Big|_0^\infty + \int_0^\infty 2e^{-x}\,dx = 0 - 2e^{-x}\Big|_0^\infty = 2$$

Instead of this method, do something completely different. Consider the integral

$$\int_0^\infty e^{-\alpha x}\,dx \tag{1.5}$$

It has the parameter α in it. The reason for this will be clear in a few lines. It is easy to evaluate, and is

$$\int_0^\infty e^{-\alpha x}\,dx = \frac{1}{-\alpha}e^{-\alpha x}\Big|_0^\infty = \frac{1}{\alpha}$$

Now differentiate this integral with respect to α,

$$\frac{d}{d\alpha}\int_0^\infty e^{-\alpha x}\,dx = \frac{d}{d\alpha}\frac{1}{\alpha} \qquad \text{or} \qquad -\int_0^\infty xe^{-\alpha x}\,dx = \frac{-1}{\alpha^2}$$

And again and again: $$+\int_0^\infty x^2e^{-\alpha x}\,dx = \frac{+2}{\alpha^3}, \qquad -\int_0^\infty x^3e^{-\alpha x}\,dx = \frac{-2\cdot 3}{\alpha^4}$$

The nth derivative is

$$\pm\int_0^\infty x^n e^{-\alpha x}\,dx = \frac{\pm n!}{\alpha^{n+1}} \tag{1.6}$$

Set $\alpha = 1$ and you see that the original integral is $n!$. This result is compatible with the standard definition for $0!$. From the equation $n! = n\cdot(n-1)!$, you take the case $n = 1$, and it requires $0! = 1$ in order to make any sense. This integral gives the same answer for $n = 0$.

The idea of this method is to change the original problem into another by introducing a parameter. Then differentiate with respect to that parameter in order to recover the problem that you really want to solve. With a little practice you'll find this easier than partial integration. Also see problem 1.47 for a variation on this theme.

Notice that I did this using definite integrals. If you try to use it for an integral without limits you can sometimes get into trouble. See for example problem 1.42.

1.3 Gaussian Integrals

Gaussian integrals are an important class of integrals that show up in kinetic theory, statistical mechanics, quantum mechanics, and any other place with a remotely statistical aspect.

$$\int dx\, x^n e^{-\alpha x^2}$$

The simplest and most common case is the definite integral from $-\infty$ to $+\infty$ or maybe from 0 to ∞.

If n is a positive odd integer, these are elementary,

$$\int_{-\infty}^{\infty} dx\, x^n e^{-\alpha x^2} = 0 \qquad (n \text{ odd}) \tag{1.7}$$

To see why this is true, sketch graphs of the integrand for a few more odd n.

For the integral over positive x and still for odd n, do the substitution $t = \alpha x^2$.

$$\int_0^{\infty} dx\, x^n e^{-\alpha x^2} = \frac{1}{2\alpha^{(n+1)/2}} \int_0^{\infty} dt\, t^{(n-1)/2} e^{-t} = \frac{1}{2\alpha^{(n+1)/2}} \big((n-1)/2\big)! \tag{1.8}$$

Because n is odd, $(n-1)/2$ is an integer and its factorial makes sense.

If n is even then doing this integral requires a special preliminary trick. Evaluate the special case $n = 0$ and $\alpha = 1$. Denote the integral by I, then

$$I = \int_{-\infty}^{\infty} dx\, e^{-x^2}, \qquad \text{and} \qquad I^2 = \left(\int_{-\infty}^{\infty} dx\, e^{-x^2}\right)\left(\int_{-\infty}^{\infty} dy\, e^{-y^2}\right)$$

In squaring the integral you must use a different label for the integration variable in the second factor or it will get confused with the variable in the first factor. Rearrange this and you have a conventional double integral.

$$I^2 = \int_{-\infty}^{\infty} dx \int_{-\infty}^{\infty} dy\, e^{-(x^2+y^2)}$$

This is something that you can recognize as an integral over the entire x-y plane. Now the trick is to switch to polar coordinates*. The element of area $dx\,dy$ now becomes $r\,dr\,d\phi$,

* See section 1.7 in this chapter

and the respective limits on these coordinates are 0 to ∞ and 0 to 2π. The exponent is just $r^2 = x^2 + y^2$.

$$I^2 = \int_0^\infty r\,dr \int_0^{2\pi} d\phi e^{-r^2}$$

The ϕ integral simply gives 2π. For the r integral substitute $r^2 = z$ and the result is 1/2. [Or use Eq. (1.8).] The two integrals together give you π.

$$I^2 = \pi, \qquad \text{so} \qquad \int_{-\infty}^{\infty} dx\, e^{-x^2} = \sqrt{\pi} \tag{1.9}$$

Now do the rest of these integrals by parametric differentiation, introducing a parameter with which to carry out the derivatives. Change e^{-x^2} to $e^{-\alpha x^2}$, then in the resulting integral change variables to reduce it to Eq. (1.9). You get

$$\int_{-\infty}^{\infty} dx\, e^{-\alpha x^2} = \sqrt{\frac{\pi}{\alpha}}, \qquad \text{so} \qquad \int_{-\infty}^{\infty} dx\, x^2 e^{-\alpha x^2} = -\frac{d}{d\alpha}\sqrt{\frac{\pi}{\alpha}} = \frac{1}{2}\left(\frac{\sqrt{\pi}}{\alpha^{3/2}}\right) \tag{1.10}$$

You can now get the results for all the higher even powers of x by further differentiation with respect to α.

1.4 erf and Gamma

What about the same integral, but with other limits? The odd-n case is easy to do in just the same way as when the limits are zero and infinity; just do the same substitution that led to Eq. (1.8). The even-n case is different because it can't be done in terms of elementary functions. It is used to define an entirely new function.

$$\operatorname{erf}(x) = \frac{2}{\sqrt{\pi}} \int_0^x dt\, e^{-t^2} \tag{1.11}$$

x	0.	0.25	0.50	0.75	1.00	1.25	1.50	1.75	2.00
erf	0.	0.276	0.520	0.711	0.843	0.923	0.967	0.987	0.995

This is called the error function. It's well studied and tabulated and even shows up as a button on some* pocket calculators, right along with the sine and cosine. (Is erf odd or even or neither?) (What is $\operatorname{erf}(\pm\infty)$?)

A related integral worthy of its own name is the Gamma function.

$$\Gamma(x) = \int_0^\infty dt\, t^{x-1} e^{-t} \tag{1.12}$$

The special case in which x is a positive integer is the one that I did as an example of parametric differentiation to get Eq. (1.6). It is

$$\Gamma(n) = (n-1)!$$

---

* See for example rpncalculator (v1.96 the latest). It is the best desktop calculator that I've found (Mac and Windows). This main site seems (2008) to have disappeared, but I did find other sources by searching the web for the pair "rpncalculator" and baker. The latter is the author's name.

The factorial is not defined if its argument isn't an integer, but the Gamma function is perfectly well defined for any argument as long as the integral converges. One special case is notable: $x = 1/2$.

$$\Gamma(1/2) = \int_0^\infty dt\, t^{-1/2} e^{-t} = \int_0^\infty 2u\, du\, u^{-1} e^{-u^2} = 2\int_0^\infty du\, e^{-u^2} = \sqrt{\pi} \qquad (1.13)$$

I used $t = u^2$ and then the result for the Gaussian integral, Eq. (1.9). You can use parametric differentiation to derive a simple and useful recursion relation. (See problem 1.14 or 1.47.)

$$x\Gamma(x) = \Gamma(x+1) \qquad (1.14)$$

From this you can get the value of $\Gamma(1\tfrac{1}{2})$, $\Gamma(2\tfrac{1}{2})$, etc. In fact, if you know the value of the function in the interval between one and two, you can use this relationship to get it anywhere else on the axis. You already know that $\Gamma(1) = 1 = \Gamma(2)$. (You do? How?) As x approaches zero, use the relation $\Gamma(x) = \Gamma(x+1)/x$ and because the numerator for small x is approximately 1, you immediately have that

$$\Gamma(x) \sim 1/x \qquad \text{for small } x \qquad (1.15)$$

The integral definition, Eq. (1.12), for the Gamma function is defined only for the case that $x > 0$. [The behavior of the integrand near $t = 0$ is approximately t^{x-1}. Integrate *this* from zero to something and see how it depends on x.] Even though the original definition of the Gamma function fails for negative x, you can extend the definition by using Eq. (1.14) to define Γ for negative arguments. What is $\Gamma(-\tfrac{1}{2})$ for example? Put $x = -\tfrac{1}{2}$ in Eq. (1.14).

$$-\frac{1}{2}\Gamma(-1/2) = \Gamma(-(1/2)+1) = \Gamma(1/2) = \sqrt{\pi}, \qquad \text{so} \qquad \Gamma(-1/2) = -2\sqrt{\pi} \qquad (1.16)$$

The same procedure works for other negative x, though it can take several integer steps to get to a positive value of x for which you can use the integral definition Eq. (1.12).

The reason for introducing these two functions now is not that they are so much more important than a hundred other functions that I could use, though they are among the more common ones. The point is that the world doesn't end with polynomials, sines, cosines, and exponentials. There are an infinite number of other functions out there waiting for you and some of them are useful. These functions can't be expressed in terms of the elementary functions that you've grown to know and love. They're different and have their distinctive behaviors.

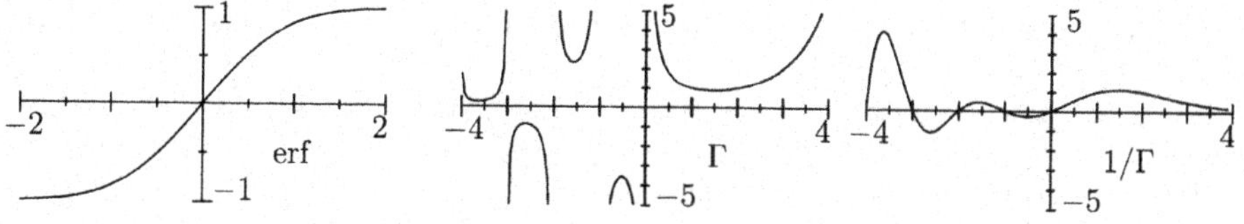

There are zeta functions and Fresnel integrals and Legendre functions and Exponential integrals and Mathieu functions and Confluent Hypergeometric functions and ...

you get the idea. When one of these shows up, you learn to look up its properties and to use them. If you're interested you may even try to understand how some of these properties are derived, but probably not the first time that you confront them. That's why there are tables, and the "Handbook of Mathematical Functions" by Abramowitz and Stegun is a premier example of such a tabulation, and it's reprinted by Dover Publications. There's also a copy on the internet* www.math.sfu.ca/~cbm/aands/ as a set of scanned page images.

Why erf?

What can you do with this function? The most likely application is probably to probability. If you flip a coin 1000 times, you expect it to come up heads *about* 500 times. But just how close to 500 will it be? If you flip it twice, you wouldn't be surprised to see two heads or two tails, in fact the equally likely possibilities are

TT HT TH HH

This says that in 1 out of $2^2 = 4$ such experiments you expect to see two heads and in 1 out of 4 you expect two tails. For just 2 out of 4 times you do the double flip do you expect exactly one head. *All this is an average. You have to try the experiment many times to see your expectation verified, and then only by averaging many experiments.*

It's easier to visualize the counting if you flip N coins at once and see how they come up. The number of coins that come up heads won't always be $N/2$, but it should be close. If you repeat the process, flipping N coins again and again, you get a distribution of numbers of heads that will vary around $N/2$ in a characteristic pattern. The result is that the fraction of the time it will come up with k heads and $N-k$ tails is, to a good approximation

$$\sqrt{\frac{2}{\pi N}}e^{-2\delta^2/N}, \qquad \text{where} \qquad \delta = k - \frac{N}{2} \tag{1.17}$$

The derivation of this can wait until section 2.6, Eq. (2.26). It is an accurate result if the number of coins that you flip in each trial is large, but try it anyway for the preceding example where $N = 2$. This formula says that the fraction of times predicted for k heads is

$$k = 0: \ \sqrt{1/\pi}e^{-1} = 0.208 \qquad k = 1 = N/2: \ 0.564 \qquad k = 2: \ 0.208$$

The exact answers are 1/4, 2/4, 1/4, but as two is not all that big a number, the fairly large error shouldn't be distressing.

If you flip three coins, the equally likely possibilities are

TTT TTH THT HTT THH HTH HHT HHH

There are 8 possibilities here, 2^3, so you expect (on average) one run out of 8 to give you 3 heads. Probability 1/8.

To see how accurate this claim is for modest values, take $N = 10$. The possible outcomes are anywhere from zero heads to ten. The exact fraction of the time that you get k heads as compared to this approximation is

---

* online books at University of Pennsylvania, onlinebooks.library.upenn.edu

$k =$	0	1	2	3	4	5
exact:	.000977	.00977	.0439	.117	.205	.246
approximate:	.0017	.0103	.0417	.113	.206	.252

For the more interesting case of big N, the exponent, $e^{-2\delta^2/N}$, varies slowly and smoothly as δ changes in integer steps away from zero. This is a key point; it allows you to approximate a sum by an integral. If $N = 1000$ and $\delta = 10$, the exponent is 0.819. It has dropped only gradually from one. For the same $N = 1000$, the fraction of the time to get exactly 500 heads is 0.025225, and this approximation is $\sqrt{2/1000\pi} = 0.025231$.

Flip N coins, then do it again and again. In what fraction of the trials will the result be between $N/2 - \Delta$ and $N/2 + \Delta$ heads? This is the sum of the fractions corresponding to $\delta = 0$, $\delta = \pm 1, \ldots, \delta = \pm\Delta$. Because the approximate function is smooth, I can replace this sum with an integral. This substitution becomes more accurate the larger N is.

$$\int_{-\Delta}^{\Delta} d\delta\, \sqrt{\frac{2}{\pi N}}\, e^{-2\delta^2/N}$$

Make the substitution $2\delta^2/N = x^2$ and you have

$$\sqrt{\frac{2}{\pi N}} \int_{-\Delta\sqrt{2/N}}^{\Delta\sqrt{2/N}} \sqrt{\frac{N}{2}}\, dx\, e^{-x^2} = \frac{1}{\sqrt{\pi}} \int_{-\Delta\sqrt{2/N}}^{\Delta\sqrt{2/N}} dx\, e^{-x^2} = \operatorname{erf}\left(\Delta\sqrt{2/N}\right) \tag{1.18}$$

The error function of one is 0.84, so if $\Delta = \sqrt{N/2}$ then in 84% of the trials heads will come up between $N/2 - \sqrt{N/2}$ and $N/2 + \sqrt{N/2}$ times. For $N = 1000$, this is between 478 and 522 heads.

1.5 Differentiating

When you differentiate a function in which the independent variable shows up in several places, how do you carry out the derivative? For example, what is the derivative with respect to x of x^x? The answer is that you treat each instance of x one at a time, ignoring the others; differentiate with respect to *that* x and add the results. For a proof, use the definition of a derivative and differentiate the function $f(x, x)$. Start with the finite difference quotient:

$$\begin{aligned} &\frac{f(x+\Delta x, x+\Delta x) - f(x,x)}{\Delta x} \\ &\qquad = \frac{f(x+\Delta x, x+\Delta x) - f(x, x+\Delta x) + f(x, x+\Delta x) - f(x,x)}{\Delta x} \\ &\qquad = \frac{f(x+\Delta x, x+\Delta x) - f(x, x+\Delta x)}{\Delta x} + \frac{f(x, x+\Delta x) - f(x,x)}{\Delta x} \end{aligned} \tag{1.19}$$

The first quotient in the last equation is, in the limit that $\Delta x \to 0$, the derivative of f with respect to its first argument. The second quotient becomes the derivative with respect to the second argument. The prescription is clear, but to remember it you may prefer a mathematical formula. A notation more common in mathematics than in physics is just what's needed:

$$\frac{d}{dt} f(t, t) = D_1 f(t, t) + D_2 f(t, t) \tag{1.20}$$

where D_1 means "differentiate with respect to the first argument." The standard "product rule" for differentiation is a special case of this.

For example,

$$\frac{d}{dx}\int_0^x dt\, e^{-xt^2} = e^{-x^3} - \int_0^x dt\, t^2 e^{-xt^2} \tag{1.21}$$

The resulting integral in this example is related to an error function, see problem 1.13, so it's not as bad as it looks.

Another example,

$$\begin{aligned}\frac{d}{dx}x^x &= x\,x^{x-1} + \frac{d}{dx}k^x \qquad \text{at } k = x \\ &= x\,x^{x-1} + \frac{d}{dx}e^{x\ln k} = x\,x^{x-1} + \ln k\, e^{x\ln k} \\ &= x^x + x^x \ln x\end{aligned}$$

1.6 Integrals

What is an integral? You've been using them for some time. I've been using the concept in this introductory chapter as if it's something that everyone knows. But what *is* it?

If your answer is something like "the function whose derivative is the given function" or "the area under a curve" then No. Both of these answers express an aspect of the subject but neither is a complete answer. The first actually refers to *the fundamental theorem of calculus,* and I'll describe that shortly. The second is a good picture that applies to some special cases, but it won't tell you how to compute it and it won't allow you to generalize the idea to the many other subjects in which it is needed. There are several different definitions of the integral, and every one of them requires more than a few lines to explain. I'll use the most common definition, the *Riemann Integral.*

An integral is a sum, obeying all the usual rules of addition and multiplication, such as $1+2+3+4 = (1+2)+(3+4)$ or $5\cdot(6+7) = (5\cdot 6)+(5\cdot 7)$. *When you've read this section, come back and translate these bits of arithmetic into statements about integrals.*

A standard way to picture the definition is to try to find the area under a curve. You can get successively better and better approximations to the answer by dividing the area into smaller and smaller rectangles — ideally, taking the limit as the number of rectangles goes to infinity.

To codify this idea takes a sequence of steps:

1. Pick an integer $N > 0$. This is the number of subintervals into which the whole interval between a and b is to be divided.

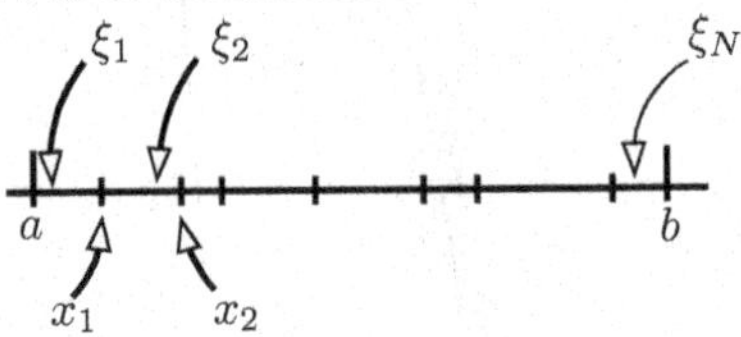

2. Pick $N-1$ points between a and b. Call them x_1, x_2, etc.

$$a = x_0 < x_1 < x_2 < \cdots < x_{N-1} < x_N = b$$

and for convenience label the endpoints as x_0 and x_N. For the sketch , $N = 8$.

3. Let $\Delta x_k = x_k - x_{k-1}$. That is,

$$\Delta x_1 = x_1 - x_0, \qquad \Delta x_2 = x_2 - x_1, \cdots$$

4. In each of the N subintervals, pick one point at which the function will be evaluated. I'll label these points by the Greek letter ξ. (That's the Greek version of "x.")

$$x_{k-1} \le \xi_k \le x_k$$
$$x_0 \le \xi_1 \le x_1, \qquad x_1 \le \xi_2 \le x_2, \cdots$$

5. Form the sum that is an approximation to the final answer.

$$f(\xi_1)\Delta x_1 + f(\xi_2)\Delta x_2 + f(\xi_3)\Delta x_3 + \cdots$$

6. Finally, take the limit as all the $\Delta x_k \to 0$ and necessarily then, as $N \to \infty$. These six steps form the definition

$$\boxed{\lim_{\Delta x_k \to 0} \sum_{k=1}^{N} f(\xi_k)\Delta x_k = \int_a^b f(x)\,dx} \tag{1.22}$$

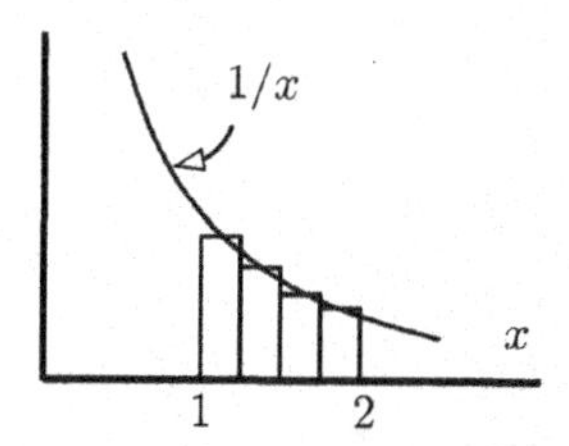

To demonstrate this numerically, pick a function and do the first five steps explicitly. Pick $f(x) = 1/x$ and integrate it from 1 to 2. The exact answer is the natural log of 2: $\ln 2 = 0.69315\ldots$

(1) Take $N = 4$ for the number of intervals

(2) Choose to divide the distance from 1 to 2 evenly, at $x_1 = 1.25$, $x_2 = 1.5$, $x_3 = 1.75$

$$a = x_0 = 1. < 1.25 < 1.5 < 1.75 < 2. = x_4 = b$$

(3) All the Δx's are equal to 0.25.

(4) Choose the midpoint of each subinterval. This is the best choice when you use a finite number of divisions without taking the limit.

$$\xi_1 = 1.125 \qquad \xi_2 = 1.375 \qquad \xi_3 = 1.625 \qquad \xi_4 = 1.875$$

(5) The sum approximating the integral is then

$$f(\xi_1)\Delta x_1 \;+\; f(\xi_2)\Delta x_2 \;+\; f(\xi_3)\Delta x_3 \;+\; f(\xi_4)\Delta x_4 \;=$$
$$\frac{1}{1.125}\times .25+\frac{1}{1.375}\times .25+\frac{1}{1.625}\times .25+\frac{1}{1.875}\times .25 = .69122$$

For such a small number of divisions, this is a very good approximation — about 0.3% error. (What do you get if you take $N = 1$ or $N = 2$ or $N = 10$ divisions?)

Fundamental Thm. of Calculus

If the function that you're integrating is complicated or if the function is itself not known to perfect accuracy then a numerical approximation just like this one for $\int_1^2 dx/x$ is often the best way to go. How can a function not be known completely? If it is experimental data. When you have to resort to this arithmetic way to do integrals, are there more efficient ways to do it than simply using the definition of the integral? Yes. That's part of the subject of numerical analysis, and there's a short introduction to the subject in chapter 11, section 11.4.

The fundamental theorem of calculus unites the subjects of differentiation and integration. The integral is defined as the limit of a sum, and the derivative is defined as the limit of a quotient of two differences. The relation between them is

IF f has an integral from a to b, that is, if $\int_a^b f(x)\,dx$ exists,
AND IF f has an anti-derivative, that is, there is a function F such that $dF/dx = f$,
THEN

$$\int_a^b f(x)\,dx = F(b) - F(a) \tag{1.23}$$

Are there cases where one of these exists without the other? Yes, though I'll admit that you are not likely to come across such functions without hunting through some advanced math books. Check out www.wikipedia.org for Volterra's function to see what it involves.

Notice an important result that follows from Eq. (1.23). Differentiate both sides with respect to b

$$\frac{d}{db}\int_a^b f(x)\,dx = \frac{d}{db}F(b) = f(b) \tag{1.24}$$

and with respect to a

$$\frac{d}{da}\int_a^b f(x)\,dx = -\frac{d}{da}F(a) = -f(a) \tag{1.25}$$

Differentiating an integral with respect to one or the other of its limits results in plus or minus the integrand. Combine this with the chain rule and you can do such calculations as

$$\frac{d}{dx}\int_{x^2}^{\sin x} e^{xt^2}\,dt = e^{x\sin^2 x}\cos x - e^{x^5}2x + \int_{x^2}^{\sin x} t^2 e^{xt^2}\,dt \tag{1.26}$$

All this requires is that you differentiate every x that is present and add the results, just as

$$\frac{d}{dx}x^2 = \frac{d}{dx}x\cdot x = \frac{dx}{dx}x + x\frac{dx}{dx} = 1\cdot x + x\cdot 1 = 2x$$

You may well ask why anyone would want to do such a thing as Eq. (1.26), but there are more reasonable examples that show up in real situations.

Riemann-Stieltjes Integrals

Are there other useful definitions of the word integral? Yes, there are many, named after various people who developed them, with Lebesgue being the most famous. His definition* is most useful in much more advanced mathematical contexts, and I won't go into it here, except to say that *very* roughly where Riemann divided the x-axis into intervals Δx_i, Lebesgue divided the y-axis into intervals Δy_i. Doesn't sound like much of a change does it? It is. There is another definition that is worth knowing about, not because it helps you to do integrals, but because it unites a couple of different types of computation into one. This is the *Riemann-Stieltjes* integral. You won't need it for any of the later work in this book, but it is a fairly simple extension of the Riemann integral and I'm introducing it mostly for its cultural value — to show you that there *are* other ways to define an integral. If you take the time to understand it, you will be able to look back at some subjects that you already know and to realize that they can be manipulated in a more compact form (e.g. center of mass).

When you try to evaluate the moment of inertia you are doing the integral

$$\int r^2\,dm$$

When you evaluate the position of the center of mass even in one dimension the integral is

$$\frac{1}{M}\int x\,dm$$

and even though you may not yet have encountered this, the electric dipole moment is

$$\int \vec{r}\,dq$$

How do you integrate x with respect to m? What exactly are you doing? A possible answer is that you can express this integral in terms of the linear density function and then $dm = \lambda(x)dx$. But if the masses are a mixture of continuous densities and point masses, this starts to become awkward. Is there a better way?

Yes

On the interval $a \le x \le b$ assume there are *two* functions, f and α. Don't assume that either of them must be continuous, though they can't be too badly behaved or nothing will converge. This starts the same way the Riemann integral does: partition the interval into a finite number (N) of sub-intervals at the points

$$a = x_0 < x_1 < x_2 < \ldots < x_N = b \tag{1.27}$$

Form the sum

$$\sum_{k=1}^{N} f(x'_k)\Delta\alpha_k, \qquad \text{where} \qquad x_{k-1} \le x'_k \le x_k \qquad \text{and} \qquad \Delta\alpha_k = \alpha(x_k) - \alpha(x_{k-1}) \tag{1.28}$$

* One of the more notable PhD theses in history

To improve the sum, keep adding more and more points to the partition so that in the limit all the intervals $x_k - x_{k-1} \to 0$. This limit is called the Riemann-Stieltjes integral,

$$\int f\,d\alpha \tag{1.29}$$

What's the big deal? Doesn't $d\alpha = \alpha' dx$? Use that and you have just the ordinary integral

$$\int f(x)\alpha'(x)\,dx?$$

Sometimes you can, but what if α isn't differentiable? Suppose that it has a step or several steps? The derivative isn't defined, but this Riemann-Stieltjes integral still makes perfectly good sense.

An example. A very thin rod of length L is placed on the x-axis with one end at the origin. It has a uniform linear mass density λ *and* an added point mass m_0 at $x = 3L/4$. (a piece of chewing gum?) Let $m(x)$ be the function defined as

$$m(x) = \big(\text{the amount of mass at coordinates} \le x\big)$$
$$= \begin{cases} \lambda x & (0 \le x < 3L/4) \\ \lambda x + m_0 & (3L/4 \le x \le L) \end{cases}$$

This is of course discontinuous.

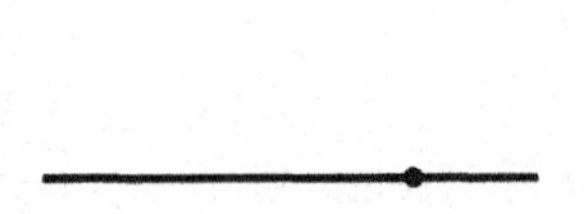

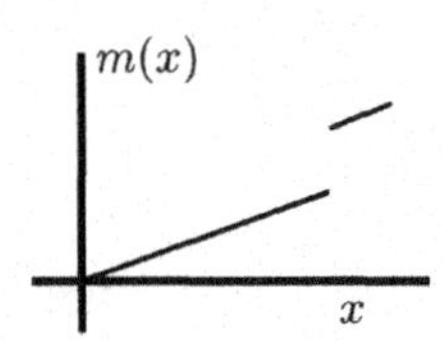

The coordinate of the center of mass is $\int x\,dm / \int dm$. The total mass in the denominator is $m_0 + \lambda L$, and I'll go through the details to evaluate the numerator, attempting to solidify the ideas that form this integral. Suppose you divide the length L into 10 equal pieces, then

$$x_k = kL/10, \quad (k = 0, 1, \ldots, 10) \qquad \text{and} \qquad \Delta m_k = \begin{cases} \lambda L/10 & (k \ne 8) \\ \lambda L/10 + m_0 & (k = 8) \end{cases}$$

$\Delta m_8 = m(x_8) - m(x_7) = (\lambda x_8 + m_0) - \lambda x_7 = \lambda L/10 + m_0$.

Choose the positions x'_k anywhere in the interval; for no particular reason I'll take the right-hand endpoint, $x'_k = kL/10$. The approximation to the integral is now

$$\sum_{k=1}^{10} x'_k \Delta m_k = \sum_{k=1}^{7} x'_k \lambda L/10 + x'_8(\lambda L/10 + m_0) + \sum_{k=9}^{10} x'_k \lambda L/10$$
$$= \sum_{k=1}^{10} x'_k \lambda L/10 + x'_8 m_0$$

As you add division points (more intervals) to the whole length this sum obviously separates into two parts. One is the ordinary integral and the other is the discrete term from the point mass.

$$\int_0^L x\lambda\,dx + m_0 3L/4 = \lambda L^2/2 + m_0 3L/4$$

The center of mass is then at

$$x_{\text{cm}} = \frac{\lambda L^2/2 + m_0 3L/4}{m_0 + \lambda L}$$

If $m_0 \ll \lambda L$, this is approximately $L/2$. In the reverse case is is approximately $3L/4$. Both are just what you should expect.

The discontinuity in $m(x)$ simply gives you a discrete added term in the overall result.

Did you need the Stieltjes integral to do this? Probably not. You would likely have simply added the two terms from the two parts of the mass and gotten the same result as with this more complicated method. The point of this is not that it provides an easier way to do computations. It doesn't. It is however a unifying notation and language that lets you avoid writing down a lot of special cases. (Is it discrete? Is it continuous?) You can even write sums as integrals: Let α be a set of steps:

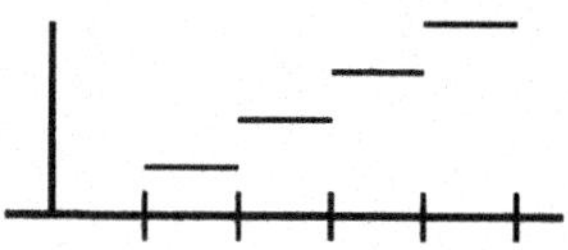

$$\alpha(x) = \begin{cases} 0 & x < 1 \\ 1 & 1 \le x < 2 \\ 2 & 2 \le x < 3 \\ \text{etc.} & \end{cases} \qquad = [x] \quad \text{for } x \ge 0$$

Where that last bracketed symbol means "greatest integer less than or equal to x." It's a notation more common in mathematics than in physics. Now in this notation the sum can be written as a Stieltjes integral.

$$\int f\,d\alpha = \int_{x=0}^{\infty} f\,d[x] = \sum_{k=1}^{\infty} f(k) \tag{1.30}$$

At every integer, where $[x]$ makes a jump by one, there is a contribution to the Riemann-Stieltjes sum, Eq. (1.28). That makes this integral just another way to write the sum over integers. This won't help you to sum the series, but it is another way to look at the subject.

The method of integration by parts works perfectly well here, though as with all the rest of this material I'll leave the proof to advanced calculus texts. If $\int f\,d\alpha$ exists then so does $\int \alpha\,df$ and

$$\int f\,d\alpha = f\alpha - \int \alpha\,df \tag{1.31}$$

This relates one Stieltjes integral to another one, and because you can express summation as an integral now, you can even do summation by parts on the equation (1.30). That's something that you are not likely to think of if you restrict yourself to the more elementary notation, and it's even occasionally useful.

1.7 Polar Coordinates

When you compute an integral in the plane, you need the element of area appropriate to the coordinate system that you're using. In the most common case, that of rectangular coordinates, you find the element of area by drawing the two lines at constant coordinates x and $x + dx$. Then you draw the two lines at constant coordinates y and $y + dy$. The little rectangle that they circumscribe has an area $dA = dx\,dy$.

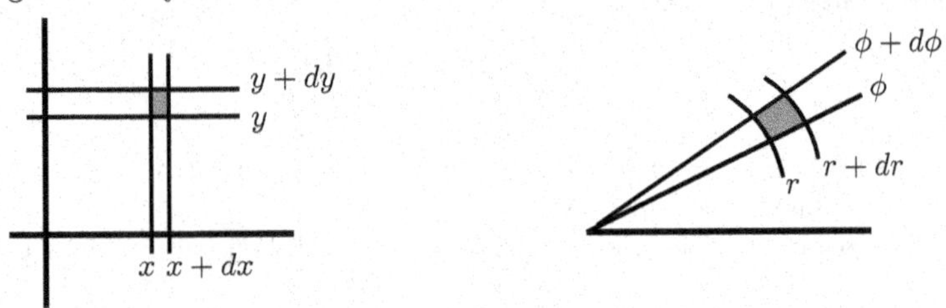

In polar coordinates you do exactly the same thing! The coordinates are r and ϕ, and the line at constant radius r and at constant $r + dr$ define two neighboring circles. The lines at constant angle ϕ and at constant angle $\phi + d\phi$ form two closely spaced rays from the origin. These four lines circumscribe a tiny area that is, for small enough dr and $d\phi$, a rectangle. You then know its area is the product of its two sides*: $dA = (dr)(r\,d\phi)$. This is the basic element of area for polar coordinates.

The area of a circle is the sum of all the pieces of area within it

$$\int dA = \int_0^R r\,dr \int_0^{2\pi} d\phi$$

I find it more useful to write double integrals in this way, so that the limits of integration are next to the differential. The other notation can put the differential a long distance from where you show the limits of integration. I get less confused my way. In either case, and to no one's surprise, you get

$$\int_0^R r\,dr \int_0^{2\pi} d\phi = \int_0^R r\,dr\,2\pi = 2\pi R^2/2 = \pi R^2$$

For the preceding example you can do the double integral in either order with no special care. If the area over which you're integrating is more complicated you will have to look more closely at the limits of integration. I'll illustrate with an example of this in rectangular coordinates: the area of a triangle. Take the triangle to have vertices $(0,0)$, $(a,0)$, and $(0,b)$. The area is

$$\int dA = \int_0^a dx \int_0^{b(a-x)/a} dy \qquad \text{or} \qquad \int_0^b dy \int_0^{a(b-y)/b} dx \tag{1.32}$$

They should both yield $ab/2$. See problem 1.25.

---

* If you're tempted to say that the area is $dA = dr\,d\phi$, *look at the dimensions.* This expression is a length, not an area.

1.8 Sketching Graphs

How do you sketch the graph of a function? This is one of the most important tool can use to understand the behavior of functions, and unless you practice it you will yourself at a loss in anticipating the outcome of many calculations. There are a ha of rules that you can follow to do this and you will find that it's not as painful as may think.

You are confronted with a function and have to sketch its graph.

1. What is the domain? That is, what is the set of values of the indepen variable that you need to be concerned with? Is it $-\infty$ to $+\infty$ or is it $0 < x < L$ or $-\pi < \phi < \pi$ or what?

2. Plot any obvious points. If you can immediately see the value of the fun at one or more points, do them right away.

3. Is the function even or odd? If the behavior of the function is the same o left as it is on the right (or perhaps inverted on the left) then you have half as much to do. Concentrate on one side and you can then make a mirror image on the left if even or an upside-down mirror image if it's odd.

4. Is the function singular anywhere? Does it go to infinity at some point w the denominator vanishes? Note these points on the axis for future examination.

5. What is the behavior of the function *near* any of the obvious points that plotted? Does it behave like x? Like x^2? If you concluded that it is even, then the s is either zero or there's a kink in the curve, such as with the absolute value function

6. At one of the singular points that you found, how does it behave as you appr the point from the right? From the left? Does the function go toward $+\infty$ or toward in each case?

7. How does the function behave as you approach the ends of the domain? I domain extends from $-\infty$ to $+\infty$, how does the function behave as you approach t regions?

8. Is the function the sum or difference of two other much simpler functions? you may find it easier to sketch the two functions and then graphically add or sub them. Similarly if it is a product.

9. Is the function related to another by translation? The function $f(x) = (x$ - is related to x^2 by translation of 2 units. Note that it is translated to the *right* fron You can see why because $(x - 2)^2$ vanishes at $x = +2$.

10. After all this, you will have a good idea of the shape of the function, sc can interpolate the behavior between the points that you've found.

Example: Sketch $f(x) = x/(a^2 - x^2)$.

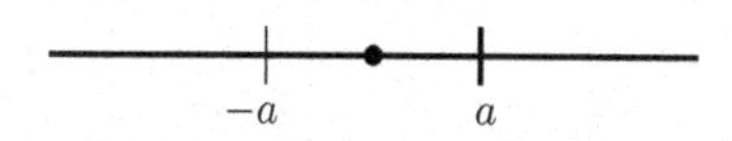

1. The domain for independent variable wasn't given, so take it to be $-\infty < x$
2. The point $x = 0$ obviously gives the value $f(0) = 0$.
4. The denominator becomes zero at the two points $x = \pm a$.
3. If you replace x by $-x$, the denominator is unchanged, and the nume changes sign. The function is odd about zero.

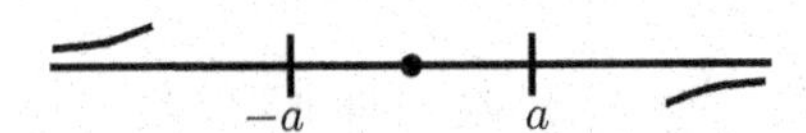

7. When x becomes very large ($|x| \gg a$), the denominator is mostly $-x^2$, so $f(x)$ behaves like $x/(-x^2) = -1/x$ for large x. It approaches zero for large x. Moreover, when x is positive, it approaches zero through negative values and when x is negative, it goes to zero through positive values.

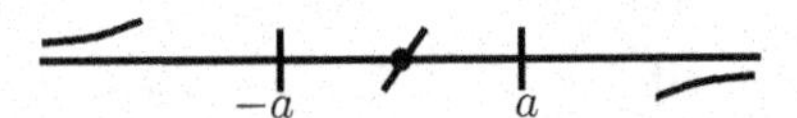

5. Near the point $x = 0$, the x^2 in the denominator is much smaller than the constant a^2 ($x^2 \ll a^2$). That means that near this point, the function f behaves like x/a^2

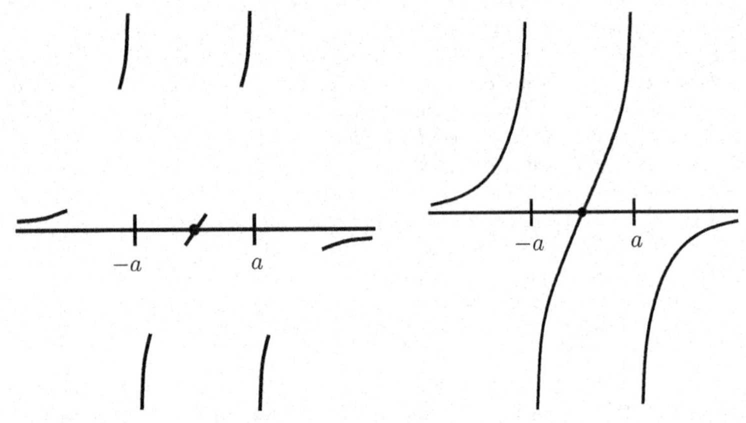

6. Go back to the places that it blows up, and ask what happens near there. If x is a little greater than a, the x^2 in the denominator is a little larger than the a^2 in the denominator. This means that the denominator is negative. When x is a little less than a, the reverse is true. Near $x = a$, The numerator is close to a. Combine these, and you see that the function approaches $-\infty$ as $x \to a$ from the right. It approaches $+\infty$ on the left side of a. I've already noted that the function is odd, so don't repeat the analysis near $x = -a$, just turn this behavior upside down.

With all of these pieces of the graph, you can now interpolate to see the whole picture.

OR, if you're clever with partial fractions, you might realize that you can rearrange f as

$$\frac{x}{a^2 - x^2} = \frac{-1/2}{x - a} + \frac{-1/2}{x + a},$$

and then follow the ideas of techniques 8 and 9 to sketch the graph. It's not obvious that this is any easier; it's just different.

Exercises

1 Express e^x in terms of hyperbolic functions.

2 If $\sinh x = 4/3$, what is $\cosh x$? What is $\tanh x$?

3 If $\tanh x = 5/13$, what is $\sinh x$? What is $\cosh x$?

4 Let n and m be positive integers. Let $a = n^2 - m^2$, $b = 2nm$, $c = n^2 + m^2$. S that a-b-c form the integer sides of a right triangle. What are the first three indepen "Pythagorean triples?" By that I mean ones that aren't just a multiple of one o others.

5 Evaluate the integral $\int_0^a dx\, x^2 \cos x$. Use parametric differentiation starting with co

6 Evaluate $\int_0^a dx\, x \sinh x$ by parametric differentiation.

7 Differentiate $xe^x \sin x \cosh x$ with respect to x.

8 Differentiate $\int_0^{x^2} dt\, \sin(xt)$ with respect to x.

9 Differentiate $\int_{-x}^{+x} dt\, e^{-xt^4}$ with respect to x.

10 Differentiate $\int_{-x}^{+x} dt\, \sin(xt^3)$ with respect to x.

11 Differentiate $\int_0^{\sqrt[3]{\sin(kx)}} dt\, e^{-\alpha t^3} J_0(\beta t)$ with respect to x. J_0 is a Bessel function.

12 Sketch the function $y = v_0 t - gt^2/2$. (First step: set all constants to one. $v_0 =$ $2 = 1$. Except exponents)

13 Sketch the function $U = -mgy + ky^2/2$. (Again: set the constant factors to one

14 Sketch $U = mg\ell(1 - \cos\theta)$.

15 Sketch $V = -V_0 e^{-x^2/a^2}$.

16 Sketch $x = x_0 e^{-\alpha t} \sin \omega t$.

17 Is it all right in Eq. (1.22) to replace "$\Delta x_k \to 0$" with "$N \to \infty$?" [No.]

18 Draw a graph of the curve parametrized as $x = \cos\theta$, $y = \sin\theta$.
Draw a graph of the curve parametrized as $x = \cosh\theta$, $y = \sinh\theta$.

19 What is the integral $\int_a^b dx\, e^{-x^2}$?

20 Given that $\int_{-\infty}^{\infty} dx/(1 + x^2) = \pi$, *i.e.* you don't have to derive this, what th $\int_{-\infty}^{\infty} dx/(\alpha + x^2)$? Now differentiate the result and find the two integrals $\int_{-\infty}^{\infty} dx/(1+$ and $\int_{-\infty}^{\infty} dx/(1 + x^2)^3$.

21 Derive the product rule as a special case of Eq. (1.20).

22 The third paragraph of section 1.6 has two simple equations in arithmetic. \ common identities about the integral do these correspond to?

Problems

1.1 What is the tangent of an angle in terms of its sine? Draw a triangle and do this in one line.

1.2 Derive the identities for $\cosh^2\theta - \sinh^2\theta$ and for $1 - \tanh^2\theta$, Equation (1.3).

1.3 Derive the expressions in Eq. (1.4) for $\cosh^{-1} y$, $\tanh^{-1} y$, and $\coth^{-1} y$. Pay particular attention to the domains and explain why these are valid for the set of y that you claim. What is $\sinh^{-1}(y) + \sinh^{-1}(-y)$?

1.4 The inverse function has a graph that is the mirror image of the original function in the 45° line $y = x$. Draw the graphs of all six of the hyperbolic functions and all six of the inverse hyperbolic functions, comparing the graphs you should get to the functions derived in the preceding problem.

1.5 Evaluate the derivatives of $\cosh x$, $\tanh x$, and $\coth x$.

1.6 What are the derivatives, $d\sinh^{-1} y/dy$ and $d\cosh^{-1} y/dy$?

1.7 Find formulas for $\sinh 2y$ and $\cosh 2y$ in terms of hyperbolic functions of y. The first one of these should take only a couple of lines. Maybe the second one too, so if you find yourself filling a page, start over.

1.8 Do a substitution to evaluate the integral (**a**) simply. Now do the same for (**b**)

$$(\mathbf{a})\quad \int \frac{dt}{\sqrt{a^2 - t^2}} \qquad\qquad (\mathbf{b})\quad \int \frac{dt}{\sqrt{a^2 + t^2}}$$

1.9 Sketch the two integrands in the preceding problem. For the second integral, if the limits are 0 and z with $z \gg a$, then before having done the integral, estimate *approximately* what the value of this integral should be. (Say $z = 10^6 a$ or $z = 10^{60} a$.) Compare your estimate to the exact answer that you just found to see if they match in any way.

1.10 Fill in the steps in the derivation of the Gaussian integrals, Eqs. (1.7), (1.8), and (1.10). In particular, draw graphs of the integrands to show why Eq. (1.7) is so.

1.11 What is the integral $\int_{-\infty}^{\infty} dt\, t^n e^{-t^2}$ if $n = -1$ or $n = -2$? [*Careful!*, no conclusion-jumping allowed.] Did you draw a graph? No? Then that's why you're having trouble with this.

1.12 Sketch a graph of the error function. In particular, what is its behavior for small x and for large x, both positive and negative? Note: "small" doesn't mean zero. First draw a sketch of the integrand e^{-t^2} and from that you can (graphically) estimate $\operatorname{erf}(x)$ for small x. Compare this to the short table in Eq. (1.11).

1.13 Put a parameter α into the defining integral for the error function, Eq. (1.11), so it has $\int_0^x dt\, e^{-\alpha t^2}$. Differentiate this integral with respect to α. Next, change variables in this same integral from t to u: $u^2 = \alpha t^2$, and differentiate *that* integral (which of course has the same value as before) with respect to alpha to show

$$\int_0^x dt\, t^2 e^{-t^2} = \frac{\sqrt{\pi}}{4}\operatorname{erf}(x) - \frac{1}{2}xe^{-x^2}$$

As a check, does this agree with the previous result for $x = \infty$, Eq. (1.10)?

1.14 Use parametric differentiation to derive the recursion relation $x\Gamma(x) = \Gamma(x+1)$. Do it once by inserting a parameter in the integral for Γ, $e^{-t} \to e^{-\alpha t}$, and differentiating. Then change variables before differentiating and equate the results.

1.15 What is the Gamma function of $x = -1/2, -3/2, -5/2$? Explain why the original definition of Γ in terms of the integral won't work here. Demonstrate why Eq. (1.12) converges for all $x > 0$ but does not converge for $x \leq 0$. Ans: $\Gamma(-5/2) = -8\sqrt{\pi}/15$

1.16 What is the Gamma function for x near 1? near 0? near -1? -2? -3? Now sketch a graph of the Gamma function from -3 through positive values. Try using the recursion relation of problem 1.14. Ans: Near -3, $\Gamma(x) \approx -1/\big(6(x+3)\big)$

1.17 Show how to express the integral for arbitrary positive x

$$\int_0^\infty dt\, t^x e^{-t^2}$$

in terms of the Gamma function. Is *positive* x the best constraint here or can you do a touch better?
Ans: $\frac{1}{2}\Gamma\big((x+1)/2\big)$

1.18 The derivative of the Gamma function at $x = 1$ is $\Gamma'(1) = -0.5772 = -\gamma$. The number γ is called Euler's constant, and like π or e it's another number that simply shows up regularly. What is $\Gamma'(2)$? What is $\Gamma'(3)$? Ans: $\Gamma'(3) = 3 - 2\gamma$

1.19 Show that

$$\Gamma(n + 1/2) = \frac{\sqrt{\pi}}{2^n}(2n-1)!!$$

The "double factorial" symbol mean the product of every other integer up to the given one. E.g. $5!! = 15$. The double factorial of an even integer can be expressed in terms of the single factorial. Do so. What about odd integers?

1.20 Evaluate this integral. Just find the right substitution. $\displaystyle\int_0^\infty dt\, e^{-t^a} \qquad (a > 0)$

1.21 A triangle has sides a, b, c, and the angle opposite c is γ. Express the area of the triangle in terms of a, b, and γ. Write the law of cosines for this triangle and then use $\sin^2\gamma + \cos^2\gamma = 1$ to express the area of a triangle solely in terms of the lengths of its three sides. The resulting formula is not especially pretty or even clearly symmetrical in the sides, but if you introduce the semiperimeter, $s = (a+b+c)/2$, you can rearrange the answer into a neat, symmetrical form. Check its validity in a couple of special cases. Ans: $\sqrt{s(s-a)(s-b)(s-c)}$ (Hero's formula)

1.22 An arbitrary linear combination of the sine and cosine, $A\sin\theta + B\cos\theta$, is a phase-shifted cosine: $C\cos(\theta+\delta)$. Solve for C and δ in terms of A and B, deriving an identity in θ.

1.23 Solve the two simultaneous linear equations

$$ax + by = e, \qquad cx + dy = f$$

and do it solely by elementary manipulation (+, −, ×, ÷), not by any special formulas. Analyze all the *qualitatively different* cases and draw graphs to describe each. In every case, how many if any solutions are there? Because of its special importance later, look at the case $e = f = 0$ and analyze it as if it's a separate problem. You should be able to discern and to classify the circumstances under which there is one solution, no solution, or many solutions. Ans: Sometimes a unique solution. Sometimes no solution. Sometimes many solutions. Draw two lines in the plane; how many qualitatively different pictures are there?

1.24 Use parametric differentiation to evaluate the integral $\int x^2 \sin x\, dx$. Find a table of integrals if you want to verify your work.

1.25 Derive all the limits on the integrals in Eq. (1.32) and then do the integrals.

1.26 Compute the area of a circle using rectangular coordinates,

1.27 (**a**) Compute the area of a triangle using rectangular coordinates, so $dA = dx\,dy$. Make it a right triangle with vertices at $(0,0)$, $(a,0)$, and (a,b). (**b**) Do it again, but reversing the order of integration. (**c**) Now compute the area of this triangle using *polar* coordinates. Examine this carefully to see which order of integration makes the problem easier.

1.28 Start from the definition of a derivative, $\lim \big(f(x+\Delta x) - f(x)\big)/\Delta x$, and derive the chain rule.

$$f(x) = g\big(h(x)\big) \Longrightarrow \frac{df}{dx} = \frac{dg}{dh}\frac{dh}{dx}$$

Now pick special, fairly simple cases for g and h to test whether your result really works. That is, choose functions so that you can do the differentiation explicitly and compare the results, but also functions with enough structure that they aren't trivial.

1.29 Starting from the definitions, derive how to do the derivative,

$$\frac{d}{dx}\int_0^{f(x)} g(t)\,dt$$

Now pick special, fairly simple cases for f and g to test whether your result really works. That is, choose functions so that you can do the integration and differentiation explicitly, but ones such the result isn't trivial.

1.30 Sketch these graphs, working by hand only, no computers:

$$\frac{x}{a^2+x^2}, \qquad \frac{x^2}{a^2-x^2}, \qquad \frac{x}{a^3+x^3}, \qquad \frac{x-a}{a^2-(x-a)^2}, \qquad \frac{x}{L^2-x^2}+\frac{x}{L}$$

1.31 Sketch by hand only, graphs of

$$\sin x \;(-3\pi < x < +4\pi), \qquad \frac{1}{\sin x}\;(-3\pi < x < +4\pi), \qquad \sin(x-\pi/2)\;(-3\pi < x < +4\pi)$$

1.32 Sketch by hand only, graphs of

$$f(\phi) = 1 + \frac{1}{2}\sin^2\phi \;(0 \le \phi \le 2\pi), \qquad f(\phi) = \begin{cases} \phi & (0 < \phi < \pi) \\ \phi - 2\pi & (\pi < \phi < 2\pi) \end{cases}$$

$$f(x) = \begin{cases} x^2 & (0 \le x < a) \\ (x-2a)^2 & (a \le x \le 2a) \end{cases}, \qquad f(r) = \begin{cases} Kr/R^3 & (0 \le r \le R) \\ K/r^2 & (R < r < \infty) \end{cases}$$

1.33 From the definition of the Riemann integral make a numerical calculation of the integral

$$\int_0^1 dx\, \frac{4}{1+x^2}$$

Use 1 interval, then 2 intervals, then 4 intervals. If you choose to write your own computer program for an arbitrary number of intervals, by all means do so. As with the example in the text, choose the midpoints of the intervals to evaluate the function. To check your answer, do a trig substitution and evaluate the integral exactly. What is the % error from the exact answer in each case? [100×(wrong − right)/right] Ans: π

1.34 Evaluate erf(1) numerically. Use 4 intervals. Ans: 0.842700792949715 (more or less)

1.35 Evaluate $\int_0^\pi dx\, \sin x/x$ numerically. Ans: 1.85193705198247 or so.

1.36 x and y are related by the equation $x^3 - 4xy + 3y^3 = 0$. You can easily check that $(x, y) = (1, 1)$ satisfies it, now what is dy/dx at that point? Unless you choose to look up and plug in to the cubic formula, I suggest that you differentiate the whole equation with respect to x and solve for dy/dx.
Generalize this to finding dy/dx if $f(x, y) = 0$. Ans: 1/5

1.37 When flipping a coin N times, what fraction of the time will the number of heads in the run lie between $\big(N/2 - 2\sqrt{N/2}\big)$ and $\big(N/2 + 2\sqrt{N/2}\big)$? What are these numbers for $N = 1000$? Ans: 99.5%

1.38 For $N = 4$ flips of a coin, count the number of times you get 0, 1, 2, *etc.* heads out of $2^4 = 16$ cases. Compare these results to the exponential approximation of Eq. (1.17). Ans: $2 \to 0.375$ and 0.399

1.39 Is the integral of Eq. (1.17) over all δ equal to one?

1.40 If there are 100 molecules of a gas bouncing around in a room, about how long will you have to wait to find that all of them are in the left half of the room? Assume that you make a new observation every microsecond and that the observations are independent of each other. Ans: A million times the age of the universe. [Care to try 10^{23} molecules?]

1.41 If you flip 1000 coins 1000 times, about how many times will you get exactly 500 heads and 500 tails? What if it's 100 coins and 100 trials, getting 50 heads? Ans: 25, 8

1.42 (**a**) Use parametric differentiation to evaluate $\int x\,dx$. Start with $\int e^{\alpha x}dx$. Differentiate and then let $\alpha \to 0$.
(**b**) Now that the problem has blown up in your face, change the integral from an indefinite to a definite integral such as $\int_a^b$ and do it again. There are easier ways to do this integral, but the point is that this method is really designed for definite integrals. It may not work on indefinite ones.

1.43 The Gamma function satisfies the identity

$$\Gamma(x)\Gamma(1-x) = \pi/\sin\pi x$$

What does this tell you about the Gamma function of 1/2? What does it tell you about its behavior near the negative integers? Compare this result to that of problem 1.16.

1.44 Start from the definition of a derivative, manipulate some terms: (**a**) derive the rule for differentiating the function h, where $h(x) = f(x)g(x)$ is the product of two other functions.
(**b**) Integrate the resulting equation with respect to x and derive the formula for integration by parts.

1.45 Show that in polar coordinates the equation $r = 2a\cos\phi$ is a circle. Now compute its area in this coordinate system.

1.46 The cycloid* has the parametric equations $x = a\theta - a\sin\theta$, and $y = a - a\cos\theta$. Compute the area, $\int y\,dx$ between one arc of this curve and the x-axis. Ans: $3\pi a^2$

1.47 An alternate approach to the problem 1.13: Change variables in the integral definition of erf to $t = \alpha u$. Now differentiate with respect to α and of course the derivative must be zero and there's your answer. Do the same thing for problem 1.14 and the Gamma function.

1.48 Recall section 1.5 and compute this second derivative to show

$$\frac{d^2}{dt^2}\int_0^t dt'\,(t-t')F(t') = F(t)$$

1.49 From the definition of a derivative show that

$$\text{If } x = f(\theta) \qquad \text{and} \qquad t = g(\theta) \qquad \text{then} \qquad \frac{dx}{dt} = \frac{df/d\theta}{dg/d\theta}$$

Make up a couple of functions that let you test this explicitly.

1.50 Redo problem 1.6 another way: $x = \sinh^{-1} y \leftrightarrow y = \sinh x$. Differentiate the second of these with respect to y and solve for dx/dy. Ans: $d\sinh^{-1} y/dy = 1/\sqrt{1+y^2}$.

* www-groups.dcs.st-and.ac.uk/~history/Curves/Cycloid.html

Infinite Series

Infinite series are among the most powerful and useful tools that you've encountered in your introductory calculus course. It's easy to get the impression that they are simply a clever exercise in manipulating limits and in studying convergence, but they are among the majors tools used in analyzing differential equations, in developing methods of numerical analysis, in defining new functions, in estimating the behavior of functions, and more.

2.1 The Basics

There are a handful of infinite series that you should memorize and should know just as well as you do the multiplication table. The first of these is the geometric series,

$$1 + x + x^2 + x^3 + x^4 + \cdots = \sum_0^\infty x^n = \frac{1}{1-x} \qquad \text{for } |x| < 1. \tag{2.1}$$

It's very easy derive because in this case you can sum the finite form of the series and then take a limit. Write the series out to the term x^N and multiply it by $(1-x)$.

$$\begin{aligned}(1 + x + x^2 + x^3 + \cdots + x^N)(1-x) &= \\ (1 + x + x^2 + x^3 + \cdots + x^N) - (x + x^2 + x^3 + x^4 + \cdots + x^{N+1}) &= 1 - x^{N+1}\end{aligned} \tag{2.2}$$

If $|x| < 1$ then as $N \to \infty$ this last term, x^{N+1}, goes to zero and you have the answer. If x is outside this domain the terms of the infinite series don't even go to zero, so there's no chance for the series to converge to anything.

The finite sum up to x^N is useful on its own. For example it's what you use to compute the payments on a loan that's been made at some specified interest rate. You use it to find the pattern of light from a diffraction grating.

$$\sum_0^N x^n = \frac{1 - x^{N+1}}{1-x} \tag{2.3}$$

Some other common series that you need to know are power series for elementary functions:

$$\begin{aligned}
e^x &= 1 + x + \frac{x^2}{2!} + \cdots &&= \sum_0^\infty \frac{x^k}{k!} \\
\sin x &= x - \frac{x^3}{3!} + \cdots &&= \sum_0^\infty (-1)^k \frac{x^{2k+1}}{(2k+1)!} \\
\cos x &= 1 - \frac{x^2}{2!} + \cdots &&= \sum_0^\infty (-1)^k \frac{x^{2k}}{(2k)!} \\
\ln(1+x) &= x - \frac{x^2}{2} + \frac{x^3}{3} - \cdots &&= \sum_1^\infty (-1)^{k+1} \frac{x^k}{k} \qquad (|x| < 1) \\
(1+x)^\alpha &= 1 + \alpha x + \frac{\alpha(\alpha-1)x^2}{2!} + \cdots &&= \sum_{k=0}^\infty \frac{\alpha(\alpha-1)\cdots(\alpha-k+1)}{k!} x^k \qquad (|x| < 1)
\end{aligned} \tag{2.4}$$

Of course, even better than memorizing them is to understand their derivations so well that you can derive them as fast as you can write them down. For example, the cosine is the derivative of the sine, so if you know the latter series all you have to do is to differentiate it term by term to get the cosine series. The logarithm of $(1+x)$ is an integral of $1/(1+x)$ so you can get its series from that of the geometric series. The geometric series is a special case of the binomial series for $\alpha=-1$, but it's easier to remember the simple case separately. You can express all of them as special cases of the general Taylor series.

What is the sine of 0.1 radians? Just use the series for the sine and you have the answer, 0.1, or to more accuracy, $0.1-0.001/6=0.099833$

What is the square root of 1.1? $\sqrt{1.1}=(1+.1)^{1/2}=1+\frac{1}{2}\cdot 0.1=1.05$

What is 1/1.9? $1/(2-.1)=1/[2(1-.05)]=\frac{1}{2}(1+.05)=.5+.025=.525$ from the first terms of the geometric series.

What is $\sqrt[3]{1024}$? $\sqrt[3]{1024}=\sqrt[3]{1000+24}=\sqrt[3]{1000(1+24/1000)}=$ $10(1+24/1000)^{1/3}=10(1+8/1000)=10.08$

As you see from the last two examples you have to cast the problem into a form fitting the expansion that you know. When you want to use the binomial series, rearrange and factor your expression so that you have

$$\left(1+\text{something small}\right)^{\alpha}$$

2.2 Deriving Taylor Series

How do you derive these series? The simplest way to get any of them is to assume that such a series exists and then to deduce its coefficients in sequence. Take the sine for example, assume that you can write

$$\sin x=A+Bx+Cx^2+Dx^3+Ex^4+\cdots$$

Evaluate this at $x=0$ to get

$$\sin 0=0=A+B0+C0^2+D0^3+E0^4+\cdots=A$$

so the first term, $A=0$. Now differentiate the series, getting

$$\cos x=B+2Cx+3Dx^2+4Ex^3+\cdots$$

Again set $x=0$ and all the terms on the right except the first one vanish.

$$\cos 0=1=B+2C0+3D0^2+4E0^3+\cdots=B$$

Keep repeating this process, evaluating in turn all the coefficients of the assumed series.

$$\begin{aligned}
\sin x&=A+Bx+Cx^2+Dx^3+Ex^4+\cdots &\qquad \sin 0&=0=A\\
\cos x&=B+2Cx+3Dx^2+4Ex^3+\cdots &\qquad \cos 0&=1=B\\
-\sin x&=2C+6Dx+12Ex^2+\cdots &\qquad -\sin 0&=0=2C\\
-\cos x&=6D+24Ex+60Fx^2+\cdots &\qquad -\cos 0&=-1=6D\\
\sin x&=24E+120Fx+\cdots &\qquad \sin 0&=0=24E\\
\cos x&=120F+\cdots &\qquad \cos 0&=1=120F
\end{aligned}$$

This shows the terms of the series for the sine as in Eq. (2.4).

Does this show that the series converges? If it converges does it show that it converges to the sine? No to both. Each statement requires more work, and I'll leave the second one to advanced calculus books. Even better, when you understand the subject of complex variables, these questions about series become much easier to understand.

The generalization to any function is obvious. You match the coefficients in the assumed expansion, and get

$$f(x) = f(0) + xf'(0) + \frac{x^2}{2!}f''(0) + \frac{x^3}{3!}f'''(0) + \frac{x^4}{4!}f''''(0) + \cdots$$

You don't have to do the expansion about the point zero. Do it about another point instead.

$$f(t) = f(t_0) + (t - t_0)f'(t_0) + \frac{(t - t_0)^2}{2!}f''(t_0) + \cdots \tag{2.5}$$

What good are infinite series?
This is sometimes the way that a new function is introduced and developed, typically by determining a series solution to a new differential equation. (Chapter 4)
This is a tool for the numerical evaluation of functions.
This is an essential tool to understand and invent numerical algorithms for integration, differentiation, interpolation, and many other common numerical methods. (Chapter 11)
To understand the behavior of complex-valued functions of a complex variable you will need to understand these series for the case that the variable is a complex number. (Chapter 14)

All the series that I've written above are power series (Taylor series), but there are many other possibilities.

$$\zeta(z) = \sum_1^\infty \frac{1}{n^z} \tag{2.6}$$

$$x^2 = \frac{L^2}{3} + \frac{4L^2}{\pi^2}\sum_1^\infty (-1)^n \frac{1}{n^2}\cos\left(\frac{n\pi x}{L}\right) \qquad (-L \le x \le L) \tag{2.7}$$

The first is a Dirichlet series defining the Riemann zeta function, a function that appears in statistical mechanics among other places.
The second is an example of a Fourier series. See chapter five for more of these.
Still another type of series is the Frobenius series, useful in solving differential equations: Its form is $\sum_k a_k x^{k+s}$. The number s need not be either positive or an integer. Chapter four has many examples of this form.

There are a few technical details about infinite series that you have to go through. In introductory calculus courses there can be a tendency to let these few details overwhelm the subject so that you are left with the impression that that's all there is, not realizing that this stuff is useful. Still, you do need to understand it.*

---

* For animations showing how fast some of these power series converge, check out www.physics.miami.edu/nearing/mathmethods/power-animations.html

2.3 Convergence

Does an infinite series converge? Does the limit as $N \to \infty$ of the sum, $\sum_1^N u_k$, exist? There are a few common and useful ways to answer this. The first and really the foundation for the others is the comparison test.

Let u_k and v_k be sequences of real numbers, positive at least after some value of k. Also assume that for all k greater than some finite value, $u_k \le v_k$. Also assume that the sum, $\sum_k v_k$ *does* converge.
The other sum, $\sum_k u_k$ then converges too. This is almost obvious, but it's worth the little effort that a proof takes.

The required observation is that an increasing sequence of real numbers, bounded above, has a limit.

After some point, $k = M$, all the u_k and v_k are positive and $u_k \le v_k$. The sum $a_n = \sum_M^n v_k$ then forms an increasing sequence of real numbers, so by assumption this has a limit (the series converges). The sum $b_n = \sum_M^n u_k$ is an increasing sequence of real numbers also. Because $u_k \le v_k$ you immediately have $b_n \le a_n$ for all n.

$$b_n \le a_n \le \lim_{n\to\infty} a_n$$

this simply says that the increasing sequence b_n has an upper bound, so it has a limit and the theorem is proved.

Ratio Test

To apply this comparison test you need a stable of known convergent series. One that you do have is the geometric series, $\sum_k x^k$ for $|x| < 1$. Let this x^k be the v_k of the comparison test. Assume at least after some point $k = K$ that all the $u_k > 0$.
Also that $u_{k+1} \le x u_k$.

$$\text{Then } u_{K+2} \le x u_{K+1} \qquad \text{and} \qquad u_{K+1} \le x u_K \qquad \text{gives} \qquad u_{K+2} \le x^2 u_K$$

You see the immediate extension is

$$u_{K+n} \le x^n u_K$$

As long as $x < 1$ this is precisely set up for the comparison test using $\sum_n u_K x^n$ as the series that dominates the $\sum_n u_n$. This test, the *ratio test* is more commonly stated for positive u_k as

$$\text{If for large } k,\ \frac{u_{k+1}}{u_k} \le x < 1 \qquad \text{then the series} \qquad \sum u_k \qquad \text{converges} \tag{2.8}$$

This is one of the more commonly used convergence tests, not because it's the best, but because it's simple and it works a lot of the time.

Integral Test

The integral test is another way to check for convergence or divergence. If f is a *decreasing positive* function and you want to determine the convergence of $\sum_n f(n)$, you can look at the integral $\int^\infty dx\, f(x)$ and check *it* for convergence. The series and the integral converge or diverge together.

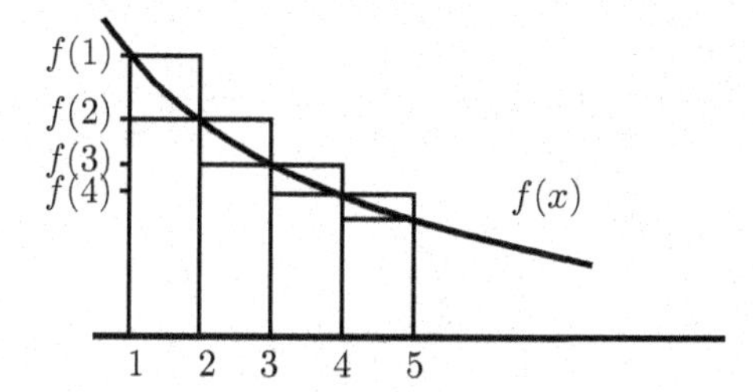

From the graph you see that the function f lies between the tops of the upper and the lower rectangles. The area under the curve of f between n and $n+1$ lies between the areas of the two rectangles. That's the reason for the assumption that f is decreasing and positive.

$$f(n)\cdot 1 > \int_n^{n+1} dx\, f(x) > f(n+1)\cdot 1$$

Add these inequalities from $n=k$ to $n=\infty$ and you get

$$\begin{aligned} f(k)+f(k+1)+\cdots > \int_k^{k+1} + \int_{k+1}^{k+2} + \cdots = \int_k^\infty dx\, f(x) \\ > f(k+1)+f(k+2)+\cdots > \int_{k+1}^\infty dx\, f(x) > f\cdots \end{aligned} \tag{2.9}$$

The only difference between the infinite series on the left and on the right is one term, so either everything converges or everything diverges.

You can do better than this and use these inequalities to get a quick estimate of the sum of a series that would be too tedious to sum by itself. For example

$$\sum_1^\infty \frac{1}{n^2} = 1 + \frac{1}{2^2} + \frac{1}{3^2} + \sum_4^\infty \frac{1}{n^2}$$

This last sum lies between two integrals.

$$\int_3^\infty dx\,\frac{1}{x^2} > \sum_4^\infty \frac{1}{n^2} > \int_4^\infty dx\,\frac{1}{x^2} \tag{2.10}$$

that is, between $1/3$ and $1/4$. Now I'll estimate the whole sum by adding the first three terms explicitly and taking the arithmetic average of these two bounds.

$$\sum_1^\infty \frac{1}{n^2} \approx 1 + \frac{1}{2^2} + \frac{1}{3^2} + \frac{1}{2}\left(\frac{1}{3}+\frac{1}{4}\right) = 1.653 \tag{2.11}$$

The exact sum is more nearly 1.644934066848226, but if you use brute-force addition of the original series to achieve accuracy equivalent to this 1.653 estimation you will need to take about 120 terms. This series converges, but not very fast. See also problem 2.24.

Quicker Comparison Test

There is another way to handle the comparison test that works very easily and quickly (if it's applicable). Look at the terms of the series for large n and see what the approximate behavior of the nth term is. That provides a comparison series. This is better shown by an example:

$$\sum_1^\infty \frac{n^3 - 2n + 1/n}{5n^5 + \sin n}$$

For large n, the numerator is essentially n^3 and the denominator is essentially $5n^5$, so for large n this series is approximately like

$$\sum^\infty \frac{1}{5n^2}$$

More precisely, the ratio of the nth term of this approximate series to that of the first series goes to one as $n \to \infty$. This comparison series converges, so the first one does too. If one of the two series diverges, then the other does too.

Apply the ratio test to the series for e^x.

$$e^x = \sum_0^\infty x^k/k! \qquad \text{so} \qquad \frac{u_{k+1}}{u_k} = \frac{x^{k+1}/(k+1)!}{x^k/k!} = \frac{x}{k+1}$$

As $k \to \infty$ this quotient approaches zero no matter the value of x. This means that the series converges for all x.

Absolute Convergence

If a series has terms of varying signs, that should help the convergence. A series is absolutely convergent if it converges when you replace each term by its absolute value. If it's absolutely convergent then it will certainly be convergent when you reinstate the signs. An example of a series that is convergent but not absolutely convergent is

$$\sum_{k=1}^\infty (-1)^{k+1}\frac{1}{k} = 1 - \frac{1}{2} + \frac{1}{3} - \ldots = \ln(1+1) = \ln 2 \tag{2.12}$$

Change all the minus signs to plus and the series is divergent. (Use the integral test.)

Can you rearrange the terms of an infinite series? Sometimes yes and sometimes no. If a series is convergent but not *absolutely* convergent, then each of the two series, the positive terms and the negative terms, is separately divergent. In this case you can rearrange the terms of the series to converge to anything you want! Take the series above that converges to $\ln 2$. I want to rearrange the terms so that it converges to $\sqrt{2}$. Easy. Just start adding the positive terms until you've passed $\sqrt{2}$. Stop and now start adding negative ones until you're below that point. Stop and start adding positive terms again. Keep going and you can get to any number you want.

$$1 + \frac{1}{3} + \frac{1}{5} - \frac{1}{2} + \frac{1}{7} + \frac{1}{9} + \frac{1}{11} + \frac{1}{13} - \frac{1}{3} \quad etc.$$

2.4 Series of Series

When you have a function whose power series you need, there are sometimes easier ways to the result than a straight-forward attack. Not always, but you should look first. If you need the expansion of e^{ax^2+bx} about the origin you can do a lot of derivatives, using the general form of the Taylor expansion. Or you can say

$$e^{ax^2+bx} = 1 + (ax^2+bx) + \frac{1}{2}(ax^2+bx)^2 + \frac{1}{6}(ax^2+bx)^3 + \cdots \tag{2.13}$$

and if you need the individual terms, expand the powers of the binomials and collect like powers of x:

$$1 + bx + (a + b^2/2)x^2 + (ab + b^3/6)x^3 + \cdots$$

If you're willing to settle for an expansion about another point, complete the square in the exponent

$$\begin{aligned} e^{ax^2+bx} &= e^{a(x^2+bx/a)} = e^{a(x^2+bx/a+b^2/4a^2)-b^2/4a} = e^{a(x+b/2a)^2-b^2/4a} = e^{a(x+b/2a)^2}e^{-b^2/4a} \\ &= e^{-b^2/4a}\big[1 + a(x+b/2a)^2 + a^2(x+b/2a)^4/2 + \cdots\big] \end{aligned}$$

and this is a power series expansion about the point $x_0 = -b/2a$.

What is the power series expansion of the secant? You can go back to the general formulation and differentiate a lot or you can use a combination of two known series, the cosine and the geometric series.

$$\begin{aligned} \sec x &= \frac{1}{\cos x} = \frac{1}{1 - \frac{1}{2!}x^2 + \frac{1}{4!}x^4 + \cdots} = \frac{1}{1 - \big[\frac{1}{2!}x^2 - \frac{1}{4!}x^4 + \cdots\big]} \\ &= 1 + [\,] + [\,]^2 + [\,]^3 + \cdots \\ &= 1 + \big[\tfrac{1}{2!}x^2 - \tfrac{1}{4!}x^4 + \cdots\big] + \big[\tfrac{1}{2!}x^2 - \tfrac{1}{4!}x^4 + \ldots\big]^2 + \cdots \\ &= 1 + \tfrac{1}{2!}x^2 + \big(-\tfrac{1}{4!} + (\tfrac{1}{2!})^2\big)x^4 + \cdots \\ &= 1 + \tfrac{1}{2!}x^2 + \tfrac{5}{24}x^4 + \cdots \end{aligned} \tag{2.14}$$

This is a geometric series, each of whose terms is itself an infinite series. It still beats plugging into the general formula for the Taylor series Eq. (2.5).

What is $1/\sin^3 x$?

$$\begin{aligned} \frac{1}{\sin^3 x} &= \frac{1}{\big(x - x^3/3! + x^5/5! - \cdots\big)^3} = \frac{1}{x^3\big(1 - x^2/3! + x^4/5! - \cdots\big)^3} \\ &= \frac{1}{x^3\big(1+z\big)^3} = \frac{1}{x^3}(1 - 3z + 6z^2 - \cdots) \\ &= \frac{1}{x^3}(1 - 3(-x^2/3! + x^4/5! - \ldots) + 6(-x^2/3! + x^4/5! - \ldots)^2 \\ &= \frac{1}{x^3} + \frac{1}{2x} + \frac{51x}{360} + \cdots \end{aligned}$$

which is a Frobenius series.

2.5 Power series, two variables

The idea of a power series can be extended to more than one variable. One way to develop it is to use exactly the same sort of brute-force approach that I used for the one-variable case. Assume that there is some sort of infinite series and successively evaluate its terms.

$$f(x,y) = A + Bx + Cy + Dx^2 + Exy + Fy^2 + Gx^3 + Hx^2y + Ixy^2 + Jy^3 \cdots$$

Include all the possible linear, quadratic, cubic, and higher order combinations. Just as with the single variable, evaluate it at the origin, the point $(0,0)$.

$$f(0,0) = A + 0 + 0 + \cdots$$

Now differentiate, but this time you have to do it twice, once with respect to x while y is held constant and once with respect to y while x is held constant.

$$\frac{\partial f}{\partial x}(x,y) = B + 2Dx + Ey + \cdots \qquad \text{then} \qquad \frac{\partial f}{\partial x}(0,0) = B$$

$$\frac{\partial f}{\partial y}(x,y) = C + Ex + 2Fy + \cdots \qquad \text{then} \qquad \frac{\partial f}{\partial y}(0,0) = C$$

Three more partial derivatives of these two equations gives the next terms.

$$\frac{\partial^2 f}{\partial x^2}(x,y) = 2D + 6Gx + 2Hy \cdots$$

$$\frac{\partial^2 f}{\partial x\,\partial y}(x,y) = E + 2Hx + 2Iy \cdots$$

$$\frac{\partial^2 f}{\partial y^2}(x,y) = 2F + 2Ix + 6Jy \cdots$$

Evaluate these at the origin and you have the values of D, E, and F. Keep going and you have all the coefficients.

This is awfully cumbersome, but mostly because the crude notation that I've used. You can make it look less messy simply by choosing a more compact notation. If you do it neatly it's no harder to write the series as an expansion about any point, not just the origin.

$$f(x,y) = \sum_{m,n=0}^{\infty} A_{mn}(x-a)^m(y-b)^n \tag{2.15}$$

Differentiate this m times with respect to x and n times with respect to y, then set $x=a$ and $y=b$. Only one term survives and that is

$$\frac{\partial^{m+n} f}{\partial x^m \partial y^n}(a,b) = m!n!A_{mn}$$

I can use subscripts to denote differentiation so that $\frac{\partial f}{\partial x}$ is f_x and $\frac{\partial^3 f}{\partial x^2 \partial y}$ is f_{xxy}. Then the two-variable Taylor expansion is

$$\begin{aligned} f(x,y) = f(0) + &f_x(0)x + f_y(0)y + \\ &\frac{1}{2}\left[f_{xx}(0)x^2 + 2f_{xy}(0)xy + f_{yy}(0)y^2\right] + \\ &\frac{1}{3!}\left[f_{xxx}(0)x^3 + 3f_{xxy}(0)x^2y + 3f_{xyy}(0)xy^2 + f_{yyy}(0)y^3\right] + \cdots \end{aligned} \tag{2.16}$$

Again put more order into the notation and rewrite the general form using A_{mn} as

$$A_{mn} = \frac{1}{(m+n)!}\left(\frac{(m+n)!}{m!n!}\right)\frac{\partial^{m+n} f}{\partial x^m \partial y^n}(a,b) \tag{2.17}$$

That factor in parentheses is variously called the binomial coefficient or a combinatorial factor. Standard notations for it are

$$\frac{m!}{n!(m-n)!} = {}_mC_n = \binom{m}{n} \tag{2.18}$$

The binomial series, Eq. (2.4), for the case of a positive integer exponent is

$$(1+x)^m = \sum_{n=0}^{m} \binom{m}{n} x^n, \qquad \text{or more symmetrically}$$

$$(a+b)^m = \sum_{n=0}^{m} \binom{m}{n} a^n b^{m-n} \tag{2.19}$$

$$(a+b)^2 = a^2 + 2ab + b^2, \qquad (a+b)^3 = a^3 + 3a^2 b + 3ab^2 + b^3,$$
$$(a+b)^4 = a^4 + 4a^3 b + 6a^2 b^2 + 4ab^3 + b^4, \quad \text{etc.}$$

Its relation to combinatorial analysis is that if you ask how many different ways can you choose n objects from a collection of m of them, ${}_mC_n$ is the answer.

2.6 Stirling's Approximation

The Gamma function for positive integers is a factorial. A clever use of infinite series and Gaussian integrals provides a useful approximate value for the factorial of large n.

$$n! \sim \sqrt{2\pi n}\, n^n e^{-n} \qquad \text{for large } n \tag{2.20}$$

Start from the Gamma function of $n+1$.

$$n! = \Gamma(\,n+1) = \int_0^\infty dt\, t^n e^{-t} = \int_0^\infty dt\, e^{-t+n\ln t}$$

The integrand starts at zero, increases, and drops back down to zero as $t \to \infty$. The graph roughly resembles a Gaussian, and I can make this more precise by expanding the exponent around the point where it is a maximum. The largest contribution to the whole integral comes from the region near this point. Differentiate the exponent to find the maximum:

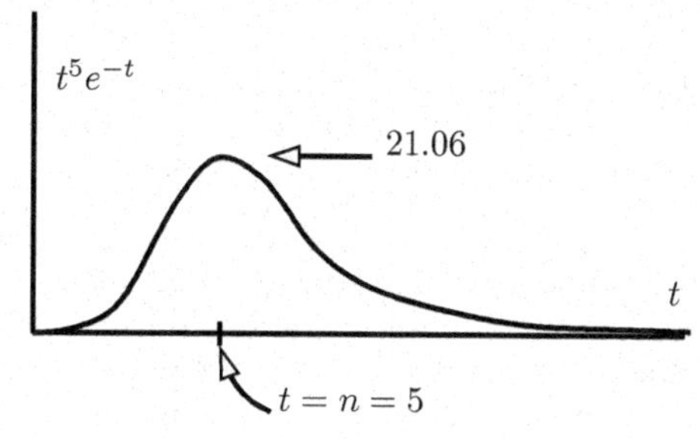

$$\frac{d}{dt}\big(-t+n\ln t\big) = -1+\frac{n}{t} = 0 \qquad \text{gives} \qquad t=n$$

Expand about this point

$$\begin{aligned} f(t) = -t+n\ln t = f(n) \qquad &+ (t-n)f'(n) \;+ (t-n)^2 f''(n)/2 \quad + \cdots \\ = -n+n\ln n \;+ 0 \qquad &\qquad\qquad + (t-n)^2(-n/n^2)/2 \;+ \cdots \end{aligned}$$

Keep terms to the second order and the integral is approximately

$$n! \sim \int_0^\infty dt\, e^{-n+n\ln n-(t-n)^2/2n} = n^n e^{-n} \int_0^\infty dt\, e^{-(t-n)^2/2n} \tag{2.21}$$

At the lower limit of the integral, at $t=0$, this integrand is $e^{-n/2}$, so if n is even moderately large then extending the range of the integral to the whole line $-\infty$ to $+\infty$ won't change the final answer much.

$$n^n e^{-n} \int_{-\infty}^\infty dt\, e^{-(t-n)^2/2n} = n^n e^{-n}\sqrt{2\pi n}$$

where the final integral is just the simplest of the Gaussian integrals in Eq. (1.10).

To see how good this is, try a few numbers

n	n!	Stirling	ratio	difference
1	1	0.922	0.922	0.078
2	2	1.919	0.960	0.081
5	120	118.019	0.983	1.981
10	3628800	3598695.619	0.992	30104.381

You can see that the *ratio* of the exact to the approximate result is approaching one even though the difference is getting very large. This is not a handicap, as there are many circumstances for which this is all you need. This derivation assumed that n is large, but notice that the result is not too bad even for modest values. The error is less than 2% for $n=5$. There are even some applications, especially in statistical mechanics, in which you can make a still cruder approximation and drop the factor $\sqrt{2\pi n}$. That is because in that context it is the logarithm of $n!$ that appears, and the ratio of the *logarithms* of the exact and even this cruder approximate number goes to one for large n. Try it.

Although I've talked about Stirling's approximation in terms of factorials, it started with the Gamma function, so Eq. (2.20) works just as well for $\Gamma(n+1)$ for any real n: $\Gamma(11.34 = 10.34+1) = 8\,116\,833.918$ and Stirling gives $8\,051\,701$.

Asymptotic

You may have noticed the symbol that I used in Eqs. (2.20) and (2.21). "$\sim$" doesn't mean "approximately equal to" or "about," because as you see here the difference between $n!$ and the Stirling approximation *grows* with n. That the ratio goes to one is the important point here and it gets this special symbol, "asymptotic to."

Probability Distribution

In section 1.4 the equation (1.17) describes the distribution of the results when you toss a coin. It's straight-forward to derive this from Stirling's formula. In fact it is just as easy to do a version of it for which the coin is biased, or more generally, for any case that one of the choices is more likely than the other.

Suppose that the two choices will come up at random with fractions a and b, where $a+b=1$. You can still picture it as a coin toss, but using a very unfair coin. Perhaps $a=1/3$ of the time it comes up tails and $b=2/3$ of the time it comes up heads. If you toss two coins, the possibilities are

TT HT TH HH

and the fractions of the time that you get each pair are respectively

$$a^2 \qquad ba \qquad ab \qquad b^2$$

This says that the fraction of the time that you get no heads, one head, or two heads are

$$a^2 = 1/9, \qquad 2ab = 4/9, \qquad b^2 = 4/9 \qquad \text{with total} \qquad (a+b)^2 = a^2 + 2ab + b^2 = 1 \tag{2.22}$$

Generalize this to the case where you throw N coins at a time and determine how often you expect to see 0, 1, ..., N heads. Equation (2.19) says

$$(a+b)^N = \sum_{k=0}^{N} \binom{N}{k} a^k b^{N-k} \qquad \text{where} \qquad \binom{N}{k} = \frac{N!}{k!(N-k)!}$$

When you make a trial in which you toss N coins, you expect that the "a" choice will come up N times only the fraction a^N of the trials. All tails and no heads. Compare problem 2.27.

The problem is now to use Stirling's formula to find an approximate result for the terms of this series. This is the fraction of the trials in which you turn up k tails and $N-k$ heads.

$$\begin{aligned} a^k b^{N-k} \frac{N!}{k!(N-k)!} &\sim a^k b^{N-k} \frac{\sqrt{2\pi N}\, N^N e^{-N}}{\sqrt{2\pi k}\, k^k e^{-k} \sqrt{2\pi(N-k)}\,(N-k)^{N-k} e^{-(N-k)}} \\ &= a^k b^{N-k} \frac{1}{\sqrt{2\pi}} \sqrt{\frac{N}{k(N-k)}}\, \frac{N^N}{k^k (N-k)^{N-k}} \end{aligned} \tag{2.23}$$

The complicated parts to manipulate are the factors with all the exponentials of k in them. Pull them out from the denominator for separate handling, leaving the square roots behind.

$$k^k (N-k)^{N-k} a^{-k} b^{-(N-k)}$$

The next trick is to take a logarithm and to do all the manipulations on it.

$$\ln \to k \ln k + (N-k)\ln(N-k) - k\ln a - (N-k)\ln b = f(k) \tag{2.24}$$

The original function is a maximum when this denominator is a minimum. When the numbers N and k are big, you can treat k as a continuous variable and differentiate with respect to it. Then set this derivative to zero and finally, expand in a power series about that point.

$$\frac{d}{dk}f(k) = \ln k + 1 - \ln(N-k) - 1 - \ln a + \ln b = 0$$

$$\ln\frac{k}{N-k} = \ln\frac{a}{b}, \qquad \frac{k}{N-k} = \frac{a}{b}, \qquad k = aN$$

This should be no surprise; a is the fraction of the time the first choice occurs, and it says that the most likely number of times that it occurs is that fraction times the number of trials. At this point, what is the second derivative?

$$\frac{d^2}{dk^2}f(k) = \frac{1}{k} + \frac{1}{N-k}$$

$$\text{when } k = aN, \qquad f''(k) = \frac{1}{k} + \frac{1}{N-k} = \frac{1}{aN} + \frac{1}{N-aN} = \frac{1}{aN} + \frac{1}{bN} = \frac{1}{abN}$$

About this point the power series for $f(k)$ is

$$\begin{aligned} f(k) &= f(aN) + (k-aN)f'(aN) + \frac{1}{2}(k-aN)^2 f''(aN) + \cdots \\ &= N\ln N + \frac{1}{2abN}(k-aN)^2 + \cdots \end{aligned} \tag{2.25}$$

To substitute this back into Eq. (2.23), take its exponential. Then because this will be a fairly sharp maximum, only the values of k near to aN will be significant. That allows me to use this central value of k in the slowly varying square root coefficient of that equation, and I can also neglect higher order terms in the power series expansion there. Let $\delta = k - aN$. The result is the Gaussian distribution.

$$\frac{1}{\sqrt{2\pi}}\sqrt{\frac{N}{aN(N-aN)}}\cdot\frac{N^N}{N^N e^{\delta^2/2abN}} = \frac{1}{\sqrt{2abN\pi}}e^{-\delta^2/2abN} \tag{2.26}$$

When $a = b = 1/2$, this reduces to Eq. (1.17).

When you accumulate N trials at a time (large N) and then look for the distribution in these cumulative results, you will commonly get a Gaussian. This is the central limit theorem, which says that whatever set of probabilities that you start with, not just a coin toss, you will get a Gaussian by averaging the data. (*Not really true.* There are some requirements* on the probabilities that aren't always met, but if as here the variable has a bounded domain then it's o.k. See problems 17.24 and 17.25 for a hint of where a naïve assumption that all distributions behave the same way that Gaussians do can be misleading.) If you listen to the clicks of a counter that records radioactive decays, they sound (and are) random, and the time interval between the clicks varies greatly. If you set the electronics to click at every tenth count, the result will sound regular, and the time interval between clicks will vary only slightly.

---

* finite mean and variance

2.7 Useful Tricks

There are a variety of ways to manipulate series, and while some of them are simple they are probably not the sort of thing you'd think of until you've seen them once. Example: What is the sum of

$$1 - \frac{1}{3} + \frac{1}{5} - \frac{1}{7} + \frac{1}{9} - \cdots ?$$

Introduce a parameter that you can manipulate, like the parameter you sometimes introduce to do integrals as in Eq. (1.5). Consider the series with the parameter x in it.

$$f(x) = x - \frac{x^3}{3} + \frac{x^5}{5} - \frac{x^7}{7} + \frac{x^9}{9} - \cdots \tag{2.27}$$

Differentiate this with respect to x to get

$$f'(x) = 1 - x^2 + x^4 - x^6 + x^8 - \cdots$$

That looks a bit like the geometric series except that it has only even powers and the signs alternate. Is that too great an obstacle? As $1/(1-x)$ has only plus signs, then change x to $-x$, and $1/(1+x)$ alternates in sign. Instead of x as a variable, use x^2, then you get exactly what you're looking for.

$$f'(x) = 1 - x^2 + x^4 - x^6 + x^8 - \cdots = \frac{1}{1+x^2}$$

Now to get back to the original series, which is $f(1)$ recall, all that I need to do is integrate this expression for $f'(x)$. The lower limit is zero, because $f(0) = 0$.

$$f(1) = \int_0^1 dx\, \frac{1}{1+x^2} = \tan^{-1} x \Big|_0^1 = \frac{\pi}{4}$$

This series converges so slowly however that you would never dream of computing π this way. If you take 100 terms, the next term is $1/201$ and you can get a better approximation to π by using $22/7$.

The geometric series is $1 + x + x^2 + x^3 + \cdots$, but what if there's an extra factor in front of each term?

$$f(x) = 2 + 3x + 4x^2 + 5x^3 + \cdots$$

Multiply this by x and it is $2x + 3x^2 + 4x^3 + 5x^4 + \cdots$, starting to look like a derivative.

$$xf(x) = 2x + 3x^2 + 4x^3 + 5x^4 + \cdots = \frac{d}{dx}\left(x^2 + x^3 + x^4 + \cdots\right)$$

Again, the geometric series pops up, though missing a couple of terms.

$$xf(x) = \frac{d}{dx}\left(1 + x + x^2 + x^3 + \cdots - 1 - x\right) = \frac{d}{dx}\left[\frac{1}{1-x} - 1 - x\right] = \frac{1}{(1-x)^2} - 1$$

The final result is then

$$f(x) = \frac{1}{x}\left[\frac{1-(1-x)^2}{(1-x)^2}\right] = \frac{2-x}{(1-x)^2}$$

2.8 Diffraction

When light passes through a very small opening it will be diffracted so that it will spread out in a characteristic pattern of higher and lower intensity. The analysis of the result uses many of the tools that you've looked at in the first two chapters, so it's worth showing the derivation first.

The light that is coming from the left side of the figure has a wavelength λ and wave number $k = 2\pi/\lambda$. The light passes through a narrow slit of width $= a$. The Huygens construction for the light that comes through the slit says that you can effectively treat each little part of the slit as if it is a source of part of the wave that comes through to the right. (As a historical note, the mathematical justification for this procedure didn't come until about 150 years after Huygens proposed it, so if you think it isn't obvious why it works, you're right.)

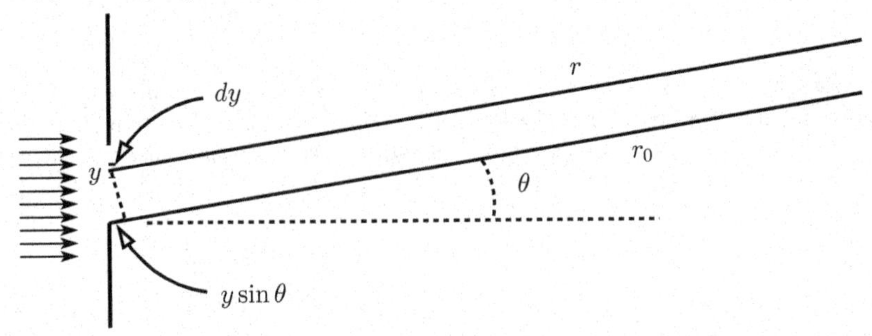

Call the coordinate along the width of the slit y, where $0 < y < a$. I want to find the total light wave that passes through the slit and that heads at the angle θ away from straight ahead. The light that passes through between coordinates y and $y + dy$ is a wave

$$A\,dy\,\cos(kr - \omega t)$$

Its amplitude is proportional to the amplitude of the incoming wave, A, and to the width dy that I am considering. The coordinate along the direction of the wave is r. The total wave that will head in this direction is the sum (integral) over all these little pieces of the slit.

Let r_0 be the distance measured from the bottom of the slit to where the light is received far away. Find the value of r by doing a little trigonometry, getting

$$r = r_0 - y\sin\theta$$

The total wave to be received is now the integral

$$\int_0^a A\,dy\,\cos\big(k(r_0 - y\sin\theta) - \omega t\big) = A\frac{\sin\big(k(r_0 - y\sin\theta) - \omega t\big)}{-k\sin\theta}\bigg|_0^a$$

Put in the limits to get

$$\frac{A}{-k\sin\theta}\big[\sin\big(k(r_0 - a\sin\theta) - \omega t\big) - \sin\big(kr_0 - \omega t\big)\big]$$

I need a trigonometric identity here, one that you can easily derive with the techniques of complex algebra in chapter 3.

$$\sin x - \sin y = 2\sin\left(\frac{x-y}{2}\right)\cos\left(\frac{x+y}{2}\right) \tag{2.28}$$

Use this and the light amplitude is

$$\frac{2A}{-k\sin\theta}\sin\left(-\frac{ka}{2}\sin\theta\right)\cos\left(k(r_0 - \frac{a}{2}\sin\theta) - \omega t\right) \tag{2.29}$$

The *wave* is the cosine factor. It is a cosine of $(k \cdot \text{distance} - \omega t)$, and the distance in question is the distance to the center of the slit. This is then a wave that appears to be coming from the middle of the slit, but with an amplitude that varies strongly with angle. That variation comes from the other factors in Eq. (2.29).

It's the variation with angle that's important. The intensity of the wave, the power per area, is proportional to the square of the wave's amplitude. I'm going to ignore all the constant factors, so there's no need to worry about the constant of proportionality. The intensity is then (up to a factor)

$$I = \frac{\sin^2\left((ka/2)\sin\theta\right)}{\sin^2\theta} \tag{2.30}$$

For light, the wavelength is about 400 to 700 nm, and the slit may be a millimeter or a tenth of a millimeter. The size of $ka/2$ is then about

$$ka/2 = \pi a/\lambda \approx 3 \cdot 0.1\,\text{mm}/500\,\text{nm} \approx 1000$$

When you plot this intensity versus angle, the numerator vanishes when the argument of $\sin^2()$ is $n\pi$, with n an integer, +, −, or 0. This says that the intensity vanishes in these directions *except* for $\theta = 0$. In that case the denominator vanishes too, so you have to look closer. For the simpler case that $\theta \neq 0$, these angles are

$$n\pi = \frac{ka}{2}\sin\theta \approx \frac{ka}{2}\theta \qquad n = \pm 1,\ \pm 2, \ldots$$

Because ka is big, you have many values of n before the approximation that $\sin\theta = \theta$ becomes invalid. You can rewrite this in terms of the wavelength because $k = 2\pi/\lambda$.

$$n\pi = \frac{2\pi a}{2\lambda}\theta, \qquad \text{or} \qquad \theta = n\lambda/a$$

What happens at zero? Use power series expansions to evaluate this indeterminate form. The first term in the series expansion of the sine is θ itself, so

$$I = \frac{\sin^2\left((ka/2)\sin\theta\right)}{\sin^2\theta} \longrightarrow \frac{\left((ka/2)\theta\right)^2}{\theta^2} = \left(\frac{ka}{2}\right)^2 \tag{2.31}$$

What is the behavior of the intensity *near* $\theta = 0$? Again, use power series expansions, but keep another term

$$\sin\theta = \theta - \frac{1}{6}\theta^3 + \cdots, \qquad \text{and} \qquad (1+x)^\alpha = 1 + \alpha x + \cdots$$

Remember, $ka/2$ is big! This means that it makes sense to keep just one term of the sine expansion for $\sin\theta$ itself, but you'd better keep an extra term in the expansion of the $\sin^2(ka\ldots)$.

$$\begin{aligned} I \approx \frac{\sin^2\big((ka/2)\theta\big)}{\theta^2} &= \frac{1}{\theta^2}\left[\left(\frac{ka}{2}\theta\right) - \frac{1}{6}\left(\frac{ka}{2}\theta\right)^3 + \cdots\right]^2 \\ &= \frac{1}{\theta^2}\left(\frac{ka}{2}\theta\right)^2\left[1 - \frac{1}{6}\left(\frac{ka}{2}\theta\right)^2 + \cdots\right]^2 \\ &= \left(\frac{ka}{2}\right)^2\left[1 - \frac{1}{3}\left(\frac{ka}{2}\theta\right)^2 + \cdots\right] \end{aligned}$$

When you use the binomial expansion, put the binomial in the standard form, $(1+x)$ as in the second line of these equations. What is the shape of this function? Forget all the constants, and it looks like $1 - \theta^2$. That's a parabola.

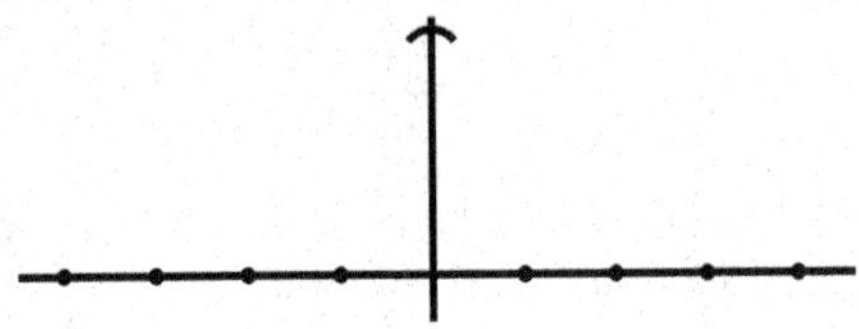

The dots are the points where the intensity goes to zero, $n\lambda/a$. Between these directions it reaches a maximum. How big is it there ? These maxima are about halfway between the points where $(ka\sin\theta)/2 = n\pi$. This is

$$\frac{ka}{2}\sin\theta = (n + {}^1\!/\!_2)\pi, \qquad n = \pm 1,\, \pm 2,\, \ldots$$

At these angles the value of I is, from Eq. (2.30),

$$I = \left(\frac{ka}{2}\right)^2\left(\frac{1}{(2n+1)\pi/2}\right)^2$$

The intensity at $\theta = 0$ is by Eq. (2.31), $(ka/2)^2$, so the maxima off to the side have intensities that are smaller than this by factors of

$$\frac{1}{9\pi^2/4} = 0.045, \qquad \frac{1}{25\pi^2/4} = 0.016, \ldots$$

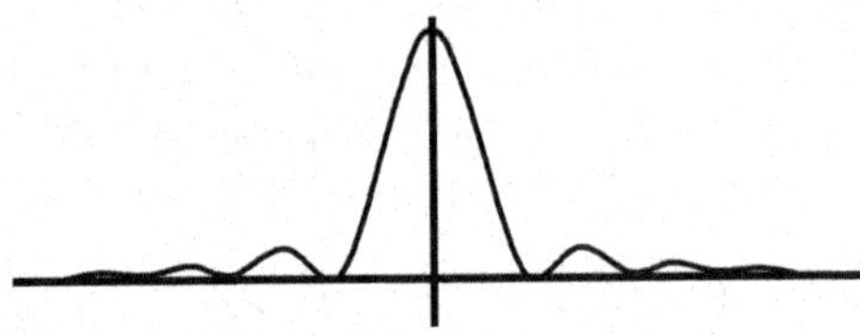

2.9 Checking Results

When you solve any problem, or at least think that you've solved it, you're not done. You still have to check to see whether your result makes any sense. If you are dealing with a problem whose solution is in the back of the book then do you think that the author is infallible? If there is no back of the book and you're working on something that you would like to publish, do you think that *you're* infallible? Either way you can't simply assume that you've made no mistakes; you have to look at your answer skeptically.

There's a second reason, at least as important, to examine your results: that's where you can learn some physics and gain some intuition. Solving a complex problem and getting a complicated answer may involve a lot of mathematics but you don't usually gain any physical insight from doing it. When you analyze your results you can gain an understanding of how the mathematical symbols are related to physical reality. Often an approximate answer to a complicated problem can give you more insight than an exact one, especially if the approximate answer is easier to analyze.

The first tool that you have to use at every opportunity is dimensional analysis. If you are computing a length and your result is a velocity then you are wrong. If you have something in your result that involves adding a time to an acceleration or an angle to a distance, then you've made a mistake; go back and find it. You can do this sort of analysis everywhere, and it is one technique that provides an automatic error finding mechanism. If an equation is dimensionally inconsistent, backtrack a few lines and see whether the units are wrong there too. If they are correct then you know that your error occurred between those two lines; then further narrow the region where the mistake happened by looking for the place at which the dimensions changed from consistent to inconsistent and that's where the mistake happened.

The second tool in your analysis is to examine all the parameters that occur in the result and to see what happens when you vary them. Especially see what happens when you push them to an extreme value. This is best explained by some examples. Start with some simple mechanics to see the procedure.

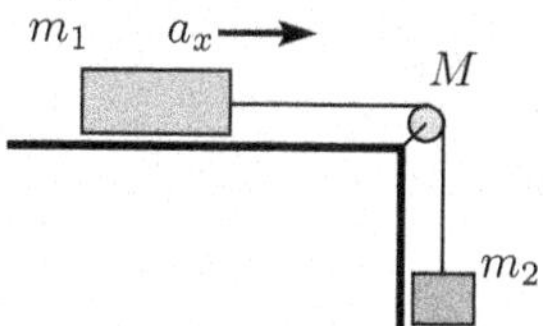

Two masses are attached by a string of negligible mass and that is wrapped around a pulley of mass M so that it can't slip on the pulley. Analyze them to determine what is wrong with each. Assume that there is no friction between m_1 and the table and that the string does not slip on the pulley.

$$\text{(a)}\ a_x = \frac{m_2 - m_1}{m_2 + m_1} g \qquad \text{(b)}\ a_x = \frac{m_2}{m_2 + m_1 - M/2} g \qquad \text{(c)}\ a_x = \frac{m_2 - M/2}{m_2 + m_1 + M/2} g$$

(a) If $m_1 \gg m_2$, this is negative, meaning that the motion of m_1 is being slowed down. But there's no friction or other such force to do this.
OR If $m_1 = m_2$, this is zero, but there are still unbalanced forces causing these masses to accelerate.

(b) If the combination of masses is just right, for example $m_1 = 1\,\text{kg}$, $m_2 = 1\,\text{kg}$, and $M = 2\,\text{kg}$, the denominator is zero. The expression for a_x blows up — a very serious problem.
OR If M is very large compared to the other masses, the denominator is negative, meaning that a_x is negative and the acceleration is a braking. Without friction, this is impossible.
(c) If $M \gg m_1$ and m_2, the numerator is mostly $-M/2$ and the denominator is mostly $+M/2$. This makes the whole expression negative, meaning that m_1 and m_2 are slowing down. There is no friction to do this, and all the forces are the direction to cause acceleration toward positive x.
OR If $m_2 = M/2$, this equals zero, saying that there is no acceleration, but in this system, a_x will always be positive.

The same picture, but *with* friction μ_k between m_1 and the table.

$$\text{(a) } a_x = \frac{m_2}{m_2 + \mu_\text{k} m_1 + M/2} g \qquad \text{(b) } a_x = \frac{m_2 - \mu_\text{k} m_1}{m_2 - M/2} g \qquad \text{(c) } a_x = \frac{m_2}{m_2 + \mu_\text{k} m_1 - M/2} g$$

(a) If μ_k is very large, this approaches zero. Large friction should cause m_1 to brake to a halt quickly with very large negative a_x.
OR If there is no friction, $\mu_\text{k} = 0$, then m_1 plays no role in this result but if it is big then you know that it will decrease the downward acceleration of m_2.
(b) The denominator can vanish. If $m_2 = M/2$ this is nonsense.
(c) This suffers from both of the difficulties of (a) and (b).

Trajectory Example
When you toss an object straight up with an initial speed v_0, you may expect an answer for the motion as a function of time to be something like

$$v_y(t) = v_0 - gt, \qquad y(t) = v_0 t - \frac{1}{2} g t^2 \tag{2.32}$$

Should you expect this? Not if you remember that there's air resistance. If I claim that the answers are

$$v_y(t) = -v_\text{t} + (v_0 + v_\text{t}) e^{-gt/v_\text{t}}, \qquad y(t) = -v_\text{t} t + (v_0 + v_\text{t}) \frac{v_\text{t}}{g} \left[1 - e^{-gt/v_\text{t}}\right] \tag{2.33}$$

then this claim has to be inspected to see if it makes sense. And I never bothered to tell you what the expression "v_t" means anyway. You have to figure that out. Fortunately that's not difficult in this case. What happens to these equations for very large time? The exponentials go to zero, so

$$v_y \longrightarrow -v_\text{t} + (v_0 + v_\text{t}) \cdot 0 = -v_\text{t}, \qquad \text{and} \qquad y \longrightarrow -v_\text{t} t + (v_0 + v_\text{t}) \frac{v_\text{t}}{g}$$

v_t is the terminal speed. After a long enough time a falling object will reach a speed for which the force by gravity and the force by the air will balance each other and the velocity then remains constant.

Do they satisfy the initial conditions? Yes:

$$v_y(0) = -v_t + (v_0 + v_t)e^0 = v_0, \qquad y(0) = 0 + (v_0 + v_t)\frac{v_t}{g}\cdot(1-1) = 0$$

What do these behave like for small time? They ought to reduce to something like the expressions in Eq. (2.32), but just as important is to determine what the deviation from that simple form is. Keep some extra terms in the series expansion. How many extra terms? If you're not certain, then keep one more than you think you will need. After some experience you will usually be able to anticipate what to do. Expand the exponential:

$$\begin{aligned} v_y(t) &= -v_t + (v_0 + v_t)\left[1 + \frac{-gt}{v_t} + \frac{1}{2}\left(\frac{-gt}{v_t}\right)^2 + \cdots\right] \\ &= v_0 - \left(1 + \frac{v_0}{v_t}\right)gt + \frac{1}{2}\left(1 + \frac{v_0}{v_t}\right)\frac{g^2t^2}{v_t} + \cdots \end{aligned}$$

The coefficient of t says that the object is slowing down more rapidly than it would have without air resistance. So far, so good. Is the factor right? Not yet clear, so keep going. Did I need to keep terms to order t^2? Probably not, but there wasn't much algebra involved in doing it, so it was harmless.

Look at the other equation, for y.

$$\begin{aligned} y(t) &= -v_t t + (v_0 + v_t)\frac{v_t}{g}\left[1 - \left[1 - \frac{gt}{v_t} + \frac{1}{2}\left(\frac{gt}{v_t}\right)^2 - \frac{1}{6}\left(\frac{gt}{v_t}\right)^3 + \cdots\right]\right] \\ &= v_0 t - \frac{1}{2}\left(1 + \frac{v_0}{v_t}\right)gt^2 - \frac{1}{6}\left(1 + \frac{v_0}{v_t}\right)\frac{g^2t^3}{v_t} + \cdots \end{aligned}$$

Now differentiate this approximate expression for y with respect to time and you get the approximate expression for v_y. That means that everything appears internally consistent, and I haven't *introduced* any obvious error in the process of approximation.

What if the terminal speed is infinite, so there's no air resistance. The work to answer this is already done. Expanding e^{-gt/v_t} for small time is the same as for large v_t, so you need only look back at the preceding two sets of equations and let $v_t \longrightarrow \infty$. The result is precisely the equations (2.32), just as you should expect.

You can even determine something about the *force* that I assumed for the air resistance: $F_y = ma_y = m\,dv_y/dt$. Differentiate the approximate expression that you already have for v_y, then at least for small t

$$\begin{aligned} F_y &= m\frac{d}{dt}\left[v_0 - \left(1 + \frac{v_0}{v_t}\right)gt + \frac{1}{2}\left(1 + \frac{v_0}{v_t}\right)\frac{g^2t^2}{v_t} + \cdots\right] \\ &= -m\left(1 + \frac{v_0}{v_t}\right)g + \cdots = -mg - mgv_0/v_t + \cdots \end{aligned} \tag{2.34}$$

This says that the force appears to be (1) gravity plus (2) a force proportional to the initial velocity. The last fact comes from the factor v_0 in the second term of the force equation, and at time zero, that *is* the velocity. Does this imply that I assumed a force acting as $F_y = -mg-$ (a constant times)v_y? To this approximation that's the best guess. (It happens to be correct.) To verify it though, you would have to go back to the original un-approximated equations (2.33) and compute the force from them.

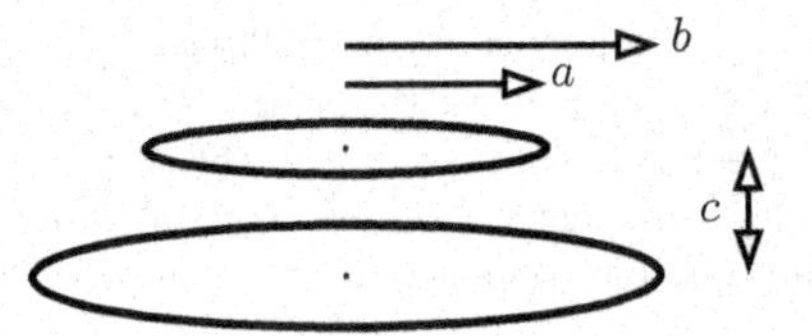

Electrostatics Example

Still another example, but from electrostatics this time: Two thin circular rings have radii a and b and carry charges Q_1 and Q_2 distributed uniformly around them. The rings are positioned in two parallel planes a distance c apart and with axes coinciding. The problem is to compute the force of one ring on the other, and for the single non-zero component the answer is (perhaps)

$$F_z = \frac{Q_1Q_2c}{2\pi^2\epsilon_0}\int_0^{\pi/2}\frac{d\theta}{\big[c^2+(b-a)^2+4ab\sin^2\theta\big]^{3/2}}. \tag{2.35}$$

Is this plausible? *First check the dimensions!* The integrand is (dimensionally) $1/(c^2)^{3/2}$ $= 1/c^3$, where c is one of the lengths. Combine this with the factors in front of the integral and one of the lengths (c's) cancels, leaving Q_1Q_2/ϵ_0c^2. This is (again dimensionally) the same as Coulomb's law, $q_1q_2/4\pi\epsilon_0r^2$, so it passes this test.

When you've done the dimensional check, start to consider the parameters that control the result. The numbers a, b, and c can be anything: small, large, or equal in any combination. For some cases you should be able to say what the answer will be, either approximately or exactly, and then check whether this complicated expression agrees with your expectation.

If the rings shrink to zero radius this has $a = b = 0$, so F_z reduces to

$$F_z \to \frac{Q_1Q_2c}{2\pi^2\epsilon_0}\int_0^{\pi/2}d\theta\frac{1}{c^3} = \frac{Q_1Q_2c}{2\pi^2\epsilon_0}\frac{\pi}{2c^3} = \frac{Q_1Q_2}{4\pi\epsilon_0c^2}$$

and this is the correct expression for two point charges a distance c apart.

If $c \gg a$ and b then this is really not very different from the preceding case, where a and b are zero.

If $a = 0$ this is

$$F_z \to \frac{Q_1Q_2c}{2\pi^2\epsilon_0}\int_0^{\pi/2}\frac{d\theta}{\big[c^2+b^2\big]^{3/2}} = \frac{Q_1Q_2c}{2\pi^2\epsilon_0}\frac{\pi/2}{\big[c^2+b^2\big]^{3/2}} = \frac{Q_1Q_2c}{4\pi\epsilon_0\big[c^2+b^2\big]^{3/2}} \tag{2.36}$$

The electric field on the axis of a ring is something that you can compute easily. The only component of the electric field at a point on the axis is itself along the axis. You can prove this by assuming that it's false. Suppose that there's a lateral component of $\vec{E}$ and say that it's to the right. Rotate everything by 180° about the axis and this component of $\vec{E}$ will now be pointing in the opposite direction. The ring of charge has not changed however, so $\vec{E}$ must be pointing in the original direction. This supposed sideways component is equal to minus itself, and something that's equal to minus itself is zero.

All the contributions to $\vec{E}$ except those parallel the axis add to zero. Along the axis each piece of charge dq contributes the component

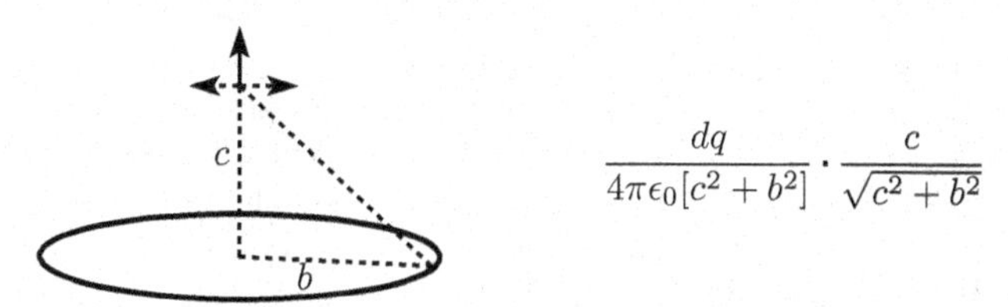

$$\frac{dq}{4\pi\epsilon_0[c^2+b^2]}\cdot\frac{c}{\sqrt{c^2+b^2}}$$

The first factor is the magnitude of the field of the point charge at a distance $r=\sqrt{c^2+b^2}$ and the last factor is the cosine of the angle between the axis and r. Add all the dq together and you get Q_1. Multiply that by Q_2 and you have the force on Q_2 and it agrees with the expressions Eq. (2.36)

If $c \to 0$ then $F_z \to 0$ in Eq. (2.35). The rings are concentric and the outer ring doesn't push the inner ring either up or down.

But wait. In this case, where $c \to 0$, what if $a=b$? Then the force should approach infinity instead of zero because the two rings are being pushed into each other. If $a=b$ then

$$F_z = \frac{Q_1Q_2c}{2\pi^2\epsilon_0}\int_0^{\pi/2}\frac{d\theta}{\left[c^2+4a^2\sin^2\theta\right]^{3/2}} \tag{2.37}$$

If you simply set $c=0$ in this equation you get

$$F_z = \frac{Q_1Q_2 0}{2\pi^2\epsilon_0}\int_0^{\pi/2}\frac{d\theta}{\left[4a^2\sin^2\theta\right]^{3/2}}$$

The numerator is zero, but look at the integral. The variable θ goes from 0 to $\pi/2$, and at the end near zero the integrand looks like

$$\frac{1}{\left[4a^2\sin^2\theta\right]^{3/2}} \approx \frac{1}{\left[4a^2\theta^2\right]^{3/2}} = \frac{1}{8a^3\theta^3}$$

Here I used the first term in the power series expansion of the sine. The integral near the zero end is then approximately

$$\int_0^{\cdots}\frac{d\theta}{\theta^3} = \frac{-1}{2\theta^2}\bigg|_0^{\cdots}$$

and that's infinite. This way to evaluate F_z is indeterminate: $0\cdot\infty$ can be anything. It doesn't show that this F_z gives the right answer, but it doesn't show that it's wrong either.

Estimating a tough integral

Although this is more difficult, even tricky, I'm going to show you how to examine this case for *small* values of c and not for $c = 0$. The problem is in figuring out how to estimate the integral (2.37) for small c, and the key is to realize that the only place the integrand gets big is in the neighborhood of $\theta = 0$. The trick then is to divide the range of integration into two pieces

$$\int_0^{\pi/2} \frac{d\theta}{\left[c^2 + 4a^2 \sin^2\theta\right]^{3/2}} = \int_0^{\Lambda} + \int_{\Lambda}^{\pi/2}$$

For any positive value of Λ the second piece of the integral will remain finite even as $c \to 0$. This means that in trying to estimate the way that the whole integral approaches infinity I can ignore the second part of the integral. Now choose Λ small enough that for $0 < \theta < \Lambda$ I can use the approximation $\sin\theta = \theta$, the first term in the series for sine. (Perhaps $\Lambda = 0.1$ or so.)

$$\text{for small } c,\quad \int_0^{\pi/2} \frac{d\theta}{\left[c^2 + 4a^2 \sin^2\theta\right]^{3/2}} \approx \int_0^{\Lambda} \frac{d\theta}{\left[c^2 + 4a^2\theta^2\right]^{3/2}} + \text{lower order terms}$$

This is an elementary integral. Let $\theta = (c/2a)\tan\phi$.

$$\int_0^{\Lambda} \frac{d\theta}{\left[c^2 + 4a^2\theta^2\right]^{3/2}} = \int_0^{\Lambda'} \frac{(c/2a)\sec^2\phi\, d\phi}{[c^2 + c^2\tan^2\phi]^{3/2}} = \frac{1}{2ac^2}\int_0^{\Lambda'} \cos\phi = \frac{1}{2ac^2}\sin\Lambda'$$

The limit Λ' comes from $\Lambda = (c/2a)\tan\Lambda'$, so this implies $\tan\Lambda' = 2a\Lambda/c$. Now given the tangent of an angle, I want the sine — that's the first page of chapter one.

$$\sin\Lambda' = \frac{2a\Lambda/c}{\sqrt{1 + (2a\Lambda/c)^2}} = \frac{2a\Lambda}{\sqrt{c^2 + 4a^2\Lambda^2}}$$

As $c \to 0$, this approaches one. Put all of this together and you have the behavior of the integral in Eq. (2.37) for small c.

$$\int_0^{\pi/2} \frac{d\theta}{\left[c^2 + 4a^2 \sin^2\theta\right]^{3/2}} \sim \frac{1}{2ac^2} + \text{lower order terms}$$

Insert this into Eq. (2.37) to get

$$F_z \sim \frac{Q_1 Q_2 c}{2\pi^2\epsilon_0} \cdot \frac{1}{2ac^2} = \frac{Q_1 Q_2}{4\pi^2\epsilon_0 ac}$$

Now why should I believe this any more than I believed the original integral? When you are very close to one of the rings, it will look like a long, straight line charge and the linear charge density on it is then $\lambda = Q_1/2\pi a$. What is the electric field of an infinitely long uniform line charge? $E_r = \lambda/2\pi\epsilon_0 r$. So now at the distance c from this line charge you know the E-field and to get the force on Q_2 you simply multiply this field by Q_2.

$$F_z \text{ should be } \frac{\lambda}{2\pi\epsilon_0 c}Q_2 = \frac{Q_1/2\pi a}{2\pi\epsilon_0 c}Q_2 \tag{2.38}$$

and that's exactly what I found in the preceding equation. After all these checks I think that I may believe the result, and more than that you begin to get an intuitive idea of what the result ought to look like. That's at least as valuable. It's what makes the difference between understanding the physics underlying a subject and simply learning how to manipulate the mathematics.

Exercises

1 Evaluate by hand $\cos 0.1$ to four places.

2 In the same way, evaluate $\tan 0.1$ to four places.

3 Use the first two terms of the binomial expansion to estimate $\sqrt{2} = \sqrt{1+1}$. What is the relative error? [(wrong−right)/right]

4 Same as the preceding exercise, but for $\sqrt{1.2}$.

5 What is the domain of convergence for $x - x^4 + x^9 - x^{4^2} + x^{5^2} - \cdots$

6 Does $\displaystyle\sum_{n=0}^{\infty} \cos(n) - \cos(n+1)$ converge?

7 Does $\displaystyle\sum_{n=1}^{\infty} \frac{1}{\sqrt{n}}$ converge?

8 Does $\displaystyle\sum_{n=1}^{\infty} \frac{n!}{n^2}$ converge?

9 What is the domain of convergence for $\dfrac{x}{1\cdot 2} - \dfrac{x^2}{2\cdot 2^2} + \dfrac{x^3}{3\cdot 3^3} - \dfrac{x^4}{4\cdot 4^4} + \cdots$?

10 From Eq. (2.1), find a series for $\dfrac{1}{(1-x)^2}$.

11 If x is positive, sum the series $1 + e^{-x} + e^{-2x} + e^{-3x} + \cdots$

12 What is the ratio of the exact value of 20! to Stirling's approximation for it?

13 For the example in Eq. (2.22), what are the approximate values that would be predicted from Eq. (2.26)?

14 Do the algebra to evaluate Eq. (2.25).

15 Translate this into a question about infinite series and evaluate the two repeating decimal numbers: 0.444444..., 0.987987987...

16 What does the integral test tell you about the convergence of the infinite series $\sum_1^\infty n^{-p}$?

17 What would the power series expansion for the sine look like if you require it to be valid in arbitrary units, not just radians? This requires using the constant "C" as in section 1.1.

Problems

2.1 (**a**) If you borrow \$200,000 to buy a house and will pay it back in monthly installments over 30 years at an annual interest rate of 6%, what is your monthly payment and what is the total money that you have paid (neglecting inflation)? To start, you have N payments p with monthly interest i and after all N payments your unpaid balance must reach zero. The initial loan is L and you pay at the end of each month.

$$((L(1+i)-p)(1+i)-p)(1+i)-p\ \cdots\ N \text{ times } = 0$$

Now carry on and find the general expression for the monthly payment. Also find the total paid.
(**b**) Does your general result for arbitrary N reduce to the correct value if you pay everything back at the end of one month? $[L(1+i)=p]$
(**c**) For general N, what does your result become if the interest rate is zero?
Ans: \$1199.10, \$431676

2.2 In the preceding problem, suppose that there is an annual inflation of 2%. Now what is the total amount of money that you've paid *in constant dollars?* That is, one hundred dollars in the year 2010 is worth just $\$100/1.02^{10} = \$82.03$ as expressed in year-2000 dollars. Each payment is paid with dollars of gradually decreasing value. Ans: \$324211

2.3 Derive all the power series that you're supposed to memorize, Eq. (2.4).

2.4 Sketch graphs of the functions

$$e^{-x^2} \qquad xe^{-x^2} \qquad x^2e^{-x^2} \qquad e^{-|x|} \qquad xe^{-|x|} \qquad x^2e^{-|x|} \qquad e^{-1/x} \qquad e^{-1/x^2}$$

2.5 The sample series in Eq. (2.7) has a simple graph (x^2 between $-L$ and $+L$) Sketch graphs of one, two, three terms of this series to see if the graph is headed toward the supposed answer.

2.6 Evaluate this same Fourier series for x^2 at $x = L$; the answer is supposed to be L^2. Rearrange the result from the series and show that you can use it to evaluate $\zeta(2)$, Eq. (2.6). Ans: $\pi^2/6$

2.7 Determine the domain of convergence for all the series in Eq. (2.4).

2.8 Determine the Taylor series for $\cosh x$ and $\sinh x$.

2.9 *Working strictly by hand,* evaluate $\sqrt[7]{0.999}$. Also $\sqrt{50}$. Ans: Here's where a calculator can tell you better than I can.

2.10 Determine the next, x^6, term in the series expansion of the secant. Ans: $61x^6/720$

2.11 The power series for the tangent is not as neat and simple as for the sine and cosine. You can derive it by taking successive derivatives as done in the text or you can use your knowledge of the series for the sine and cosine, and the geometric series.

$$\tan x = \frac{\sin x}{\cos x} = \frac{x - x^3/3! + \cdots}{1 - x^2/2! + \cdots} = \left[x - x^3/3! + \cdots\right]\left[1 + (-x^2/2! + \cdots)\right]^{-1}$$

Use the expansion for the geometric series to place all the x^2, x^4, etc. terms into the numerator, treating every term after the "1" as a single small thing. Then collect the like powers to obtain the series at least through x^5.
Ans: $x + x^3/3 + 2x^5/15 + 17x^7/315 + \cdots$

2.12 What is the series expansion for $\csc x = 1/\sin x$? As in the previous problem, use your knowledge of the sine series and the geometric series to get this result at least through x^5. Note: the first term in *this* series is $1/x$. Ans: $1/x + x/6 + 7x^3/360 + 31x^5/15120 + \cdots$

2.13 The exact relativistic expression for the kinetic energy of an object with non-zero mass is

$$K = mc^2(\gamma - 1) \qquad \text{where} \qquad \gamma = \left(1 - v^2/c^2\right)^{-1/2}$$

and c is the speed of light in vacuum. If the speed v is small compared to the speed of light, find an approximate expression for K to show that it reduces to the Newtonian expression for the kinetic energy, but include the next term in the expansion to determine how large the speed v must be in order that this correction term is 10% of the Newtonian expression for the kinetic energy? Ans: $v \approx 0.36\,c$

2.14 Use series expansions to evaluate

$$\lim_{x\to 0} \frac{1 - \cos x}{1 - \cosh x} \qquad \text{and} \qquad \lim_{x\to 0} \frac{\sin kx}{x}$$

2.15 Evaluate using series; you will need both the sine series and the binomial series.

$$\lim_{x\to 0} \left(\frac{1}{\sin^2 x} - \frac{1}{x^2}\right)$$

Now do it again, setting up the algebra differently and finding an easier (or harder) way.
Ans: 1/3

2.16 For some more practice with series, evaluate

$$\lim_{x\to 0} \left(\frac{2}{x} + \frac{1}{1 - \sqrt{1+x}}\right)$$

Ans: Check experimentally with a few values of x on a pocket calculator.

2.17 Expand the integrand and find the power series expansion for

$$\ln(1 + x) = \int_0^x dt\,(1 + t)^{-1}$$

Ans: Eq. (2.4)

2.18 (**a**) The error function $\operatorname{erf}(x)$ is defined by an integral. Expand the integrand, integrate term by term, and develop a power series representation for erf. For what values of x does it converge? Evaluate $\operatorname{erf}(1)$ from this series and compare it to the result of problem 1.34. (**b**) Also, as further validation of the integral in problem 1.13, do the power series expansion of both sides of the equation and verify the expansions of the two sides of the equation agree .

2.19 Verify that the combinatorial factor ${}_mC_n$ is really what results for the coefficients when you specialize the binomial series Eq. (2.4) to the case that the exponent is an integer.

2.20 Determine the double power series representation about $(0,0)$ of $1/(1-x/a)(1-y/b)$

2.21 Determine the double power series representation about $(0,0)$ of $1/(1-x/a-y/b)$

2.22 Use a pocket calculator that can handle 100! and find the ratio of Stirling's approximation to the exact value. You may not be able to find the difference of two such large numbers. An improvement on the basic Stirling's formula is

$$\sqrt{2\pi n}\, n^n e^{-n}\left(1+\frac{1}{12n}\right)$$

What is the ratio of approximate to exact for $n = 1,\ 2,\ 10$?
Ans: 0.99898, 0.99948, ...

2.23 Evaluate the sum $\sum_1^\infty 1/n(n+1)$. To do this, write the single term $1/n(n+1)$ as a combination of two fractions with denominator n and $(n+1)$ respectively, then start to write out the stated infinite series to a few terms to see the pattern. When you do this you may be tempted to separate it into two series, of positive and of negative terms. Examine the problem of convergence and show why this is wrong. Ans: 1

2.24 (**a**) You can sometimes use the result of the previous problem to improve the convergence of a slow-converging series. The sum $\sum_1^\infty 1/n^2$ converges, but not very fast. If you add zero to it you don't change the answer, but if you're clever about how you add it you can change this into a much faster converging series. Add $1-\sum_1^\infty 1/n(n+1)$ to this series and combine the sums. (**b**) After Eq. (2.11) it says that it takes 120 terms to get the stated accuracy. Verify this. For the same accuracy, how many terms does this improved sum take? Ans: about 8 terms

2.25 The electric potential from one point charge is kq/r. For two point charges, you add the potentials of each: $kq_1/r_1 + kq_2/r_2$. Place a charge $-q$ at the origin; place a charge $+q$ at position $(x,y,z) = (0,0,a)$. Write the total potential from these at an arbitrary position P with coordinates (x,y,z). Now suppose that a is small compared to the distance of P to the origin $\left(r=\sqrt{x^2+y^2+z^2}\right)$ and expand your result to the first non-vanishing power of a, or really of a/r. This is the potential of an electric dipole. Also express your answer in spherical coordinates. See section 8.8 if you need. Ans: $kqa\cos\theta/r^2$

2.26 Do the previous problem, but with charge $-2q$ at the origin and charges $+q$ at each of the two points $(0,0,a)$ and $(0,0,-a)$. Again, you are looking for the potential at a point far away from the charges, and up to the lowest non-vanishing power of a. In effect you're doing a series expansion in a/r and keeping the first surviving term. Also express the result in spherical coordinates. The angular dependence should be proportional to $P_2(\cos\theta) = \frac{3}{2}\cos^2\theta - \frac{1}{2}$, a "Legendre polynomial." The r dependence will have a $1/r^3$ in it. This potential is that of a linear quadrupole.

2.27 The combinatorial factor Eq. (2.18) is supposed to be the number of different ways of choosing n objects out of a set of m objects. Explicitly verify that this gives the correct number of ways for $m = 1, 2, 3, 4.$ and all n from zero to m.

2.28 Pascal's triangle is a visual way to compute the values of ${}_mC_n$. Start with the single digit 1 on the top line. Every new line is computed by adding the two neighboring digits on the line above. (At the end of the line, treat the empty space as a zero.)

$$\begin{array}{ccccccc} & & & 1 & & & \\ & & 1 & & 1 & & \\ & 1 & & 2 & & 1 & \\ 1 & & 3 & & 3 & & 1 \end{array}$$

Write the next couple of lines of the triangle and then prove that this algorithm works, that is that the mth row is the ${}_mC_n$, where the top row has $m = 0$. Mathematical induction is the technique that I recommend.

2.29 Sum the series and show

$$\frac{1}{2!} + \frac{2}{3!} + \frac{3}{4!} + \cdots = 1$$

2.30 You know the power series representation for the exponential function, but now apply it in a slightly different context. Write out the power series for the exponential, but with an argument that is a differential operator. The letter h represents some fixed number; interpret the square of d/dx as d^2/dx^2 and find

$$e^{h\frac{d}{dx}} f(x)$$

Interpret the terms of the series and show that the value of this is $f(x+h)$.

2.31 The Doppler effect for sound with a moving source and for a moving observer have different formulas. The Doppler effect for light, including relativistic effects is different still. Show that for low speeds they are all about the same.

$$f' = f\frac{v - v_o}{v} \qquad f' = f\frac{v}{v + v_s} \qquad f' = f\sqrt{\frac{1 - v/c}{1 + v/c}}$$

The symbols have various meanings: v is the speed of sound in the first two, with the other terms being the velocity of the observer and the velocity of the source. In the third equation c is the speed of light and v is the velocity of the observer. And no, $1 = 1$ isn't good enough; you should get these at least to first order in the speed.

2.32 In the equation (2.30) for the light diffracted through a narrow slit, the width of the central maximum is dictated by the angle at the first dark region. How does this angle vary as you vary the width of the slit, a? What is this angle if $a = 0.1\,\text{mm}$ and $\lambda = 700\,\text{nm}$? And how wide will the central peak be on a wall 5 meters from the slit? Take this width to be the distance between the first dark regions on either side of the center.

2.33 An object is a distance d below the surface of a medium with index of refraction n. (For example, water.) When viewed from directly above the object in air (i.e. use small angle approximation), the object appears to be a distance below the surface given by (maybe) one of the following expressions. Show why most of these expressions are implausible; that is, give reasons for eliminating the wrong ones *without* solving the problem explicitly.

(1) $d\sqrt{1+n^2}/n$ (2) $dn/\sqrt{1+n^2}$ (3) nd (4) d/n (5) $dn^2/\sqrt{1+n^2}$

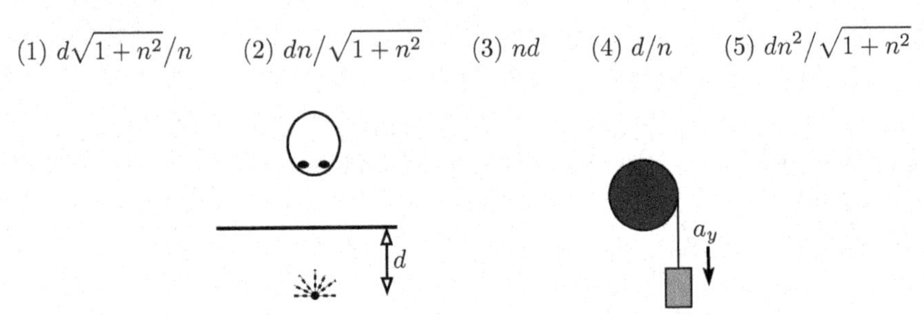

2.34 A mass m_1 hangs from a string that is wrapped around a pulley of mass M. As the mass m_1 falls with acceleration a_y, the pulley rotates. An anonymous source claims that the acceleration of m_1 is one of the following answers. Examine them to determine if any is plausible. That is, examine each and show why it could not be correct. NOTE: solving the problem and then seeing if any of these agree is *not* what this is about.

(1) $a_y = Mg/(m_1 - M)$ (2) $a_y = Mg/(m_1 + M)$ (3) $a_y = m_1 g/M$

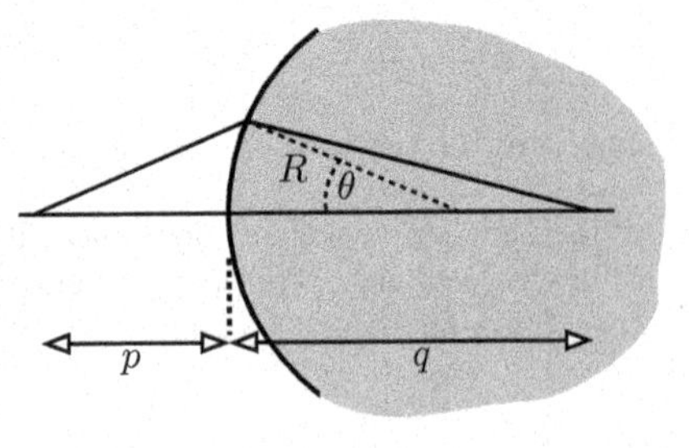

2.35 Light travels from a point on the left (p) to a point on the right (q), and on the left it is in vacuum while on the right of the spherical surface it is in glass with an index of refraction n. The radius of the spherical surface is R and you can parametrize the point on the surface by the angle θ from the center of the sphere. Compute the time it takes light to travel on the indicated path (two straight line segments) as a function of the angle θ. Expand the time through second order in a power series in θ and show that the function $T(\theta)$ has a minimum if the distance

q is small enough, but that it switches to a maximum when q exceeds a particular value. This position is the focus.

2.36 Combine two other series to get the power series in θ for $\ln(\cos\theta)$.
Ans: $-\frac{1}{2}\theta^2 - \frac{1}{12}\theta^4 - \frac{1}{45}\theta^6 + \cdots$

2.37 Subtract the series for $\ln(1-x)$ and $\ln(1+x)$. For what range of x does this series converge? For what range of arguments of the logarithm does it converge?
Ans: $-1 < x < 1, \quad 0 < \arg < \infty$

2.38 A function is defined by the integral

$$f(x) = \int_0^x \frac{dt}{1-t^2}$$

Expand the integrand with the binomial expansion and derive the power (Taylor) series representation for f about $x = 0$. Also make a hyperbolic substitution to evaluate it in closed form.

2.39 Light travels from a point on the right (p), hits a spherically shaped mirror and goes to a point (q). The radius of the spherical surface is R and you can parametrize the point on the surface by the angle θ from the center of the sphere. Compute the time it takes light to travel on the indicated path (two straight line segments) as a function of the angle θ.

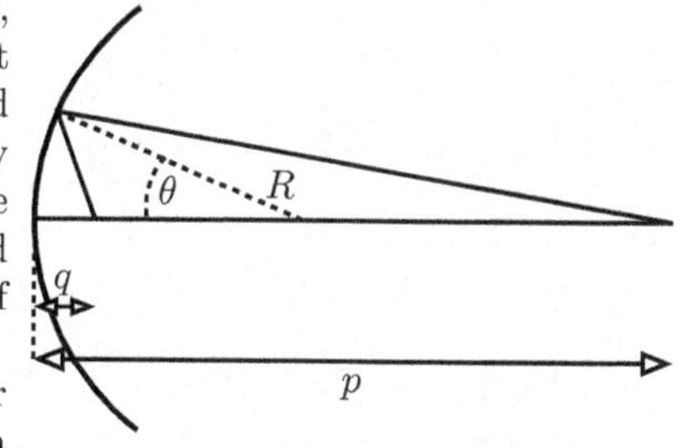

Expand the time through second order in a power series in θ and show that the function $T(\theta)$ has a minimum if the distance q is small enough, but that it switches to a maximum when q exceeds a particular value. This is the focus.

2.40 (**a**) The quadratic equation $ax^2 + bx + c = 0$ is almost a linear equation if a is small enough: $bx + c = 0 \Rightarrow x = -c/b$. You can get a more accurate solution iteratively by rewriting the equation as

$$x = -\frac{c}{b} - \frac{a}{b}x^2$$

Solve this by neglecting the second term, then with this approximate x_1 get an improved value of the root by

$$x_2 = -\frac{c}{b} - \frac{a}{b}x_1^2$$

and you can repeat the process. For comparison take the exact solution and do a power series expansion on it for small a. See if the results agree.
(**b**) Where does the other root come from? That value of x is very large, so the first two terms in the quadratic are the big ones and must nearly cancel. $ax^2 + bx = 0$ so $x = -b/a$.

Rearrange the equation so that you can iterate it, and compare the iterated solution to the series expansion of the exact solution.

$$x = -\frac{b}{a} - \frac{c}{ax}$$

Solve $0.001x^2 + x + 1 = 0$. Ans: Solve it exactly and compare.

2.41 Evaluate the limits

$$\text{(a)}\ \lim_{x\to 0} \frac{\sin x - \tan x}{x}, \qquad \text{(b)}\ \lim_{x\to 0} \frac{\sin x - \tan x}{x^2}, \qquad \text{(c)}\ \lim_{x\to 0} \frac{\sin x - \tan x}{x^3}$$

Ans: Check with a pocket calculator for $x = 1.0,\ 0.1,\ 0.01$

2.42 Fill in the missing steps in the derivation of Eq. (2.26).

2.43 Is the result in Eq. (2.26) normalized properly? What is its integral $d\delta$ over all δ? Ans: 1

2.44 A political survey asks 1500 people randomly selected from the entire country whom they will vote for as dog-catcher-in-chief. The results are 49.0% for T.I. Hulk and 51.0% for T.A. Spiderman. Assume that these numbers are representative, an unbiased sample of the electorate. The number $0.49 \times 1500 = aN$ is now your best estimate for the number of votes Mr. Hulk will get in a sample of 1500. Given this estimate, what is the probability that Mr. Hulk will win the final vote anyway? (**a**) Use Eq. (2.26) to represent this estimate of the probability of his getting various possible outcomes, where the center of the distribution is at $k = aN$. Using $\delta = k - aN$, this probability function is proportional to $\exp\big(-\delta^2/2abN\big)$, and the probability of winning is the sum of all the probabilities of having $k > N/2$, that is, $\int_{N/2}^{\infty} dk$. (**b**) What would the answer be if the survey had asked 150 or 15000 people with the same 49-51 results? Ans: (**a**) $\frac{1}{2}\big[1 - \operatorname{erf}\big(\sqrt{N/2ab}\,(\frac{1}{2} - a)\big)\big]$. 22%, (**b**) 40%, 0.7%

2.45 For the function defined in problem 2.38, what is its behavior near $x = 1$? Compare this result to equation (1.4). Note: the integral is $\int_0^\Lambda + \int_\Lambda^x$. Also, $1 - t^2 = (1+t)(1-t)$, and this $\approx 2(1-t)$ near 1.

2.46 (**a**) What is the expansion of $1/(1+t^2)$ in powers of t for small t. (**b**) That was easy, now what is it for large t? In each case, what is the domain of convergence?

2.47 The "average" of two numbers a and b commonly means $(a+b)/2$, the arithmetic mean. There are many other averages however. $(a, b > 0)$

$$M_n(a,b) = \big[(a^n + b^n)/2\big]^{1/n}$$

is the nth mean, also called the power mean, and it includes many others as special cases. $n = 2$: root-mean-square, $n = -1$: harmonic mean. Show that this includes the geometric mean too: $\sqrt{ab} = \lim_{n\to 0} M_n(a,b)$. It can be shown that $dM_n/dn > 0$; what inequalities does this imply for various means? Ans: harmonic $\le$ geometric $\le$ arithmetic $\le$ rms

2.48 Using the definition in the preceding problem, show that $dM_n/dn > 0$. [Tough!]

2.49 In problem 2.18 you found the power series expansion for the error function — good for small arguments. Now what about large arguments?

$$\operatorname{erf}(x) = \frac{2}{\sqrt{\pi}}\int_0^x dt\, e^{-t^2} = 1 - \frac{2}{\sqrt{\pi}}\int_x^\infty dt\, e^{-t^2} = 1 - \frac{2}{\sqrt{\pi}}\int_x^\infty dt\, \frac{1}{t}\cdot te^{-t^2}$$

Notice that you *can* integrate the te^{-t^2} factor explicitly, so integrate by parts. Then do it again and again. This provides a series in inverse powers that allows you evaluate the error function for large arguments. What is erf(3)? Ans: 0.9999779095 See Abramowitz and Stegun: 7.1.23.

2.50 A friend of mine got a different result for Eq. (2.35). Instead of $\sin^2\theta$ in the denominator, he got a $\sin\theta$. Analyze his answer for plausibility.

2.51 Find the minimum of the function $f(r) = ar + b/r$ for a, b, $r > 0$. Then find the series expansion of f about that point, at least as far as the first non-constant term.

2.52 In problem 2.15 you found the limit of a function as $x \to 0$. Now find the behavior of the same function as a series expansion for small x, through terms in x^2. Ans: $\frac{1}{3} + \frac{1}{15}x^2$. To test whether this answer or yours or neither is likely to be correct, evaluate the exact and approximate values of this for moderately small x on a pocket calculator.

2.53 Following Eq. (2.34) the tentative conclusion was that the force assumed for the air resistance was a constant times the velocity. Go back the the exact equations (2.33) and compute this force without approximation, showing that it is in fact a constant times the velocity. And of course find the constant.

2.54 An object is thrown straight up with speed v_0. There is air resistance and the resulting equation for the velocity is claimed to be (only while it's going up)

$$v_y(t) = v_t \frac{v_0 - v_t\tan(gt/v_t)}{v_t + v_0\tan(gt/v_t)}$$

where v_t is the terminal speed of the object after it turns around and has then been falling long enough. (**a**) Check whether this equation is plausible by determining if it reduces to the correct result if there is no air resistance and the terminal speed goes to infinity. (**b**) Now, what is the velocity for small time and then use $F_y = ma_y$ to infer the probable speed dependence of what I assumed for the air resistance in deriving this expression. See problem 2.11 for the tangent series. (**c**) Use the exact $v_y(t)$ to show that no matter how large the initial speed is, it will stop in no more than some maximum time. For a bullet that has a terminal speed of 100 m/s, this is about 16 s.

2.55 Under the same circumstances as problem 2.54, the equation for position versus time is

$$y(t) = \frac{v_t^2}{g}\ln\left(\frac{v_t\cos(gt/v_t) + v_0\sin(gt/v_t)}{v_t}\right)$$

(**a**) What is the behavior of this for small time? Analyze and interpret what it says and whether it behaves as it should. (**b**) At the time that it reaches its maximum height ($v_y = 0$), what is its position? Note that you don't have to have an explicit value of t for which this happens; you use the equation that t satisfies.

2.56 You can get the individual terms in the series Eq. (2.13) another way: multiply the two series:

$$e^{ax^2+bx} = e^{ax^2}e^{bx}$$

Do so and compare it to the few terms found after (2.13).

Complex Algebra

When the idea of negative numbers was broached a couple of thousand years ago, they were considered suspect, in some sense not "real." Later, when probably one of the students of Pythagoras discovered that numbers such as $\sqrt{2}$ are irrational and cannot be written as a quotient of integers, legends have it that the discoverer suffered dire consequences. Now both negatives and irrationals are taken for granted as ordinary numbers of no special consequence. Why should $\sqrt{-1}$ be any different? Yet it was not until the middle 1800's that complex numbers were accepted as fully legitimate. Even then, it took the prestige of Gauss to persuade some. How can this be, because the general solution of a quadratic equation had been known for a long time? When it gave complex roots, the response was that those are meaningless and you can discard them.

3.1 Complex Numbers

As soon as you learn to solve a quadratic equation, you are confronted with complex numbers, but what is a complex number? If the answer involves $\sqrt{-1}$ then an appropriate response might be "What is *that*?" Yes, we can manipulate objects such as $-1 + 2i$ and get consistent results with them. We just have to follow certain rules, such as $i^2 = -1$. But is that an answer to the question? You can go through the entire subject of complex algebra and even complex calculus without learning a better answer, but it's nice to have a more complete answer once, if then only to relax* and forget it.

An answer to this question is to define complex numbers as pairs of real numbers, (a, b). These pairs are made subject to rules of addition and multiplication:

$$(a, b) + (c, d) = (a + c, b + d) \qquad \text{and} \qquad (a, b)(c, d) = (ac - bd, ad + bc)$$

An algebraic system has to have something called zero, so that it plus any number leaves that number alone. Here that role is taken by $(0, 0)$

$$(0, 0) + (a, b) = (a + 0, b + 0) = (a, b) \qquad \text{for all values of } (a, b)$$

What is the identity, the number such that it times any number leaves that number alone?

$$(1, 0)(c, d) = (1 \cdot c - 0 \cdot d, 1 \cdot d + 0 \cdot c) = (c, d)$$

so $(1, 0)$ has this role. Finally, where does $\sqrt{-1}$ fit in?

$$(0, 1)(0, 1) = (0 \cdot 0 - 1 \cdot 1, 0 \cdot 1 + 1 \cdot 0) = (-1, 0)$$

and the sum $(-1, 0) + (1, 0) = (0, 0)$ so $(0, 1)$ is the representation of $i = \sqrt{-1}$, that is $i^2 + 1 = 0$. $\big[(0, 1)^2 + (1, 0) = (0, 0)\big]$.

---

* If you think that this question is an easy one, you can read about some of the difficulties that the greatest mathematicians in history had with it: "An Imaginary Tale: The Story of $\sqrt{-1}$" by Paul J. Nahin. I recommend it.

This set of pairs of real numbers satisfies all the desired properties that you want for complex numbers, so having shown that it is possible to express complex numbers in a precise way, I'll feel free to ignore this more cumbersome notation and to use the more conventional representation with the symbol i:

$$(a, b) \longleftrightarrow a + ib$$

That complex number will in turn usually be represented by a single letter, such as $z = x + iy$.

The graphical interpretation of complex numbers is the Cartesian geometry of the plane. The x and y in $z = x + iy$ indicate a point in the plane, and the operations of addition and multiplication can be interpreted as operations in the plane. Addition of complex numbers is simple to interpret; it's nothing more than common vector addition where you think of the point as being a vector from the origin. It reproduces the parallelogram law of vector addition.

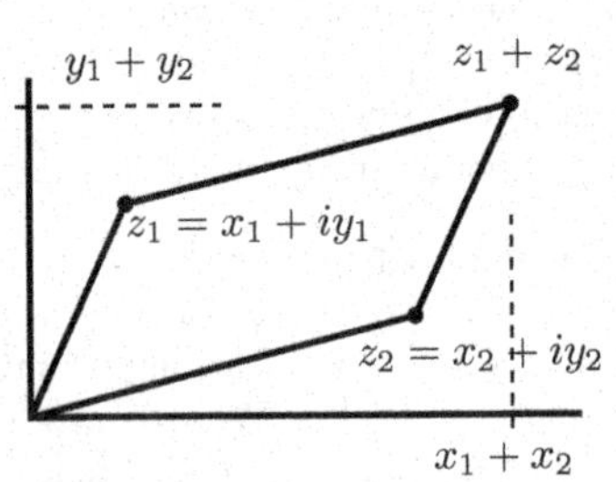

The *magnitude* of a complex number is defined in the same way that you define the magnitude of a vector in the plane. It is the distance to the origin using the Euclidean idea of distance.

$$|z| = |x + iy| = \sqrt{x^2 + y^2} \tag{3.1}$$

The multiplication of complex numbers doesn't have such a familiar interpretation in the language of vectors. (And why should it?)

3.2 Some Functions

For the algebra of complex numbers I'll start with some simple looking questions of the sort that you know how to handle with real numbers. If z is a complex number, what are z^2 and $\sqrt{z}$? Use x and y for real numbers here.

$$z = x + iy, \qquad \text{so} \qquad z^2 = (x + iy)^2 = x^2 - y^2 + 2ixy$$

That was easy, what about the square root? A little more work:

$$\sqrt{z} = w \Longrightarrow z = w^2$$

If $z = x + iy$ and the unknown is $w = u + iv$ (u and v real) then

$$x + iy = u^2 - v^2 + 2iuv, \qquad \text{so} \qquad x = u^2 - v^2 \qquad \text{and} \qquad y = 2uv$$

These are two equations for the two unknowns u and v, and the problem is now to solve them.

$$v = \frac{y}{2u}, \qquad \text{so} \qquad x = u^2 - \frac{y^2}{4u^2}, \qquad \text{or} \qquad u^4 - xu^2 - \frac{y^2}{4} = 0$$

This is a quadratic equation for u^2.

$$u^2 = \frac{x \pm \sqrt{x^2+y^2}}{2}, \qquad \text{then} \qquad u = \pm\sqrt{\frac{x \pm \sqrt{x^2+y^2}}{2}} \tag{3.2}$$

Use $v = y/2u$ and you have four roots with the four possible combinations of plus and minus signs. You're supposed to get only two square roots, so something isn't right yet; which of these four have to be thrown out? See problem 3.2.

What is the reciprocal of a complex number? You can treat it the same way as you did the square root: Solve for it.

$$(x+iy)(u+iv) = 1, \qquad \text{so} \qquad xu - yv = 1, \qquad xv + yu = 0$$

Solve the two equations for u and v. The result is

$$\frac{1}{z} = \frac{x-iy}{x^2+y^2} \tag{3.3}$$

See problem 3.3. At least it's obvious that the dimensions are correct even before you verify the algebra. In both of these cases, the square root and the reciprocal, there is another way to do it, a much simpler way. That's the subject of the next section.

Complex Exponentials

A function that is central to the analysis of differential equations and to untold other mathematical ideas: the exponential, the familiar e^x. What is this function for complex values of the exponent?

$$e^z = e^{x+iy} = e^x e^{iy} \tag{3.4}$$

This means that all that's necessary is to work out the value for the purely imaginary exponent, and the general case is then just a product. There are several ways to work this out, and I'll pick what is probably the simplest. Use the series expansions Eq. (2.4) for the exponential, the sine, and the cosine and apply it to this function.

$$\begin{aligned} e^{iy} &= 1 + iy + \frac{(iy)^2}{2!} + \frac{(iy)^3}{3!} + \frac{(iy)^4}{4!} + \cdots \\ &= 1 - \frac{y^2}{2!} + \frac{y^4}{4!} - \cdots + i\left[y - \frac{y^3}{3!} + \frac{y^5}{5!} - \cdots\right] = \cos y + i \sin y \end{aligned} \tag{3.5}$$

A few special cases of this are worth noting: $e^{i\pi/2} = i$, also $e^{i\pi} = -1$ and $e^{2i\pi} = 1$. In fact, $e^{2n\pi i} = 1$ so the exponential is a periodic function in the imaginary direction.

The magnitude or absolute value of a complex number $z = x + iy$ is $r = \sqrt{x^2+y^2}$. Combine this with the complex exponential and you have another way to represent complex numbers.

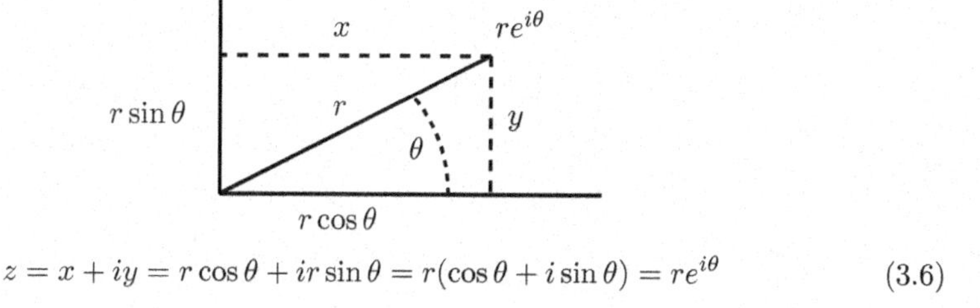

$$z = x + iy = r\cos\theta + ir\sin\theta = r(\cos\theta + i\sin\theta) = re^{i\theta} \tag{3.6}$$

This is the polar form of a complex number and $x+iy$ is the rectangular form of the same number. The magnitude is $|z| = r = \sqrt{x^2+y^2}$. What is $\sqrt{i}$? Express it in polar form: $\left(e^{i\pi/2}\right)^{1/2}$, or better, $\left(e^{i(2n\pi+\pi/2)}\right)^{1/2}$. This is

$$e^{i(n\pi+\pi/4)} = \left(e^{i\pi}\right)^n e^{i\pi/4} = \pm(\cos\pi/4 + i\sin\pi/4) = \pm\frac{1+i}{\sqrt{2}}$$

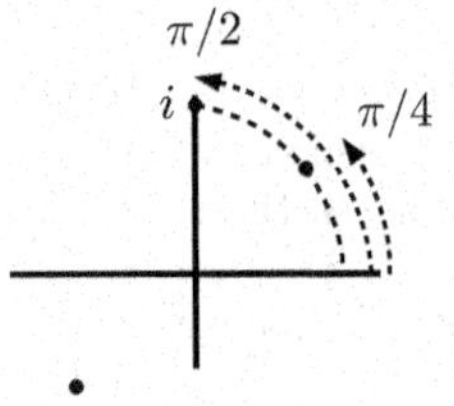

$\pi/2$ i $\pi/4$

3.3 Applications of Euler's Formula

When you are adding or subtracting complex numbers, the rectangular form is more convenient, but when you're multiplying or taking powers the polar form has advantages.

$$z_1 z_2 = r_1 e^{i\theta_1} r_2 e^{i\theta_2} = r_1 r_2 e^{i(\theta_1+\theta_2)} \tag{3.7}$$

Putting it into words, you multiply the magnitudes and add the angles in polar form.

From this you can immediately deduce some of the common trigonometric identities. Use Euler's formula in the preceding equation and write out the two sides.

$$r_1(\cos\theta_1 + i\sin\theta_1) r_2(\cos\theta_2 + i\sin\theta_2) = r_1 r_2\big[\cos(\theta_1+\theta_2) + i\sin(\theta_1+\theta_2)\big]$$

The factors r_1 and r_2 cancel. Now multiply the two binomials on the left and match the real and the imaginary parts to the corresponding terms on the right. The result is the pair of equations

$$\begin{aligned}\cos(\theta_1+\theta_2) &= \cos\theta_1\cos\theta_2 - \sin\theta_1\sin\theta_2\\ \sin(\theta_1+\theta_2) &= \cos\theta_1\sin\theta_2 + \sin\theta_1\cos\theta_2\end{aligned} \tag{3.8}$$

and you have a much simpler than usual derivation of these common identities. You can do similar manipulations for other trigonometric identities, and in some cases you will encounter relations for which there's really no other way to get the result. That is why you will find that in physics applications where you might use sines or cosines (oscillations, waves) no one uses anything but complex exponentials. Get used to it.

The trigonometric functions of complex argument follow naturally from these.

$$e^{i\theta} = \cos\theta + i\sin\theta, \qquad \text{so, for negative angle} \qquad e^{-i\theta} = \cos\theta - i\sin\theta$$

Add these and subtract these to get

$$\cos\theta = \frac{1}{2}\left(e^{i\theta} + e^{-i\theta}\right) \qquad \text{and} \qquad \sin\theta = \frac{1}{2i}\left(e^{i\theta} - e^{-i\theta}\right) \tag{3.9}$$

What is this if $\theta = iy$?

$$\cos iy = \frac{1}{2}\left(e^{-y} + e^{+y}\right) = \cosh y \qquad \text{and} \qquad \sin iy = \frac{1}{2i}\left(e^{-y} - e^{+y}\right) = i\sinh y \tag{3.10}$$

Apply Eq. (3.8) for the addition of angles to the case that $\theta = x + iy$.

$$\cos(x+iy) = \cos x \cos iy - \sin x \sin iy = \cos x \cosh y - i \sin x \sinh y \qquad \text{and}$$
$$\sin(x+iy) = \sin x \cosh y + i \cos x \sinh y \tag{3.11}$$

You can see from this that the sine and cosine of complex angles can be real and larger than one. The hyperbolic functions and the circular trigonometric functions are now the same functions. You're just looking in two different directions in the complex plane. It's as if you are changing from the equation of a circle, $x^2 + y^2 = R^2$, to that of a hyperbola, $x^2 - y^2 = R^2$. Compare this to the hyperbolic functions at the beginning of chapter one.

Equation (3.9) doesn't require that θ itself be real; call it z. Then what is $\sin^2 z + \cos^2 z$?

$$\cos z = \frac{1}{2}\left(e^{iz} + e^{-iz}\right) \qquad \text{and} \qquad \sin z = \frac{1}{2i}\left(e^{iz} - e^{-iz}\right)$$
$$\cos^2 z + \sin^2 z = \frac{1}{4}\left[e^{2iz} + e^{-2iz} + 2 - e^{2iz} - e^{-2iz} + 2\right] = 1$$

This polar form shows a geometric interpretation for the periodicity of the exponential. $e^{i(\theta+2\pi)} = e^{i\theta} = e^{i(\theta+2k\pi)}$. In the picture, you're going around a circle and coming back to the same point. If the angle θ is negative you're just going around in the opposite direction. An angle of $-\pi$ takes you to the same point as an angle of $+\pi$.

Complex Conjugate

The complex conjugate of a number $z = x + iy$ is the number $z^* = x - iy$. Another common notation is $\bar{z}$. The product z^*z is $(x - iy)(x + iy) = x^2 + y^2$ and that is $|z|^2$, the square of the magnitude of z. You can use this to rearrange complex fractions, combining the various terms with i in them and putting them in one place. This is best shown by some examples.

$$\frac{3+5i}{2+3i} = \frac{(3+5i)(2-3i)}{(2+3i)(2-3i)} = \frac{21+i}{13}$$

What happens when you add the complex conjugate of a number to the number, $z + z^*$?
What happens when you subtract the complex conjugate of a number from the number?
If one number is the complex conjugate of another, how do their squares compare?
What about their cubes?
What about $z + z^2$ and $z^* + z^{*2}$?
What about comparing $e^z = e^{x+iy}$ and e^{z^*}?
What is the product of a number and its complex conjugate written in polar form?
Compare $\cos z$ and $\cos z^*$.
What is the quotient of a number and its complex conjugate?
What about the magnitude of the preceding quotient?

Examples

Simplify these expressions, making sure that you can do all of these manipulations yourself.

$$\frac{3-4i}{2-i} = \frac{(3-4i)(2+i)}{(2-i)(2+i)} = \frac{10-5i}{5} = 2 - i.$$

$$(3i+1)^2\left[\frac{1}{2-i} + \frac{3i}{2+i}\right] = (-8+6i)\left[\frac{(2+i)+3i(2-i)}{(2-i)(2+i)}\right] = (-8+6i)\frac{5+7i}{5} = \frac{2-26i}{5}.$$

$$\frac{i^3 + i^{10} + i}{i^2 + i^{137} + 1} = \frac{(-i) + (-1) + i}{(-1) + (i) + (1)} = \frac{-1}{i} = i.$$

Manipulate these using the polar form of the numbers, though in some cases you can do it either way.

$$\sqrt{i} = \left(e^{i\pi/2}\right)^{1/2} = e^{i\pi/4} = \frac{1+i}{\sqrt{2}}.$$

$$\left(\frac{1-i}{1+i}\right)^3 = \left(\frac{\sqrt{2}e^{-i\pi/4}}{\sqrt{2}e^{i\pi/4}}\right)^3 = \left(e^{-i\pi/2}\right)^3 = e^{-3i\pi/2} = i.$$

$$\left(\frac{2i}{1+i\sqrt{3}}\right)^{25} = \left(\frac{2e^{i\pi/2}}{2\left(\frac{1}{2}+i\frac{1}{2}\sqrt{3}\right)}\right)^{25} = \left(\frac{2e^{i\pi/2}}{2e^{i\pi/3}}\right)^{25} = \left(e^{i\pi/6}\right)^{25} = e^{i\pi(4+1/2)} = i$$

Roots of Unity

What is the cube root of one? One of course, but not so fast; there are three cube roots, and you can easily find all of them using complex exponentials.

$$1 = e^{2k\pi i}, \qquad \text{so} \qquad 1^{1/3} = \left(e^{2k\pi i}\right)^{1/3} = e^{2k\pi i/3} \tag{3.12}$$

and k is any integer. $k = 0, 1, 2$ give

$$1^{1/3} = 1, \qquad e^{2\pi i/3} = \cos(2\pi/3) + i\sin(2\pi/3), \qquad e^{4\pi i/3} = \cos(4\pi/3) + i\sin(4\pi/3)$$
$$= -\frac{1}{2} + i\frac{\sqrt{3}}{2} \qquad\qquad = -\frac{1}{2} - i\frac{\sqrt{3}}{2}$$

and other positive or negative integers k just keep repeating these three values.

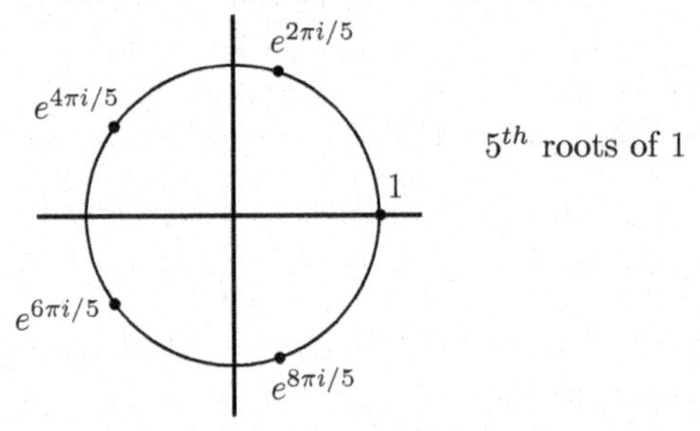

The roots are equally spaced around the unit circle. If you want the nth root, you do the same sort of calculation: the $1/n$ power and the integers $k = 0, 1, 2, \ldots, (n-1)$. These are n points, and the angles between adjacent ones are equal.

3.4 Geometry

Multiply a number by 2 and you change its length by that factor.
Multiply it by i and you rotate it counterclockwise by 90° about the origin.
Multiply is by $i^2 = -1$ and you rotate it by 180° about the origin. (Either direction: $i^2 = (-i)^2$)

The Pythagorean Theorem states that if you construct three squares from the three sides of a right triangle, the sum of the two areas on the shorter sides equals the area of the square constructed on the hypotenuse. What happens if you construct four squares on the four sides of an arbitrary quadrilateral?

Represent the four sides of the quadrilateral by four complex numbers that add to zero. Start from the origin and follow the complex number a. Then follow b, then c, then d. The result brings you back to the origin. Place four squares on the four sides and locate the centers of those squares: P_1, P_2,... Draw lines between these points as shown.

These lines are orthogonal and have the same length. Stated in the language of complex numbers, this is

$$P_1 - P_3 = i(P_2 - P_4) \tag{3.13}$$

$$a+b+c+d=0$$

$$\tfrac{1}{2}a + \tfrac{1}{2}ia = P_1$$

$$a + \tfrac{1}{2}b + \tfrac{1}{2}ib = P_2$$

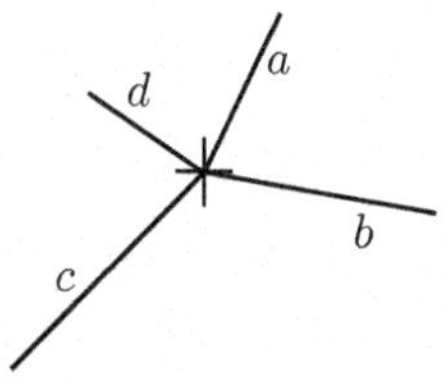

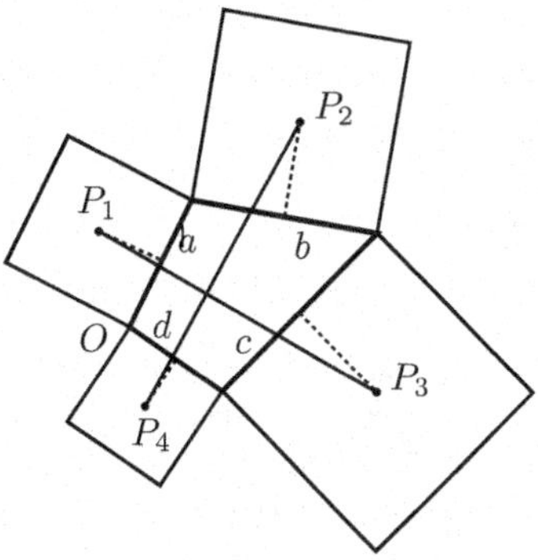

Pick the origin at one corner, then construct the four center points $P_{1,2,3,4}$ as complex numbers, following the pattern shown above for the first two. *E.g.* , you get to P_1 from the origin by going halfway along a, turning left, then going the distance $|a|/2$. Now write out the two complex number $P_1 - P_3$ and $P_2 - P_4$ and finally manipulate them by using the defining equation for the quadrilateral, $a + b + c + d = 0$. The result is the stated theorem. See problem 3.54.

3.5 Series of cosines

There are standard identities for the cosine and sine of the sum of angles and less familiar ones for the sum of two cosines or sines. You can derive that latter sort of equations using Euler's formula and a little manipulation. The sum of two cosines is the real part of $e^{ix} + e^{iy}$, and you can use simple identities to manipulate these into a useful form.

$$x = \tfrac{1}{2}(x+y) + \tfrac{1}{2}(x-y) \qquad \text{and} \qquad y = \tfrac{1}{2}(x+y) - \tfrac{1}{2}(x-y)$$

See problems 3.34 and 3.35 to complete these.

What if you have a sum of many cosines or sines? Use the same basic ideas of the preceding manipulations, and combine them with some of the techniques for manipulating series.

$$1 + \cos\theta + \cos 2\theta + \cdots + \cos N\theta = 1 + e^{i\theta} + e^{2i\theta} + \cdots e^{Ni\theta} \qquad \text{(Real part)}$$

The last series is geometric, so it is nothing more than Eq. (2.3).

$$1 + e^{i\theta} + \left(e^{i\theta}\right)^2 + \left(e^{i\theta}\right)^3 + \cdots \left(e^{i\theta}\right)^N = \frac{1 - e^{i(N+1)\theta}}{1 - e^{i\theta}}$$

$$= \frac{e^{i(N+1)\theta/2}\left(e^{-i(N+1)\theta/2} - e^{i(N+1)\theta/2}\right)}{e^{i\theta/2}\left(e^{-i\theta/2} - e^{i\theta/2}\right)} = e^{iN\theta/2}\frac{\sin\left[(N+1)\theta/2\right]}{\sin\theta/2} \tag{3.14}$$

From this you now extract the real part and the imaginary part, thereby obtaining the series you want (plus another one, the series of sines). These series appear when you analyze the behavior of a diffraction grating. Naturally you have to check the plausibility of these results; do the answers work for small θ?

3.6 Logarithms

The logarithm is the inverse function for the exponential. If $e^w = z$ then $w = \ln z$. To determine what this is, let

$$w = u + iv \qquad \text{and} \qquad z = re^{i\theta}, \qquad \text{then} \qquad e^{u+iv} = e^u e^{iv} = re^{i\theta}$$

This implies that $e^u = r$ and so $u = \ln r$, but it doesn't imply $v = \theta$. Remember the periodic nature of the exponential function? $e^{i\theta} = e^{i(\theta+2n\pi)}$, so you can conclude instead that $v = \theta + 2n\pi$.

$$\ln z = \ln\left(re^{i\theta}\right) = \ln r + i(\theta + 2n\pi) \tag{3.15}$$

has an infinite number of possible values. Is this bad? You're already familiar with the square root function, and that has *two* possible values, $\pm$. This just carries the idea farther. For example $\ln(-1) = i\pi$ or $3i\pi$ or $-7i\pi$ etc. As with the square root, the specific problem that you're dealing with will tell you which choice to make.

A sample graph of the logarithm in the complex plane is $\ln(1 + it)$ as t varies from $-\infty$ to $+\infty$.

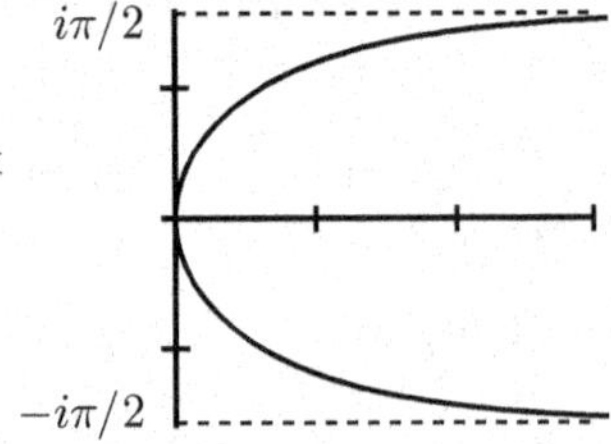

3.7 Mapping

When you apply a complex function to a region in the plane, it takes that region into another region. When you look at this as a geometric problem you start to get some very pretty and occasionally useful results. Start with a simple example,

$$w = f(z) = e^z = e^{x+iy} = e^x e^{iy} \tag{3.16}$$

If $y = 0$ and x goes from $-\infty$ to $+\infty$, this function goes from 0 to ∞.
If y is $\pi/4$ and x goes over this same range of values, f goes from 0 to infinity along the ray at angle $\pi/4$ above the axis.
At any fixed y, the horizontal line parallel to the x-axis is mapped to the ray that starts at the origin and goes out to infinity.
The strip from $-\infty < x < +\infty$ and $0 < y < \pi$ is mapped into the upper half plane.

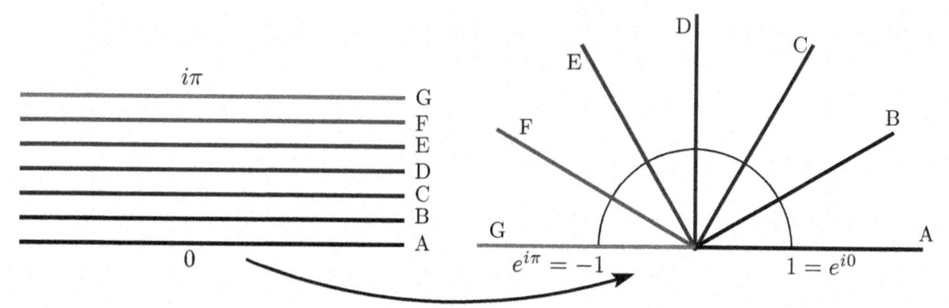

The line B from $-\infty + i\pi/6$ to $+\infty + i\pi/6$ is mapped onto the ray B from the origin along the angle $\pi/6$.

For comparison, what is the image of the same strip under a different function? Try

$$w = f(z) = z^2 = x^2 - y^2 + 2ixy$$

The image of the line of fixed y is a parabola. The real part of w has an x^2 in it while the imaginary part is linear in x. That is the representation of a parabola. The image of the strip is the region among the lines below.

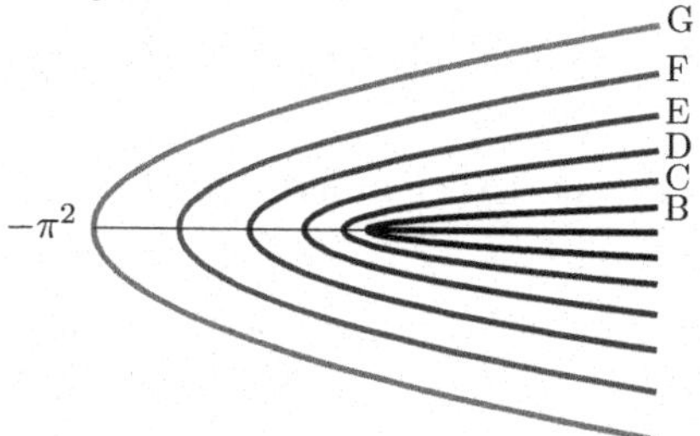

Pretty yes, but useful? In certain problems in electrostatics and in fluid flow, it is possible to use complex algebra to map one region into another, with the accompanying electric fields and potentials or respectively fluid flows mapped from a complicated problem into a simple one. Then you can map the simple solution back to the original problem and you have your desired solution to the original problem. Easier said than done. It's the sort of method that you can learn about when you find that you need it.

Exercises

1 Express in the form $a+ib$: $(3-i)^2$, $(2-3i)(3+4i)$. Draw the geometric representation for each calculation.

2 Express in polar form, $re^{i\theta}$: -2, $3i$, $3+3i$. Draw the geometric representation for each.

3 Show that $(1+2i)(3+4i)(5+6i)$ satisfies the associative law of multiplication. *I.e.* multiply first pair first or multiply the second pair first, no matter.

4 Solve the equation $z^2 - 2z + c = 0$ and plot the roots as points in the complex plane. Do this as the real number c moves from $c = 0$ to $c = 2$

5 Now show that $(a + bi)\big[(c + di)(e + fi)\big] = \big[(a + bi)(c + di)\big](e + fi)$. After all, just because real numbers satisfy the associative law of multiplication it isn't immediately obvious that complex numbers do too.

6 Given $z_1 = 2e^{i60^\circ}$ and $z_2 = 4e^{i120^\circ}$, evaluate z_1^2, $z_1 z_2$, z_2/z_1. Draw pictures too.

7 Evaluate $\sqrt{i}$ using the rectangular form, Eq. (3.2), and compare it to the result you get by using the polar form.

8 Given $f(z) = z^2 + z + 1$, evaluate $f(3 + 2i)$, $f(3 - 2i)$.

9 For the same f as the preceding exercise, what are $f'(3 + 2i)$ and $f'(3 - 2i)$?

10 Do the arithmetic and draw the pictures of these computations:

$$(3 + 2i) + (-1 + i), \qquad (3 + 2i) - (-1 + i), \qquad (-4 + 3i) - (4 + i), \qquad -5 + (3 - 5i)$$

11 Show that the real part of z is $(z + z^*)/2$. Find a similar expression for the imaginary part of z.

12 What is i^n for integer n? Draw the points in the complex plane for a variety of positive and negative n.

13 What is the magnitude of $(4 + 3i)/(3 - 4i)$? What is its polar angle?

14 Evaluate $(1 + i)^{19}$.

15 What is $\sqrt{1 - i}$? Do this by the method of Eq. (3.2).

16 What is $\sqrt{1 - i}$? Do this by the method of Eq. (3.6).

17 Sketch a plot of the curve $z = \alpha e^{i\alpha}$ as the real parameter α varies from zero to infinity. Does the behavior of your sketch conform to the small α behavior of the function? (And when no one's looking you can plug in a few numbers for α to see what this behavior is.)

18 Verify the graph following Eq. (3.15).

Problems

3.1 Pick a pair of complex numbers and plot them in the plane. Compute their product and plot that point. Do this for several pairs, trying to get a feel for how complex multiplication works. When you do this, be sure that you're not simply repeating yourself. Place the numbers in qualitatively different places.

3.2 In the calculation of the square root of a complex number,Eq. (3.2), I found four roots instead of two. Which ones don't belong? Do the other two expressions have any meaning?

3.3 Finish the algebra in computing the reciprocal of a complex number, Eq. (3.3).

3.4 Pick a complex number and plot it in the plane. Compute its reciprocal and plot it. Compute its square and square root and plot them. Do this for several more (qualitatively different) examples.

3.5 Plot e^{ct} in the plane where c is a complex constant of your choosing and the parameter t varies over $0 \leq t < \infty$. Pick another couple of values for c to see how the resulting curves change. Don't pick values that simply give results that are qualitatively the same; pick values sufficiently varied so that you can get different behavior. If in doubt about how to plot these complex numbers as functions of t, pick a few numerical values: *e.g.* $t = 0.01, 0.1$, 0.2, 0.3, *etc.* Ans: Spirals or straight lines, depending on where you start

3.6 Plot $\sin ct$ in the plane where c is a complex constant of your choosing and the parameter t varies over $0 \leq t < \infty$. Pick another couple of qualitatively different values for c to see how the resulting curves change.

3.7 Solve the equation $z^2 + iz + 1 = 0$

3.8 Just as Eq. (3.11) presents the circular functions of complex arguments, what are the hyperbolic functions of complex arguments?

3.9 From $\left(e^{ix}\right)^3$, deduce trigonometric identities for the cosine and sine of triple angles in terms of single angles. Ans: $\cos 3x = \cos x - 4\sin^2 x \cos x = 4\cos^3 x - 3\cos x$

3.10 For arbitrary integer $n > 1$, compute the sum of all the nth roots of one. (When in doubt, try $n = 2$, 3, 4 first.)

3.11 Either solve for z in the equation $e^z = 0$ or prove that it can't be done.

3.12 Evaluate z/z^* in polar form.

3.13 From the geometric picture of the magnitude of a complex number, the set of points z defined by $|z - z_0| = R$ is a circle. Write it out in rectangular components to see what this is in conventional Cartesian coordinates.

3.14 An ellipse is the set of points z such that the sum of the distances to two fixed points is a constant: $|z - z_1| + |z - z_2| = 2a$. Pick the two points to be $z_1 = -f$ and $z_2 = +f$ on the real axis ($f < a$). Write z as $x + iy$ and manipulate this equation for the ellipse into a simple standard form. I suggest that you leave everything in terms of complex numbers (z, z^*, z_1, z_1^*, *etc.*) until some distance into the problem. Use $x+iy$ only after it becomes truly useful to do so.

3.15 Repeat the previous problem, but for the set of points such that the *difference* of the distances from two fixed points is a constant.

3.16 There is a vertical line $x = -f$ and a point on the x-axis $z_0 = +f$. Find the set of points z so that the distance to z_0 is the same as the perpendicular distance to the line $x = -f$.

3.17 Sketch the set of points $|z - 1| < 1$.

3.18 Simplify the numbers

$$\frac{1+i}{1-i}, \qquad \frac{-1+i\sqrt{3}}{+1+i\sqrt{3}}, \qquad \frac{i^5+i^3}{\sqrt{3\sqrt{i}-7\sqrt[3]{17-4i}}}, \qquad \left(\frac{\sqrt{3}+i}{1+i}\right)^2$$

3.19 Express in polar form; include a sketch in each case.

$$2-2i, \qquad \sqrt{3}+i, \qquad -\sqrt{5}i, \qquad -17-23i$$

3.20 Take two complex numbers; express them in polar form, and subtract them.

$$z_1 = r_1 e^{i\theta_1}, \qquad z_2 = r_2 e^{i\theta_2}, \qquad \text{and} \qquad z_3 = z_2 - z_1$$

Compute $z_3^* z_3$, the magnitude squared of z_3, and so derive the law of cosines. You *did* draw a picture didn't you?

3.21 What is i^i? Ans: If you'd like to check your result, type $i \wedge i$ into Google. Or use a calculator such as the one mentioned on page 6.

3.22 For what argument does $\sin\theta = 2$? Next: $\cos\theta = 2$?
Ans: $\sin^{-1} 2 = 1.5708 \pm i1.3170$

3.23 What are the other trigonometric functions, $\tan(ix)$, $\sec(ix)$, etc. What are tan and sec for the general argument $x + iy$.
Ans: $\tan(x+iy) = (\tan x + i\tanh y)/(1 - i\tan x \tanh y)$

3.24 The diffraction pattern from a grating involves the sum of waves from a large number of parallel slits. For light observed at an angle θ away from directly ahead, this sum is, for $N + 1$ slits,

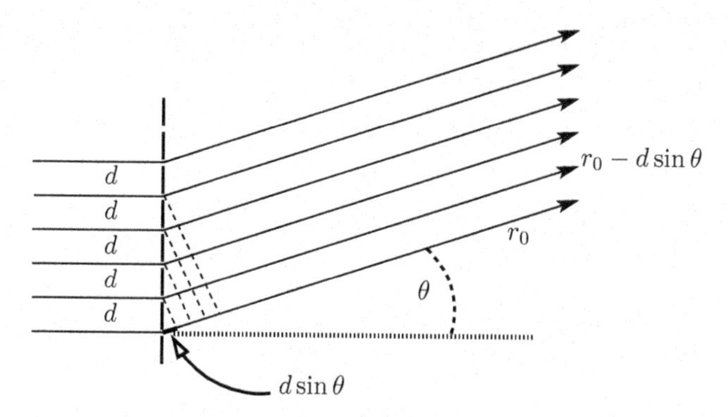

$$\cos\big(kr_0 - \omega t\big) + \cos\big(k(r_0 - d\sin\theta) - \omega t\big) + \cos\big(k(r_0 - 2d\sin\theta) - \omega t\big) + \\ \ldots + \cos\big(k(r_0 - Nd\sin\theta) - \omega t\big)$$

Express this as the real part of complex exponentials and sum the finite series. Show that the resulting wave is

$$\frac{\sin\big(\frac{1}{2}(N+1)kd\sin\theta\big)}{\sin\big(\frac{1}{2}kd\sin\theta\big)} \cos\big(k(r_0 - \tfrac{1}{2}Nd\sin\theta) - \omega t\big)$$

Interpret this result as a wave that appears to be coming from some particular point (where?) and with an intensity pattern that varies strongly with θ.

3.25 (**a**) If the coefficients in a quadratic equation are real, show that if z is a complex root of the equation then so is z^*. If you do this by reference to the quadratic formula, you'd better find another way too, because the second part of this problem is
(**b**) Generalize this to the roots of an arbitrary polynomial with real coefficients.

3.26 You can represent the motion of a particle in two dimensions by using a time-dependent complex number with $z = x + iy = re^{i\theta}$ showing its rectangular or polar coordinates. Assume that r and θ are functions of time and differentiate $re^{i\theta}$ to get the velocity. Differentiate it again to get the acceleration. You can interpret $e^{i\theta}$ as the unit vector along the radius and $ie^{i\theta}$ as the unit vector perpendicular to the radius and pointing in the direction of increasing theta. Show that

$$\frac{d^2z}{dt^2} = e^{i\theta}\left[\frac{d^2r}{dt^2} - r\left(\frac{d\theta}{dt}\right)^2\right] + ie^{i\theta}\left[r\frac{d^2\theta}{dt^2} + 2\frac{dr}{dt}\frac{d\theta}{dt}\right] \tag{3.17}$$

and translate this into the usual language of components of vectors, getting the radial ($\hat{r}$) component of acceleration and the angular component of acceleration as in section 8.9.

3.27 Use the results of the preceding problem, and examine the case of a particle moving directly away from the origin. (**a**) What is its acceleration? (**b**) If instead, it is moving at

$r =$constant, what is its acceleration? (**c**) If instead, $x = x_0$ and $y = v_0 t$, what are $r(t)$ and $\theta(t)$? Now compute $d^2 z/dt^2$ from Eq. (3.17).

3.28 Was it really legitimate simply to substitute $x + iy$ for $\theta_1 + \theta_2$ in Eq. (3.11) to get $\cos(x + iy)$? Verify the result by substituting the expressions for $\cos x$ and for $\cosh y$ as exponentials to see if you can reconstruct the left-hand side.

3.29 The roots of the quadratic equation $z^2 + bz + c = 0$ are functions of the parameters b and c. For real b and c and for both cases $c > 0$ and $c < 0$ (say ± 1 to be specific) plot the trajectories of the roots in the complex plane as b varies from $-\infty$ to $+\infty$. You should find various combinations of straight lines and arcs of circles.

3.30 In integral tables you can find the integrals for such functions as

$$\int dx\, e^{ax} \cos bx, \qquad \text{or} \qquad \int dx\, e^{ax} \sin bx$$

Show how easy it is to do these by doing both integrals at once. Do the first plus i times the second and then separate the real and imaginary parts.

3.31 Find the sum of the series

$$\sum_{1}^{\infty} \frac{i^n}{n}$$

Ans: $i\pi/4 - \frac{1}{2}\ln 2$

3.32 Evaluate $|\cos z|^2$. Evaluate $|\sin z|^2$.

3.33 Evaluate $\sqrt{1 + i}$. Evaluate $\ln(1 + i)$. Evaluate $\tan(1 + i)$.

3.34 (**a**) Beats occur in sound when two sources emit two frequencies that are almost the same. The perceived wave is the sum of the two waves, so that at your ear, the wave is a sum of two cosines of $\omega_1 t$ and of $\omega_2 t$. Use complex algebra to evaluate this. The sum is the real part of

$$e^{i\omega_1 t} + e^{i\omega_2 t}$$

Notice the two identities

$$\omega_1 = \frac{\omega_1 + \omega_2}{2} + \frac{\omega_1 - \omega_2}{2}$$

and the difference of these for ω_2. Use the complex exponentials to derive the results; don't just look up some trig identity. Factor the resulting expression and sketch a graph of the resulting real part, interpreting the result in terms of beats if the two frequencies are close to each other. (**b**) In the process of doing this problem using complex exponentials, what is the trigonometric identity for the sum of two cosines? While you're about it, what is the difference of two cosines?
Ans: $\cos\omega_1 t + \cos\omega_2 t = 2\cos\frac{1}{2}(\omega_1 + \omega_2)t \cos\frac{1}{2}(\omega_1 - \omega_2)t$.

3.35 Derive using complex exponentials: $\sin x - \sin y = 2\sin\left(\frac{x-y}{2}\right)\cos\left(\frac{x+y}{2}\right)$

3.36 The equation (3.4) assumed that the usual rule for multiplying exponentials still holds when you are using complex numbers. Does it? You can prove it by looking at the infinite series representation for the exponential and showing that

$$e^a e^b = \left[1 + a + \frac{a^2}{2!} + \frac{a^3}{3!} + \cdots\right]\left[1 + b + \frac{b^2}{2!} + \frac{b^3}{3!} + \cdots\right] = \left[1 + (a+b) + \frac{(a+b)^2}{2!} + \cdots\right]$$

You may find Eq. (2.19) useful.

3.37 Look at the vertical lines in the z-plane as mapped by Eq. (3.16). I drew the images of lines $y =$constant, now you draw the images of the straight line segments $x =$constant from $0 < y < \pi$. The two sets of lines in the original plane intersect at right angles. What is the angle of intersection of the corresponding curves in the image?

3.38 Instead of drawing the image of the lines $x =$constant as in the previous problem, draw the image of the line $y = x \tan\alpha$, the line that makes an angle α with the horizontal lines. The image of the horizontal lines were radial lines. At a point where this curve intersects one of the radial lines, what angle does the curve make with the radial line? Show that the answer is α, the same angle of intersection as in the original picture.

3.39 Write each of these functions of z as two real functions u and v such that $f(z) = u(x, y) + iv(x, y)$.

$$z^3, \qquad \frac{1+z}{1-z}, \qquad \frac{1}{z^2}, \qquad \frac{z}{z^*}$$

3.40 Evaluate z^i where z is an arbitrary complex number, $z = x + iy = re^{i\theta}$.

3.41 What is the image of the domain $-\infty < x < +\infty$ and $0 < y < \pi$ under the function $w = \sqrt{z}$? Ans: One boundary is a hyperbola.

3.42 What is the image of the disk $|z - a| < b$ under the function $w = cz + d$? Allow c and d to be complex. Take a real.

3.43 What is the image of the disk $|z - a| < b$ under the function $w = 1/z$? Assume $b < a$. Ans: Another disk, centered at $a/(a^2 - b^2)$.

3.44 (**a**) Multiply $(2 + i)(3 + i)$ and deduce the identity

$$\tan^{-1}(1/2) + \tan^{-1}(1/3) = \pi/4$$

(**b**) Multiply $(5 + i)^4(-239 + i)$ and deduce

$$4\tan^{-1}(1/5) - \tan^{-1}(1/239) = \pi/4$$

For (**b**) a sketch will help sort out some signs.
(**c**) Using the power series representation of the $\tan^{-1}$, Eq. (2.27), how many terms would it take to compute 100 digits of π as $4\tan^{-1} 1$? How many terms would it take using each of these two representations, (**a**) and (**b**), for π? Ans: Almost a googol versus respectively about 540 and a few more than 180 terms.

3.45 Use Eq. (3.9) and look back at the development of Eq. (1.4) to find the $\sin^{-1}$ and $\cos^{-1}$ in terms of logarithms.

3.46 Evaluate the integral $\int_{-\infty}^{\infty} dx\, e^{-\alpha x^2} \cos\beta x$ for fixed real α and β. Sketch a graph of the result versus β. Sketch a graph of the result versus α, and why does the graph behave as it does? Notice the rate at which the result approaches zero as either $\alpha \to 0$ or $\alpha \to \infty$. The behavior is very different in the two cases. Ans: $e^{-\beta^2/4\alpha}\sqrt{\pi/\alpha}$

3.47 Does the equation $\sin z = 0$ have any roots other than the real ones? How about the cosine? The tangent?

3.48 Compute (**a**) $\sin^{-1} i$. (**b**) $\cos^{-1} i$. (**c**) $\tan^{-1} i$. (**d**) $\sinh^{-1} i$. Ans: $\sin^{-1} i = 0{+}0.881\, i$, $\cos^{-1} i = \pi/2 - 0.881\, i$.

3.49 By writing

$$\frac{1}{1+x^2} = \frac{i}{2}\left[\frac{1}{x+i} - \frac{1}{x-i}\right]$$

and integrating, check the equation

$$\int_0^1 \frac{dx}{1+x^2} = \frac{\pi}{4}$$

3.50 Solve the equations (**a**) $\cosh u = 0$ (**b**) $\tanh u = 2$ (**c**) $\operatorname{sech} u = 2i$
Ans: $\operatorname{sech}^{-1} 2i = 0.4812 - i1.5707$

3.51 Solve the equations (**a**) $z - 2z^* = 1$ (**b**) $z^3 - 3z^2 + 4z = 2i$ after verifying that $1+i$ is a root. Compare the result of problem 3.25.

3.52 Confirm the plot of $\ln(1+iy)$ following Eq. (3.15). Also do the corresponding plots for $\ln(10+iy)$ and $\ln(100+iy)$. And what do these graphs look like if you take the other branches of the logarithm, with the $i(\theta + 2n\pi)$?

3.53 Check that the results of Eq. (3.14) for cosines and for sines give the correct results for small θ? What about $\theta \to 2\pi$?

3.54 Finish the calculation leading to Eq. (3.13), thereby proving that the two indicated lines have the same length and are perpendicular.

3.55 In the same spirit as Eq. (3.13) concerning squares drawn on the sides of an arbitrary quadrilateral, start with an arbitrary triangle and draw equilateral triangles on each side. Find the centroids of each of the equilateral triangles and connect them. The result is an equilateral triangle. Recall: the centroid is one third the distance from the base to the vertex. [This one requires more algebra than the one in the text.] (Napoleon's Theorem)

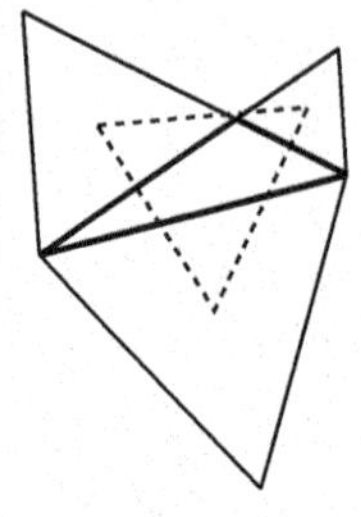

Differential Equations

The subject of ordinary differential equations encompasses such a large field that you can make a profession of it. There are however a small number of techniques in the subject that you *have* to know. These are the ones that come up so often in physical systems that you not need both the skills to use them and the intuition about what they will do. That small group of methods is what I'll concentrate on in this chapter.

4.1 Linear Constant-Coefficient

A differential equation such as

$$\left(\frac{d^2x}{dt^2}\right)^3 + t^2x^4 + 1 = 0$$

relating acceleration to position and time, is not one that I'm especially eager to solve, and one of the things that makes it difficult is that it is non-linear. This means that starting with two solutions $x_1(t)$ and $x_2(t)$, the sum $x_1 + x_2$ is not a solution; look at all the cross-terms you get if you try to plug the sum into the equation and have to cube the sum of the second derivatives. Also if you multiply $x_1(t)$ itself by 2 you no longer have a solution.

An equation such as

$$e^t\frac{d^3x}{dt^3} + t^2\frac{dx}{dt} - x = 0$$

may be a mess to solve, but if you have two solutions, $x_1(t)$ and $x_2(t)$ then the sum $\alpha x_1 + \beta x_2$ is also a solution. Proof? Plug in:

$$\begin{aligned} &e^t\frac{d^3(\alpha x_1 + \beta x_2)}{dt^3} + t^2\frac{d(\alpha x_1 + \beta x_2)}{dt} - (\alpha x_1 + \beta x_2) \\ &= \alpha\left(e^t\frac{d^3x_1}{dt^3} + t^2\frac{dx_1}{dt} - x_1\right) + \beta\left(e^t\frac{d^3x_2}{dt^3} + t^2\frac{dx_2}{dt} - x_2\right) = 0 \end{aligned}$$

This is called a linear, homogeneous equation because of this property. A similar-looking equation,

$$e^t\frac{d^3x}{dt^3} + t^2\frac{dx}{dt} - x = t$$

does not have this property, though it's close. It is called a linear, inhomogeneous equation. If $x_1(t)$ and $x_2(t)$ are solutions to this, then if I try their sum as a solution I get $2t = t$, and that's no solution, but it misses working only because of the single term on the right, and that will make it not too far removed from the preceding case.

One of the most common sorts of differential equations that you see is an especially simple one to solve. That's part of the reason it's so common. This is the linear, constant-coefficient, differential equation. If you have a mass tied to the end of a spring and the other end of the spring is fixed, the force applied to the mass by the spring is to a good approximation proportional to the distance that the mass has moved from its equilibrium position.

If the coordinate x is measured from the mass's equilibrium position, the equation $\vec{F} = m\vec{a}$ says

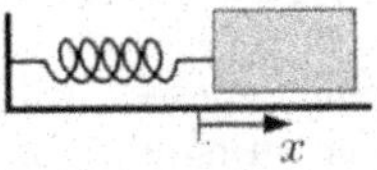

$$m\frac{d^2x}{dt^2} = -kx \tag{4.1}$$

If there's friction (and there's *always* friction), the force has another term. Now how do you describe friction mathematically? The common model for dry friction is that the magnitude of the force is independent of the magnitude of the mass's velocity and opposite to the direction of the velocity. If you try to write that down in a compact mathematical form you get something like

$$\vec{F}_{\text{friction}} = -\mu_{\text{k}} F_{\text{N}} \frac{\vec{v}}{|\vec{v}|} \tag{4.2}$$

This is hard to work with. It can be done, but I'm going to do something different. (See problem 4.31 however.) Wet friction is easier to handle mathematically because when you lubricate a surface, the friction becomes velocity dependent in a way that is, for low speeds, proportional to the velocity.

$$\vec{F}_{\text{friction}} = -b\vec{v} \tag{4.3}$$

Neither of these two representations is a completely accurate description of the way friction works. That's far more complex than either of these simple models, but these approximations are good enough for many purposes and I'll settle for them.

Assume "wet friction" and the differential equation for the motion of m is

$$m\frac{d^2x}{dt^2} = -kx - b\frac{dx}{dt} \tag{4.4}$$

This is a second order, linear, homogeneous differential equation, which simply means that the highest derivative present is the second, the sum of two solutions is a solution, and a constant multiple of a solution is a solution. That the coefficients are constants makes this an easy equation to solve.

All you have to do is to recall that the derivative of an exponential is an exponential. $de^t/dt = e^t$. Substitute this exponential for $x(t)$, and of course it can't work as a solution; it doesn't even make sense dimensionally. What is e to the power of a day? You need something in the exponent to make it dimensionless, $e^{\alpha t}$. Also, the function x is supposed to give you a position, with dimensions of length. Use another constant: $x(t) = Ae^{\alpha t}$. Plug *this* into the differential equation (4.4) to find

$$mA\alpha^2 e^{\alpha t} + bA\alpha e^{\alpha t} + kAe^{\alpha t} = Ae^{\alpha t}\big[m\alpha^2 + b\alpha + k\big] = 0$$

The product of factors is zero, and the only way that a product of two numbers can be zero is if one of the numbers is zero. The exponential never vanishes, and for a non-trivial solution $A \neq 0$, so all that's left is the polynomial in α.

$$m\alpha^2 + b\alpha + k = 0, \qquad \text{with solutions} \qquad \alpha = \frac{-b \pm \sqrt{b^2 - 4km}}{2m} \tag{4.5}$$

The position function is then

$$x(t) = Ae^{\alpha_1 t} + Be^{\alpha_2 t} \tag{4.6}$$

where A and B are arbitrary constants and α_1 and α_2 are the two roots.

Isn't this supposed to be oscillating? It is a harmonic oscillator after all, but the exponentials don't look very oscillatory. If you have a mass on the end of a spring and the entire system is immersed in honey, it won't do much oscillating! Translated into mathematics, this says that if the constant b is too large, there is no oscillation. In the equation for α, if b is large enough the argument of the square root is positive, and both α's are real — no oscillation. Only if b is small enough does the argument of the square root become negative; then you get complex values for the α's and hence oscillations.

Push this to the extreme case where the damping vanishes: $b = 0$. Then $\alpha_1 = i\sqrt{k/m}$ and $\alpha_2 = -i\sqrt{k/m}$. Denote $\omega_0 = \sqrt{k/m}$.

$$x(t) = Ae^{i\omega_0 t} + Be^{-i\omega_0 t} \tag{4.7}$$

You can write this in other forms using sines and cosines, see problem 4.10. To determine the arbitrary constant A and B you need two equations. They come from some additional information about the problem, typically some initial conditions. Take a specific example in which you start from the origin with a kick, $x(0) = 0$ and $\dot{x}(0) = v_0$.

$$x(0) = 0 = A + B, \qquad \dot{x}(0) = v_0 = i\omega_0 A - i\omega_0 B$$

Solve for A and B to get $A = -B = v_0/(2i\omega_0)$. Then

$$x(t) = \frac{v_0}{2i\omega_0}\left[e^{i\omega_0 t} - e^{-i\omega_0 t}\right] = \frac{v_0}{\omega_0}\sin\omega_0 t$$

As a check on the algebra, use the first term in the power series expansion of the sine function to see how x behaves for small t. The sine factor is $\sin\omega_0 t \approx \omega_0 t$, and then $x(t)$ is approximately $v_0 t$, just as it should be. Also notice that despite all the complex numbers, the final answer is real. This is another check on the algebra.

Damped Oscillator

If there is damping, but not too much, then the α's have an imaginary part *and* a negative real part. (Is it important whether it's negative or not?)

$$\alpha = \frac{-b \pm i\sqrt{4km - b^2}}{2m} = -\frac{b}{2m} \pm i\omega', \qquad \text{where} \qquad \omega' = \sqrt{\frac{k}{m} - \frac{b^2}{4m^2}} \tag{4.8}$$

This represents a damped oscillation and has frequency a bit lower than the one in the undamped case. Use the same initial conditions as above and you will get similar results (let $\gamma = b/2m$)

$$x(t) = Ae^{(-\gamma + i\omega')t} + Be^{(-\gamma - i\omega')t}$$

$$x(0) = A + B = 0, \qquad v_x(0) = (-\gamma + i\omega')A + (-\gamma - i\omega')B = v_0 \tag{4.9}$$

The two equations for the unknowns A and B imply $B = -A$ and

$$2i\omega' A = v_0, \qquad \text{so} \qquad x(t) = \frac{v_0}{2i\omega'} e^{-\gamma t}\left[e^{i\omega' t} - e^{-i\omega' t}\right] = \frac{v_0}{\omega'} e^{-\gamma t} \sin \omega' t \tag{4.10}$$

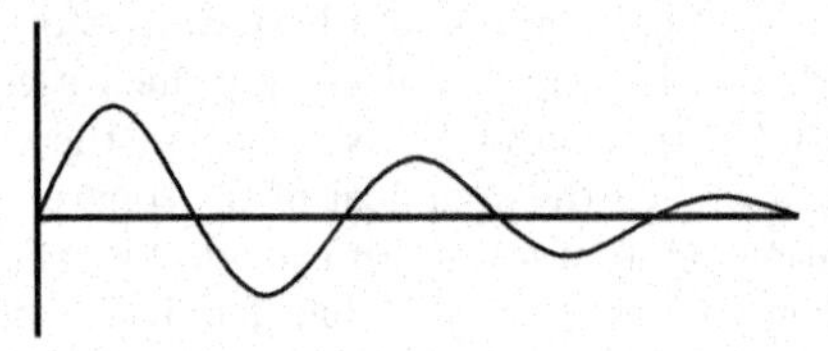

For small values of t, the first terms in the power series expansion of this result are

$$x(t) = \frac{v_0}{\omega'}[1 - \gamma t + \gamma^2 t^2/2 - \ldots][\omega' t - \omega'^3 t^3/6 + \ldots] = v_0 t - v_0 \gamma t^2 + \ldots$$

The first term is what you should expect, as the initial velocity is $v_x = v_0$. The negative sign in the next term says that it doesn't move as far as it would without the damping, but analyze it further. Does it have the right size as well as the right sign? It is $-v_0\gamma t^2 = -v_0(b/2m)t^2$. But that's an acceleration: $a_x t^2/2$. It says that the acceleration just after the motion starts is $a_x = -bv_0/m$. Is that what you should expect? As the motion starts, the mass hasn't gone very far so the spring doesn't yet exert much force. The viscous friction is however $-bv_x$. Set that equal to ma_x and you see that $-v_0\gamma t^2$ has precisely the right value:

$$x(t) \approx v_0 t - v_0 \gamma t^2 = v_0 t - v_0 \frac{b}{2m} t^2 = v_0 t + \frac{1}{2}\frac{-bv_0}{m} t^2$$

The last term says that the acceleration starts as $a_x = -bv_0/m$, as required.

In Eq. (4.8) I assumed that the two roots of the quadratic, the two α's, are different. What if they aren't? Then you have just one value of α to use in defining the solution $e^{\alpha t}$ in Eq. (4.9). You now have just one arbitrary constant with which to match two initial conditions. You're stuck. See problem 4.11 to understand how to handle this case (critical damping). It's really a special case of what I've already done.

What is the energy for this damped oscillator? The kinetic energy is $mv^2/2$ and the potential energy for the spring is $kx^2/2$. Is the sum constant? No.

$$\text{If} \quad F_x = ma_x = -kx + F_{x,\text{frict}}, \quad \text{then}$$
$$\frac{dE}{dt} = \frac{d}{dt}\frac{1}{2}\left(mv^2 + kx^2\right) = mv\frac{dv}{dt} + kx\frac{dx}{dt} = v_x\left(ma_x + kx\right) = F_{x,\text{frict}} v_x \tag{4.11}$$

"Force times velocity" is a common expression for power, and this says that the total energy is decreasing according to this formula. For the wet friction used here, this is $dE/dt = -bv_x^2$, and the energy decreases exponentially on average.

4.2 Forced Oscillations

What happens if the equation is inhomogeneous? That is, what if there is a term that doesn't involve x or its derivatives at all. In this harmonic oscillator example, apply an extra external force. Maybe it's a constant; maybe it's an oscillating force; it can be anything you want not involving x.

$$m\frac{d^2x}{dt^2} = -kx - b\frac{dx}{dt} + F_{\text{ext}}(t) \tag{4.12}$$

The key result that you need for this class of equations is very simple to state and not too difficult to implement. It is a procedure for attacking any linear inhomogeneous differential equation and consists of three steps.

1. Temporarily throw out the inhomogeneous term [here $F_{\text{ext}}(t)$] and completely solve the resulting homogeneous equation. In the current case that's what you just saw when I worked out the solution to the differential equation $md^2x/dt^2 + bdx/dt + kx = 0$. $[x_{\text{hom}}(t)]$
2. Find any *one* solution to the full inhomogeneous equation. Note that for step one you have to have all the arbitrary constants present; for step two you do not. $[x_{\text{inh}}(t)]$
3. Add the results of steps one and two. $[x_{\text{hom}}(t) + x_{\text{inh}}(t)]$

I've already done step one. To carry out the next step I'll start with a particular case of the forcing function. If $F_{\text{ext}}(t)$ is simple enough, you should be able to *guess* the answer to step two. If it's a constant, then a constant will work for x. If it's a sine or cosine, then you can guess that a sine or cosine or a combination of the two should work. If it's an exponential, then guess an exponential — remember that the derivative of an exponential is an exponential. If it's the sum of two terms, such as a constant and an exponential, it's easy to verify that you add the results that you get for the two cases separately. If the forcing function is too complicated for you to guess a solution then there's a general method using Green's functions that I'll get to in section 4.6.

Choose a specific example

$$F_{\text{ext}}(t) = F_0\big[1 - e^{-\beta t}\big] \tag{4.13}$$

This starts at zero and builds up to a final value of F_0. It does it slowly or quickly depending on β.

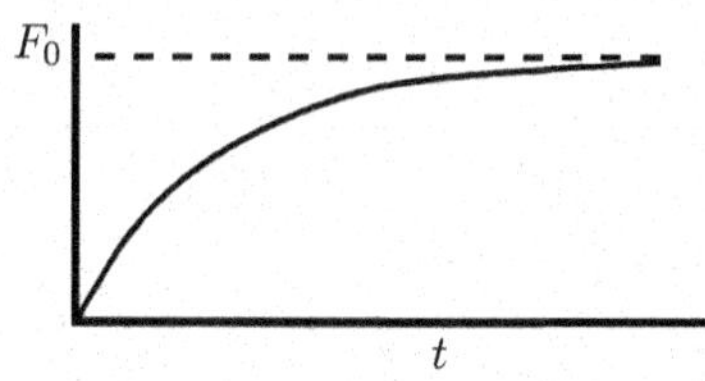

Start with the first term, F_0, for external force in Eq. (4.12). Try $x(t) = C$ and plug into that equation to find

$$kC = F_0$$

This is simple and determines C.

Next, use the second term as the forcing function, $-F_0e^{-\beta t}$. Guess a solution $x(t) = C'e^{-\beta t}$ and plug in. The exponential cancels, leaving

$$mC'\beta^2 - bC'\beta + kC' = -F_0 \qquad \text{or} \qquad C' = \frac{-F_0}{m\beta^2 - b\beta + k}$$

The total solution for the inhomogeneous part of the equation is then the sum of these two expressions.

$$x_{\text{inh}}(t) = F_0\left(\frac{1}{k} - \frac{1}{m\beta^2 - b\beta + k}e^{-\beta t}\right)$$

The homogeneous part of Eq. (4.12) has the solution found in Eq. (4.6) and the total is

$$x(t) = x_{\text{hom}}(t) + x_{\text{inh}}(t) = x(t) = Ae^{\alpha_1 t} + Be^{\alpha_2 t} + F_0\left(\frac{1}{k} - \frac{1}{m\beta^2 - b\beta + k}e^{-\beta t}\right) \tag{4.14}$$

There are two arbitrary constants here, and this is what you need because you have to be able to specify the initial position and the initial velocity independently; this is a second order differential equation after all. Take for example the conditions that the initial position is zero and the initial velocity is zero. Everything is at rest until you start applying the external force. This provides two equations for the two unknowns.

$$x(0) = 0 = A + B + F_0\frac{m\beta^2 - b\beta}{k(m\beta^2 - b\beta + k)}$$
$$\dot{x}(0) = 0 = A\alpha_1 + B\alpha_2 + F_0\frac{\beta}{m\beta^2 - b\beta + k}$$

Now all you have to do is solve the two equations in the two unknowns A and B. Take the first, multiply it by α_2 and subtract the second. This gives A. Do the same with α_1 instead of α_2 to get B. The results are

$$A = \frac{1}{\alpha_1 - \alpha_2}F_0\frac{\alpha_2(m\beta^2 - b\beta) - k\beta}{k(m\beta^2 - b\beta + k)}$$

Interchange α_1 and α_2 to get B.

The final result is

$$x(t) = \frac{F_0}{\alpha_1 - \alpha_2}\frac{\big(\alpha_2(m\beta^2 - b\beta) - k\beta\big)e^{\alpha_1 t} - \big(\alpha_1(m\beta^2 - b\beta) - k\beta\big)e^{\alpha_2 t}}{k(m\beta^2 - b\beta + k)}$$
$$+ F_0\left(\frac{1}{k} - \frac{1}{m\beta^2 - b\beta + k}e^{-\beta t}\right) \tag{4.15}$$

If you think this is messy and complicated, you haven't seen messy and complicated. When it takes 20 pages to write out the equation, then you're entitled say that it is starting to become involved.

Why not start with a simpler example, one without all the terms? The reason is that a complex expression is often easier to analyze than a simple one. There are more things that you can do to it, and so more opportunities for it to go wrong. The problem isn't finished until you've analyzed the supposed solution. After all, I may have made some errors in algebra along the way. Also, analyzing the solution is the way you learn how these functions *work*.

1. Everything in the solution is proportional to F_0 and that's not surprising.
2. I'll leave it as an exercise to check the dimensions.
3. A key parameter to vary is β. What should happen if it is either very large or very small? In the former case the exponential function in the force drops to zero quickly so the force jumps from zero to F_0 in a very short time — a step in the limit that $\beta \to 0$.
4. If β is very small the force turns on very gradually and gently, as though you are being very careful not to disturb the system.

Take point 3 above: For large β the dominant terms in both numerator and denominator everywhere are the $m\beta^2$ terms. This result is then very nearly

$$x(t) \approx \frac{F_0}{\alpha_1 - \alpha_2} \frac{\big(\alpha_2(m\beta^2)\big)e^{\alpha_1 t} - \big(\alpha_1(m\beta^2)\big)e^{\alpha_2 t}}{km\beta^2} + F_0\left(\frac{1}{k} - \frac{1}{(m\beta^2)}e^{-\beta t}\right)$$
$$\approx \frac{F_0}{k(\alpha_1 - \alpha_2)}\big[(\alpha_2 e^{\alpha_1 t} - \alpha_1 e^{\alpha_2 t}\big] + F_0\frac{1}{k}$$

Use the notation of Eq. (4.9) and you have

$$x(t) \approx \frac{F_0}{k\big(-\gamma + i\omega' - (-\gamma - i\omega')\big)}\big[((-\gamma - i\omega')e^{(-\gamma+i\omega')t} - (-\gamma + i\omega')e^{(-\gamma-i\omega')t}\big] + F_0\frac{1}{k}$$
$$= \frac{F_0 e^{-\gamma t}}{k(2i\omega')}\big[-2i\gamma\sin\omega' t - 2i\omega'\cos\omega' t\big] + F_0\frac{1}{k}$$
$$= \frac{F_0 e^{-\gamma t}}{k}\big[-\frac{\gamma}{\omega'}\sin\omega' t - \cos\omega' t\big] + F_0\frac{1}{k} \tag{4.16}$$

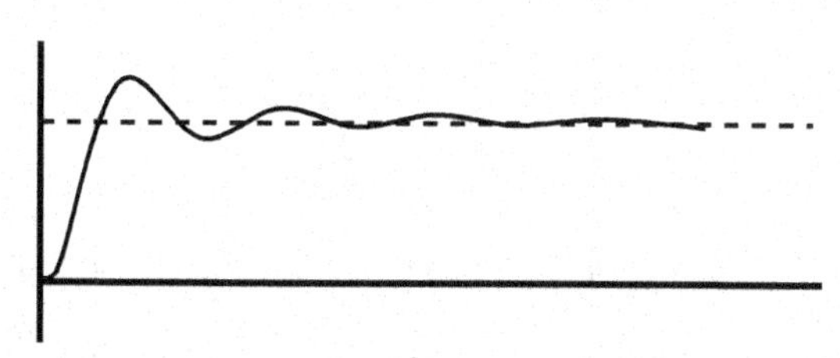

At time $t = 0$ this is still zero even with the approximations. That's comforting, but if it hadn't happened it's not an insurmountable disaster. This is an approximation to the exact answer after all, so it could happen that the initial conditions are obeyed only approximately. The exponential terms have oscillations and damping, so the mass oscillates about its eventual equilibrium position and after a long enough time the oscillations die out and you are left with the equilibrium solution $x = F_0/k$.

Look at point 4 above: For small β the β^2 terms in Eq. (4.15) are small compared to the β terms to which they are added or subtracted. The numerators of the terms with

$e^{\alpha t}$ are then proportional to β. The denominator of the same terms has a $k - b\beta$ in it. That means that as $\beta \to 0$, the numerator of the homogeneous term approaches zero and its denominator doesn't. The last terms, that came from the inhomogeneous part, don't have any β in the numerator so they don't vanish in this limit. The approximate final result then comes solely from the $x_{\text{inh}}(t)$ term.

$$x(t) \approx F_0 \frac{1}{k}\left(1 - e^{-\beta t}\right)$$

It doesn't oscillate at all and just gradually moves from equilibrium to equilibrium as time goes on. It's what you get if you go back to the differential equation (4.12) and say that the acceleration and the velocity are negligible.

$$m\frac{d^2x}{dt^2}[\approx 0] = -kx - b\frac{dx}{dt}[\approx 0] + F_{\text{ext}}(t) \qquad \Longrightarrow \qquad x \approx \frac{1}{k}F_{\text{ext}}(t)$$

The spring force nearly balances the external force at all times; this is "quasi-static," in which the external force is turned on so slowly that it doesn't cause any oscillations.

4.3 Series Solutions

A linear, second order differential equation can always be rearranged into the form

$$y'' + P(x)y' + Q(x)y = R(x) \tag{4.17}$$

If at some point x_0 the functions P and Q are well-behaved, if they have convergent power series expansions about x_0, then this point is called a "*regular point*" and you can expect good behavior of the solutions there — at least if R is also regular there.

I'll look just at the case for which the inhomogeneous term $R = 0$. If P or Q has a singularity at x_0, perhaps something such as $1/(x - x_0)$ or $\sqrt{x - x_0}$, then x_0 is called a "*singular point*" of the differential equation.

Regular Singular Points

The most important special case of a singular point is the "*regular singular point*" for which the behaviors of P and Q are not too bad. Specifically this requires that $(x-x_0)P(x)$ and $(x - x_0)^2Q(x)$ have no singularity at x_0. For example

$$y'' + \frac{1}{x}y' + \frac{1}{x^2}y = 0 \qquad \text{and} \qquad y'' + \frac{1}{x^2}y' + xy = 0$$

have singular points at $x = 0$, but the first one is a regular singular point and the second one is not. The importance of a regular singular point is that there is a procedure guaranteed to find a solution near a regular singular point (Frobenius series). For the more general singular point there is no guaranteed procedure (though there are a few tricks* that sometimes work).

---

* The book by Bender and Orszag: "Advanced mathematical methods for scientists and engineers" is a very readable source for this and many other topics.

Examples of equations that show up in physics problems are

$$\begin{aligned} y'' + y &= 0 \\ (1-x^2)y'' - 2xy' + \ell(\ell+1)y &= 0 \qquad \text{regular singular points at } \pm 1 \\ x^2y'' + xy' + (x^2 - n^2)y &= 0 \qquad \text{regular singular point at zero} \\ xy'' + (\alpha + 1 - x)y' + ny &= 0 \qquad \text{regular singular point at zero} \end{aligned} \tag{4.18}$$

These are respectively the classical simple harmonic oscillator, Legendre equation, Bessel equation, generalized Laguerre equation.

A standard procedure to solve these equations is to use series solutions, but not just the standard power series such as those in Eq. (2.4). Essentially, you assume that there is a solution in the form of an infinite series and you systematically compute the terms of the series. I'll pick the Bessel equation from the above examples, as the other three equations are done the same way. The parameter n in that equation is often an integer, but it can be anything. It's common for it to be $^1\!/_2$ or $^3\!/_2$ or sometimes even imaginary, but there's no need to make any assumptions about it for now.

Assume a solution in the form :

$$\text{Frobenius Series:} \qquad y(x) = \sum_0^\infty a_k x^{k+s} \qquad (a_0 \neq 0) \tag{4.19}$$

If $s = 0$ or a positive integer, this is just the standard Taylor series you saw so much of in chapter two, but this simple-looking extension makes it much more flexible and suited for differential equations. It often happens that s is a fraction or negative, but this case is no harder to handle than the Taylor series. For example, what is the series expansion of $(\cos x)/x$ about the origin? This is singular at zero, but it's easy to write the answer anyway because you already know the series for the cosine.

$$\frac{\cos x}{x} = \frac{1}{x} - \frac{x}{2} + \frac{x^3}{24} - \frac{x^5}{720} + \cdots$$

It starts with the term $1/x$ corresponding to $s = -1$ in the Frobenius series.

Always assume that $a_0 \neq 0$, because that just defines the coefficient of the most negative power, x^s. If you allow it be zero, that's just the same as redefining s and it gains nothing except confusion. Plug this into the Bessel differential equation.

$$x^2y'' + xy' + (x^2 - n^2)y = 0$$

$$x^2 \sum_{k=0}^\infty a_k(k+s)(k+s-1)x^{k+s-2} + x\sum_{k=0}^\infty a_k(k+s)x^{k+s-1} + (x^2 - n^2)\sum_{k=0}^\infty a_k x^{k+s} = 0$$

$$\sum_{k=0}^\infty a_k(k+s)(k+s-1)x^{k+s} + \sum_{k=0}^\infty a_k(k+s)x^{k+s} + \sum_{k=0}^\infty a_k x^{k+s+2} - n^2\sum_0^\infty a_k x^{k+s} = 0$$

$$\sum_{k=0}^\infty a_k\big[(k+s)(k+s-1) + (k+s) - n^2\big]x^{k+s} + \sum_{k=0}^\infty a_k x^{k+s+2} = 0$$

The coefficients of all the like powers of x must match, and in order to work out the matches efficiently, and so as not to get myself confused in a mess of indices, I'll make an explicit change of the index in the sums. *Do this trick every time. It keeps you out of trouble.*

Let $\ell = k$ in the first sum. Let $\ell = k + 2$ in the second. *Explicitly* show the limits of the index on the sums, or you're bound to get it wrong.

$$\sum_{\ell=0}^{\infty} a_\ell\big[(\ell+s)^2 - n^2\big]x^{\ell+s} + \sum_{\ell=2}^{\infty} a_{\ell-2}x^{\ell+s} = 0$$

The lowest power of x in this equation comes from the $\ell = 0$ term in the first sum. That coefficient of x^s must vanish. $(a_0 \neq 0)$

$$a_0\big[s^2 - n^2\big] = 0 \tag{4.20}$$

This is called the *indicial* equation. It determines s, or in this case, maybe two s's. After this, set to zero the coefficient of $x^{\ell+s}$.

$$a_\ell\big[(\ell+s)^2 - n^2\big] + a_{\ell-2} = 0 \tag{4.21}$$

This determines a_2 in terms of a_0; it determines a_4 in terms of a_2 *etc.*

$$a_\ell = -a_{\ell-2}\frac{1}{(\ell+s)^2 - n^2}, \qquad \ell = 2,\ 4,\ \ldots$$

For example, if $n = 0$, the indicial equation says $s = 0$.

$$a_2 = -a_0\frac{1}{2^2}, \qquad a_4 = -a_2\frac{1}{4^2} = +a_0\frac{1}{2^2 4^2}, \qquad a_6 = -a_4\frac{1}{6^2} = -a_0\frac{1}{2^2 4^2 6^2}$$

$$a_{2k} = (-1)^k a_0\frac{1}{2^{2k}k!^2} \qquad \text{then} \qquad y(x) = a_0\sum_{k=0}^{\infty}(-1)^k\frac{(x/2)^{2k}}{(k!)^2} = a_0 J_0(x) \tag{4.22}$$

and in the last equation I rearranged the factors and used the standard notation for the Bessel function, $J_n(x)$.

This is a second order differential equation. What about the other solution? This Frobenius series method is guaranteed to find one solution near a regular singular point. Sometimes it gives both but not always, and in this example it produces only one. There are procedures that will let you find the second solution to this sort of second order differential equation. See problem 4.49 for one such method.

For the case $n = 1/2$ the calculations just above will produce two solutions. The indicial equation gives $s = \pm 1/2$. After that, the recursion relation for the coefficients give

$$a_\ell = -a_{\ell-2}\frac{1}{(\ell+s)^2 - n^2} = -a_{\ell-2}\frac{1}{\ell^2 + 2\ell s} = -a_{\ell-2}\frac{1}{\ell(\ell+2s)} = -a_{\ell-2}\frac{1}{\ell(\ell\pm 1)}$$

For the $s = +1/2$ result

$$a_2 = -a_0 \frac{1}{2 \cdot 3} \qquad a_4 = -a_2 \frac{1}{4 \cdot 5} = +a_0 \frac{1}{2 \cdot 3 \cdot 4 \cdot 5}$$
$$a_{2k} = (-1)^k a_0 \frac{1}{(2k+1)!}$$

This solution is then

$$y(x) = a_0 x^{1/2} \left[1 - \frac{x^2}{3!} + \frac{x^4}{5!} - \ldots \right]$$

This series looks suspiciously like the series for the sine function, but is has some of the x's or some of the factorials in the wrong place. You can fix that if you multiply the series in brackets by x. You then have

$$y(x) = a_0 x^{-1/2} \left[x - \frac{x^3}{3!} + \frac{x^5}{5!} - \ldots \right] = a_0 \frac{\sin x}{x^{1/2}} \tag{4.23}$$

I'll leave it to problem 4.15 for you to find the other solution.

Do you need to use a Frobenius series instead of just a power series for all differential equations? No, but I recommend it. If you are expanding about a regular point of the equation then a power series will work, but I find it more systematic to use the same method for all cases. It's less prone to error.

4.4 Some General Methods

It is important to be familiar with the arsenal of special methods that work on special types of differential equations. What if you encounter an equation that doesn't fit these special methods? There are some techniques that you should be familiar with, even if they are mostly not ones that you will want to use often. Here are a couple of methods that can get you started, and there's a much broader set of approaches under the heading of numerical analysis; you can explore those in section 11.5.

If you have a first order differential equation, $dx/dt = f(x,t)$, with initial condition $x(t_0) = x_0$ then you can follow the spirit of the series method, computing successive orders in the expansion. Assume for now that the function f is smooth, with as many derivatives as you want, then use the chain rule a lot to get the higher derivatives of x

$$\begin{aligned}
\frac{dx}{dt} &= f(x,t) \\
\frac{d^2x}{dt^2} &= \frac{\partial f}{\partial t} + \frac{\partial f}{\partial x}\frac{dx}{dt} = \ddot{x} = f_t + f_x \dot{x} \\
\dddot{x} &= f_{tt} + 2f_{xt}\dot{x} + f_{xx}\dot{x}^2 + f_x \ddot{x} = f_{tt} + 2f_{xt}\dot{x} + f_{xx}\dot{x}^2 + f_x[f_t + f_x \dot{x}] \\
x(t) &= x_0 + f(x_0,t_0)(t-t_0) + \tfrac{1}{2}\ddot{x}(t_0)(t-t_0)^2 + \tfrac{1}{6}\dddot{x}(t_0)x(t_0)(t-t_0)^3 + \cdots
\end{aligned} \tag{4.24}$$

Here the dot-notation ($\dot{x}$ *etc.*) is a standard shorthand for derivative with respect to time. This is unlike using a prime for derivative, which is with respect to anything you want. These equations show that once you have the initial data (t_0, x_0), you can compute the next derivatives from them and from the properties of f. Of course if f is complicated this will quickly become a mess, but even then it can be useful to compute the first few terms in the power series expansion of x.

For example, $\dot{x} = f(x,t) = Ax^2(1+\omega t)$ with $t_0 = 0$ and $x_0 = \alpha$.

$$\dot{x}_0 = A\alpha^2, \qquad \ddot{x}_0 = A\alpha^2\omega + 2A^2\alpha^3, \qquad \dddot{x}_0 = 4A^2\alpha^3\omega + 2A^3\alpha^4 + 2A\alpha\big[A\alpha^2\omega + 2A^2\alpha^3\big] \tag{4.25}$$

If $A = 1/\,\mathrm{m\cdot s}$ and $\omega = 1/\mathrm{s}$ with $\alpha = 1\,\mathrm{m}$ this is

$$x(t) = 1 + t + \tfrac{3}{2}t^2 + 2t^3 + \cdots$$

You can also solve this example exactly and compare the results to check the method.

What if you have a second order differential equation? Pretty much the same thing, though it is sometimes convenient to make a slight change in the appearance of the equations when you do this.

$$\ddot{x} = f(x,\dot{x},t) \qquad \text{can be written} \qquad \dot{x} = v, \qquad \dot{v} = f(x,v,t) \tag{4.26}$$

so that it looks like two simultaneous first order equations. Either form will let you compute the higher derivatives, but the second one often makes for a less cumbersome notation. You start by knowing t_0, x_0, and now $v_0 = \dot{x}_0$.

Some of the numerical methods you will find in chapter 11 start from the ideas of these expansions, but then develop them along different lines.

There is an iterative methods that of more theoretical than practical importance, but it's easy to understand. I'll write it for a first order equation, but you can rewrite it for the second (or higher) order case by doing the same thing as in Eq. (4.26).

$$\dot{x} = f(x,t) \qquad \text{with} \qquad x(t_0) = x_0 \qquad \text{generates} \qquad x_1(t) = \int_{t_0}^{t} dt'\, f(x_0,t')$$

This is not a solution of the differential equation, but it forms the starting point to find one because you can iterate this approximate solution x_1 to form an improved approximation.

$$x_k(t) = \int_{t_0}^{t} dt'\, f\big(x_{k-1}(t'),t'\big), \quad k = 2,\, 3,\, \ldots \tag{4.27}$$

This will form a sequence that is usually different from that of the power series approach, though the end result better be the same. This iterative approach is used in one proof that shows under just what circumstances this differential equation $\dot{x} = f$ has a unique solution.

4.5 Trigonometry via ODE's

The differential equation $u'' = -u$ has two independent solutions. The point of this exercise is to derive all (or at least some) of the standard relationships for sines and cosines *strictly from the differential equation.* The reasons for spending some time on this are twofold. First, it's neat. Second, you have to get used to manipulating a differential equation in order to find properties of its solutions. This is essential in the study of Fourier series as you will see in section 5.3.

Two solutions can be defined when you specify boundary conditions. Call the functions $c(x)$ and $s(x)$, and specify their respective boundary conditions to be

$$c(0) = 1, \quad c'(0) = 0, \qquad \text{and} \qquad s(0) = 0, \quad s'(0) = 1 \tag{4.28}$$

What is $s'(x)$? First observe that s' satisfies the same differential equation as s and c:

$$u'' = -u \implies (u')'' = (u'')' = -u', \qquad \text{and that shows the desired result.}$$

This in turn implies that s' is a linear combination of s and c, as that is the most general solution to the original differential equation.

$$s'(x) = Ac(x) + Bs(x)$$

Use the boundary conditions:

$$s'(0) = 1 = Ac(0) + Bs(0) = A$$

From the differential equation you also have

$$s''(0) = -s(0) = 0 = Ac'(0) + Bs'(0) = B$$

Put these together and you have

$$s'(x) = c(x) \qquad \text{And a similar calculation shows} \qquad c'(x) = -s(x) \tag{4.29}$$

What is $c(x)^2 + s(x)^2$? Differentiate this expression to get

$$\frac{d}{dx}[c(x)^2 + s(x)^2] = 2c(x)c'(x) + 2s(x)s'(x) = -2c(x)s(x) + 2s(x)c(x) = 0$$

This combination is therefore a constant. What constant? Just evaluate it at $x = 0$ and you see that the constant is one. There are many more such results that you can derive, but that's left for the exercises, Eq. 4.21 *et seq.*

4.6 Green's Functions

Is there a general way to find the solution to the whole harmonic oscillator inhomogeneous differential equation? One that does not require guessing the form of the solution and applying initial conditions? Yes there is. It's called the method of Green's functions. The idea behind it is that you can think of any force as a sequence of short, small kicks. In fact, because of the atomic nature of matter, that's not so far from the truth. If you can figure out the result of an impact by one molecule, you can add the results of many such kicks to get the answer for 10^{23} molecules.

I'll start with the simpler case where there's no damping, $b = 0$ in the harmonic oscillator equation.

$$m\ddot{x} + kx = F_{\text{ext}}(t) \tag{4.30}$$

Suppose that everything is at rest at the origin and then at time t' the external force provides a small impulse. The motion from that point on will be a sine function starting at t',

$$A \sin\left(\omega_0(t - t')\right) \qquad (t > t') \tag{4.31}$$

The amplitude will depend on the strength of the kick. A constant force F applied for a very short time, $\Delta t'$, will change the momentum of the mass by $m\Delta v_x = F\Delta t'$. If this time interval is short enough the mass doesn't have a chance to move very far before the force is turned off, then from that time on it's subject only to the $-kx$ force. This kick gives m a velocity $F\Delta t'/m$, and that's what determines the unknown constant A.

Just after $t = t'$, $v_x = A\omega_0 = F\Delta t'/m$. This determines A, so the position of m is

$$x(t) = \begin{cases} \frac{F\Delta t'}{m\omega_0} \sin\left(\omega_0(t - t')\right) & (t > t') \\ 0 & (t \le t') \end{cases} \tag{4.32}$$

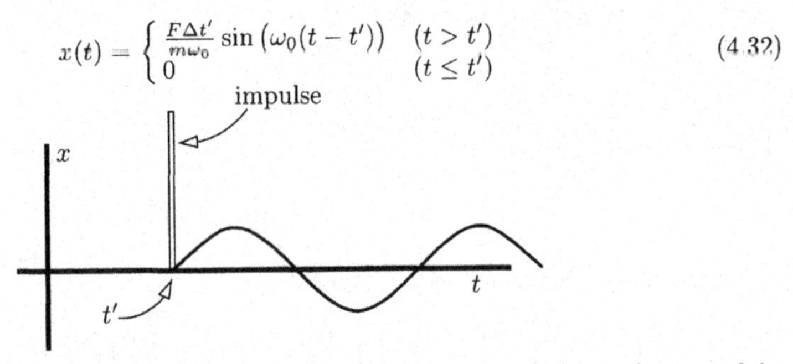

When the external force is the sum of two terms, the total solution is the sum of the solutions for the individual forces. If an impulse at one time gives a solution Eq. (4.32), an impulse at a later time gives a solution that starts its motion at that later time. The key fact about the equation that you're trying to solve is that it is linear, so you can get the solution for two impulses simply by adding the two simpler solutions.

$$m\frac{d^2x_1}{dt^2} + kx_1 = F_1(t) \qquad \text{and} \qquad m\frac{d^2x_2}{dt^2} + kx_2 = F_2(t)$$

then

$$m\frac{d^2(x_1 + x_2)}{dt^2} + k(x_1 + x_2) = F_1(t) + F_2(t)$$

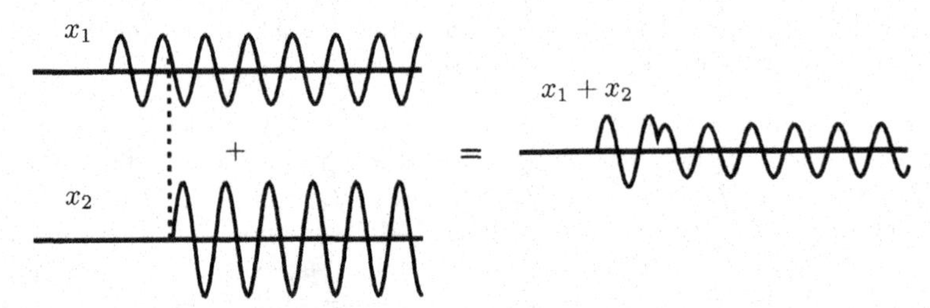

The way to make use of this picture is to take a sequence of contiguous steps. One step follows immediately after the preceding one. If two such impulses are two steps

$$F_0 = \begin{cases} F(t_0) & (t_0 < t < t_1) \\ 0 & \text{(elsewhere)} \end{cases} \qquad \text{and} \qquad F_1 = \begin{cases} F(t_1) & (t_1 < t < t_2) \\ 0 & \text{(elsewhere)} \end{cases}$$

$$m\ddot{x} + kx = F_0 + F_1 \tag{4.33}$$

then if x_0 is the solution to Eq. (4.30) with only the F_0 on its right, and x_1 is the solution with only F_1, then the full solution to Eq. (4.33) is the sum, $x_0 + x_1$.

Think of a general forcing function $F_{x,\text{ext}}(t)$ in the way that you would set up an integral. Approximate it as a sequence of very short steps as in the picture. Between t_k and t_{k+1} the force is essentially $F(t_k)$. The response of m to this piece of the total force is then Eq. (4.32).

$$x_k(t) = \begin{cases} \frac{F(t_k)\Delta t_k}{m\omega_0} \sin\big(\omega_0(t - t_k)\big) & (t > t_k) \\ 0 & (t \le t_k) \end{cases}$$

where $\Delta t_k = t_{k+1} - t_k$.

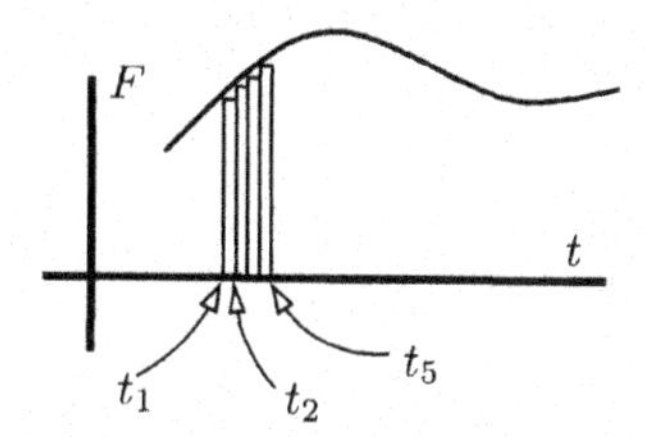

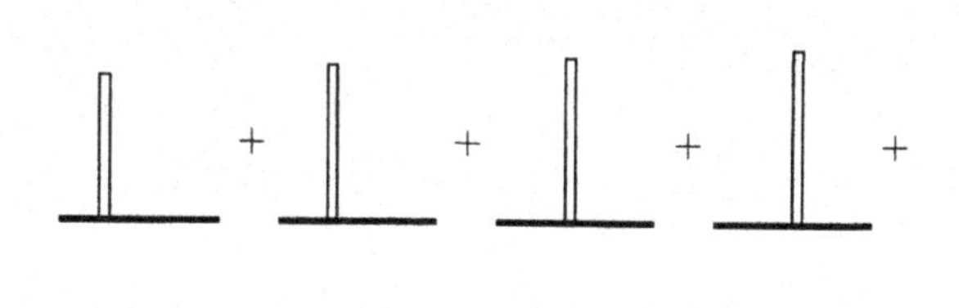

To complete this idea, the external force is the sum of a lot of terms, the force between t_1 and t_2, that between t_2 and t_3 etc. The total response is the sum of all these individual responses.

$$x(t) = \sum_k \begin{cases} \frac{F(t_k)\Delta t_k}{m\omega_0} \sin\big(\omega_0(t - t_k)\big) & (t > t_k) \\ 0 & (t \le t_k) \end{cases}$$

For a specified time t, only the times t_k before and up to t contribute to this sum. The impulses occurring at the times after the time t can't change the value of $x(t)$; they haven't happened yet. In the limit that $\Delta t_k \to 0$, this sum becomes an integral.

$$x(t) = \int_{-\infty}^{t} dt' \, \frac{F(t')}{m\omega_0} \sin\big(\omega_0(t - t')\big) \tag{4.34}$$

Apply this to an example. The simplest is to start at rest and begin applying a constant force from time zero on.

$$F_{\text{ext}}(t) = \begin{cases} F_0 & (t > 0) \\ 0 & (t \le 0) \end{cases} \qquad\qquad x(t) = \int_0^t dt' \, \frac{F_0}{m\omega_0} \sin\big(\omega_0(t - t')\big)$$

and the last expression applies only for $t > 0$. It is

$$x(t) = \frac{F_0}{m\omega_0^2}\big[1 - \cos(\omega_0 t)\big] \tag{4.35}$$

As a check for the plausibility of this result, look at the special case of small times. Use the power series expansion of the cosine, keeping a couple of terms, to get

$$x(t) \approx \frac{F_0}{m\omega_0^2}\big[1 - \big(1 - (\omega_0 t)^2/2\big)\big] = \frac{F_0}{m\omega_0^2}\frac{\omega_0^2 t^2}{2} = \frac{F_0}{m}\frac{t^2}{2}$$

and this is just the result you'd get for constant acceleration F_0/m. In this short time, the position hasn't changed much from zero, so the spring hasn't had a chance to stretch very far, so it can't apply much force, and you have nearly constant acceleration.

This is a sufficiently important subject that it will be repeated elsewhere in this text. A completely different approach to Green's functions will appear is in section 15.5, and chapter 17 is largely devoted to the subject.

4.7 Separation of Variables

If you have a first order differential equation — I'll be more specific for an example, in terms of x and t — and if you are able to move the variables around until everything involving x and dx is on one side of the equation and everything involving t and dt is on the other side, then you have "separated variables." Now all you have to do is integrate.

For example, the total energy in the undamped harmonic oscillator is $E = mv^2/2 + kx^2/2$. Solve for dx/dt and

$$\frac{dx}{dt} = \sqrt{\frac{2}{m}(E - kx^2/2)} \tag{4.36}$$

To separate variables, multiply by dt and divide by the right-hand side.

$$\frac{dx}{\sqrt{\frac{2}{m}(E - kx^2/2)}} = dt$$

Now it's just manipulation to put this into a convenient form to integrate.

$$\sqrt{\frac{m}{k}}\frac{dx}{\sqrt{(2E/k) - x^2}} = dt, \qquad \text{or} \qquad \int \frac{dx}{\sqrt{(2E/k) - x^2}} = \int \sqrt{\frac{k}{m}}dt$$

Make the substitution $x = a\sin\theta$ and you see that if $a^2 = 2E/k$ then the integral on the left simplifies.

$$\int \frac{a\cos\theta d\theta}{a\sqrt{1-\sin^2\theta}} = \int \sqrt{\frac{k}{m}}dt \qquad \text{so} \qquad \theta = \sin^{-1}\frac{x}{a} = \omega_0 t + C$$

$$\text{or} \qquad x(t) = a\sin(\omega_0 t + C) \qquad \text{where} \qquad \omega_0 = \sqrt{k/m}$$

An electric circuit with an inductor, a resistor, and a battery has a differential equation for the current flow:

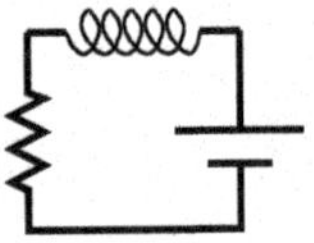

$$L\frac{dI}{dt} + IR = V_0 \tag{4.37}$$

Manipulate this into

$$L\frac{dI}{dt} = V_0 - IR, \qquad \text{then} \qquad L\frac{dI}{V_0 - IR} = dt$$

Now integrate this to get

$$L\int\frac{dI}{V_0 - IR} = t + C, \qquad \text{or} \qquad -\frac{L}{R}\ln(V_0 - IR) = t + C$$

Solve for the current I to get

$$RI(t) = V_0 - e^{-(L/R)(t+C)} \tag{4.38}$$

Now does this make sense? Look at the dimensions and you see that it *doesn't*, at least not yet. The problem is the logarithm on the preceding line where you see that its units don't make sense either. How can this be? The differential equation that you started with is correct, so how did the units get messed up? It goes back the the standard equation for integration,

$$\int dx/x = \ln x + C$$

If x is a length for example, then the left side is dimensionless, but this right side is the logarithm of a length. It's a peculiarity of the logarithm that leads to this anomaly. You can write the constant of integration as $C = -\ln C'$ where C' is another arbitrary constant, then

$$\int dx/x = \ln x + C = \ln x - \ln C' = \ln\frac{x}{C'}$$

If C' is a length this is perfectly sensible dimensionally. To see that the dimensions in Eq. (4.38) will work themselves out (this time), put on some initial conditions. Set $I(0) = 0$ so that the circuit starts with zero current.

$$\begin{aligned} R\cdot 0 &= V_0 - e^{-(L/R)(0+C)} & \text{implies} \qquad & e^{-(L/R)(C)} = V_0 \\ RI(t) &= V_0 - V_0 e^{-Lt/R} & \text{or} \qquad & I(t) = (1 - e^{-Lt/R})V_0/R \end{aligned}$$

and somehow the units have worked themselves out. Logarithms do this, but you still better check. The current in the circuit starts at zero and climbs gradually to its final value $I = V_0/R$.

4.8 Circuits

The methods of section 4.1 apply to simple linear circuits, and the use of complex algebra as in that section leads to powerful and simple ways to manipulate such circuit equations. You probably remember the result of putting two resistors in series or in parallel, but what about combinations of capacitors or inductors under the same circumstances? And what if you have some of each? With the right tools, all of these questions become the same question, so it's not several different techniques, but one.

If you have an oscillating voltage source (a wall plug), and you apply it to a resistor or to a capacitor or to an inductor, what happens? In the first case, $V = IR$ of course, but what about the others? The voltage equation for a capacitor is $V = q/C$ and for an inductor it is $V = L\,dI/dt$. A voltage that oscillates at frequency ω is $V = V_0\cos\omega t$, but using this trigonometric function forgoes all the advantages that complex exponentials provide. Instead, assume that your voltage source is $V = V_0 e^{i\omega t}$ with the real part understood. Carry this exponential through the calculation, and take the real part only at the end — often you won't even need to do that.

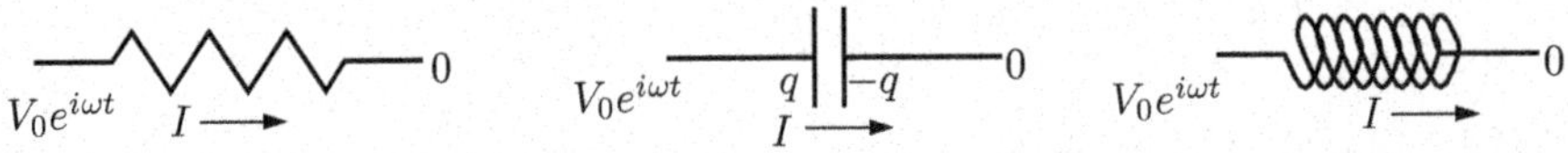

These are respectively

$$V_0e^{i\omega t} = IR = I_0e^{i\omega t}R$$
$$V_0e^{i\omega t} = q/C \implies i\omega V_0e^{i\omega t} = \dot{q}/C = I/C = I_0e^{i\omega t}/C$$
$$V_0e^{i\omega t} = L\dot{I} = i\omega LI = i\omega LI_0e^{i\omega t}$$

In each case the exponential factor is in common, and you can cancel it. These equations are then

$$V = IR \qquad V = I/i\omega C \qquad V = i\omega L\, I$$

All three of these have the same form: $V =$ (something times)I, and in each case the size of the current is proportional to the applied voltage. The factors of i implies that in the second and third cases the current is $\pm 90^\circ$ out of phase with the voltage cycle.

The coefficients in these equations generalize the concept of resistance, and they are called "impedance," respectively resistive impedance, capacitive impedance, and inductive impedance.

$$V = Z_RI = RI \qquad V = Z_CI = \frac{1}{i\omega C}I \qquad V = Z_LI = i\omega LI \tag{4.39}$$

Impedance appears in the same place as does resistance in the direct current situation, and this implies that it can be manipulated in the same way. The left figure shows two impedances in series.

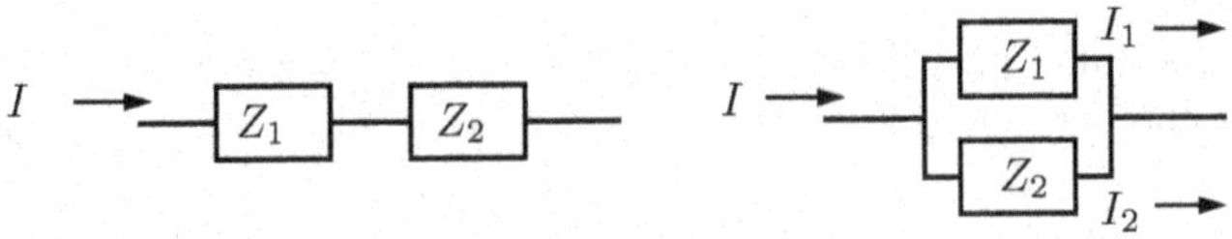

The total voltage from left to right in the left picture is

$$V = Z_1I + Z_2I = (Z_1 + Z_2)I = Z_{\text{total}}I \tag{4.40}$$

It doesn't matter if what's inside the box is a resistor or some more complicated impedance, it matters only that each box obeys $V = ZI$ and that the total voltage from left to right is the sum of the two voltages. Impedances in series add. You don't need the common factor $e^{i\omega t}$.

For the second picture, for which the components are in parallel, the voltage is the same on each impedance and charge is conserved, so the current entering the circuit obeys

$$I = I_1 + I_2, \qquad \text{then} \qquad \frac{V}{Z_{\text{total}}} = \frac{V}{Z_1} + \frac{V}{Z_2} \qquad \text{or} \qquad \frac{1}{Z_{\text{total}}} = \frac{1}{Z_1} + \frac{1}{Z_2} \tag{4.41}$$

Impedances in parallel add as reciprocals, so both of these formulas generalize the common equations for resistors in series and parallel. They also include as a special case the formula you may have seen before for adding capacitors in series and parallel.

In the example Eq. (4.37), if you replace the constant voltage by an oscillating voltage, you have two impedances in series.

$$Z_{\text{tot}} = Z_R + Z_L = R + i\omega L \implies I = V/(R + i\omega L)$$

What happened to the $e^{-Lt/R}$ term of the previous solution? This impedance manipulation tells you the *inhomogeneous* solution; you still must solve the homogeneous part of the differential equation and add that.

$$L\frac{dI}{dt} + IR = 0 \implies I(t) = Ae^{-Rt/L}$$

The total solution is the sum

$$\begin{aligned} I(t) &= Ae^{-Rt/L} + V_0 e^{i\omega t}\frac{1}{R + i\omega L} \\ \text{real part} &= Ae^{-Rt/L} + V_0\frac{\cos(\omega t - \phi)}{\sqrt{R^2 + \omega^2 L^2}} \qquad \text{where} \qquad \phi = \tan^{-1}\frac{\omega L}{R} \end{aligned} \tag{4.42}$$

How did that last manipulation come about? Change the complex number $R + i\omega L$ in the denominator from rectangular to polar form. Then the division of the complex numbers becomes easy. The dying exponential is called the "transient" term, and the other term is the "steady-state" term.

The denominator is

$$R + i\omega L = \alpha + i\beta = \sqrt{\alpha^2 + \beta^2}\,\frac{\alpha + i\beta}{\sqrt{\alpha^2 + \beta^2}} \tag{4.43}$$

The reason for this multiplication and division by the same factor is that it makes the final fraction have magnitude one. That allows me to write it as an exponential, $e^{i\phi}$. From the picture, the cosine and the sine of the angle ϕ are the two terms in the fraction.

$$\alpha + i\beta = \sqrt{\alpha^2 + \beta^2}\,(\cos\phi + i\sin\phi) = \sqrt{\alpha^2 + \beta^2}\, e^{i\phi} \qquad \text{and} \qquad \tan\phi = \beta/\alpha$$

In summary,

$$V = IZ \longrightarrow I = \frac{V}{Z} \longrightarrow Z = |Z|e^{i\phi} \longrightarrow I = \frac{V}{\sqrt{R^2 + \omega^2 L^2}e^{i\phi}}$$

To satisfy initial conditions, you need the parameter A, but you also see that it gives a dying exponential. After some time this transient term will be negligible, and only the oscillating steady-state term is left. That is what this impedance idea provides.

In even the simplest circuits such as these, that fact that Z is complex implies that the applied voltage is out of phase with the current. $Z = |Z|e^{i\phi}$, so $I = V/Z$ has a phase change of $-\phi$ from V.

What if you have more than one voltage source, perhaps the second having a different frequency from the first? Remember that you're just solving an inhomogeneous differential equation, and you are using the methods of section 4.2. If the external force in Eq. (4.12) has two terms, you can handle them separately then add the results.

4.9 Simultaneous Equations

What's this doing in a chapter on differential equations? Patience. Solve two equations in two unknowns:

$$\begin{array}{lcr} \text{(X)}\ ax + by = e & & adx + bdy - bcx - bdy = ed - fb \\ & d{\times}(\text{X}) - b{\times}(\text{Y}): & \\ \text{(Y)}\ cx + dy = f & & (ad - bc)x = ed - fb \end{array}$$

Similarly, multiply (Y) by a and (X) by c and subtract:

$$\begin{aligned} acx + ady - acx - cby &= fa - ec \\ (ad - bc)y &= fa - ec \end{aligned}$$

Divide by the factor on the left side and you have

$$x = \frac{ed - fb}{ad - bc}, \qquad y = \frac{fa - ec}{ad - bc} \tag{4.44}$$

provided that $ad - bc \neq 0$. This expression appearing in both denominators is the determinant of the equations.

Classify all the essentially different cases that can occur with this simple-looking set of equations and draw graphs to illustrate them. If this looks like problem 1.23, it should.

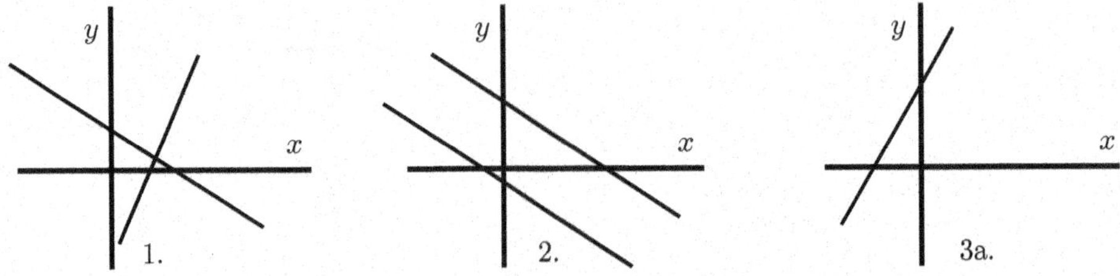

1. The solution is just as in Eq. (4.44) above and nothing goes wrong. There is exactly one solution. The two graphs of the two equations are two intersecting straight lines.

2. The denominator, the determinant, is zero and the numerator isn't. This is impossible and there are no solutions. When the determinant vanishes, the two straight lines are parallel and the fact that the numerator isn't zero implies that the two lines are

distinct and never intersect. (This could also happen if in one of the equations, say (X), $a = b = 0$ and $e \neq 0$. For example $0 = 1$. This obviously makes no sense.)

3a. The determinant is zero and so are both numerators. In this case the two lines are not only parallel, they are the same line. The two equations are not really independent and you have an infinite number of solutions.

3b. You can get zero over zero another way. Both equations (X) and (Y) are $0 = 0$. This sounds trivial, but it can really happen. *Every* x and y will satisfy the equation.

4. Not strictly a different case, but sufficiently important to discuss it separately: Suppose the the right-hand sides of (X) and (Y) are zero, $e = f = 0$. If the determinant is non-zero, there is a unique solution and it is $x = 0$, $y = 0$.

5. With $e = f = 0$, if the determinant is zero, the two equations are the same equation and there are an infinite number of non-zero solutions.

In the important case for which $e = f = 0$ and the determinant is zero, there are two cases: (3b) and (5). In the latter case there is a one-parameter family of solutions and in the former case there is a two-parameter family. Put another way, for case (5) the set of all solutions is a straight line, a one-dimensional set. For case (3b) the set of all solutions is the whole plane, a two-dimensional set.

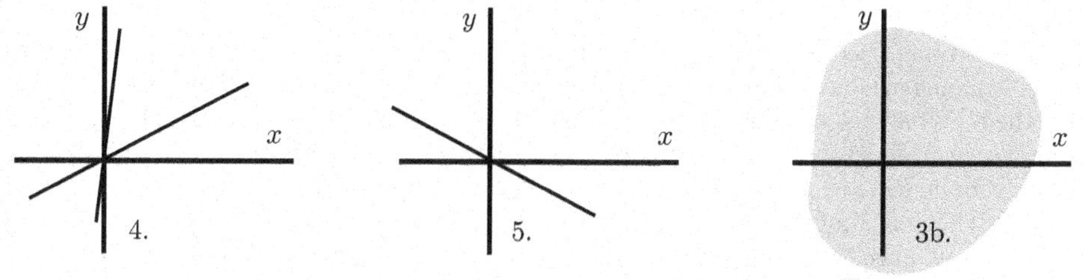

Example: Consider the two equations

$$kx + (k-1)y = 0, \qquad (1-k)x + (k-1)^2 y = 0$$

For whatever reason, I would like to get a non-zero solution for x and y. Can I? The condition depends on the determinant, so take the determinant and set it equal to zero.

$$k(k-1)^2 - (1-k)(k-1) = 0, \qquad \text{or} \qquad (k+1)(k-1)^2 = 0$$

There are two roots, $k = -1$ and $k = +1$. In the $k = -1$ case the two equations become

$$-x - 2y = 0, \qquad \text{and} \qquad 2x + 4y = 0$$

The second is just -2 times the first, so it isn't a separate equation. The family of solutions is all those x and y satisfying $x = -2y$, a straight line.

In the $k = +1$ case you have

$$x + 0y = 0, \qquad \text{and} \qquad 0 = 0$$

The solution to this is $x = 0$ and $y =$anything and it is again a straight line (the y-axis).

4.10 Simultaneous ODE's

Single point masses are an idealization that has some application to the real world, but there are many more cases for which you need to consider the interactions among many masses. To approach this, take the first step, from one mass to two masses.

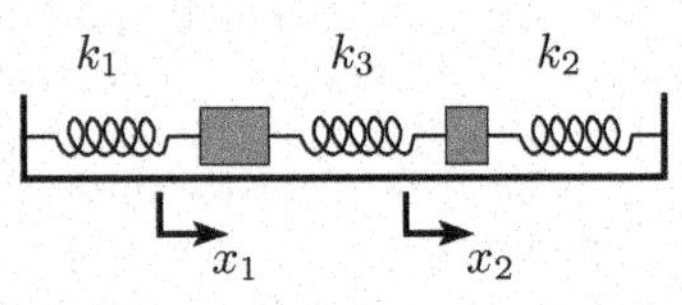

Two masses are connected to a set of springs and fastened between two rigid walls as shown. The coordinates for the two masses (moving along a straight line) are x_1 and x_2, and I'll pick the zero point for these coordinates to be the positions at which everything is at equilibrium — no total force on either. When a mass moves away from its equilibrium position there is a force on it. On m_1, the two forces are proportional to the distance by which the two springs k_1 and k_3 are stretched. These two distances are x_1 and $x_1 - x_2$ respectively, so $F_x = ma_x$ applied to each mass gives the equations

$$m_1\frac{d^2x_1}{dt^2} = -k_1x_1 - k_3(x_1 - x_2), \qquad \text{and} \qquad m_2\frac{d^2x_2}{dt^2} = -k_2x_2 - k_3(x_2 - x_1) \tag{4.45}$$

I'm neglecting friction simply to keep the algebra down. These are linear, constant coefficient, homogeneous equations, just the same sort as Eq. (4.4) except that there are two of them. What made the solution of (4.4) easy is that the derivative of an exponential is an exponential, so that when you substituted $x(t) = Ae^{\alpha t}$ all that you were left with was an algebraic factor — a quadratic equation in α. Exactly the same method works here.

The only way to find out if this is true is to try it. The big difference is that there are two unknowns instead of one, and the amplitude of the two motions will probably not be the same. If one mass is a lot bigger than the other, you expect it to move less.

Try the solution

$$x_1(t) = Ae^{\alpha t}, \qquad x_2(t) = Be^{\alpha t} \tag{4.46}$$

When you plug this into the differential equations for the masses, all the factors of $e^{\alpha t}$ cancel, just the way it happens in the one variable case.

$$m_1\alpha^2 A = -k_1A - k_3(A - B), \qquad \text{and} \qquad m_2\alpha^2 B = -k_2B - k_3(B - A) \tag{4.47}$$

Rearrange these to put them into a neater form.

$$\begin{aligned} (k_1 + k_3 + m_1\alpha^2)A + \quad & (-k_3)B = 0 \\ (-k_3)A + \quad & (k_2 + k_3 + m_2\alpha^2)B = 0 \end{aligned} \tag{4.48}$$

The results of problem 1.23 and of section 4.9 tell you all about such equations. In particular, for the pair of equations $ax + by = 0$ and $cx + dy = 0$, the only way to have a non-zero solution for x and y is for the determinant of the coefficients to be zero: $ad - bc = 0$. Apply this result to the problem at hand. Either $A = 0$ and $B = 0$ with a trivial solution *or* the determinant is zero.

$$(k_1 + k_3 + m_1\alpha^2)(k_2 + k_3 + m_2\alpha^2) - (k_3)^2 = 0 \tag{4.49}$$

This is a quadratic equation for α^2, and it determines the frequencies of the oscillation. Note the plural in the word frequencies.

Equation (4.49) is just a quadratic, but it's still messy. For a first example, try a special, symmetric case: $m_1 = m_2 = m$ and $k_1 = k_2$. There's a lot less algebra.

$$(k_1 + k_3 + m\alpha^2)^2 - (k_3)^2 = 0 \tag{4.50}$$

You could use the quadratic formula on this, but why? It's already set up to be factored.

$$(k_1 + k_3 + m\alpha^2 - k_3)(k_1 + k_3 + m\alpha^2 + k_3) = 0$$

The product is zero, so one or the other factors is zero. These determine the αs.

$$\alpha_1^2 = -\frac{k_1}{m} \qquad \text{and} \qquad \alpha_2^2 = -\frac{k_1 + 2k_3}{m} \tag{4.51}$$

These are negative, and that's what you should expect. There's no damping and the springs provide restoring forces that should give oscillations. That's just what these imaginary α's provide.

When you examine the equations $ax + by = 0$ and $cx + dy = 0$ the condition that the determinant vanishes is the condition that the two equations are really just one equation, and that the other is not independent of it; it is actually a multiple of the first. You must solve that equation for x and y. Here, arbitrarily pick the first of the equations (4.48) and find the relation between A and B.

$$\alpha_1^2 = -\frac{k_1}{m} \implies \big(k_1 + k_3 + m(-(k_1/m))\big)A + (-k_3)B = 0 \implies B = A$$

$$\alpha_2^2 = -\frac{k_1 + 2k_3}{m} \implies \big(k_1 + k_3 + m(-(k_1 + 2k_3/m))\big)A + (-k_3)B = 0 \implies B = -A$$

For the first case, $\alpha_1 = \pm i\omega_1 = \pm i\sqrt{k_1/m}$, there are two solutions to the original differential equations. These are called "normal modes."

$$\begin{aligned} x_1(t) &= A_1 e^{i\omega_1 t} \\ x_2(t) &= A_1 e^{i\omega_1 t} \end{aligned} \qquad \text{and} \qquad \begin{aligned} x_1(t) &= A_2 e^{-i\omega_1 t} \\ x_2(t) &= A_2 e^{-i\omega_1 t} \end{aligned}$$

The other frequency has the corresponding solutions

$$\begin{aligned} x_1(t) &= A_3 e^{i\omega_2 t} \\ x_2(t) &= -A_3 e^{i\omega_2 t} \end{aligned} \qquad \text{and} \qquad \begin{aligned} x_1(t) &= A_4 e^{-i\omega_2 t} \\ x_2(t) &= -A_4 e^{-i\omega_2 t} \end{aligned}$$

The total solution to the differential equations is the sum of all four of these.

$$\begin{aligned} x_1(t) &= A_1 e^{i\omega_1 t} + A_2 e^{-i\omega_1 t} + A_3 e^{i\omega_2 t} + A_4 e^{-i\omega_2 t} \\ x_2(t) &= A_1 e^{i\omega_1 t} + A_2 e^{-i\omega_1 t} - A_3 e^{i\omega_2 t} - A_4 e^{-i\omega_2 t} \end{aligned} \tag{4.52}$$

The two second order differential equations have four arbitrary constants in their solution. You can specify the initial values of two positions and of two velocities this way. As a specific example suppose that all initial velocities are zero and that the first mass is pushed to coordinate x_0 and released.

$$\begin{aligned} x_1(0) &= x_0 = A_1 + A_2 + A_3 + A_4 \\ x_2(0) &= 0 = A_1 + A_2 - A_3 - A_4 \\ v_{x1}(0) &= 0 = i\omega_1 A_1 - i\omega_1 A_2 + i\omega_2 A_3 - i\omega_2 A_4 \\ v_{x2}(0) &= 0 = i\omega_1 A_1 - i\omega_1 A_2 - i\omega_2 A_3 + i\omega_2 A_4 \end{aligned} \tag{4.53}$$

With a little thought (i.e. don't plunge blindly ahead) you can solve these easily.

$$A_1 = A_2 = A_3 = A_4 = \frac{x_0}{4}$$

$$\begin{aligned} x_1(t) &= \frac{x_0}{4}\left[e^{i\omega_1 t} + e^{-i\omega_1 t} + e^{i\omega_2 t} + e^{-i\omega_2 t}\right] = \frac{x_0}{2}\left[\cos\omega_1 t + \cos\omega_2 t\right] \\ x_2(t) &= \frac{x_0}{4}\left[e^{i\omega_1 t} + e^{-i\omega_1 t} - e^{i\omega_2 t} - e^{-i\omega_2 t}\right] = \frac{x_0}{2}\left[\cos\omega_1 t - \cos\omega_2 t\right] \end{aligned}$$

From the results of problem 3.34, you can rewrite these as

$$\begin{aligned} x_1(t) &= x_0 \cos\left(\frac{\omega_2 + \omega_1}{2}t\right)\cos\left(\frac{\omega_2 - \omega_1}{2}t\right) \\ x_2(t) &= x_0 \sin\left(\frac{\omega_2 + \omega_1}{2}t\right)\sin\left(\frac{\omega_2 - \omega_1}{2}t\right) \end{aligned} \tag{4.54}$$

As usual you have to draw some graphs to understand what these imply. If the center spring k_3 is a lot weaker than the outer ones, then Eq. (4.51) implies that the two frequencies are close to each other and so $|\omega_1 - \omega_2| \ll \omega_1 + \omega_2$. Examine Eq. (4.54) and you see that one of the two oscillating factors oscillate at a much higher frequency than the other. To sketch the graph of x_2 for example you should draw one factor $\left[\sin\left((\omega_2 + \omega_1)t/2\right)\right]$ and the other factor $\left[\sin\left((\omega_2 - \omega_1)t/2\right)\right]$ and graphically multiply them.

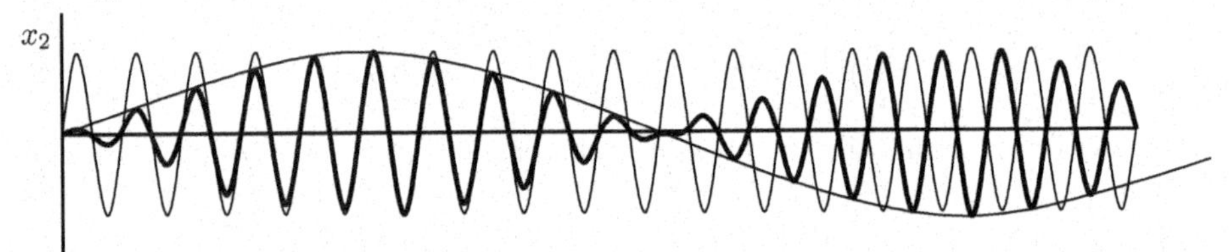

The mass m_2 starts without motion and its oscillations gradually build up. Later they die down and build up again (though with reversed phase). Look at the other mass, governed by the equation for $x_1(t)$ and you see that the low frequency oscillation from the $(\omega_2 - \omega_1)/2$ part is big where the one for x_2 is small and vice versa. The oscillation energy moves back and forth from one mass to the other.

4.11 Legendre's Equation

This equation and its solutions appear when you solve electric and gravitational potential problems in spherical coordinates [problem 9.20]. They appear when you study Gauss's method of numerical integration [Eq. (11.27)] and they appear when you analyze orthogonal functions [problem 6.7]. Because it shows up so often it is worth the time to go through the details in solving it.

$$\big[(1-x^2)y'\big]' + Cy = 0, \qquad \text{or} \qquad (1-x^2)y'' - 2xy' + Cy = 0 \tag{4.55}$$

Assume a Frobenius solutions about $x = 0$

$$y = \sum_0^\infty a_k x^{k+s}$$

and substitute into (4.55). Could you use an ordinary Taylor series instead? Yes, the point $x = 0$ is not a singular point at all, but it is just as easy (and more systematic and less prone to error) to use the same method in all cases.

$$(1-x^2)\sum_0^\infty a_k(k+s)(k+s-1)x^{k+s-2} - 2x\sum_0^\infty a_k(k+s)x^{k+s-1} + C\sum_0^\infty a_k x^{k+s} = 0$$

$$\sum_0^\infty a_k(k+s)(k+s-1)x^{k+s-2} +$$
$$\sum_0^\infty a_k\big[-2(k+s) - (k+s)(k+s-1)\big]x^{k+s} + C\sum_0^\infty a_k x^{k+s} = 0$$

$$\sum_{n=-2}^\infty a_{n+2}(n+s+2)(n+s+1)x^{n+s} -$$
$$\sum_{n=0}^\infty a_n\big[(n+s)^2 + (n+s)\big]x^{n+s} + C\sum_{n=0}^\infty a_n x^{n+s} = 0$$

In the last equation you see the usual substitution $k = n+2$ for the first sum and $k = n$ for the rest. That makes the exponents match across the equation. In the process, I simplified some of the algebraic expressions.

The indicial equation comes from the $n = -2$ term, which appears only once.

$$a_0 s(s-1) = 0, \qquad \text{so} \qquad s = 0,\ 1$$

Now set the coefficient of x^{n+s} to zero, and solve for a_{n+2} in terms of a_n. Also note that s is a non-negative integer, which says that the solution is non-singular at $x = 0$, consistent with the fact that zero is a regular point of the differential equation.

$$a_{n+2} = a_n\frac{(n+s)(n+s+1) - C}{(n+s+2)(n+s+1)} \tag{4.56}$$

$$a_2 = a_0 \frac{s(s+1) - C}{(s+2)(s+1)}, \qquad \text{then} \qquad a_4 = a_2 \frac{(s+2)(s+3) - C}{(s+4)(s+3)}, \qquad \textit{etc.} \tag{4.57}$$

This looks messier than it is. Notice that the only combination of indices that shows up is $n+s$. The index s is 0 or 1, and n is an even number, so the combination $n+s$ covers the non-negative integers: 0, 1, 2, ...

The two solutions to the Legendre differential equation come from the two cases, $s = 0, 1$.

$$\begin{aligned} s = 0: \quad & a_0 \left[1 + \left(\frac{-C}{2}\right) x^2 + \left(\frac{-C}{2}\right)\left(\frac{2\cdot 3 - C}{4\cdot 3}\right) x^4 + \right. \\ & \qquad \left. \left(\frac{-C}{2}\right)\left(\frac{2\cdot 3 - C}{4\cdot 3}\right)\left(\frac{4\cdot 5 - C}{6\cdot 5}\right) x^6 \cdots \right] \\ s = 1: \quad & a_0' \left[x + \left(\frac{1\cdot 2 - C}{3\cdot 2}\right) x^3 + \left(\frac{1\cdot 2 - C}{3\cdot 2}\right)\left(\frac{3\cdot 4 - C}{5\cdot 4}\right) x^5 + \cdots \right] \end{aligned} \tag{4.58}$$

and the general solution is a sum of these.

This procedure gives both solutions to the differential equation, one with even powers and one with odd powers. Both are infinite series and are called Legendre Functions. An important point about both of them is that they blow up as $x \to \pm 1$. This fact shouldn't be too surprising, because the differential equation (4.55) has a singular point there.

$$y'' - \frac{2x}{(1+x)(1-x)} y' + \frac{C}{(1+x)(1-x)} y = 0 \tag{4.59}$$

It's a regular singular point, but it is still singular. A detailed calculation in the next section shows that these solutions behave as $\ln(1-x)$ near $x = 1$.

There is an exception! If the constant C is for example $C = 6$, then with $s = 0$ the equations (4.57) are

$$a_2 = a_0 \frac{-6}{2}, \qquad a_4 = a_2 \frac{6-6}{12} = 0, \qquad a_6 = a_8 = \ldots = 0$$

The infinite series terminates in a polynomial

$$a_0 + a_2 x^2 = a_0 [1 - 3x^2]$$

This (after a conventional rearrangement) is a Legendre Polynomial,

$$P_2(x) = \frac{3}{2} x^2 - \frac{1}{2}$$

The numerator in Eq. (4.56) for a_{n+2} is $[(n+s)(n+s+1) - C]$. If this happen to equal zero for some value of $n = N$, then $a_{N+2} = 0$ and so then all the rest of $a_{N+4} \cdots$ are zero too. The series is a polynomial. This will happen only for special values of C, such as the value $C = 6$ above. The values of C that have this special property are

$$C = \ell(\ell+1), \qquad \text{for} \qquad \ell = 0,\ 1,\ 2,\ \ldots \tag{4.60}$$

This may be easier to see in the explicit representation, Eq. (4.58). When a numerator equals zero, all the rest that follow are zero too. When $C = \ell(\ell+1)$ for even ℓ, the first series terminates in a polynomial. Similarly for odd ℓ the second series is a polynomial. These are the Legendre polynomials, denoted $P_\ell(x)$, and the conventional normalization is to require that their value at $x = 1$ is one.

$$\begin{aligned} &P_0(x) = 1 \qquad P_1(x) = x \qquad P_2(x) = \tfrac{3}{2}x^2 - \tfrac{1}{2} \\ &P_3(x) = \tfrac{5}{2}x^3 - \tfrac{3}{2}x \qquad P_4(x) = \tfrac{35}{8}x^4 - \tfrac{30}{8}x^2 + \tfrac{3}{8} \end{aligned} \tag{4.61}$$

The special case for which the series terminates in a polynomial is by far the most commonly used solution to Legendre's equation. You seldom encounter the general solutions as in Eq. (4.58).

A few properties of the P_ℓ are

$$\begin{aligned}
&(a) \qquad \int_{-1}^{1} dx\, P_n(x)P_m(x) = \frac{2}{2n+1}\delta_{nm} \qquad \text{where} \qquad \delta_{nm} = \begin{cases} 1 & \text{if } n = m \\ 0 & \text{if } n \neq m \end{cases} \\
&(b) \qquad (n+1)P_{n+1}(x) = (2n+1)xP_n(x) - nP_{n-1}(x) \\
&(c) \qquad P_n(x) = \frac{(-1)^n}{2^n n!}\frac{d^n}{dx^n}(1-x^2)^n \\
&(d) \qquad P_n(1) = 1 \qquad P_n(-x) = (-1)^n P_n(x) \\
&(e) \qquad \left(1 - 2tx + t^2\right)^{-1/2} = \sum_{n=0}^{\infty} t^n P_n(x)
\end{aligned} \tag{4.62}$$

4.12 Asymptotic Behavior

This is a slightly technical subject, but it will come up occasionally in electromagnetism when you dig into the details of boundary value problems. It will come up in quantum mechanics when you solve some of the standard eigenvalue problems that you face near the beginning of the subject. If you haven't come to these yet then you can skip this part for now.

You solve a differential equation using a Frobenius series and now you need to know something about the solution. In particular, how does the solution behave for large values of the argument? All you have in front of you is an infinite series, and it isn't obvious how it will behave far away from the origin. In the line just after Eq. (4.59) it says that these Legendre functions behave as $\ln(1-x)$. How can you tell this from the series in Eq. (4.58)?

There is a theorem that addresses this. Take two functions described by two series:

$$f(x) = \sum^{\infty} a_k x^k \qquad \text{and} \qquad g(x) = \sum^{\infty} b_k x^k$$

It does not matter where the sums start because you are concerned just with the large values of k. The lower limit could as easily be -14 or $+27$ with no change in the result. The ratio test, Eq. (2.8), will determine the radius of convergence of these series, and

$$\left|\frac{a_{k+1}x^{k+1}}{a_k x^k}\right| < C < 1 \quad \text{for large enough } k$$

is enough to insure convergence. The largest x for which this holds defines the radius of convergence, maybe 1, maybe ∞.... Call it R.

Assume that (after some value of k) all the a_k and b_k are positive, then look at the ratio of the ratios,

$$\frac{a_{k+1}/a_k}{b_{k+1}/b_k}$$

If this approaches one, that will tell you only that the radii of convergence of the two series are the same. If it approaches one *very fast*, and if either one of the functions goes to infinity as x approaches the radius of convergence, then it says that the asymptotic behaviors of the functions defined by the series are the same.

$$\text{If } \frac{a_{k+1}/a_k}{b_{k+1}/b_k} - 1 \longrightarrow 0 \text{ as fast as } \frac{1}{k^2}, \text{ and if either } f(x) \text{ or } g(x) \longrightarrow \infty \text{ as } x \to R$$

$$\text{Then } \frac{f(x)}{g(x)} \longrightarrow \text{a constant as } x \to R$$

There are more general ways to state this, but this handles most cases of interest.

Compare these series near $x = 1$.

$$\frac{1}{1-x} = \sum_0^\infty x^k, \qquad \text{or} \qquad \ln(1-x) = \sum_1^\infty \frac{x^k}{k}, \qquad \text{or}$$

$$(1-x)^{-1/2} = \sum_{k=0}^\infty \frac{\alpha(\alpha-1)\cdots(\alpha-k+1)}{k!}(-x)^k \quad (\alpha = -1/2)$$

Even in the third case, the signs of the terms are the same after a while, so this is relevant to the current discussion. The ratio of ratios for the first and second series is

$$\frac{a_{k+1}/a_k}{b_{k+1}/b_k} = \frac{1}{(k+1)/k} = \frac{1}{1+1/k} = 1 - \frac{1}{k} + \cdots$$

These series behave differently as x approaches the radius of convergence ($x \to 1$). But you knew that. The point is to compare an unknown series to a known one.

Applying this theorem requires some fussy attention to detail. You must make sure that the indices in one series correspond exactly to the indices in the other. Take the Legendre series, Eq. (4.56) and compare it to a logarithmic series. Choose $s = 0$ to be specific; then only even powers of x appear. That means that I want to compare it to a series with even powers, and with radius of convergence $= 1$. First try a binomial series such as for $(1-x^2)^\alpha$, but that doesn't work. See for yourself. The logarithm $\ln(1-x^2)$ turns out to be right. From Eq. (4.56) and from the logarithm series,

$$f(x) = \sum_{n \text{ even}}^\infty a_n x^n \qquad \text{with} \qquad a_{n+2} = a_n \frac{(n+s)(n+s+1) - C}{(n+s+2)(n+s+1)}$$

$$g(x) = -\ln(1-x^2) = \sum_1^\infty \frac{x^{2k}}{k} = \sum b_k x^{2k}$$

To make the indices match, let $n = 2k$ in the second series.

$$g(x) = \sum_{n \text{ even}} \frac{x^n}{n/2} = \sum c_n x^n$$

Now look at the ratios.

$$\frac{a_{n+2}}{a_n} = \frac{n(n+1) - C}{(n+2)(n+1)} = \frac{1 + \frac{1}{n} - \frac{C}{n^2}}{1 + \frac{3}{n} + \frac{2}{n^2}} = 1 - \frac{2}{n} + \cdots$$

$$\frac{c_{n+2}}{c_n} = \frac{2/(n+2)}{2/n} = \frac{n}{n+2} = \frac{1}{1 + \frac{2}{n}} = 1 - \frac{2}{n} + \cdots$$

These agree to order $1/n$, so the ratio of the ratios differs from one only in order $1/n^2$, satisfying the requirements of the test. This says the the Legendre functions (the ones where the series does not terminate in a polynomial) are logarithmically infinite near $x = \pm 1$. It's a mild infinity, but it is still an infinity. Is this bad? Not by itself, after all the electric potential of a line charge has a logarithm of r in it. Singular solutions aren't necessarily wrong, it just means that you have to look closely at how you are using them.

Exercises

1 What algebraic equations do you have to solve to find the solutions of these differential equations?

$$\frac{d^3x}{dt^3} + a\frac{dx}{dt} + bx = 0, \qquad \frac{d^{10}z}{du^{10}} - 3z = 0$$

2 These equations are separable, as in section 4.7. Separate them and integrate, solving for the dependent variable, with one arbitrary constant.

$$\frac{dN}{dt} = -\lambda N, \qquad \frac{dx}{dt} = a^2 + x^2, \qquad \frac{dv_x}{dt} = -a\left(1 - e^{-\beta v_x}\right)$$

3 From Eq. (4.40) and (4.41) what are the formulas for putting capacitors or inductors in series and parallel?

Problems

4.1 If the equilibrium position $x = 0$ for Eq. (4.4) is unstable instead of stable, this reverses the sign in front of k. Solve the problem that led to Eq. (4.10) under these circumstances. That is, the damping constant is b as before, and the initial conditions are $x(0) = 0$ and $v_x(0) = v_0$. What is the small time and what is the large time behavior?
Ans: $\left(2mv_0/\sqrt{b^2+4km}\right)e^{-bt/2m}\sinh\left(\sqrt{b^2/4m+k/m}\,t\right)$

4.2 In the damped harmonic oscillator problem, Eq. (4.4), suppose that the damping term is an *anti*-damping term. It has the sign opposite to the one that I used $(+b\,dx/dt)$. Solve the problem with the initial condition $x(0) = 0$ and $v_x(0) = v_0$ and describe the resulting behavior.
Ans: $\left(2mv_0/\sqrt{4km-b^2}\right)e^{bt/2m}\sin\left(\sqrt{4km-b^2}\,t/2m\right)$

4.3 A point mass m moves in one dimension under the influence of a force F_x that has a potential energy $V(x)$. Recall that the relation between these is

$$F_x = -\frac{dV}{dx}$$

Take the specific potential energy $V(x) = -V_0a^2/(a^2+x^2)$, where V_0 is positive. Sketch V. Write the equation $F_x = ma_x$. There is an equilibrium point at $x = 0$, and if the motion is over only small distances you can do a power series expansion of F_x about $x = 0$. What is the differential equation now? Keep just the lowest order non-vanishing term in the expansion for the force and solve that equation subject to the initial conditions that at time $t = 0$, $x(0) = x_0$ and $v_x(0) = 0$.
How does the graph of V change as you vary a from small to large values and how does this same change in a affect the behavior of your solution? Ans: $\omega = \sqrt{2V_0/ma^2}$

4.4 The same as the preceding problem except that the potential energy function is $+V_0a^2/(a^2+x^2)$. Ans: $x(t) = x_0\cosh\left(\sqrt{2V_0/ma^2}\,t\right)$ ($|x| < a/4$ or so, depending on the accuracy you want.)

4.5 For the case of the undamped harmonic oscillator and the force Eq. (4.13), start from the beginning and derive the solution subject to the initial conditions that the initial position is zero and the initial velocity is zero. At the end, compare your result to the result of Eq. (4.15) to see if they agree where they should agree.

4.6 Check the dimensions in the result for the forced oscillator, Eq. (4.15).

4.7 Fill in the missing steps in the derivation of Eq. (4.15).

4.8 For the undamped harmonic oscillator apply an extra oscillating force so that the equation to solve is

$$m\frac{d^2x}{dt^2} = -kx + F_{\text{ext}}(t)$$

where the external force is $F_{\text{ext}}(t) = F_0 \cos\omega t$. Assume that $\omega \neq \omega_0 = \sqrt{k/m}$.
Find the general solution to the homogeneous part of this problem.
Find a solution for the inhomogeneous case. You can readily guess what sort of function will give you a $\cos\omega t$ from a combination of x and its second derivative.
Add these and apply the initial conditions that at time $t = 0$ the mass is at rest at the origin. Be sure to check your results for plausibility: 0) dimensions; 1) $\omega = 0$; 2) $\omega \to \infty$; 3) t small (not zero). In each case explain why the result is as it should be.
Ans: $(F_0/m)[-\cos\omega_0 t + \cos\omega t]/(\omega_0^2 - \omega^2)$

4.9 In the preceding problem I specified that $\omega \neq \omega_0 = \sqrt{k/m}$. Having solved it, you know why this condition is needed. Now take the final result of that problem, including the initial conditions, and take the limit as $\omega \to \omega_0$. [What is the *definition* of a derivative?] You did draw a graph of your result didn't you? Ans: $(F_0/2m\omega_0)t\sin\omega_0 t$

4.10 Show explicitly that you can write the solution Eq. (4.7) in any of several equivalent ways,

$$Ae^{i\omega_0 t} + Be^{-i\omega_0 t} = C\cos\omega_0 t + D\sin\omega_0 t = E\cos(\omega_0 t + \phi)$$

I.e. , given A and B, what are C and D, what are E and ϕ? Are there any restrictions in any of these cases?

4.11 In the damped harmonic oscillator, you can have the special (critical damping) case for which $b^2 = 4km$ and for which $\omega' = 0$. Use a series expansion to take the limit of Eq. (4.10) as $\omega' \to 0$. Also graph this solution. What would happen if you took the same limit in Eqs. (4.8) and (4.9), *before* using the initial conditions?

4.12 (**a**) In the limiting solution for the forced oscillator, Eq. (4.16), what is the nature of the result for small time? Expand the solution through order t^2 and understand what you get. Be careful to be consistent in keeping terms to the same order in t.
(**b**) Part (**a**) involved letting β be very large, then examining the behavior for small t. Now reverse the order: what is the first non-vanishing order in t that you will get if you go back to Eq. (4.13), expand that to first non-vanishing order in time, use that for the external force in Eq. (4.12), and find $x(t)$ for small t. Recall that in this example $x(0) = 0$ and $\dot{x}(0) = 0$, so you can solve for $\ddot{x}(0)$ and then for $\dddot{x}(0)$. The two behaviors are very different.

4.13 The undamped harmonic oscillator equation is $d^2x/dt^2 + \omega^2 x = 0$. Solve this by Frobenius series expansion about $t = 0$.

4.14 Check the algebra in the derivation of the $n = 0$ Bessel equation. Explicitly verify that the general expression for a_{2k} in terms of a_0 is correct, Eq. (4.22).

4.15 Work out the Frobenius series solution to the Bessel equation for the $n = 1/2$, $s = -1/2$ case. Graph both solutions, this one and Eq. (4.23).

4.16 Derive the Frobenius series solution to the Bessel equation for the value of $n = 1$. Show that this method doesn't yield a second solution for this case either.

4.17 Try using a Frobenius series method on $y'' + y/x^3 = 0$ around $x = 0$.

4.18 Solve by Frobenius series $x^2u'' + 4xu' + (x^2 + 2)u = 0$. You should be able to recognize the resulting series (after a little manipulation).

4.19 The harmonic oscillator equation, $d^2y/dx^2 + k^2y = 0$, is easy in terms of the variable x. What is this equation if you change variables to $z = 1/x$, getting an equation in such things as d^2y/dz^2. What sort of singularity does this equation have at $z = 0$? And of course, write down the answer for $y(z)$ to see what this sort of singularity can lead to. Graph it.

4.20 Solve by Frobenius series solution about $x = 0$: $y'' + xy = 0$.
Ans: $1 - (x^3/3!) + (1\cdot 4x^6/6!) - (1\cdot 4\cdot 7x^9/9!) + \cdots$ is one.

4.21 From the differential equation $d^2u/dx^2 = -u$, finish the derivation for c' as in Eq. (4.29). Derive identities for the functions $c(x + y)$ and $s(x + y)$.

4.22 The chain rule lets you take the derivative of the composition of two functions. The function inverse to s is the function f that satisfies $f\big(s(x)\big) = x$. Differentiate this equation with respect to x and derive that f satisfies $df(x)/dx = 1/\sqrt{1 - x^2}$. What is the derivative of the function inverse to c?

4.23 For the differential equation $u'' = +u$ (note the sign change) use the same boundary conditions for two independent solutions that I used in Eq. (4.28). For this new example evaluate c' and s'. Does $c^2 + s^2$ have the nice property that it did in section 4.5? What about $c^2 - s^2$? What are $c(x + y)$ and $s(x + y)$? What is the derivative of the function inverse to s? to c?

4.24 Apply the Green's function method for the force $F_0\big(1 - e^{-\beta t}\big)$ on the harmonic oscillator without damping. Verify that it agrees with the previously derived result, Eq. (4.15). They should match in a special case.

4.25 An undamped harmonic oscillator with natural frequency ω_0 is at rest for time $t < 0$. Starting at time zero there is an added force $F_0 \sin\omega_0 t$. Use Green's functions to find the motion for time $t > 0$, and analyze the solution for both small and large time, determining if your results make sense. Compare the solution to problems 4.9 and 4.11.
Ans: $\big(F_0/2m\omega_0^2\big)\big[\sin(\omega_0 t) - \omega_0 t\cos(\omega_0 t)\big]$

4.26 Derive the Green's function analogous to Eq. (4.32) for the case that the harmonic oscillator is damped.

4.27 Radioactive processes have the property that the rate of decay of nuclei is proportional to the number of nuclei present. That translates into the differential equation $dN/dt = -\lambda N$, where λ is a constant depending on the nucleus. At time $t = 0$ there are N_0 nuclei; how many are present at time t later? The half-life is the time in which one-half of the nuclei decay; what is that in terms of λ? Ans: $\ln 2/\lambda$

4.28 (**a**) In the preceding problem, suppose that the result of the decay is another nucleus (the "daughter") that is itself radioactive with its own decay constant λ_2. Call the first one above λ_1. Write the differential equation for the time-derivative of the number, N_2 of this nucleus. You note that N_2 will change for two reasons, so in time dt the quantity dN_2

has two contributions, one is the decrease because of the radioactivity of the daughter, the other an increase due to the decay of the parent. Set up the differential equation for N_2 and you will be able to use the result of the previous problem as input to this; then solve the resulting differential equation for the number of daughter nuclei as a function of time. Assume that you started with none, $N_2(0) = 0$.
(**b**) Next, the "activity" is the total number of *all* types of decays per time. Compute the activity and graph it. For the plot, assume that λ_1 is substantially smaller than λ_2 and plot the total activity as a function of time. Then examine the reverse case, $\lambda_1 \gg \lambda_2$
Ans: $N_0\lambda_1\big[(2\lambda_2 - \lambda_1)e^{-\lambda_1 t} - \lambda_2 e^{-\lambda_2 t}\big]/(\lambda_2 - \lambda_1)$

4.29 The "snowplow problem" was made famous by Ralph Agnew: A snowplow starts at 12:00 Noon in a heavy and steady snowstorm. In the first hour it goes 2 miles; in the second hour it goes 1 mile. When did the snowstorm start? Ans: 11:23

4.30 Verify that the equations (4.52) really do satisfy the original differential equations.

4.31 When you use the "dry friction" model Eq. (4.2) for the harmonic oscillator, you can solve the problem by dividing it into pieces. Start at time $t = 0$ and position $x = x_0$ (positive). The initial velocity of the mass m is zero. As the mass is pulled to the left, set up the differential equation and solve it up to the point at which it comes to a halt. Where is that? You can take that as a new initial condition and solve the problem as the mass is pulled to the right until it stops. Where is that? Then keep repeating the process. Instead or further repetition, examine the case for which the coefficient of kinetic friction is small, and determine to lowest order in the coefficient of friction what is the change in the amplitude of oscillation up to the first stop. From that, predict what the amplitude will be after the mass has swung back to the original side and come to its second stop. In this small μ_k approximation, how many oscillations will it undergo until all motion stops. Let $b = \mu_\mathsf{k} F_N$ Ans: Let $t_n = \pi n/\omega_0$, then for $t_n < t < t_{n+1}$, $x(t) = [x_0 - (2n+1)b/k]\cos\omega_0 t + (-1)^n b/k$. Stops when $t \approx \pi k x_0/2\omega_0 b$ *roughly*.

4.32 A mass m is in an undamped one-dimensional harmonic oscillator and is at rest. A constant external force F_0 is applied for the time interval T and is then turned off. What is the motion of the oscillator as a function of time for all $t > 0$? For what value of T is the amplitude of the motion a maximum after the force is turned off? For what values is the motion a minimum? Of course you need an explanation of why you should have been able to anticipate these two results.

4.33 Starting from the solution Eq. (4.52) assume the initial conditions that both masses start from the equilibrium position and that the first mass is given an initial velocity $v_{x1} = v_0$. Find the subsequent motion of the two masses and analyze it.

4.34 If there is viscous damping on the middle spring of Eqs. (4.45) so that each mass feels an extra force depending on their relative velocity, then these equations will be

$$m_1\frac{d^2x_1}{dt^2} = -k_1x_1 - k_3(x_1 - x_2) - b(\dot{x}_1 - \dot{x}_2), \quad \text{and}$$

$$m_2\frac{d^2x_2}{dt^2} = -k_2x_2 - k_3(x_2 - x_1) - b(\dot{x}_2 - \dot{x}_1)$$

Solve these subject to the conditions that all initial velocities are zero and that the first mass is pushed to coordinate x_0 and released. Use the same assumption as before that $m_1 = m_2 = m$ and $k_1 = k_2$.

4.35 For the damped harmonic oscillator apply an extra oscillating force so that the equation to solve is

$$m\frac{d^2x}{dt^2} = -b\frac{dx}{dt} - kx + F_{\text{ext}}(t)$$

where the external force is $F_{\text{ext}}(t) = F_0 e^{i\omega t}$.
(**a**) Find the general solution to the homogeneous part of this problem.
(**b**) Find a solution for the inhomogeneous case. You can readily guess what sort of function will give you an $e^{i\omega t}$ from a combination of x and its first two derivatives. This problem is easier to solve than the one using $\cos\omega t$, and at the end, to get the solution for the cosine case, all you have to do is to take the real part of your result.

4.36 You can solve the circuit equation Eq. (4.37) more than one way. Solve it by the methods used earlier in this chapter.

4.37 For a second order differential equation you can pick the position and the velocity any way that you want, and the equation then determines the acceleration. Differentiate the equation and you find that the third derivative is determined too.

$$\frac{d^2x}{dt^2} = -\frac{b}{m}\frac{dx}{dt} - \frac{k}{m}x \qquad \text{implies} \qquad \frac{d^3x}{dt^3} = -\frac{b}{m}\frac{d^2x}{dt^2} - \frac{k}{m}\frac{dx}{dt}$$

Assume the initial position is zero, $x(0) = 0$ and the initial velocity is $v_x(0) = v_0$; determine the second derivative at time zero. Now determine the third derivative at time zero. Now differentiate the above equation again and determine the fourth derivative at time zero.
From this, write down the first five terms of the power series expansion of $x(t)$ about $t = 0$.
Compare this result to the power series expansion of Eq. (4.10) to this order.

4.38 Use the techniques of section 4.6, start from the equation $m\,d^2x/dt^2 = F_x(t)$ with *no* spring force or damping. (**a**) Find the Green's function for this problem, that is, what is the response of the mass to a small kick over a small time interval (the analog of Eq. (4.32))? Develop the analog of Eq. (4.34) for this case. Apply your result to the special case that $F_x(t) = F_0$, a constant for time $t > 0$.
(**b**) You know that the solution of this differential equation involves two integrals of $F_x(t)$ with respect to time, so how can this single integral do the same thing? Differentiate this Green's function integral (for arbitrary F_x) twice with respect to time to verify that it really gives what it's supposed to. This is a special case of some general results, problems 15.19 and 15.20.
Ans: $\frac{1}{m}\int_{-\infty}^{t} dt'\, F_x(t')(t - t')$

4.39 A point mass m moves in one dimension under the influence of a force F_x that has a potential energy $V(x)$. Recall that the relation between these is $F_x = -dV/dx$, and take the specific potential energy $V(x) = -V_0 e^{-x^2/a^2}$, where V_0 is positive. Sketch V. Write the equation $F_x = ma_x$. There is an equilibrium point at $x = 0$, and if the motion is over only small distances you can do a power series expansion of F_x about $x = 0$. What is the differential equation now? Keep just the lowest order non-vanishing term in the expansion for the force and solve that equation subject to the initial conditions that at time $t = 0$, $x(0) = x_0$ and $v_x(0) = 0$. As usual, analyze large and small a.

4.40 Solve by Frobenius series methods

$$\frac{d^2y}{dx^2} + \frac{2}{x}\frac{dy}{dx} + \frac{1}{x}y = 0$$

Ans: $\sum_{k=0}^{\infty}(-1)^k \frac{x^k}{n!(n+1)!}$ is one.

4.41 Find a series solution about $x = 0$ for $y'' + y \sec x = 0$, at least to a few terms.
Ans: $a_0\left[1 - \frac{1}{2}x^2 + 0x^4 + \frac{1}{720}x^6 + \cdots\right] + a_1\left[x - \frac{1}{6}x^3 - \frac{1}{60}x^5 + \cdots\right]$

4.42 Fill in the missing steps in the equations (4.55) to Eq. (4.58).

4.43 Verify the orthogonality relation Eq. (4.62)(**a**) for Legendre polynomials of order $\ell = 0,\ 1,\ 2,\ 3$.

4.44 Start with the function $\left(1 - 2xt + t^2\right)^{-1/2}$. Use the binomial expansion and collect terms to get a power series in t. The coefficients in this series are functions of x. Carry this out at least to the coefficient of t^3 and show that the coefficients are Legendre polynomials. This is called the generating function for the P_ℓ's. It is $\sum_0^\infty P_\ell(x)t^\ell$

4.45 In the equation of problem 4.17, make the change of independent variable $x = 1/z$. Without actually carrying out the solution of the resulting equation, what can you say about solving it?

4.46 Show that Eq. (4.62)(**c**) has the correct value $P_n(1) = 1$ for all n. Note: $(1 - x^2) = (1 + x)(1 - x)$ and you are examining the point $x = 1$.

4.47 Solve for the complete solution of Eq. (4.55) for the case $C = 0$. For this, don't use series methods, but get the closed form solution. Ans: $A \tanh^{-1} x + B$

4.48 Derive the condition in Eq. (4.60). Which values of s correspond to which values of ℓ?

4.49 Start with the equation $y'' + P(x)y' + Q(x)y = 0$ and assume that you have found one solution: $y = f(x)$. Perhaps you used series methods to find it. (**a**) Make the substitution $y(x) = f(x)g(x)$ and deduce a differential equation for g. Let $G = g'$ and solve the resulting first order equation for G. Finally write g as an integral. This is one method (not necessarily the best) to find the second solution to a differential equation.
(**b**) Apply this result to the $\ell = 0$ solution of Legendre's equation to find another way to solve problem 4.47. Ans: $y = f \int dx \frac{1}{f^2} \exp - \int P\, dx$

4.50 Treat the damped harmonic oscillator as a two-point boundary value problem.

$$m\ddot{x} + b\dot{x} + kx = 0, \qquad \text{with} \qquad x(0) = 0 \qquad \text{and} \qquad x(T) = d$$

[For this problem, if you want to set $b = k = T = d = 1$ that's o.k.]
(**a**) Assume that m is very small. To a first approximation neglect it and solve the problem.
(**b**) Since you failed to do part (**a**) — it blew up in your face — solve it exactly instead and examine the solution for very small m. Why couldn't you make the approximation of neglecting m? Draw graphs. Welcome to the world of boundary layer theory and singular perturbations. Ans: $x(t) \approx e^{1-t} - e^{1-t/m}$

4.51 Solve the differential equation $\dot{x} = Ax^2(1+\omega t)$ in closed form and compare the series expansion of the result to Eq. (4.25). Ans: $x = \alpha/\big[1 - A\alpha(t + \omega t^2/2)\big]$

4.52 Solve the same differential equation $\dot{x} = Ax^2(1+\omega t)$ with $x(t_0) = \alpha$ by doing a few iterations of Eq. (4.27).

4.53 Analyze the steady-state part of the solution Eq. (4.42). For the input potential $V_0 e^{i\omega t}$, find the real part of the current explicitly, writing the final form as $I_{\text{max}} \cos(\omega t - \phi)$. Plot I_{max} and ϕ versus ω. Plot $V(t)$ and $I(t)$ on a second graph with time as the axis. Recall these V and I are the real part understood.

4.54 If you have a resistor, a capacitor, and an inductor in series with an oscillating voltage source, what is the steady-state solution for the current? Write the final form as $I_{\text{max}} \cos(\omega t - \phi)$, and plot I_{max} and ϕ versus ω. See what happens if you vary some of the parameters.
Ans: $I = V_0 \cos(\omega t - \phi)/|Z|$ where $|Z| = \sqrt{R^2 + (\omega L - 1/\omega C)^2}$ and $\tan\phi = (\omega L - 1/\omega C)/R$

4.55 In the preceding problem, what if the voltage source is a combination of DC and AC, so it is $V(t) = V_0 + V_1 \cos\omega t$. Find the steady state solution now.

4.56

What is the total impedance left to right in the circuit

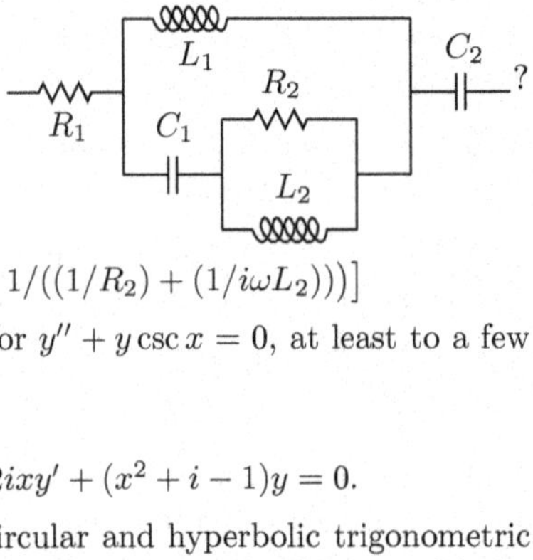

Ans: $R_1 + (1/i\omega C_2) + 1/\big[(1/i\omega L_1) + 1/\big((1/i\omega C_1) + 1/((1/R_2) + (1/i\omega L_2))\big)\big]$

4.57 Find a Frobenius series solution about $x = 0$ for $y'' + y \csc x = 0$, at least to a few terms.
Ans: $x - \frac{1}{2}x^2 + \frac{1}{12}x^3 - \frac{1}{48}x^4 + \frac{1}{192}x^5 + \cdots$

4.58 Find a series solution about $x = 0$ for $x^2y'' - 2ixy' + (x^2 + i - 1)y = 0$.

4.59 Just as you can derive the properties of the circular and hyperbolic trigonometric functions from the differential equations that they satisfy, you can do the same for the exponential. Take the equation $u' = u$ and consider that solution satisfying the boundary condition $u(0) = 1$.
(**a**) Prove that u satisfies the identity $u(x + y) = u(x)u(y)$.
(**b**) Prove that the function inverse to u has the derivative $1/x$.

4.60 Find the asymptotic behavior of the Legendre series for the $s = 1$ case.

4.61 Find Frobenius series solutions for $xy'' + y = 0$.

Fourier Series

Fourier series started life as a method to solve problems about the flow of heat through ordinary materials. It has grown so far that if you search any university's library catalog for the keyword "Fourier" you will find hundreds of entries. It is a tool in abstract analysis and electromagnetism and statistics and radio communication and .... People have even tried to use it to analyze the stock market. (It didn't help.) The representation of musical sounds as sums of waves of various frequencies is an audible example. It provides an indispensible tool in solving partial differential equations, and a later chapter will show some of these tools at work.

5.1 Examples

The power series or Taylor series is based on the idea that you can write a general function as an infinite series of powers. The idea of Fourier series is that you can write a function as an infinite series of sines and cosines. You can also use functions other than trigonometric ones, but I'll leave that generalization aside for now, except to say that Legendre polynomials are an important example of functions used for such more general expansions.

An example: On the interval $0 < x < L$ the function x^2 varies from 0 to L^2. It can be written as the series of cosines

$$\begin{aligned} x^2 &= \frac{L^2}{3} + \frac{4L^2}{\pi^2}\sum_1^\infty \frac{(-1)^n}{n^2}\cos\frac{n\pi x}{L} \\ &= \frac{L^2}{3} - \frac{4L^2}{\pi^2}\left[\cos\frac{\pi x}{L} - \frac{1}{4}\cos\frac{2\pi x}{L} + \frac{1}{9}\cos\frac{3\pi x}{L} - \cdots\right] \end{aligned} \tag{5.1}$$

To see if this is even plausible, examine successive partial sums of the series, taking one term, then two terms, etc. Sketch the graphs of these partial sums to see if they start to look like the function they are supposed to represent (left graph). The graphs of the series, using terms up to $n = 5$ do pretty well at representing the parabola.

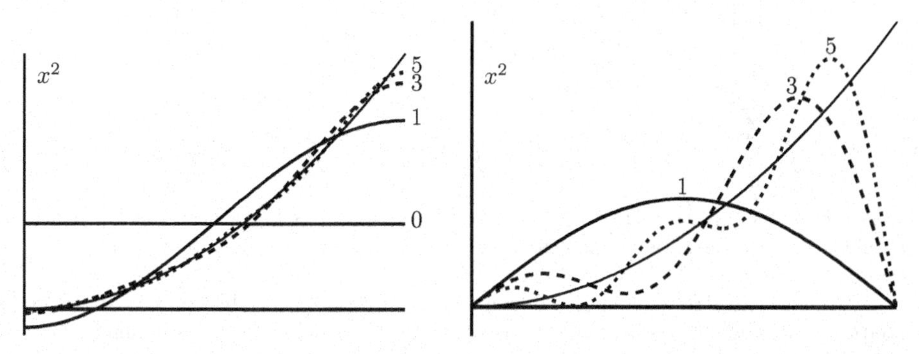

The same function can be written in terms of sines with another series:

$$x^2 = \frac{2L^2}{\pi}\sum_1^\infty\left[\frac{(-1)^{n+1}}{n} - \frac{2}{\pi^2 n^3}\big(1-(-1)^n\big)\right]\sin\frac{n\pi x}{L} \tag{5.2}$$

and again you can see how the series behaves by taking one to several terms of the series. (right graph) The graphs show the parabola $y = x^2$ and partial sums of the two series with terms up to $n = 1, 3, 5$.

The second form doesn't work as well as the first one, and there's a reason for that. The sine functions all go to zero at $x = L$ and x^2 doesn't, making it hard for the sum of sines to approximate the desired function. They can do it, but it takes a lot more terms in the series to get a satisfactory result. The series Eq. (5.1) has terms that go to zero as $1/n^2$, while the terms in the series Eq. (5.2) go to zero only as $1/n$.*

5.2 Computing Fourier Series

How do you determine the details of these series starting from the original function? For the Taylor series, the trick was to assume a series to be an infinitely long polynomial and then to evaluate it (and its successive derivatives) at a point. You require that all of these values match those of the desired function at that one point. That method won't work in this case. (Actually I've read that it can work here too, but with a ridiculous amount of labor and some mathematically suspect procedures.)

The idea of Fourier's procedure is like one that you can use to determine the components of a vector in three dimensions. You write such a vector as

$$\vec{A} = A_x\,\hat{x} + A_y\,\hat{y} + A_z\,\hat{z}$$

And then use the orthonormality of the basis vectors, $\hat{x}\cdot\hat{y} = 0$ etc. Take the scalar product of the preceding equation with $\hat{x}$.

$$\hat{x}\cdot\vec{A} = \hat{x}\cdot\big(A_x\hat{x} + A_y\hat{y} + A_z\hat{z}\big) = A_x \qquad \text{and} \qquad \hat{y}\cdot\vec{A} = A_y \qquad \text{and} \qquad \hat{z}\cdot\vec{A} = A_z \tag{5.3}$$

This lets you get all the components of $\vec{A}$. For example,

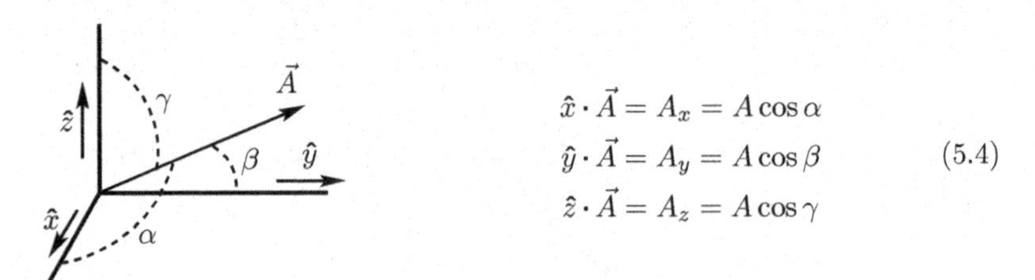

$$\begin{aligned} \hat{x}\cdot\vec{A} &= A_x = A\cos\alpha \\ \hat{y}\cdot\vec{A} &= A_y = A\cos\beta \\ \hat{z}\cdot\vec{A} &= A_z = A\cos\gamma \end{aligned} \tag{5.4}$$

This shows the three direction cosines for the vector $\vec{A}$. You will occasionally see these numbers used to describe vectors in three dimensions, and it's easy to see that $\cos^2\alpha + \cos^2\beta + \cos^2\gamma = 1$.

In order to stress the close analogy between this scalar product and what you do in Fourier series, I will introduce another notation for the scalar product. You don't typically see it in introductory courses for the simple reason that it isn't needed there. Here however it will turn out to be very useful, and in the next chapter you will see nothing *but* this

* For animated sequences showing the convergence of some of these series, see www.physics.miami.edu/nearing/mathmethods/animations.html

notation. Instead of $\hat{x}\cdot\vec{A}$ or $\vec{A}\cdot\vec{B}$ you use $\langle\hat{x},\vec{A}\,\rangle$ or $\langle\vec{A},\vec{B}\,\rangle$. The angle bracket notation will make it very easy to generalize the idea of a dot product to cover other things. In this notation the above equations will appear as

$$\langle\hat{x},\vec{A}\,\rangle = A\cos\alpha, \qquad \langle\hat{y},\vec{A}\,\rangle = A\cos\beta, \qquad \langle\hat{z},\vec{A}\,\rangle = A\cos\gamma$$

and they mean exactly the same thing as Eq. (5.4).

There are orthogonality relations similar to the ones for $\hat{x}$, $\hat{y}$, and $\hat{z}$, but for sines and cosines. Let n and m represent integers, then

$$\int_0^L dx\,\sin\left(\frac{n\pi x}{L}\right)\sin\left(\frac{m\pi x}{L}\right) = \begin{cases} 0 & n\neq m \\ L/2 & n=m \end{cases} \tag{5.5}$$

This is sort of like $\hat{x}\cdot\hat{z}=0$ and $\hat{y}\cdot\hat{y}=1$, where the analog of $\hat{x}$ is $\sin\pi x/L$ and the analog of $\hat{y}$ is $\sin 2\pi x/L$. The biggest difference is that it doesn't stop with three vectors in the basis; it keeps on with an infinite number of values of n and the corresponding different sines. There are an infinite number of very different possible functions, so you need an infinite number of basis functions in order to express a general function as a sum of them. The integral Eq. (5.5) is a continuous analog of the coordinate representation of the common dot product. The sum over three terms $A_xB_x+A_yB_y+A_zB_z$ becomes a sum (integral) over a continuous index, the integration variable. By using this integral as a generalization of the ordinary scalar product, you can say that $\sin(\pi x/L)$ and $\sin(2\pi x/L)$ are *orthogonal*. Let i be an index taking on the values x, y, and z, then the notation A_i is a function of the variable i. In this case the independent variable takes on just three possible values instead of the infinite number in Eq. (5.5).

How do you derive an identity such as Eq. (5.5)? The first method is just straight integration, using the right trigonometric identities. The easier (and more general) method can wait for a few pages.

$$\cos(x\pm y) = \cos x\cos y \mp \sin x\sin y, \qquad \text{subtract:} \qquad \cos(x-y)-\cos(x+y) = 2\sin x\sin y$$

Use this in the integral.

$$2\int_0^L dx\,\sin\left(\frac{n\pi x}{L}\right)\sin\left(\frac{m\pi x}{L}\right) = \int_0^L dx\,\left[\cos\left(\frac{(n-m)\pi x}{L}\right) - \cos\left(\frac{(n+m)\pi x}{L}\right)\right]$$

Now do the integral, assuming $n\neq m$ and that n and m are positive integers.

$$= \frac{L}{(n-m)\pi}\sin\left(\frac{(n-m)\pi x}{L}\right) - \frac{L}{(n+m)\pi}\sin\left(\frac{(n+m)\pi x}{L}\right)\bigg|_0^L = 0 \tag{5.6}$$

Why assume that the integers are positive? Aren't the negative integers allowed too? Yes, but they aren't needed. Put $n=-1$ into $\sin(n\pi x/L)$ and you get the same function as for $n=+1$, but turned upside down. It isn't an independent function, just -1 times what you already have. Using it would be sort of like using for your basis not only $\hat{x}$, $\hat{y}$, and $\hat{z}$ but $-\hat{x}$, $-\hat{y}$, and $-\hat{z}$ too. Do the $n=m$ case of the integral yourself.

Computing an Example

For a simple example, take the function $f(x) = 1$, the constant on the interval $0 < x < L$, and assume that there is a series representation for f on this interval.

$$1 = \sum_1^\infty a_n \sin\left(\frac{n\pi x}{L}\right) \qquad (0 < x < L) \tag{5.7}$$

Multiply both sides by the sine of $m\pi x/L$ and integrate from 0 to L.

$$\int_0^L dx\, \sin\left(\frac{m\pi x}{L}\right) 1 = \int_0^L dx\, \sin\left(\frac{m\pi x}{L}\right) \sum_{n=1}^\infty a_n \sin\left(\frac{n\pi x}{L}\right) \tag{5.8}$$

Interchange the order of the sum and the integral, and the integral that shows up is the orthogonality integral derived just above. When you use the orthogonality of the sines, only one term in the infinite series survives.

$$\begin{aligned}\int_0^L dx\, \sin\left(\frac{m\pi x}{L}\right) 1 &= \sum_{n=1}^\infty a_n \int_0^L dx\, \sin\left(\frac{m\pi x}{L}\right) \sin\left(\frac{n\pi x}{L}\right) \\ &= \sum_{n=1}^\infty a_n \cdot \begin{cases} 0 & (n \neq m) \\ L/2 & (n = m) \end{cases} \\ &= a_m\, L/2. \end{aligned} \tag{5.9}$$

Now all you have to do is to evaluate the integral on the left.

$$\int_0^L dx\, \sin\left(\frac{m\pi x}{L}\right) 1 = \frac{L}{m\pi}\left[-\cos\frac{m\pi x}{L}\right]_0^L = \frac{L}{m\pi}\left[1 - (-1)^m\right]$$

This is zero for even m, and when you equate it to (5.9) you get

$$a_m = \frac{4}{m\pi} \qquad \text{for } m \text{ odd}$$

You can relabel the indices so that the sum shows only odd integers $m = 2k + 1$ and the Fourier series is

$$\frac{4}{\pi} \sum_{m \text{ odd} > 0} \frac{1}{m} \sin\frac{m\pi x}{L} = \frac{4}{\pi} \sum_{k=0}^\infty \frac{1}{2k+1} \sin\frac{(2k+1)\pi x}{L} = 1, \qquad (0 < x < L) \tag{5.10}$$

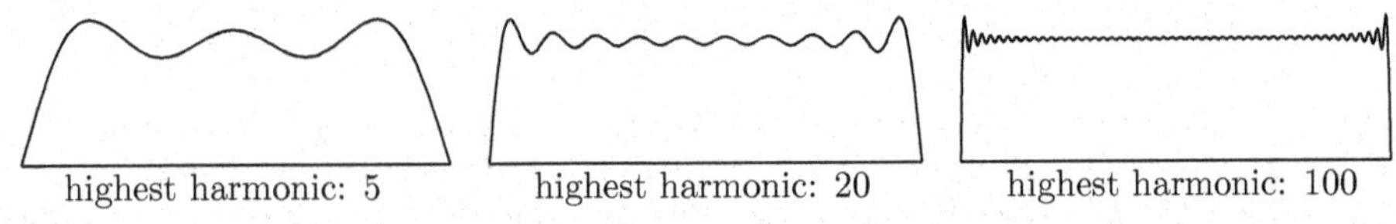

The graphs show the sum of the series up to $2k + 1 = 5,\ 19,\ 99$ respectively. It is not a very rapidly converging series, but it's a start. You can see from the graphs that near the end of the interval, where the function is discontinuous, the series has a hard time handling the jump. The resulting overshoot is called the Gibbs phenomenon, and it is analyzed in section 5.7.

Notation

The point of introducing that other notation for the scalar product comes right here. The same notation is used for these integrals. In this context define

$$\langle f, g\rangle = \int_0^L dx\, f(x)^* g(x) \tag{5.11}$$

and it will behave just the same way that $\vec{A}\cdot\vec{B}$ does. Eq. (5.5) then becomes

$$\langle u_n, u_m\rangle = \begin{cases} 0 & n \neq m \\ L/2 & n = m \end{cases} \qquad \text{where} \qquad u_n(x) = \sin\left(\frac{n\pi x}{L}\right) \tag{5.12}$$

$$\text{precisely analogous to} \qquad \langle \hat{x}, \hat{x}\rangle = 1 \qquad \text{and} \qquad \langle \hat{y}, \hat{z}\rangle = 0$$

These u_n are orthogonal to each other even though they aren't normalized to one the way that $\hat{x}$ and $\hat{y}$ are, but that turns out not to matter. $\langle u_n, u_n\rangle = L/2$ instead of $= 1$, so you simply keep track of it. (What happens to the series Eq. (5.7) if you multiply every u_n by 2? Nothing, because the coefficients a_n get multiplied by $1/2$.)

The Fourier series manipulations, Eqs. (5.7), (5.8), (5.9), become

$$1 = \sum_1^\infty a_n u_n \qquad \text{then} \qquad \langle u_m, 1\rangle = \left\langle u_m, \sum_1^\infty a_n u_n\right\rangle = \sum_{n=1}^\infty a_n \langle u_m, u_n\rangle = a_m \langle u_m, u_m\rangle \tag{5.13}$$

This is far more compact than you see in the steps between Eq. (5.7) and Eq. (5.10). You *still* have to evaluate the integrals $\langle u_m, 1\rangle$ and $\langle u_m, u_m\rangle$, but when you master this notation you'll likely make fewer mistakes in figuring out what integral you have to do. Again, you can think of Eq. (5.11) as a continuous analog of the discrete sum of three terms, $\langle \vec{A}, \vec{B}\,\rangle = A_x B_x + A_y B_y + A_z B_z$.

The analogy between the vectors such as $\hat{x}$ and functions such as sine is really far deeper, and it is central to the subject of the next chapter. In order not to to get confused by the notation, you have to distinguish between a whole function f, and the value of that function at a point, $f(x)$. The former is the whole graph of the function, and the latter is one point of the graph, analogous to saying that $\vec{A}$ is the whole vector and A_y is one of its components.

The scalar product notation defined in Eq. (5.11) is not necessarily restricted to the interval $0 < x < L$. Depending on context it can be over any interval that you happen to be considering at the time. In Eq. (5.11) there is a complex conjugation symbol. The functions here have been real, so this made no difference, but you will often deal with complex functions and then the fact that the notation $\langle f, g\rangle$ includes a conjugation is important. This notation is a special case of the general development that will come in section 6.6. The basis vectors such as $\hat{x}$ are conventionally normalized to one, $\hat{x}\cdot\hat{x} = 1$, but you don't have to require it even there, and in the context of Fourier series it would clutter up the notation to require $\langle u_n, u_n\rangle = 1$, so I don't bother.

Some Examples

To get used to this notation, try showing that these pairs of functions are orthogonal on the interval $0 < x < L$. Sketch graphs of both functions in every case.

$$\left\langle x, L - \tfrac{3}{2}x \right\rangle = 0 \qquad \left\langle \sin \pi x/L, \cos \pi x/L \right\rangle = 0 \qquad \left\langle \sin 3\pi x/L, L - 2x \right\rangle = 0$$

The notation has a complex conjugation built into it, but these examples are all real. What if they aren't? Show that these are orthogonal too. How do you graph these? Not easily.*

$$\left\langle e^{2i\pi x/L}, e^{-2i\pi x/L} \right\rangle = 0 \qquad \left\langle L - \tfrac{1}{4}(7+i)x, L + \tfrac{3}{2}ix \right\rangle = 0$$

Extending the function

In Equations (5.1) and (5.2) the original function was specified on the interval $0 < x < L$. The two Fourier series that represent it can be evaluated for *any* x. Do they equal x^2 everywhere? No. The first series involves only cosines, so it is an even function of x, but it's periodic: $f(x + 2L) = f(x)$. The second series has only sines, so it's odd, and it too is periodic with period $2L$.

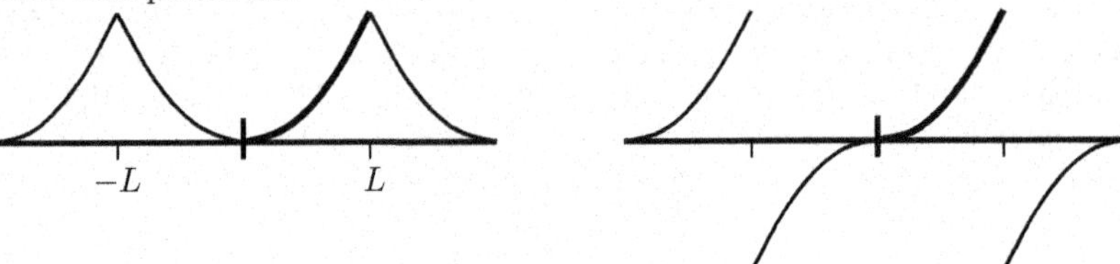

Here the discontinuity in the sine series is more obvious, a fact related to its slower convergence.

5.3 Choice of Basis

When you work with components of vectors in two or three dimensions, you will choose the basis that is most convenient for the problem you're working with. If you do a simple mechanics problem with a mass moving on an incline, you can choose a basis $\hat{x}$ and $\hat{y}$ that are arranged horizontally and vertically. OR, you can place them at an angle so that they point down the incline and perpendicular to it. The latter is often a simpler choice in that type of problem.

The same applies to Fourier series. The interval on which you're working is not necessarily from zero to L, and even on the interval $0 < x < L$ you can choose many sets of function for a basis:

$\sin n\pi x/L \quad (n = 1, 2, \ldots)$ as in equations (5.10) and (5.2), or you can choose a basis

$\cos n\pi x/L \quad (n = 0, 1, 2, \ldots)$ as in Eq. (5.1), or you can choose a basis

$\sin(n + 1/2)\pi x/L \quad (n = 0, 1, 2, \ldots)$, or you can choose a basis

* but see if you can find a copy of the book by Jahnke and Emde, published long before computers. They show examples. Also check out
www.geom.uiuc.edu/˜banchoff/script/CFGInd.html or
www.math.ksu.edu/˜bennett/jomacg/

$e^{2\pi inx/L} \quad (n = 0, \pm 1, \pm 2, \ldots)$, or an infinite number of other possibilities.

In order to use any of these you need a relation such as Eq. (5.5) for each separate case. That's a lot of integration. You need to do it for any interval that you may need and that's even more integration. Fortunately there's a way out:

Fundamental Theorem

If you want to show that each of these respective choices provides an orthogonal set of functions you can integrate every special case as in Eq. (5.6), *or* you can do all the cases at once by deriving an important theorem. This theorem starts from the fact that all of these sines and cosines and complex exponentials satisfy the same differential equation, $u'' = \lambda u$, where λ is some constant, different in each case. If you studied section 4.5, you saw how to derive properties of trigonometric functions simply by examining the differential equation that they satisfy. If you didn't, now might be a good time to look at it, because this is more of the same. (I'll wait.)

You have two functions u_1 and u_2 that satisfy

$$u_1'' = \lambda_1 u_1 \qquad \text{and} \qquad u_2'' = \lambda_2 u_2$$

Make no assumption about whether the λ's are positive or negative or even real. The u's can also be complex. Multiply the first equation by u_2^* and the second by u_1^*, then take the complex conjugate of the second product.

$$u_2^* u_1'' = \lambda_1 u_2^* u_1 \qquad \text{and} \qquad u_1 u_2^{*\prime\prime} = \lambda_2^* u_1 u_2^*$$

Subtract the equations.

$$u_2^* u_1'' - u_1 u_2^{*\prime\prime} = (\lambda_1 - \lambda_2^*) u_2^* u_1$$

Integrate from a to b

$$\int_a^b dx\,\left(u_2^* u_1'' - u_1 u_2^{*\prime\prime}\right) = (\lambda_1 - \lambda_2^*) \int_a^b dx\, u_2^* u_1 \tag{5.14}$$

Now do two partial integrations. Work on the second term on the left:

$$\int_a^b dx\, u_1 u_2^{*\prime\prime} = u_1 u_2^{*\prime}\Big|_a^b - \int_a^b dx\, u_1' u_2^{*\prime} = u_1 u_2^{*\prime}\Big|_a^b - u_1' u_2^*\Big|_a^b + \int_a^b dx\, u_1'' u_2^*$$

Put this back into the Eq. (5.14) and the integral terms cancel, leaving

$$\boxed{u_1' u_2^* - u_1 u_2^{*\prime}\Big|_a^b = (\lambda_1 - \lambda_2^*) \int_a^b dx\, u_2^* u_1} \tag{5.15}$$

This is the central identity from which all the orthogonality relations in Fourier series derive. It's even more important than that because it tells you what types of boundary conditions you can use in order to get the desired orthogonality relations. (It tells you even more than that, as it tells you how to compute the adjoint of the second derivative operator. But not now — save that for later.) The expression on the left side of the equation has a name: "bilinear concomitant."

You can see how this is related to the work with the functions $\sin(n\pi x/L)$. They satisfy the differential equation $u'' = \lambda u$ with $\lambda = -n^2\pi^2/L^2$. The interval in that case was $0 < x < L$ for $a < x < b$.

There are generalizations of this theorem that you will see in places such as problems 6.16 and 6.17 and 10.21. In those extensions these same ideas will allow you to handle Legendre polynomials and Bessel functions and Ultraspherical polynomials and many other functions in just the same way that you handle sines and cosines. That development comes under the general name Sturm-Liouville theory.

The key to using this identity will be to figure out what sort of boundary conditions will cause the left-hand side to be zero. For example if $u(a) = 0$ and $u(b) = 0$ then the left side vanishes. These are not the only possible boundary conditions that make this work; there are several other common cases soon to appear.

The first consequence of Eq. (5.15) comes by taking a special case, the one in which the two functions u_1 and u_2 are in fact the same function. If the boundary conditions make the left side zero then

$$0 = (\lambda_1 - \lambda_1^*) \int_a^b dx\, u_1^*(x) u_1(x)$$

The λ's are necessarily the same because the u's are. The only way the product of two numbers can be zero is if one of them is zero. The integrand, $u_1^*(x)u_1(x)$ is always non-negative and is continuous, so the integral can't be zero unless the function u_1 is identically zero. As that would be a trivial case, assume it's not so. This then implies that the other factor, $(\lambda_1 - \lambda_1^*)$ must be zero, and this says that the constant λ_1 is real. Yes, $-n^2\pi^2/L^2$ is real.

[To use another language that will become more familiar later, λ is an eigenvalue and d^2/dx^2 with these boundary conditions is an operator. This calculation guarantees that the eigenvalue is real.]

Now go back the the more general case of two different functions, and drop the complex conjugation on the λ's.

$$0 = (\lambda_1 - \lambda_2) \int_a^b dx\, u_2^*(x) u_1(x)$$

This says that if the boundary conditions on u make the left side zero, then for two solutions with different eigenvalues (λ's) the orthogonality integral is zero. Eq. (5.5) is a special case of the following equation.

$$\text{If} \quad \lambda_1 \neq \lambda_2, \qquad \text{then} \qquad \langle u_2, u_1 \rangle = \int_a^b dx\, u_2^*(x) u_1(x) = 0 \tag{5.16}$$

Apply the Theorem

As an example, carry out a full analysis of the case for which $a = 0$ and $b = L$, and for the boundary conditions $u(0) = 0$ and $u(L) = 0$. The parameter λ is positive, zero, or negative. If $\lambda > 0$, then set $\lambda = k^2$ and

$$u(x) = A\sinh kx + B\cosh kx, \qquad \text{then} \qquad u(0) = B = 0$$
$$\text{and so} \qquad u(L) = A\sinh kL = 0 \Rightarrow A = 0$$

No solutions there, so try $\lambda = 0$

$$u(x) = A + Bx, \qquad \text{then} \qquad u(0) = A = 0 \qquad \text{and so} \qquad u(L) = BL = 0 \Rightarrow B = 0$$

No solutions here either. Try $\lambda < 0$, setting $\lambda = -k^2$.

$$u(x) = A\sin kx + B\cos kx, \qquad \text{then} \qquad u(0) = 0 = B, \qquad \text{so} \qquad u(L) = A\sin kL = 0$$

Now there are many solutions because $\sin n\pi = 0$ allows $k = n\pi/L$ with n any integer. But, $\sin(-x) = -\sin(x)$ so negative integers just reproduce the same functions as do the positive integers; they are redundant and you can eliminate them. The complete set of solutions to the equation $u'' = \lambda u$ with these boundary conditions has $\lambda_n = -n^2\pi^2/L^2$ and reproduces the result of the explicit integration as in Eq. (5.6).

$$u_n(x) = \sin\left(\frac{n\pi x}{L}\right) \qquad n = 1,\ 2,\ 3, \ldots \qquad \text{and}$$
$$\langle u_n, u_m\rangle = \int_0^L dx\,\sin\left(\frac{n\pi x}{L}\right)\sin\left(\frac{m\pi x}{L}\right) = 0 \qquad \text{if} \qquad n \neq m \quad (5.17)$$

There are other choices of boundary condition that will make the bilinear concomitant vanish. (Verify these!) For example

$$u(0) = 0, \qquad u'(L) = 0 \qquad \text{gives} \qquad u_n(x) = \sin\left(n + 1/2\right)\pi x/L \qquad n = 0,\ 1,\ 2,\ 3, \ldots$$

and without further integration you have the orthogonality integral for non-negative integers n and m

$$\langle u_n, u_m\rangle = \int_0^L dx\,\sin\left(\frac{(n + 1/2)\pi x}{L}\right)\sin\left(\frac{(m + 1/2)\pi x}{L}\right) = 0 \qquad \text{if} \qquad n \neq m \quad (5.18)$$

A very common choice of boundary conditions is

$$u(a) = u(b), \qquad u'(a) = u'(b) \qquad \text{(periodic boundary conditions)} \qquad (5.19)$$

It is often more convenient to use complex exponentials in this case (though of course not necessary). On $0 < x < L$

$$u(x) = e^{ikx}, \qquad \text{where} \qquad k^2 = -\lambda \qquad \text{and} \qquad u(0) = 1 = u(L) = e^{ikL}$$

The periodic behavior of the exponential implies that $kL = 2n\pi$. The condition that the derivatives match at the boundaries makes no further constraint, so the basis functions are

$$u_n(x) = e^{2\pi inx/L}, \qquad (n = 0,\ \pm1, \pm2,\ \ldots) \qquad (5.20)$$

Notice that in this case the index n runs over all positive and negative numbers and zero, not just the positive integers. The functions $e^{2\pi inx/L}$ and $e^{-2\pi inx/L}$ are independent, unlike the case of the sines discussed above. Without including both of them you don't have a basis and can't do Fourier series. If the interval is symmetric about the origin as it often is, $-L < x < +L$, the conditions are

$$u(-L) = e^{-ikL} = u(+L) = e^{+ikL}, \qquad \text{or} \qquad e^{2ikL} = 1 \tag{5.21}$$

This says that $2kL = 2n\pi$, so

$$u_n(x) = e^{n\pi ix/L}, \qquad (n = 0,\ \pm1, \pm2,\ \ldots) \qquad \text{and} \qquad f(x) = \sum_{-\infty}^{\infty} c_n u_n(x)$$

The orthogonality properties determine the coefficients:

$$\big\langle u_m, f\big\rangle = \big\langle u_m, \sum_{-\infty}^{\infty} c_n u_n\big\rangle = c_m\big\langle u_m, u_m\big\rangle$$

$$\int_{-L}^{L} dx\, e^{-m\pi ix/L} f(x) = c_m\big\langle u_m, u_m\big\rangle$$

$$= c_m \int_{-L}^{L} dx\, e^{-m\pi ix/L} e^{+m\pi ix/L} = c_m \int_{-L}^{L} dx\, 1 = 2Lc_m$$

In this case, sometimes the real form of this basis is more convenient and you can use the combination of the two sets u_n and v_n, where

$$u_n(x) = \cos(n\pi x/L),\ (n = 0,\ 1,\ 2,\ \ldots)$$
$$v_n(x) = \sin(n\pi x/L),\ (n = 1,\ 2,\ \ldots)$$
$$\big\langle u_n, u_m\big\rangle = 0 \quad (n \neq m), \qquad \big\langle v_n, v_m\big\rangle = 0 \quad (n \neq m), \qquad \big\langle u_n, v_m\big\rangle = 0 \quad (\text{all } n,\, m) \tag{5.22}$$

and the Fourier series is a sum such as $f(x) = \sum_0^\infty a_n u_n + \sum_1^\infty b_n v_n$.

There are an infinite number of other choices, a few of which are even useful, *e.g.*

$$u'(a) = 0 = u'(b) \tag{5.23}$$

Take the same function as in Eq. (5.7) and try a different basis. Choose the basis for which the boundary conditions are $u(0) = 0$ and $u'(L) = 0$. This gives the orthogonality conditions of Eq. (5.18). The general structure is always the same.

$$f(x) = \sum a_n\, u_n(x), \qquad \text{and use} \qquad \big\langle u_m, u_n\big\rangle = 0\ (n \neq m)$$

Take the scalar product of this equation with u_m to get

$$\big\langle u_m, f\big\rangle = \big\langle u_m, \sum a_n\, u_n\big\rangle = a_m\big\langle u_m, u_m\big\rangle \tag{5.24}$$

This is exactly as before in Eq. (5.13), but with a different basis. To evaluate it you still have to do the integrals.

$$\langle u_m, f\rangle = \int_0^L dx\, \sin\left(\frac{(m+1/2)\pi x}{L}\right) 1 = a_m \int_0^L dx\, \sin^2\left(\frac{(m+1/2)\pi x}{L}\right) = a_m \langle u_m, u_m\rangle$$

$$\frac{L}{(m+1/2)\pi}\left[1-\cos\left((m+1/2)\pi\right)\right] = \frac{L}{2}a_m$$

$$a_m = \frac{4}{(2m+1)\pi}$$

Then the series is

$$\frac{4}{\pi}\left[\sin\frac{\pi x}{2L} + \frac{1}{3}\sin\frac{3\pi x}{2L} + \frac{1}{5}\sin\frac{5\pi x}{2L} + \cdots\right] \tag{5.25}$$

5.4 Musical Notes

Different musical instruments sound different even when playing the same note. You won't confuse the sound of a piano with the sound of a guitar, and the reason is tied to Fourier series. The note middle C has a frequency that is 261.6 Hz on the standard equal tempered scale. The angular frequency is then 2π times this, or 1643.8 radians/sec. Call it $\omega_0 = 1644$. When you play this note on any musical instrument, the result is always a combination of many frequencies, this one and many multiples of it. A pure frequency has just ω_0, but a real musical sound has many harmonics: ω_0, $2\omega_0$, $3\omega_0$, *etc.*

$$\text{Instead of} \quad e^{i\omega_0 t} \qquad \text{an instrument produces} \quad \sum_{n=1}^{?} a_n\, e^{ni\omega_0 t} \tag{5.26}$$

A pure frequency is the sort of sound that you hear from an electronic audio oscillator, and it's not very interesting. Any real musical instrument will have at least a few and usually many frequencies combined to make what you hear.

Why write this as a complex exponential? A sound wave is a real function of position and time, the pressure wave, but it's easier to manipulate complex exponentials than sines and cosines, so when I write this, I really mean to take the real part for the physical variable, the pressure variation. The imaginary part is carried along to be discarded later. Once you're used to this convention you don't bother writing the "real part understood" anywhere — it's understood.

$$p(t) = \Re\sum_{n=1}^{?} a_n\, e^{ni\omega_0 t} = \sum_{n=1}^{?} |a_n| \cos\left(n\omega_0 t + \phi_n\right) \qquad \text{where} \qquad a_n = |a_n| e^{i\phi_n} \tag{5.27}$$

I wrote this using the periodic boundary conditions of Eq. (5.19). The period is the period of the lowest frequency, $T = 2\pi/\omega_0$.

A flute produces a combination of frequencies that is mostly concentrated in a small number of harmonics, while a violin or reed instrument produces a far more complex combination of frequencies. The size of the coefficients a_n in Eq. (5.26) determines the quality of the note that you hear, though oddly enough its *phase*, ϕ_n, doesn't have an effect on your perception of the sound.

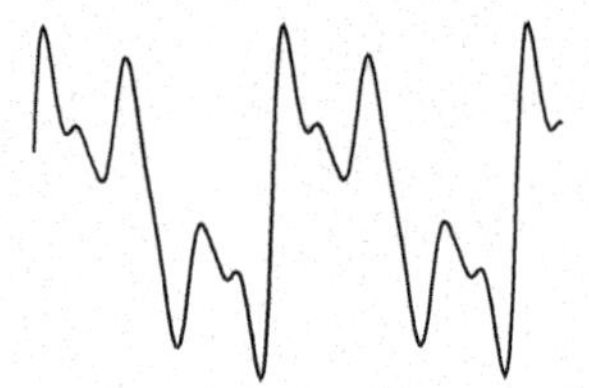 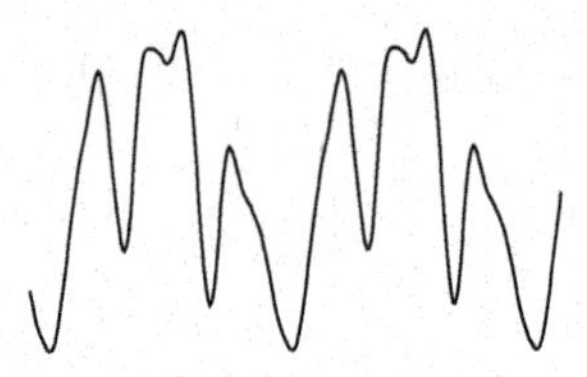

These represent a couple of cycles of the sound of a clarinet. The left graph is about what the wave output of the instrument looks like, and the right graph is what the graph would look like if I add a random phase, ϕ_n, to each of the Fourier components of the sound as in Eq. (5.27). They may look very different, but to the human ear they sound alike.

You can hear examples of the sound of Fourier series online via the web sites:

www.jhu.edu/~signals/listen/music1.html
www.educypedia.be/education/physicsjavasound.htm

You can hear the (lack of) effect of phase on sound. You can also synthesize your own series and hear what they sound like under such links as "Fourier synthese" and "Harmonics applet" found on this page. You can back up from this link to larger topics by using the links shown in the left column of the web page.

Real musical sound is of course more than just these Fourier series. At the least, the Fourier coefficients, a_n, are themselves functions of time. The time scale on which they vary is however much longer than the basic period of oscillation of the wave. That means that it makes sense to treat them as (almost) constant when you are trying to describe the harmonic structure of the sound. Even the lowest pitch notes you can hear are at least 20 Hz, and few home sound systems can produce frequencies nearly that low. Musical notes change on time scales much greater than 1/20 or 1/100 of a second, and this allows you to treat the notes by Fourier series even though the Fourier coefficients are themselves time-dependent. The attack and the decay of the note greatly affects our perception of it, and that is described by the time-varying nature of these coefficients.*

Parseval's Identity

Let u_n be the set of orthogonal functions that follow from your choice of boundary conditions.

$$f(x) = \sum_n a_n u_n(x)$$

---

* For an enlightening web page, including a complete and impressively thorough text on mathematics and music, look up the book by David Benson. It is available both in print from Cambridge Press and online.
www.maths.abdn.ac.uk/~bensondj/html/ and especially
www.maths.abdn.ac.uk/~bensondj/html/maths-music.html

Evaluate the integral of the absolute square of f over the domain.

$$\int_a^b dx\,|f(x)|^2 = \int_a^b dx \left[\sum_m a_m u_m(x)\right]^* \left[\sum_n a_n u_n(x)\right]$$
$$= \sum_m a_m^* \sum_n a_n \int_a^b dx\, u_m(x)^* u_n(x) = \sum_n |a_n|^2 \int_a^b dx\,|u_n(x)|^2$$

In the more compact notation this is

$$\langle f, f\rangle = \Big\langle \sum_m a_m u_m, \sum_n a_n u_n \Big\rangle = \sum_{m,n} a_m^* a_n \langle u_m, u_n\rangle = \sum_n |a_n|^2 \langle u_n, u_n\rangle \tag{5.28}$$

The first equation is nothing more than substituting the series for f. The second moved the integral under the summation. The third equation uses the fact that all these integrals are zero except for the ones with $m = n$. That reduces the double sum to a single sum. If you have chosen to normalize all of the functions u_n so that the integrals of $|u_n(x)|^2$ are one, then this relation takes on a simpler appearance. This is sometimes convenient.

What does this say if you apply it to a series I've just computed? Take Eq. (5.10) and see what it implies.

$$\langle f, f\rangle = \int_0^L dx\,1 = L = \sum_{k=0}^\infty |a_k|^2 \langle u_n, u_n\rangle =$$
$$\sum_{k=0}^\infty \left(\frac{4}{\pi(2k+1)}\right)^2 \int_0^L dx\,\sin^2\left(\frac{(2k+1)\pi x}{L}\right) = \sum_{k=0}^\infty \left(\frac{4}{\pi(2k+1)}\right)^2 \frac{L}{2}$$

Rearrange this to get

$$\sum_{k=0}^\infty \frac{1}{(2k+1)^2} = \frac{\pi^2}{8} \tag{5.29}$$

A bonus. You have the sum of this infinite series, a result that would be quite perplexing if you see it without knowing where it comes from. While you have it in front of you, what do you get if you simply *evaluate* the infinite series of Eq. (5.10) at $L/2$. The answer is 1, but what is the other side?

$$1 = \frac{4}{\pi}\sum_{k=0}^\infty \frac{1}{2k+1}\sin\frac{(2k+1)\pi(L/2)}{L} = \frac{4}{\pi}\sum_{k=0}^\infty \frac{1}{2k+1}(-1)^k$$
$$\text{or}\qquad 1 - \frac{1}{3} + \frac{1}{5} - \frac{1}{7} + \frac{1}{9} - \cdots = \frac{\pi}{4}$$

But does it Work?

If you are in the properly skeptical frame of mind, you may have noticed a serious omission on my part. I've done all this work showing how to get orthogonal functions and to manipulate them to derive Fourier series for a general function, but when did I show that this actually works? Never. How do I know that a general function, even a well-behaved general function, can be written as such a series? I've proved that the set of functions $\sin(n\pi x/L)$ are orthogonal on $0 < x < L$, but that's not good enough.

Maybe a clever mathematician will invent a new function that I haven't thought of and that will be orthogonal to all of these sines and cosines that I'm trying to use for a basis, just as $\hat{k}$ is orthogonal to $\hat{\imath}$ and $\hat{\jmath}$. It won't happen. There are proper theorems that specify the conditions under which all of this Fourier manipulation works. Dirichlet worked out the key results, which are found in many advanced calculus texts.

For example if the function is continuous with a continuous derivative on the interval $0 \le x \le L$ then the Fourier series will exist, will converge, and will converge to the specified function (except maybe at the endpoints). If you allow it to have a finite number of finite discontinuities but with a continuous derivative in between, then the Fourier series will converge and will (except maybe at the discontinuities) converge to the specified function. At these discontinuities it will converge to the average value taken from the left and from the right. There are a variety of other sufficient conditions that you can use to insure that all of this stuff works, but I'll leave that to the advanced calculus books.

5.5 Periodically Forced ODE's

If you have a harmonic oscillator with an added external force, such as Eq. (4.12), there are systematic ways to solve it, such as those found in section 4.2. One part of the problem is to find a solution to the inhomogeneous equation, and if the external force is simple enough you can do this easily. Suppose though that the external force is complicated *but periodic*, as for example when you're pushing a child on a swing.

$$m\frac{d^2x}{dt^2} = -kx - b\frac{dx}{dt} + F_{\text{ext}}(t)$$

That the force is periodic means $F_{\text{ext}}(t) = F_{\text{ext}}(t+T)$ for all times t. The period is T.

Pure Frequency Forcing

Before attacking the general problem, look at a simple special case. Take the external forcing function to be $F_0 \cos\omega_{\text{e}} t$ where this frequency is $\omega_{\text{e}} = 2\pi/T$. This equation is now

$$m\frac{d^2x}{dt^2} + kx + b\frac{dx}{dt} = F_0\cos\omega_{\text{e}} t = \frac{F_0}{2}\left[e^{i\omega_{\text{e}} t} + e^{-i\omega_{\text{e}} t}\right] \tag{5.30}$$

Find a solution corresponding to each term separately and add the results. To get an exponential out, put an exponential in.

$$\text{for}\qquad m\frac{d^2x}{dt^2} + kx + b\frac{dx}{dt} = e^{i\omega_{\text{e}} t} \qquad \text{assume} \qquad x_{\text{inh}}(t) = Ae^{i\omega_{\text{e}} t}$$

Substitute the assumed form and it will determine A.

$$\left[m(-\omega_{\text{e}}^2) + b(i\omega_{\text{e}}) + k\right]Ae^{i\omega_{\text{e}} t} = e^{i\omega_{\text{e}} t}$$

This tells you the value of A is

$$A = \frac{1}{-m\omega_{\text{e}}^2 + b\,i\omega_{\text{e}} + k} \tag{5.31}$$

The other term in Eq. (5.30) simply changes the sign in front of i everywhere. The total solution for Eq. (5.30) is then

$$x_{\mathsf{inh}}(t) = \frac{F_0}{2}\left[\frac{1}{-m\omega_{\mathsf{e}}^2 + b\,i\omega_{\mathsf{e}} + k}e^{i\omega_{\mathsf{e}}t} + \frac{1}{-m\omega_{\mathsf{e}}^2 - b\,i\omega_{\mathsf{e}} + k}e^{-i\omega_{\mathsf{e}}t}\right] \tag{5.32}$$

This is the sum of a number and its complex conjugate, so it's real. You can rearrange it so that it looks a lot simpler, but there's no need to do that right now. Instead I'll look at what it implies for certain values of the parameters.

Suppose that the viscous friction is small (b is small). If the forcing frequency, ω_{e} is such that $-m\omega_{\mathsf{e}}^2 + k = 0$, or is even close to zero, the denominators of the two terms become very small. This in turn implies that the response of x to the oscillating force is huge. *Resonance.* See problem 5.27. In a contrasting case, look at ω_{e} very large. Now the response of the mass is very small; it barely moves.

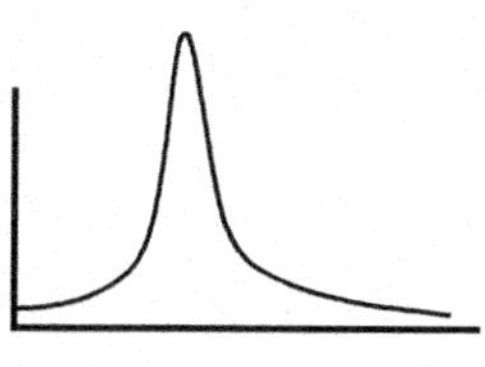

General Periodic Force

Now I'll go back to the more general case of a periodic forcing function, but not one that is simply a cosine. If a function is periodic I can use Fourier series to represent it on the whole axis. The basis to use will of course be the one with periodic boundary conditions (what else?). Use complex exponentials, then

$$u(t) = e^{i\omega t} \qquad \text{where} \qquad e^{i\omega(t+T)} = e^{i\omega t}$$

This is just like Eq. (5.20) but with t instead of x, so

$$u_n(t) = e^{2\pi int/T}, \qquad (n = 0,\ \pm 1,\ \ldots) \tag{5.33}$$

Let $\omega_{\mathsf{e}} = 2\pi/T$, and this is

$$u_n(t) = e^{in\omega_{\mathsf{e}}t}$$

The external force can now be represented by the Fourier series

$$F_{\text{ext}}(t) = \sum_{k=-\infty}^{\infty} a_k\, e^{ik\omega_{\mathsf{e}}t}, \qquad \text{where}$$

$$\left\langle e^{in\omega_{\mathsf{e}}t}, \sum_{k=-\infty}^{\infty} a_k\, e^{ik\omega_{\mathsf{e}}t}\right\rangle = a_n\,T = \left\langle e^{in\omega_{\mathsf{e}}t}, F_{\text{ext}}(t)\right\rangle = \int_0^T dt\, e^{-in\omega_{\mathsf{e}}t}\, F_{\text{ext}}(t)$$

(Don't forget the implied complex conjugation in the definition of the scalar product, $\langle\,,\rangle$, Eq. (5.11)) Because the force is periodic, any other time interval of duration T is just as good, perhaps $-T/2$ to $+T/2$ if that's more convenient.

How does this solve the differential equation? Plug in.

$$m\frac{d^2x}{dt^2} + b\frac{dx}{dt} + kx = \sum_{n=-\infty}^{\infty} a_n\, e^{in\omega_{\mathsf{e}}t} \tag{5.34}$$

All there is to do now is to solve for an inhomogeneous solution one term at a time and then to add the results. Take one term alone on the right:

$$m\frac{d^2x}{dt^2} + b\frac{dx}{dt} + kx = e^{in\omega_e t}$$

This is what I just finished solving a few lines ago, Eq. (5.31), but with $n\omega_e$ instead of simply ω_e. The inhomogeneous solution is the sum of the solutions from each term.

$$x_{\text{inh}}(t) = \sum_{n=-\infty}^{\infty} a_n \frac{1}{-m(n\omega_e)^2 + bin\omega_e + k} e^{ni\omega_e t} \tag{5.35}$$

Suppose for example that the forcing function is a simple square wave.

$$F_{\text{ext}}(t) = \begin{cases} F_0 & (0 < t < T/2) \\ -F_0 & (T/2 < t < T) \end{cases} \qquad \text{and} \qquad F_{\text{ext}}(t+T) = F_{\text{ext}}(t) \tag{5.36}$$

The Fourier series for this function is one that you can do in problem 5.12. The result is

$$F_{\text{ext}}(t) = F_0 \frac{2}{\pi i} \sum_{n \text{ odd}} \frac{1}{n} e^{ni\omega_e t} \tag{5.37}$$

The solution corresponding to Eq. (5.35) is now

$$x_{\text{inh}}(t) = F_0 \frac{1}{2\pi i} \sum_{n \text{ odd}} \frac{1}{\big(-m(n\omega_e)^2 + ibn\omega_e + k\big)} \frac{1}{n} e^{ni\omega_e t} \tag{5.38}$$

A real force ought to give a real result; does this? Yes. For every positive n in the sum, there is a corresponding negative one and the sum of those two is real. You can see this because every n that appears is either squared or is multiplied by an "i." When you add the $n = +5$ term to the $n = -5$ term it's adding a number to its own complex conjugate, and that's real.

What peculiar features does this result imply? With the simply cosine force the phenomenon of resonance occurred, in which the response to a small force at a frequency that matched the intrinsic frequency $\sqrt{k/m}$ produced a disproportionately large response. What other things happen here?

The natural frequency of the system is (for small damping) still $\sqrt{k/m}$. Look to see where a denominator in Eq. (5.38) can become very small. This time it is when $-m(n\omega_e)^2 + k = 0$. This is not only when the external frequency ω_e matches the natural frequency; it's when $n\omega_e$ matches it. If the natural frequency is $\sqrt{k/m} = 100$ radians/sec you get a big response if the forcing frequency is 100 radians/sec or 33 radians/sec or 20 radians/sec or 14 radians/sec etc. What does this mean? The square wave in Eq. (5.36) contains many frequencies. It contains more than just the main frequency $2\pi/T$, it contains 3 times this and 5 times it and many higher frequencies. When any one of these harmonics matches the natural frequency you will have the large resonant response.

Not only do you get a large response, look at the way the mass oscillates. If the force has a square wave frequency 20 radians/sec, the mass responds* with a large sinusoidal oscillation at a frequency 5 times higher — 100 radians/sec.

---

* The next time you have access to a piano, gently depress a key without making a sound, then strike the key one octave lower. Release the lower key and listen to the sound of the upper note. Then try it with an interval of an octave plus a fifth.

5.6 Return to Parseval

When you have a periodic wave such as a musical note, you can Fourier analyze it. The boundary conditions to use are naturally the periodic ones, Eq. (5.20) or (5.33), so that

$$f(t) = \sum_{-\infty}^{\infty} a_n e^{in\omega_0 t}$$

If this represents the sound of a flute, the amplitudes of the higher frequency components (the a_n) drop off rapidly with n. If you are hearing an oboe or a violin the strength of the higher components is greater.

If this function represents the sound wave as received by your ear, the power that you receive is proportional to the square of f. If f represent specifically the pressure disturbance in the air, the intensity (power per area) carried by the wave is $f(t)^2 v/B$ where v is the speed of the wave and B is the bulk modulus of the air. The key property of this is that it is proportional to the square of the wave's amplitude. That's the same relation that occurs for light or any other wave. Up to a known factor then, the power received by the ear is proportional to $f(t)^2$.

This time average of the power is (up to that constant factor that I'm ignoring)

$$\langle f^2 \rangle = \lim_{T\to\infty} \frac{1}{2T} \int_{-T}^{+T} dt\, f(t)^2$$

Now put the Fourier series representation of the sound into the integral to get

$$\lim_{T\to\infty} \frac{1}{2T} \int_{-T}^{+T} dt \left[\sum_{-\infty}^{\infty} a_n e^{in\omega_0 t} \right]^2$$

The sound $f(t)$ is real, so by problem 5.11, $a_{-n} = a_n^*$. Also, using the result of problem 5.18 the time average of $e^{i\omega t}$ is zero unless $\omega = 0$; then it's one.

$$\begin{aligned}
\langle f^2 \rangle &= \lim_{T\to\infty} \frac{1}{2T} \int_{-T}^{+T} dt \left[\sum_n a_n e^{in\omega_0 t} \right] \left[\sum_m a_m e^{im\omega_0 t} \right] \\
&= \lim_{T\to\infty} \frac{1}{2T} \int_{-T}^{+T} dt \sum_n \sum_m a_n e^{in\omega_0 t} a_m e^{im\omega_0 t} \\
&= \sum_n \sum_m a_n a_m \lim_{T\to\infty} \frac{1}{2T} \int_{-T}^{+T} dt\, e^{i(n+m)\omega_0 t} \\
&= \sum_n a_n a_{-n} \\
&= \sum_n |a_n|^2
\end{aligned} \tag{5.39}$$

Put this into words and it says that the time-average power received is the sum of many terms, each one of which can be interpreted as the amount of power coming in *at that frequency* $n\omega_0$. The Fourier coefficients squared (absolute-squared really) are then proportional to the part of the power at a particular frequency. The "power spectrum."

Other Applications

In section 10.2 Fourier series will be used to solve partial differential equations, leading to equations such as Eq. (10.15).

In quantum mechanics, Fourier series and its generalizations will manifest themselves in displaying the discrete energy levels of bound atomic and nuclear systems.

Music synthesizers are *all* about Fourier series and its generalizations.

5.7 Gibbs Phenomenon

There's a picture of the Gibbs phenomenon with Eq. (5.10). When a function has a discontinuity, its Fourier series representation will not handle it in a uniform way, and the series overshoots its goal at the discontinuity. The detailed calculation of this result is quite pretty, and it's an excuse to pull together several of the methods from the chapters on series and on complex algebra.

$$\frac{4}{\pi}\sum_{k=0}^{\infty}\frac{1}{2k+1}\sin\frac{(2k+1)\pi x}{L} = 1, \qquad (0 < x < L)$$

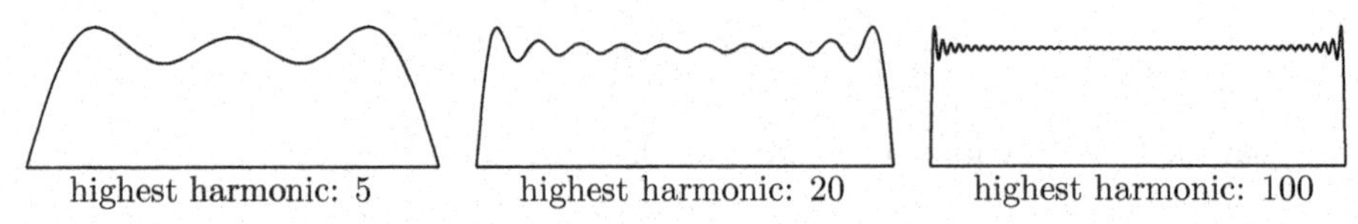

The analysis sounds straight-forward. Find the position of the first maximum. Evaluate the series there. It really is almost that clear. First however, you have to start with the a *finite* sum and find the first maximum of that. Stop the sum at $k = N$.

$$\frac{4}{\pi}\sum_{k=0}^{N}\frac{1}{2k+1}\sin\frac{(2k+1)\pi x}{L} = f_N(x) \tag{5.40}$$

For a maximum, set the derivative to zero.

$$f_N'(x) = \frac{4}{L}\sum_{0}^{N}\cos\frac{(2k+1)\pi x}{L}$$

Write this as the real part of a complex exponential and use Eq. (2.3).

$$\sum_{0}^{N} e^{i(2k+1)\pi x/L} = \sum_{0}^{N} z^{2k+1} = z\sum_{0}^{N} z^{2k} = z\frac{1-z^{2N+2}}{1-z^2}$$

Factor these complex exponentials in order to put this into a nicer form.

$$= e^{i\pi x/L}\,\frac{e^{-i\pi x(N+1)/L} - e^{i\pi x(N+1)/L}}{e^{-i\pi x/L} - e^{i\pi x/L}}\,\frac{e^{i\pi x(N+1)/L}}{e^{i\pi x/L}} = \frac{\sin(N+1)\pi x/L}{\sin\pi x/L}\,e^{i\pi x(N+1)/L}$$

The real part of this changes the last exponential into a cosine. Now you have the product of the sine and cosine of $(N+1)\pi x/L$, and that lets you use the trigonometric double angle formula.

$$f_N'(x) = \frac{4}{L}\,\frac{\sin 2(N+1)\pi x/L}{2\sin \pi x/L} \tag{5.41}$$

This is zero at the maximum. The first maximum after $x = 0$ is at $2(N+1)\pi x/L = \pi$, or $x = L/2(N+1)$.

Now for the value of f_N at this point,

$$f_N\big(L/2(N+1)\big) = \frac{4}{\pi}\sum_{k=0}^{N}\frac{1}{2k+1}\sin\frac{(2k+1)\pi L/2(N+1)}{L} = \frac{4}{\pi}\sum_{k=0}^{N}\frac{1}{2k+1}\sin\frac{(2k+1)\pi}{2(N+1)}$$

The final step is to take the limit as $N\to\infty$. As k varies over the set 0 to N, the argument of the sine varies from a little more than zero to a little less than π. As N grows you have the sum over a lot of terms, each of which is approaching zero. It's an integral. Let $t_k = k/N$ then $\Delta t_k = 1/N$. This sum is approximately

$$\frac{4}{\pi}\sum_{k=0}^{N}\frac{1}{2Nt_k}\sin t_k\pi = \frac{2}{\pi}\sum_{0}^{N}\Delta t_k\,\frac{1}{t_k}\sin t_k\pi \longrightarrow \frac{2}{\pi}\int_0^1\frac{dt}{t}\sin\pi t$$

In this limit $2k+1$ and $2k$ are the same, and $N+1$ is the same as N.

Finally, put this into a standard form by changing variables to $\pi t = x$.

$$\frac{2}{\pi}\int_0^{\pi} dx\,\frac{\sin x}{x} = \frac{2}{\pi}\,\mathrm{Si}(\pi) = 1.17898 \qquad\qquad \int_0^x dt\,\frac{\sin t}{t} = \mathrm{Si}(x) \tag{5.42}$$

The function Si is called the "sine integral." It's just another tabulated function, along with erf, Γ, and others. This equation says that as you take the limit of the series, the first part of the graph approaches a vertical line starting from the origin, but it overshoots its target by 18%.

Exercises

1 A vector is given to be $\vec A = 5\hat x + 3\hat y$. Let a new basis be $\hat e_1 = (\hat x + \hat y)/\sqrt2$, and $\hat e_2 = (\hat x - \hat y)/\sqrt2$. Use scalar products to find the components of $\vec A$ in the new basis: $\vec A = A_1\,\hat e_1 + A_2\,\hat e_2$.

2 For the same vector as the preceding problem, and another basis $\vec f_1 = 3\hat x + 4\hat y$ and $\vec f_2 = -8\hat x + 6\hat y$, express $\vec A$ in the new basis. Are these basis vectors orthogonal?

3 On the interval $0 < x < L$, sketch three graphs: The first term alone, then the second term alone, then the third. Try to get the scale of the graphs reasonable accurate. Now add the first two and graph. Then add the third also and graph. Do all this by hand, no graphing calculators, though if you want to use a calculator to calculate a few points, that's ok.

$$\sin\left(\pi x/L\right) - \tfrac{1}{9}\sin\left(3\pi x/L\right) + \tfrac{1}{25}\sin\left(5\pi x/L\right)$$

4 For what values of α are the vectors $\vec{A} = \alpha\hat{x} - 2\hat{y} + \hat{z}$ and $\vec{B} = 2\alpha\hat{x} + \alpha\hat{y} - 4\hat{z}$ orthogonal?

5 On the interval $0 < x < L$ with a scalar product defined as $\langle f, g\rangle = \int_0^L dx\, f(x)^* g(x)$, show that these are zero, making the functions orthogonal:

$$x \quad \text{and} \quad L - \tfrac{3}{2}x, \qquad \sin\pi x/L \quad \text{and} \quad \cos\pi x/L, \qquad \sin 3\pi x/L \quad \text{and} \quad L - 2x$$

6 Same as the preceding, show that these functions are orthogonal:

$$e^{2i\pi x/L} \quad \text{and} \quad e^{-2i\pi x/L}, \qquad L - \tfrac{1}{4}(7+i)x \quad \text{and} \quad L + \tfrac{3}{2}ix$$

7 With the same scalar product the last two exercises, for what values of α are the functions $f_1(x) = \alpha x - (1-\alpha)(L - \frac{3}{2}x)$ and $f_2(x) = 2\alpha x + (1+\alpha)(L - \frac{3}{2}x)$ orthogonal? What is the interpretation of the two roots?

8 Repeat the preceding exercise but use the scalar product $\langle f, g\rangle = \int_L^{2L} dx\, f(x)^* g(x)$.

9 Use the scalar product $\langle f, g\rangle = \int_{-1}^{1} dx\, f(x)^* g(x)$, and show that the Legendre polynomials P_0, P_1, P_2, P_3 of Eq. (4.61) are mutually orthogonal.

10 Change the scalar product in the preceding exercise to $\langle f, g\rangle = \int_0^1 dx\, f(x)^* g(x)$ and determine if the same polynomials are still orthogonal.

Problems

5.1 Get the results in Eq. (5.18) by explicitly calculating the integrals.

5.2 (**a**) The functions with periodic boundary conditions, Eq. (5.20), are supposed to be orthogonal on $0 < x < L$. That is, $\langle u_n, u_m \rangle = 0$ for $n \neq m$. Verify this by explicit integration. What is the result if $n = m$ or $n = -m$? The notation is defined in Eq. (5.11).
(**b**) Same calculation for the real version, $\langle u_n, u_m \rangle$, $\langle v_n, v_m \rangle$, and $\langle u_n, v_m \rangle$, Eq. (5.22)

5.3 Find the Fourier series for the function $f(x) = 1$ as in Eq. (5.10), but use as a basis the set of functions u_n on $0 < x < L$ that satisfy the differential equation $u'' = \lambda u$ with boundary conditions $u'(0) = 0$ and $u'(L) = 0$. (Eq. (5.23)) *Necessarily the first step* will be to examine all the solutions to the differential equation and to find the cases for which the bilinear concomitant vanishes.
(**b**) Graph the resulting Fourier series on $-2L < x < 2L$.
(**c**) Graph the Fourier series Eq. (5.10) on $-2L < x < 2L$.

5.4 (**a**) Compute the Fourier series for the function x^2 on the interval $0 < x < L$, using as a basis the functions with boundary conditions $u'(0) = 0$ and $u'(L) = 0$.
(**b**) Sketch the partial sums of the series for 1, 2, 3 terms. Also sketch this sum *outside* the original domain and see what this series produces for an extension of the original function.
Ans: Eq. (5.1)

5.5 (**a**) Compute the Fourier series for the function x on the interval $0 < x < L$, using as a basis the functions with boundary conditions $u(0) = 0 = u(L)$. How does the coefficient of the nth term decrease as a function of n? (**b**) Also sketch this sum within *and* outside the original domain to see what this series produces for an extension of the original function.
Ans: $\frac{2L}{\pi} \sum_1^\infty \frac{(-1)^{n+1}}{n} \sin(n\pi x/L)$

5.6 (**a**) In the preceding problem the sine functions that you used don't match the qualitative behavior of the function x on this interval because the sine is zero at $x = L$ and x isn't. The qualitative behavior is different from the basis you are using for the expansion. You should be able to get better convergence for the series if you choose functions that more closely match the function that you're expanding, so try repeating the calculation using basis functions that satisfy $u(0) = 0$ and $u'(L) = 0$. How does the coefficient of the nth term decrease as a function of n? (**b**) As in the preceding problem, sketch some partial sums of the series and its extension outside the original domain.
Ans: $\frac{8L}{\pi^2} \sum_0^\infty \left((-1)^n/(2n+1)^2\right) \sin\left((n + 1/2)\pi x/L\right)$

5.7 The function $\sin^2 x$ is periodic with period π. What is its Fourier series representation using as a basis functions that have this period? Eqs. (5.20) or (5.22).

5.8 In the two problems 5.5 and 5.6 you improved the convergence by choosing boundary conditions that better matched the function that you want. Can you do better? The function x vanishes at the origin, but its derivative isn't zero at L, so try boundary conditions $u(0) = 0$ and $u(L) = Lu'(L)$. These conditions match those of x so this ought to give even better convergence, but first you have to verify that these conditions guarantee the orthogonality of the basis functions. You have to verify that the left side of

Eq. (5.15) is in fact zero. When you set up the basis, you will examine functions of the form $\sin kx$, but you will not be able to solve explicitly for the values of k. Don't worry about it. When you use Eq. (5.24) to get the coefficients all that you need to do is to use the equation that k satisfies to do the integrals. You do not need to have *solved* it. If you do all the algebra correctly you will probably have a surprise.

5.9 (**a**) Use the periodic boundary conditions on $-L < x < +L$ and basis $e^{\pi inx/L}$ to write x^2 as a Fourier series. Sketch the sums up to a few terms. (**b**) Evaluate your result at $x = L$ where you know the answer to be L^2 and deduce from this the value of $\zeta(2)$.

5.10 On the interval $-\pi < x < \pi$, the function $f(x) = \cos x$. Expand this in a Fourier series defined by $u'' = \lambda u$ and $u(-\pi) = 0 = u(\pi)$. If you use your result for the series outside of this interval you define an extension of the original function. Graph this extension and compare it to what you normally think of as the graph of $\cos x$. As always, go back to the differential equation to get all the basis functions.
Ans: $-\frac{4}{\pi}\sum_{k=0}^{\infty}\frac{2k+1}{(2k+3)(2k-1)}\sin\big((2k+1)(x+\pi)/2\big)$

5.11 Represent a function f on the interval $-L < x < L$ by a Fourier series using periodic boundary conditions

$$f(x) = \sum_{-\infty}^{\infty} a_n\, e^{n\pi ix/L}$$

(**a**) If the function f is odd, prove that for all n, $a_{-n} = -a_n$
(**b**) If the function f is even, prove that all $a_{-n} = a_n$.
(**c**) If the function f is real, prove that all $a_{-n} = a_n^*$.
(**d**) If the function is both real and even, characterize a_n.
(**e**) If the function is imaginary and odd, characterize a_n.

5.12 Derive the series Eq. (5.37).

5.13 For the function $e^{-\alpha t}$ on $0 < t < T$, express it as a Fourier series using periodic boundary conditions $[u(0) = u(T)$ and $u'(0) = u'(T)]$. Examine for plausibility the cases of large and small α. The basis functions for periodic boundary conditions can be expressed either as cosines and sines or as complex exponentials. Unless you can analyze the problem ahead of time and determine that it has some special symmetry that matches that of the trig functions, you're usually better off with the exponentials.
Ans: $\big[(1-e^{-\alpha T})/\alpha T\big]\big[1+2\sum_1^\infty[\alpha^2\cos n\omega t + \alpha n\omega\sin n\omega t]/[\alpha^2+n^2\omega^2]\big]$

5.14 (**a**) On the interval $0 < x < L$, write $x(L-x)$ as a Fourier series, using boundary conditions that the expansion functions vanish at the endpoints. Next, evaluate the series at $x = L/2$ to see if it gives an interesting result. (**b**) Finally, what does Parseval's identity tell you?
Ans: $\sum_1^\infty \frac{4L^2}{n^3\pi^3}\big[1-(-1)^n\big]\sin(n\pi x/L)$

5.15 A full-wave rectifier takes as an input a sine wave, $\sin\omega t$ and creates the output $f(t) = |\sin\omega t|$. The period of the original wave is $2\pi/\omega$, so write the Fourier series for the output in terms of functions periodic with this period. Graph the function f first and use the graph to anticipate which terms in the Fourier series will be present.

When you're done, use the result to evaluate the infinite series $\sum_1^\infty (-1)^{k+1}/(4k^2-1)$
Ans: $\pi/4 - 1/2$

5.16 A half-wave rectifier takes as an input a sine wave, $\sin\omega t$ and creates the output

$$\sin\omega t \quad \text{if } \sin\omega t > 0 \qquad \text{and} \qquad 0 \quad \text{if } \sin\omega t \le 0$$

The period of the original wave is $2\pi/\omega$, so write the Fourier series for the output in terms of functions periodic with this period. Graph the function first. Check that the result gives the correct value at $t = 0$, manipulating it into a telescoping series. Sketch a few terms of the whole series to see if it's heading in the right direction.
Ans: ${}^4/\pi + {}^1/_2 \sin\omega t - {}^8/\pi \sum_{n \text{ even} > 0} \cos(n\omega t)/(n^2-1)$

5.17 For the undamped harmonic oscillator, apply an oscillating force (cosine). This is a simpler version of Eq. (5.30). Solve this problem and add the general solution to the homogeneous equation. Solve this subject to the initial conditions that $x(0) = 0$ and $v_x(0) = v_0$.

5.18 The average (arithmetic mean) value of a function is

$$\langle f \rangle = \lim_{T\to\infty} \frac{1}{2T} \int_{-T}^{+T} dt\, f(t) \qquad \text{or} \qquad \langle f \rangle = \lim_{T\to\infty} \frac{1}{T} \int_0^T dt\, f(t)$$

as appropriate for the problem.
What is $\langle \sin\omega t \rangle$? What is $\langle \sin^2\omega t \rangle$? What is $\langle e^{-at^2} \rangle$?
What is $\langle \sin\omega_1 t \sin\omega_2 t \rangle$? What is $\langle e^{i\omega t} \rangle$?

5.19 In the calculation leading to Eq. (5.39) I assumed that $f(t)$ is real and then used the properties of a_n that followed from that fact. Instead, make no assumption about the reality of $f(t)$ and compute

$$\langle |f(t)|^2 \rangle = \langle f(t)^* f(t) \rangle$$

Show that it leads to the same result as before, $\sum |a_n|^2$.

5.20 The series

$$\sum_{n=0}^{\infty} a^n \cos n\theta \qquad (|a| < 1)$$

represents a function. Sum this series and determine what the function is. While you're about it, sum the similar series that has a sine instead of a cosine. Don't try to do these separately; combine them and do them as one problem. And check some limiting cases of course. And graph the functions. Ans: $a\sin\theta/\big(1 + a^2 - 2a\cos\theta\big)$

5.21 Apply Parseval's theorem to the result of problem 5.9 and see what you can deduce.

5.22 If you take all the elements u_n of a basis and multiply each of them by 2, what happens to the result for the Fourier series for a given function?

5.23 In the section 5.3 several bases are mentioned. Sketch a few terms of each basis.

5.24 A function is specified on the interval $0 < t < T$ to be

$$f(t) = \begin{cases} 1 & (0 < t < t_0) \\ 0 & (t_0 < t < T) \end{cases} \qquad 0 < t_0 < T$$

On this interval, choose boundary conditions such that the left side of the basic identity (5.15) is zero. Use the corresponding choice of basis functions to write f as a Fourier series on this interval.

5.25 Show that the boundary conditions $u(0) = 0$ and $\alpha u(L) + \beta u'(L) = 0$ make the bilinear concomitant in Eq. (5.15) vanish. Are there any restrictions on α and β? Do not automatically assume that these numbers are real.

5.26 Derive a Fourier series for the function

$$f(x) = \begin{cases} Ax & (0 < x < L/2) \\ A(L-x) & (L/2 < x < L) \end{cases}$$

Choose the Fourier basis that you prefer. Evaluate the resulting series at $x = L/2$ to check the result. Sketch the sum of a couple of terms. Comment on the convergence properties of the result. Are they what you expect? What does Parseval's identity say?
Ans: $(2AL/\pi^2)\sum_{k\text{ odd}}(-1)^{(k-1)/2}\sin(k\pi x/L)/k^2$

5.27 Rearrange the solution Eq. (5.32) into a more easily understood form. (**a**) Write the first denominator as

$$-m\omega_e^2 + b\,i\omega_e + k = Re^{i\phi}$$

What are R and ϕ? The second term does not require you to repeat this calculation, just use its results, now combine everything and write the answer as an amplitude times a phase-shifted cosine.
(**b**) Assume that b is not too big and plot both R and ϕ versus the forcing frequency ω_e. Also, and perhaps more illuminating, plot $1/R$.

5.28 Find the form of Parseval's identity appropriate for power series. Assume a scalar product $\langle f, g\rangle = \int_{-1}^{1} f(x)^* g(x)\,dx$ for the series $f(x) = \sum_0^\infty a_n x^n$, and $g(x) = \sum_0^\infty b_n x^n$, expressing the result in terms of matrices. Next, test your result on a simple, low-order polynomial.
Ans: $(a_0^*\ a_1^* \ldots)M(b_0\ b_1 \ldots)\tilde{}$ where $M_{00} = 2,\ M_{02} = 2/3\ M_{04} = 2/5, \ldots$ and $\tilde{}$ is transpose.

5.29 (**a**) In the Gibbs phenomenon, after the first maximum there is a first *minimum*. Where is it? how big is the function there? What is the limit of this point? That is, repeat the analysis of section 5.7 for this minimum point.
(**b**) While you're about it, what will you get for the limit of the sine integral, $\text{Si}(\infty)$? The last result can also be derived by complex variable techniques of chapter 14, Eq. (14.16).
Ans: $(2/\pi)\,\text{Si}(2\pi) = 0.9028$

5.30 Make a blown-up copy of the graph preceding Eq. (5.40) and measure the size of the overshoot. Compare this experimental value to the theoretical limit. Same for the first minimum.

5.31 Find the power series representation about the origin for the sine integral, Si, that appeared in Eq. (5.42). What is its domain of convergence?
Ans: $\frac{2}{\pi}\sum_0^\infty(-1)^n\big(x^{2n+1}/(2n+1)(2n+1)!\big)$

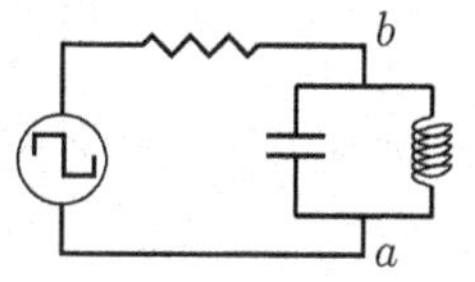

5.32 An input potential in a circuit is given to be a square wave $\pm V_0$ at frequency ω. What is the voltage between the points a and b? In particular, assume that the resistance is small, and show that you can pick values of the capacitance and the inductance so that the output potential is almost exactly a sine wave at frequency 3ω. A filter circuit. Recall section 4.8.

5.33 For the function $\sin(\pi x/L)$ on $(0 < x < 2L)$, expand it in a Fourier series using as a basis the trigonometric functions with the boundary conditions $u'(0) = 0 = u'(2L)$, the cosines. Graph the resulting series as extended outside the original domain.

5.34 For the function $\cos(\pi x/L)$ on $(0 < x < 2L)$, expand it in a Fourier series using as a basis the trigonometric functions with the boundary conditions $u(0) = 0 = u(2L)$, the sines. Graph the resulting series as extended outside the original domain.

5.35 (**a**) For the function $f(x) = x^4$, evaluate the Fourier series on the interval $-L < x < L$ using periodic boundary conditions $\big(u(-L) = u(L)$ and $u'(-L) = u'(L)\big)$. (**b**) Evaluate the series at the point $x = L$ to derive the zeta function value $\zeta(4) = \pi^4/90$. Evaluate it at $x = 0$ to get a related series.
Ans: $\frac{1}{5}L^4 + L^4\sum_1^\infty(-1)^n\left[\frac{8}{n^2\pi^2} - \frac{48}{n^4\pi^4}\right]\cos n\pi x/L$

5.36 Fourier series depends on the fact that the sines and cosines are orthogonal when integrated over a suitable interval. There are other functions that allow this too, and you've seen one such set. The Legendre polynomials that appeared in section 4.11 in the chapter on differential equations satisfied the equations (4.62). One of these is

$$\int_{-1}^{1} dx\, P_n(x)P_m(x) = \frac{2}{2n+1}\delta_{nm}$$

This is an orthogonality relation, $\langle P_n, P_m\rangle = 2\delta_{nm}/(2n+1)$, much like that for trigonometric functions. Write a function $f(x) = \sum_0^\infty a_n\, P_n(x)$ and deduce an expression for evaluating the coefficients a_n. Apply this to the function $f(x) = x^2$.

5.37 For the standard differential equation $u'' = \lambda u$, use the boundary conditions $u(0) = 0$ and $2u(L) = Lu'(L)$. This is a special case of problem 5.25, so the bilinear concomitant vanishes. If you haven't done that problem, at least do this special case. Find all the solutions that satisfy these conditions and graph a few of them. You will not be able to find an explicit solution for the λs, but you can estimate a few of them to sketch graphs. Did you get them *all*?

5.38 Examine the function on $-L < x < L$ given by

$$f(x) = \begin{cases} 0 & (-L < x < -L/2) \text{ and } (L/2 < x < L) \\ 1 & (0 < x < L/2) \\ -1 & (-L/2 < x < 0) \end{cases}$$

Draw it first. Now find a Fourier series representation for it. You may choose to do this by doing lots of integrals, OR you may prefer to start with some previous results in this chapter, change periods, add or subtract, and do no integrals at all.

5.39 In Eq. (5.30) I wrote $\cos\omega_e t$ as the sum of two exponentials, $e^{i\omega_e t} + e^{-i\omega_e t}$. Instead, write the cosine as $e^{i\omega_e t}$ with the understanding that at the end you take the real part of the result, Show that the result is the same.

5.40 From Eq. (5.41) write an approximate closed form expression for the partial sum $f_N(x)$ for the region $x \ll L$ but *not* necessarily $x \ll NL$, though that extra-special case is worth doing too.

5.41 Evaluate the integral $\int_0^L dx\, x^2$ using the series Eq. (5.1) and using the series (5.2).

5.42 The Fourier series in problem 5.5 uses the same basis as the series Eq. (5.10). What is the result of evaluating the scalar products $\langle 1, 1\rangle$ and $\langle 1, x\rangle$ with these series?

5.43 If you evaluated the $n = m$ case of Eq. (5.6) by using a different trig identity, you can do it by an alternative method: say that n and m in this equation aren't necessarily integers. Then take the limit as $n \to m$.

Vector Spaces

The idea of vectors dates back to the middle 1800's, but our current understanding of the concept waited until Peano's work in 1888. Even then it took many years to understand the importance and generality of the ideas involved. This one underlying idea can be used to describe the forces and accelerations in Newtonian mechanics and the potential functions of electromagnetism and the states of systems in quantum mechanics and the least-square fitting of experimental data and much more.

6.1 The Underlying Idea

What *is* a vector?

If your answer is along the lines "something with magnitude and direction" then you have something to unlearn. Maybe you heard this definition in a class that I taught. If so, I lied; sorry about that. At the very least I didn't tell the whole truth. Does an automobile have magnitude and direction? Does that make it a vector?

The idea of a vector is far more general than the picture of a line with an arrowhead attached to its end. That special case is an important one, but it doesn't tell the whole story, and the whole story is one that unites many areas of mathematics. The short answer to the question of the first paragraph is

A vector is an element of a vector space.

Roughly speaking, a vector space is some set of things for which the operation of addition is defined and the operation of multiplication by a scalar is defined. You don't necessarily have to be able to multiply two vectors by each other or even to be able to define the length of a vector, though those *are* very useful operations and will show up in most of the interesting cases. You can add two cubic polynomials together:

$$\left(2 - 3x + 4x^2 - 7x^3\right) + \left(-8 - 2x + 11x^2 + 9x^3\right)$$

makes sense, resulting in a cubic polynomial. You can multiply such a polynomial by* 17 and it's still a cubic polynomial. The set of all cubic polynomials in x forms a vector space and the vectors are the individual cubic polynomials.

The common example of directed line segments (arrows) in two or three dimensions fits this idea, because you can add such arrows by the parallelogram law and you can multiply them by numbers, changing their length (and reversing direction for negative numbers).

Another, equally important example consists of all ordinary real-valued functions of a real variable: two such functions can be added to form a third one; you can multiply a function by a number to get another function. The example of cubic polynomials above is then a special case of this one.

A complete definition of a vector space requires pinning down these ideas and making them less vague. In the end, the way to do that is to express the definition as a set of axioms. From these axioms the general properties of vectors will follow.

A *vector space* is a set whose elements are called "vectors" and such that there are two operations defined on them: you can add vectors to each other and you can multiply

---

* The physicist's canonical random number

them by scalars (numbers). These operations must obey certain simple rules, the axioms for a vector space.

6.2 Axioms

The precise definition of a vector space is given by listing a set of axioms. For this purpose, I'll denote vectors by arrows over a letter, and I'll denote scalars by Greek letters. These scalars will, for our purpose, be either real or complex numbers — it makes no difference which for now.*

1 There is a function, addition of vectors, denoted +, so that $\vec{v}_1 + \vec{v}_2$ is another vector.
2 There is a function, multiplication by scalars, denoted by juxtaposition, so that $\alpha\vec{v}$ is a vector.
3 $(\vec{v}_1 + \vec{v}_2) + \vec{v}_3 = \vec{v}_1 + (\vec{v}_2 + \vec{v}_3)$ (the associative law).
4 There is a zero vector, so that for each $\vec{v}$, $\vec{v} + \vec{O} = \vec{v}$.
5 There is an additive inverse for each vector, so that for each $\vec{v}$, there is another vector $\vec{v}\,'$ so that $\vec{v} + \vec{v}\,' = \vec{O}$.
6 The commutative law of addition holds: $\vec{v}_1 + \vec{v}_2 = \vec{v}_2 + \vec{v}_1$.
7 $(\alpha + \beta)\vec{v} = \alpha\vec{v} + \beta\vec{v}$.
8 $(\alpha\beta)\vec{v} = \alpha(\beta\vec{v})$.
9 $\alpha(\vec{v}_1 + \vec{v}_2) = \alpha\vec{v}_1 + \alpha\vec{v}_2$.
10 $1\vec{v} = \vec{v}$.

In axioms 1 and 2 I called these operations "functions." Is that the right use of the word? Yes. Without going into the precise definition of the word (see section 12.1), you know it means that you have one or more independent variables and you have a single output. Addition of vectors and multiplication by scalars certainly fit that idea.

6.3 Examples of Vector Spaces

Examples of sets satisfying these axioms abound:

1 The usual picture of directed line segments in a plane, using the parallelogram law of addition.
2 The set of real-valued functions of a real variable, defined on the domain $[a \le x \le b]$. Addition is defined pointwise. If f_1 and f_2 are functions, then the value of the function $f_1 + f_2$ at the point x is the number $f_1(x) + f_2(x)$. That is, $f_1 + f_2 = f_3$ means $f_3(x) = f_1(x) + f_2(x)$. Similarly, multiplication by a scalar is defined as $(\alpha f)(x) = \alpha(f(x))$. Notice a small confusion of notation in this expression. The first multiplication, (αf), multiplies the scalar α by the vector f; the second multiplies the scalar α by the number $f(x)$.
3 Like example 2, but restricted to continuous functions. The one observation beyond the previous example is that the sum of two continuous functions is continuous.
4 Like example 2, but restricted to bounded functions. The one observation beyond the previous example is that the sum of two bounded functions is bounded.

---

* For a nice introduction online see distance-ed.math.tamu.edu/Math640, chapter three.

5 The set of n-tuples of real numbers: $(a_1, a_2, \ldots, a_n)$ where addition and scalar multiplication are defined by

$$(a_1, \ldots, a_n) + (b_1, \ldots, b_n) = (a_1 + b_1, \ldots, a_n + b_n) \qquad \alpha(a_1, \ldots, a_n) = (\alpha a_1, \ldots, \alpha a_n)$$

6 The set of square-integrable real-valued functions of a real variable on the domain $[a \le x \le b]$. That is, restrict example two to those functions with $\int_a^b dx\,|f(x)|^2 < \infty$. Axiom 1 is the only one requiring more than a second to check.

7 The set of solutions to the equation $\partial^2\phi/\partial x^2 + \partial^2\phi/\partial y^2 = 0$ in any fixed domain. (Laplace's equation)

8 Like example 5, but with $n = \infty$.

9 Like example 8, but each vector has only a finite number of non-zero entries.

10 Like example 8, but restricting the set so that $\sum_1^\infty |a_k|^2 < \infty$. Again, only axiom one takes work.

11 Like example 10, but the sum is $\sum_1^\infty |a_k| < \infty$.

12 Like example 10, but $\sum_1^\infty |a_k|^p < \infty$. $(p \ge 1)$

13 Like example 6, but $\int_a^b dx\,|f(x)|^p < \infty$.

14 Any of examples 2–13, but make the scalars complex, and the functions complex valued.

15 The set of all $n \times n$ matrices, with addition being defined element by element.

16 The set of all polynomials with the obvious laws of addition and multiplication by scalars.

17 Complex valued functions on the domain $[a \le x \le b]$ with $\sum_x |f(x)|^2 < \infty$. (Whatever this means. See problem 6.18)

18 $\{\vec{O}\}$, the space consisting of the zero vector alone.

19 The set of all solutions to the equations describing small motions of the surface of a drumhead.

20 The set of solutions of Maxwell's equations without charges or currents and with finite energy. That is, $\int [E^2 + B^2] d^3x < \infty$.

21 The set of all functions of a complex variable that are differentiable everywhere and satisfy

$$\int dx\,dy\,e^{-x^2-y^2}|f(z)|^2 < \infty,$$

where $z = x + iy$.

To verify that any of these is a vector space you have to run through the ten axioms, checking each one. (Actually, in a couple of pages there's a theorem that will greatly simplify this.) To see what is involved, take the first, most familiar example, arrows that all start at one point, the origin. I'll go through the details of each of the ten axioms to show that the process of checking is very simple. There are some cases for which this checking isn't so simple, but the difficulty is usually confined to verifying axiom one.

The picture shows the definitions of addition of vectors and multiplication by scalars, the first two axioms. The commutative law, axiom 6, is clear, as the diagonal of the parallelogram doesn't depend on which side you're looking at.

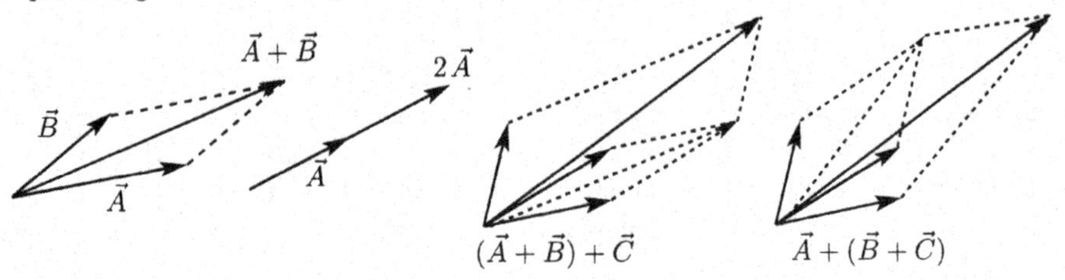

The associative law, axiom 3, is also illustrated in the picture. The zero vector, axiom 4, appears in this picture as just a point, the origin.

The definition of multiplication by a scalar is that the length of the arrow is changed (or even reversed) by the factor given by the scalar. Axioms 7 and 8 are then simply the statement that the graphical interpretation of multiplication of numbers involves adding and multiplying their lengths.

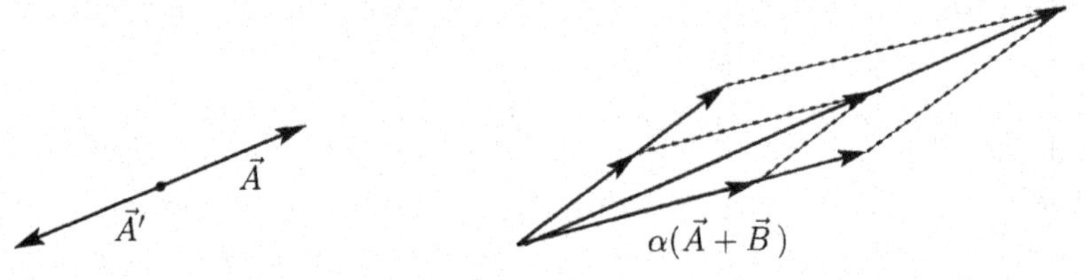

Axioms 5 and 9 appear in this picture.

Finally, axiom 10 is true because you leave the vector alone when you multiply it by one.

This process looks almost *too* easy. Some of the axioms even look as though they are trivial and unnecessary. The last one for example: why do you have to *assume* that multiplication by one leaves the vector alone? For an answer, I will show an example of something that satisfies all of axioms one through nine but *not* the tenth. These processes, addition of vectors and multiplication by scalars, are functions. I could write "$f(\vec{v}_1, \vec{v}_2)$" instead of "$\vec{v}_1 + \vec{v}_2$" and write "$g(\alpha, \vec{v})$" instead of "$\alpha\vec{v}$". The standard notation is just that — a common way to write a vector-valued function of two variables. I can define any function that I want and then see if it satisfies the required properties.

On the set of arrows just above, redefine multiplication by a scalar (the function g of the preceding paragraph) to be the zero vector for all scalars and vectors. That is, $\alpha\vec{v} = \vec{O}$ for all α and $\vec{v}$. Look back and you see that this definition satisfies all the assumptions 1–9 but not 10. For example, 9: $\alpha(\vec{v}_1 + \vec{v}_2) = \alpha\vec{v}_1 + \alpha\vec{v}_2$ because both sides of the equation are the zero vector. This observation proves that the tenth axiom is independent of the others. If you could derive the tenth axiom from the first nine, then this example couldn't exist. This construction is of course not a vector space.

Function Spaces

Is example 2 a vector space? How can a function be a vector? This comes down to your understanding of the word "function." Is $f(x)$ a function or is $f(x)$ a number? Answer: It's a number. This is a confusion caused by the conventional notation for functions. We routinely call $f(x)$ a function, but it is really the result of feeding the particular value, x, to the function f in order to get the number $f(x)$. This confusion in notation is so ingrained that it's hard to change, though in more sophisticated mathematics books it *is* changed.

In a better notation, the symbol f is the function, expressing the relation between all the possible inputs and their corresponding outputs. Then $f(1)$, or $f(\pi)$, or $f(x)$ are the results of feeding f the particular inputs, and the results are (at least for example 2) real numbers. Think of the function f as the whole graph relating input to output; the pair $\big(x, f(x)\big)$ is then just one point on the graph. Adding two functions is adding their graphs. For a precise, set theoretic definition of the word function, see section 12.1. Reread the statement of example 2 in light of these comments.

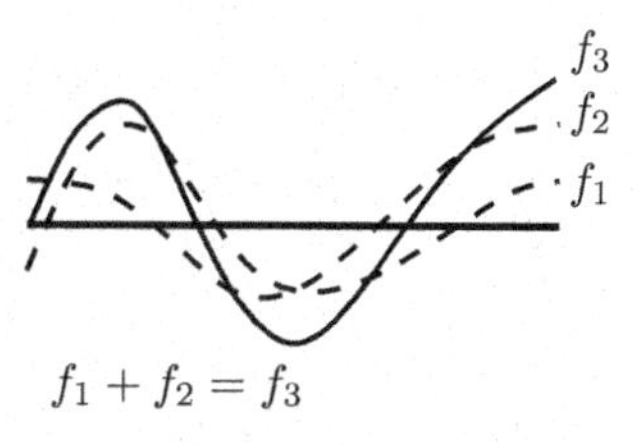

$f_1 + f_2 = f_3$

Special Function Space

Go through another of the examples of vector spaces written above. Number 6, the square-integrable real-valued functions on the interval $a \le x \le b$. The single difficulty here is the first axiom: Is the sum of two square-integrable functions itself square-integrable? The other nine axioms are yours to check.

Suppose that

$$\int_a^b f(x)^2\,dx < \infty \qquad \text{and} \qquad \int_a^b g(x)^2\,dx < \infty.$$

simply note the combination

$$\big(f(x) + g(x)\big)^2 + \big(f(x) - g(x)\big)^2 = 2f(x)^2 + 2g(x)^2$$

The integral of the right-hand side is by assumption finite, so the same must hold for the left side. This says that the sum (and difference) of two square-integrable functions is square-integrable. For this example then, it isn't very difficult to show that it satisfies the axioms for a vector space, but it requires more than just a glance.

There are a few properties of vector spaces that seem to be missing. There is the somewhat odd notation $\vec{v}\,'$ for the additive inverse in axiom 5. Isn't that just $-\vec{v}$? Isn't the zero vector simply the number zero times a vector? Yes in both cases, but these are theorems that follow easily from the ten axioms listed. See problem 6.20. I'll do part (a) of that exercise as an example here:

Theorem: The vector $\vec{O}$ is unique.

Proof: Assume it is not, then there are two such vectors, $\vec{O}_1$ and $\vec{O}_2$.
By [4], $\vec{O}_1 + \vec{O}_2 = \vec{O}_1$ ($\vec{O}_2$ is a zero vector)
By [6], the left side is $\vec{O}_2 + \vec{O}_1$
By [4], this is $\vec{O}_2$ ($\vec{O}_1$ is a zero vector)
Put these together and $\vec{O}_1 = \vec{O}_2$.

Theorem: If a subset of a vector space is closed under addition and multiplication by scalars, then it is itself a vector space. This means that if you add two elements of this subset to each other they remain in the subset and multiplying any element of the subset

by a scalar leaves it in the subset. It is a "subspace."
Proof: The assumption of the theorem is that axioms 1 and 2 are satisfied as regards the subset. That axioms 3 through 10 hold follows because the elements of the subset inherit their properties from the larger vector space of which they are a part. Is this all there is to it? Not quite. Axioms 4 and 5 take a little more thought, and need the results of the problem 6.20, parts (b) and (d).

6.4 Linear Independence

A set of non-zero vectors is linearly dependent if one element of the set can be written as a linear combination of the others. The set is linearly independent if this cannot be done.

Bases, Dimension, Components

A basis for a vector space is a linearly independent set of vectors such that any vector in the space can be written as a linear combination of elements of this set. The *dimension* of the space is the number of elements in this basis.

If you take the usual vector space of arrows that start from the origin and lie in a plane, the common basis is denoted $\hat{\imath}$, $\hat{\jmath}$. If I propose a basis consisting of

$$\hat{\imath}, \qquad -\tfrac{1}{2}\hat{\imath} + \tfrac{\sqrt{3}}{2}\hat{\jmath}, \qquad -\tfrac{1}{2}\hat{\imath} - \tfrac{\sqrt{3}}{2}\hat{\jmath}$$

these will certainly span the space. Every vector can be written as a linear combination of them. They are however, redundant; the sum of all three is zero, so they aren't linearly independent and aren't a basis. If you use them as if they are a basis, the components of a given vector won't be unique. Maybe that's o.k. and you want to do it, but either be careful or look up the mathematical subject called "frames."

Beginning with the most elementary problems in physics and mathematics, it is clear that the choice of an appropriate coordinate system can provide great computational advantages. In dealing with the usual two and three dimensional vectors it is useful to express an arbitrary vector as a sum of unit vectors. Similarly, the use of Fourier series for the analysis of functions is a very powerful tool in analysis. These two ideas are essentially the same thing when you look at them as aspects of vector spaces.

If the elements of the basis are denoted $\vec{e}_i$, and a vector $\vec{a}$ is

$$\vec{a} = \sum_i a_i \vec{e}_i,$$

the numbers $\{a_i\}$ are called the *components* of $\vec{a}$ in the specified basis. Note that you don't have to talk about orthogonality or unit vectors or any other properties of the basis vectors save that they span the space and they're independent.

Example 1 is the prototype for the subject, and the basis usually chosen is the one designated $\hat{x}$, $\hat{y}$, (and $\hat{z}$ for three dimensions). Another notation for this is $\hat{\imath}$, $\hat{\jmath}$, $\hat{k}$ — I'll use $\hat{x}$-$\hat{y}$. In any case, the two (or three) arrows are at right angles to each other.

In example 5, the simplest choice of basis is

$$\begin{aligned} \vec{e}_1 &= (\,1 \quad 0 \quad 0 \quad \ldots \quad 0\,) \\ \vec{e}_2 &= (\,0 \quad 1 \quad 0 \quad \ldots \quad 0\,) \\ &\;\;\vdots \\ \vec{e}_n &= (\,0 \quad 0 \quad 0 \quad \ldots \quad 1\,) \end{aligned} \tag{6.1}$$

In example 6, if the domain of the functions is from $-\infty$ to $+\infty$, a possible basis is the set of functions

$$\psi_n(x) = x^n e^{-x^2/2}.$$

The major distinction between this and the previous cases is that the dimension here is infinite. There is a basis vector corresponding to each non-negative integer. It's not obvious that this is a basis, but it's true.

If two vectors are equal to each other and you express them in the same basis, the corresponding components must be equal.

$$\sum_i a_i \vec{e}_i = \sum_i b_i \vec{e}_i \qquad \Longrightarrow \qquad a_i = b_i \quad \text{for all } i \tag{6.2}$$

Suppose you have the relation between two functions of time

$$A - B\omega + \gamma t = \beta t \tag{6.3}$$

that is, that the two *functions* are the same, think of this in terms of vectors: On the vector space of polynomials in t a basis is

$$\vec{e}_0 = 1, \qquad \vec{e}_1 = t, \qquad \vec{e}_2 = t^2, \ etc.$$

Translate the preceding equation into this notation.

$$(A - B\omega)\vec{e}_0 + \gamma\vec{e}_1 = \beta\vec{e}_1 \tag{6.4}$$

For this to be valid the corresponding components must match:

$$A - B\omega = 0, \qquad \text{and} \qquad \gamma = \beta$$

Differential Equations

When you encounter differential equations such as

$$m\frac{d^2x}{dt^2} + b\frac{dx}{dt} + kx = 0, \qquad \text{or} \qquad \gamma\frac{d^3x}{dt^3} + kt^2\frac{dx}{dt} + \alpha e^{-\beta t}x = 0, \tag{6.5}$$

the sets of solutions to each of these equations form vector spaces. All you have to do is to check the axioms, and because of the theorem in section 6.3 you don't even have to do all of that. The solutions are functions, and as such they are elements of the vector space of example 2. All you need to do now is to verify that the sum of two solutions is a solution and that a constant times a solution is a solution. That's what the phrase "linear, homogeneous" means.

Another common differential equation is

$$\frac{d^2\theta}{dt^2} + \frac{g}{\ell}\sin\theta = 0$$

This describes the motion of an undamped pendulum, and the set of its solutions do *not* form a vector space. The sum of two solutions is not a solution.

The first of Eqs. (6.5) has two independent solutions,

$$x_1(t) = e^{-\gamma t}\cos\omega' t, \qquad \text{and} \qquad x_2(t) = e^{-\gamma t}\sin\omega' t \tag{6.6}$$

where $\gamma = -b/2m$ and $\omega' = \sqrt{\frac{k}{m} - \frac{b^2}{4m^2}}$. This is from Eq. (4.8). Any solution of this differential equation is a linear combination of these functions, and I can restate that fact in the language of this chapter by saying that x_1 and x_2 form a basis for the vector space of solutions of the damped oscillator equation. It has dimension two.

The second equation of the pair (6.5) is a third order differential equation, and as such you will need to specify three conditions to determine the solution and to determine all the three arbitrary constants. In other words, the dimension of the solution space of this equation is three.

In chapter 4 on the subject of differential equations, one of the topics was simultaneous differential equations, coupled oscillations. The simultaneous differential equations, Eq. (4.45), are

$$m_1\frac{d^2x_1}{dt^2} = -k_1x_1 - k_3(x_1 - x_2), \qquad \text{and} \qquad m_2\frac{d^2x_2}{dt^2} = -k_2x_2 - k_3(x_2 - x_1)$$

and have solutions that are pairs of functions. In the development of section 4.10 (at least for the equal mass, symmetric case), I found four pairs of functions that satisfied the equations. Now translate that into the language of this chapter, using the notation of column matrices for the functions. The solution is the vector

$$\begin{pmatrix} x_1(t) \\ x_2(t) \end{pmatrix}$$

and the four basis vectors for this four-dimensional vector space are

$$\vec e_1 = \begin{pmatrix} e^{i\omega_1 t} \\ e^{i\omega_1 t} \end{pmatrix}, \qquad \vec e_2 = \begin{pmatrix} e^{-i\omega_1 t} \\ e^{-i\omega_1 t} \end{pmatrix}, \qquad \vec e_3 = \begin{pmatrix} e^{i\omega_2 t} \\ -e^{i\omega_2 t} \end{pmatrix}, \qquad \vec e_4 = \begin{pmatrix} e^{-i\omega_2 t} \\ -e^{-i\omega_2 t} \end{pmatrix}$$

Any solution of the differential equations is a linear combination of these. In the original notation, you have Eq. (4.52). In the current notation you have

$$\begin{pmatrix} x_1 \\ x_2 \end{pmatrix} = A_1\,\vec e_1 + A_2\,\vec e_2 + A_3\,\vec e_3 + A_4\,\vec e_4$$

6.5 Norms

The "norm" or length of a vector is a particularly important type of function that can be defined on a vector space. It is a function, usually denoted by $\|\ \ \|$, and that satisfies

1. $\|\vec v\| \geq 0$; $\|\vec v\| = 0$ if and only if $\vec v = \vec O$
2. $\|\alpha\vec v\| = |\alpha|\,\|\vec v\|$
3. $\|\vec v_1 + \vec v_2\| \leq \|\vec v_1\| + \|\vec v_2\|$ (the triangle inequality) The distance between two vectors $\vec v_1$ and $\vec v_2$ is taken to be $\|\vec v_1 - \vec v_2\|$.

6.6 Scalar Product

The scalar product of two vectors is a scalar valued function of *two* vector variables. It could be denoted as $f(\vec{u},\vec{v})$, but a standard notation for it is $\langle\vec{u},\vec{v}\rangle$. It must satisfy the requirements

1. $\langle\vec{w},(\vec{u}+\vec{v})\rangle = \langle\vec{w},\vec{u}\rangle + \langle\vec{w},\vec{v}\rangle$
2. $\langle\vec{w},\alpha\vec{v}\rangle = \alpha\langle\vec{w},\vec{v}\rangle$
3. $\langle\vec{u},\vec{v}\rangle^* = \langle\vec{v},\vec{u}\rangle$
4. $\langle\vec{v},\vec{v}\rangle \geq 0$; and $\langle\vec{v},\vec{v}\rangle = 0$ if and only if $\vec{v} = \vec{O}$

When a scalar product exists on a space, a norm naturally does too:

$$\|\vec{v}\| = \sqrt{\langle\vec{v},\vec{v}\rangle}. \tag{6.7}$$

That this *is* a norm will follow from the Cauchy-Schwartz inequality. Not all norms come from scalar products.

Examples

Use the examples of section 6.3 to see what these are. The numbers here refer to the numbers of that section.

1 A norm is the usual picture of the length of the line segment. A scalar product is the usual product of lengths times the cosine of the angle between the vectors.

$$\langle\vec{u},\vec{v}\rangle = \vec{u}\cdot\vec{v} = u\,v\cos\vartheta. \tag{6.8}$$

4 A norm can be taken as the least upper bound of the magnitude of the function. This is distinguished from the "maximum" in that the function may not actually achieve a maximum value. Since it is bounded however, there is an upper bound (many in fact) and we take the smallest of these as the norm. On $-\infty < x < +\infty$, the function $|\tan^{-1}x|$ has $\pi/2$ for its least upper bound, though it never equals that number.

5 A possible scalar product is

$$\big\langle(a_1,\ldots,a_n),(b_1,\ldots,b_n)\big\rangle = \sum_{k=1}^{n} a_k^*\,b_k. \tag{6.9}$$

There are other scalar products for the same vector space, for example

$$\big\langle(a_1,\ldots,a_n),(b_1,\ldots,b_n)\big\rangle = \sum_{k=1}^{n} k\,a_k^*\,b_k \tag{6.10}$$

In fact any other positive function can appear as the coefficient in the sum and it still defines a valid scalar product. It's surprising how often something like this happens in real situations. In studying normal modes of oscillation the masses of different particles will appear as coefficients in a natural scalar product.

I used complex conjugation on the first factor here, but example 5 referred to real numbers only. The reason for leaving the conjugation in place is that when you jump to example 14 you want to allow for complex numbers, and it's harmless to put it in for the real case because in that instance it leaves the number alone.

For a norm, there are many possibilities:

$$\begin{aligned} &(1)\ \|(a_1,\ldots,a_n)\| = \sqrt{\sum\nolimits_{k=1}^{n} |a_k|^2} \\ &(2)\ \|(a_1,\ldots,a_n)\| = \sum\nolimits_{k=1}^{n} |a_k| \\ &(3)\ \|(a_1,\ldots,a_n)\| = \max\nolimits_{k=1}^{n} |a_k| \\ &(4)\ \|(a_1,\ldots,a_n)\| = \max\nolimits_{k=1}^{n} k|a_k|. \end{aligned} \tag{6.11}$$

The United States Postal Service prefers a variation on the second of these norms, see problem 8.45.

6 A possible choice for a scalar product is

$$\langle f, g\rangle = \int_a^b dx\, f(x)^*\, g(x). \tag{6.12}$$

9 Scalar products and norms used here are just like those used for example 5. The difference is that the sums go from 1 to infinity. The problem of convergence doesn't occur because there are only a finite number of non-zero terms.

10 Take the norm to be

$$\|(a_1, a_2, \ldots)\| = \sqrt{\sum\nolimits_{k=1}^{\infty} |a_k|^2}, \tag{6.13}$$

and this by assumption will converge. The natural scalar product is like that of example 5, but with the sum going out to infinity. It requires a small amount of proof to show that this will converge. See problem 6.19.

11 A norm is $\|\vec v\,\| = \sum_{i=1}^{\infty} |a_i|$. There is no scalar product that will produce this norm, a fact that you can prove by using the results of problem 6.13.

13 A natural norm is

$$\|f\| = \left[\int_a^b dx\, |f(x)|^p\right]^{1/p}. \tag{6.14}$$

To demonstrate that this *is* a norm requires the use of some special inequalities found in advanced calculus books.

15 If A and B are two matrices, a scalar product is $\langle A, B\rangle = \mathrm{Tr}(A^\dagger B)$, where $\dagger$ is the transpose complex conjugate of the matrix and Tr means the trace, the sum of the diagonal elements. Several possible norms can occur. One is $\|A\| = \sqrt{\mathrm{Tr}(A^\dagger A)}$. Another is the maximum value of $\|A\vec u\|$, where $\vec u$ is a unit vector and the norm of $\vec u$ is taken to be $\left[|u_1|^2 + \cdots + |u_n|^2\right]^{1/2}$.

19 A valid definition of a norm for the motions of a drumhead is its total energy, kinetic plus potential. How do you describe this mathematically? It's something like

$$\int dx\, dy\, \frac{1}{2}\left[\left(\frac{\partial f}{\partial t}\right)^2 + (\nabla f)^2\right]$$

I've left out all the necessary constants, such as mass density of the drumhead and tension in the drumhead. You can perhaps use dimensional analysis to surmise where they go.

There is an example in criminal law in which the distinctions between some of these norms have very practical consequences. If you're caught selling drugs in New York there is a longer sentence if your sale is within 1000 feet of a school. If you are an attorney defending someone accused of this crime, which of the norms in Eq. (6.11) would you argue for? The legislators who wrote this law didn't know linear algebra, so they didn't specify which norm they intended. The prosecuting attorney argued for norm #1, "as the crow flies," but the defense argued that "crows don't sell drugs" and humans move along city streets, so norm #2 is more appropriate.

The New York Court of Appeals decided that the Pythagorean norm (#1) is the appropriate one and they rejected the use of the pedestrian norm that the defendant advocated (#2).
www.courts.state.ny.us/ctapps/decisions/nov05/162opn05.pdf

6.7 Bases and Scalar Products

When there is a scalar product, a most useful type of basis is the orthonormal one, satisfying

$$\langle \vec{v}_i, \vec{v}_j \rangle = \delta_{ij} = \begin{cases} 1 & \text{if } i = j \\ 0 & \text{if } i \neq j \end{cases} \tag{6.15}$$

The notation δ_{ij} represents the very useful Kronecker delta symbol.

In the example of Eq. (6.1) the basis vectors are orthonormal with respect to the scalar product in Eq. (6.9). It is orthogonal with respect to the other scalar product mentioned there, but it is not in that case normalized to magnitude one.

To see how the choice of even an orthonormal basis depends on the scalar product, try a different scalar product on this space. Take the special case of two dimensions. The vectors are now pairs of numbers. Think of the vectors as 2×1 matrix column and use the 2×2 matrix

$$\begin{pmatrix} 2 & 1 \\ 1 & 2 \end{pmatrix}$$

Take the scalar product of two vectors to be

$$\big\langle (a_1, a_2), (b_1, b_2) \big\rangle = \begin{pmatrix} a_1^* & a_2^* \end{pmatrix} \begin{pmatrix} 2 & 1 \\ 1 & 2 \end{pmatrix} \begin{pmatrix} b_1 \\ b_2 \end{pmatrix} = 2a_1^* b_1 + a_1^* b_2 + a_2^* b_1 + 2a_2^* b_2 \tag{6.16}$$

To show that this satisfies all the defined requirements for a scalar product takes a small amount of labor. The vectors that you may expect to be orthogonal, (1 0) and (0 1), are not.

In example 6, if we let the domain of the functions be $-L < x < +L$ and the scalar product is as in Eq. (6.12), then the set of trigonometric functions can be used as a basis.

$$\sin\frac{n\pi x}{L} \quad \text{and} \quad \cos\frac{m\pi x}{L}$$
$$n = 1, 2, 3, \ldots \quad \text{and} \quad m = 0, 1, 2, 3, \ldots.$$

That a function can be written as a series

$$f(x)=\sum_{1}^{\infty}a_n\sin\frac{n\pi x}{L}+\sum_{0}^{\infty}b_m\cos\frac{m\pi x}{L} \tag{6.17}$$

on the domain $-L<x<+L$ is just an example of Fourier series, and the components of f in this basis are Fourier coefficients $a_1,\ldots,b_0,\ldots$. An equally valid and more succinctly stated basis is

$$e^{n\pi ix/L},\qquad n=0,\ \pm1,\ \pm2,\ \ldots$$

Chapter 5 on Fourier series shows many other choices of bases, all orthogonal, but not necessarily normalized.

To emphasize the relationship between Fourier series and the ideas of vector spaces, this picture represents three out of the infinite number of basis vectors and part of a function that uses these vectors to form a Fourier series.

$$f(x)=\frac{1}{2}\sin\frac{\pi x}{L}+\frac{2}{3}\sin\frac{2\pi x}{L}+\frac{1}{3}\sin\frac{3\pi x}{L}+\cdots$$

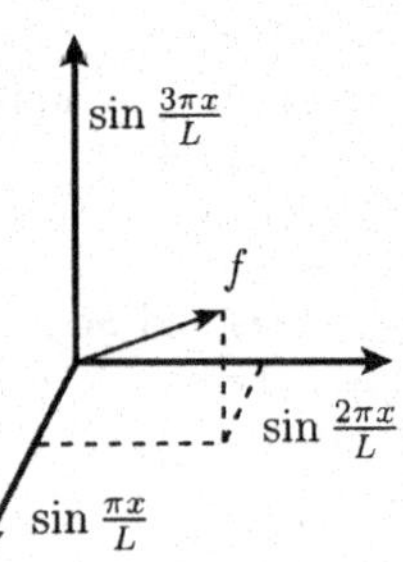

The orthogonality of the sines becomes the geometric term "perpendicular," and if you look at section 8.11, you will see that the subject of least square fitting of data to a sum of sine functions leads you right back to Fourier series, and to the same picture as here.

6.8 Gram-Schmidt Orthogonalization

From a basis that is not orthonormal, it is possible to construct one that is. This device is called the Gram-Schmidt procedure. Suppose that a basis is known (finite or infinite), $\vec v_1,\ \vec v_2,\ldots$

Step 1: Normalize $\vec v_1$: $\quad \vec e_1=\vec v_1\big/\sqrt{\langle\vec v_1,\vec v_1\rangle}$.

Step 2: Construct a linear combination of $\vec v_1$ and $\vec v_2$ that is orthogonal to $\vec v_1$:
Let $\vec e_{20}=\vec v_2-\vec e_1\langle\vec e_1,\vec v_2\rangle$ and then normalize it.

$$\vec e_2=\vec e_{20}\big/\langle\vec e_{20},\vec e_{20}\rangle^{1/2}. \tag{6.18}$$

Step 3: Let $\vec e_{30}=\vec v_3-\vec e_1\langle\vec e_1,\vec v_3\rangle-\vec e_2\langle\vec e_2,\vec v_3\rangle$ *etc.* repeating step 2.
What does this look like? See problem 6.3.

6.9 Cauchy-Schwartz inequality

For common three-dimensional vector geometry, it is obvious that for any real angle, $\cos^2\theta\le1$. In terms of a dot product, this is $|\vec A\cdot\vec B|\le AB$. This can be generalized to any scalar product on any vector space:

$$\big|\langle\vec u,\vec v\rangle\big|\le\|\vec u\|\,\|\vec v\|. \tag{6.19}$$

The proof starts from a simple but not-so-obvious point. The scalar product of a vector with itself is by definition positive, so for any two vectors $\vec u$ and $\vec v$ you have the inequality

$$\langle\vec u-\lambda\vec v,\vec u-\lambda\vec v\rangle\ge0. \tag{6.20}$$

where λ is any complex number. This expands to

$$\langle \vec{u}, \vec{u} \rangle + |\lambda|^2 \langle \vec{v}, \vec{v} \rangle - \lambda \langle \vec{u}, \vec{v} \rangle - \lambda^* \langle \vec{v}, \vec{u} \rangle \geq 0. \tag{6.21}$$

How much bigger than zero the left side is will depend on the parameter λ. To find the smallest value that the left side can have you simply differentiate. Let $\lambda = x + iy$ and differentiate with respect to x and y, setting the results to zero. This gives (see problem 6.5)

$$\lambda = \langle \vec{v}, \vec{u} \rangle / \langle \vec{v}, \vec{v} \rangle. \tag{6.22}$$

Substitute this value into the above inequality (6.21)

$$\langle \vec{u}, \vec{u} \rangle + \frac{\left|\langle \vec{u}, \vec{v} \rangle\right|^2}{\langle \vec{v}, \vec{v} \rangle} - \frac{\left|\langle \vec{u}, \vec{v} \rangle\right|^2}{\langle \vec{v}, \vec{v} \rangle} - \frac{\left|\langle \vec{u}, \vec{v} \rangle\right|^2}{\langle \vec{v}, \vec{v} \rangle} \geq 0. \tag{6.23}$$

This becomes

$$\left|\langle \vec{u}, \vec{v} \rangle\right|^2 \leq \langle \vec{u}, \vec{u} \rangle \langle \vec{v}, \vec{v} \rangle \tag{6.24}$$

This isn't quite the result needed, because Eq. (6.19) is written differently. It refers to a norm and I haven't established that the square root of $\langle \vec{v}, \vec{v} \rangle$ *is* a norm. When I do, then the square root of this is the desired inequality (6.19).

For a couple of examples of this inequality, take specific scalar products. First the common directed line segments:

$$\langle \vec{u}, \vec{v} \rangle = \vec{u} \cdot \vec{v} = uv \cos\theta, \qquad \text{so} \qquad |uv \cos\theta|^2 \leq |u|^2 |v|^2$$

$$\left| \int_a^b dx\, f(x)^* g(x) \right|^2 \leq \left[\int_a^b dx\, |f(x)|^2 \right] \left[\int_a^b dx\, |g(x)|^2 \right]$$

The first of these is familiar, but the second is not, though when you look at it from the general vector space viewpoint they are essentially the same.

Norm from a Scalar Product

The equation (6.7), $\|\vec{v}\| = \sqrt{\langle \vec{v}, \vec{v} \rangle}$, defines a norm. Properties one and two for a norm are simple to check. (Do so.) The third requirement, the triangle inequality, takes a bit of work and uses the inequality Eq. (6.24).

$$\begin{aligned}
\langle \vec{v}_1 + \vec{v}_2, \vec{v}_1 + \vec{v}_2 \rangle &= \langle \vec{v}_1, \vec{v}_1 \rangle + \langle \vec{v}_2, \vec{v}_2 \rangle + \langle \vec{v}_1, \vec{v}_2 \rangle + \langle \vec{v}_2, \vec{v}_1 \rangle \\
&\leq \langle \vec{v}_1, \vec{v}_1 \rangle + \langle \vec{v}_2, \vec{v}_2 \rangle + \left|\langle \vec{v}_1, \vec{v}_2 \rangle\right| + \left|\langle \vec{v}_2, \vec{v}_1 \rangle\right| \\
&= \langle \vec{v}_1, \vec{v}_1 \rangle + \langle \vec{v}_2, \vec{v}_2 \rangle + 2\left|\langle \vec{v}_1, \vec{v}_2 \rangle\right| \\
&\leq \langle \vec{v}_1, \vec{v}_1 \rangle + \langle \vec{v}_2, \vec{v}_2 \rangle + 2\sqrt{\langle \vec{v}_1, \vec{v}_1 \rangle \langle \vec{v}_2, \vec{v}_2 \rangle} \\
&= \left(\sqrt{\langle \vec{v}_1, \vec{v}_1 \rangle} + \sqrt{\langle \vec{v}_2, \vec{v}_2 \rangle} \right)^2
\end{aligned}$$

The first inequality is a property of complex numbers. The second one is Eq. (6.24). The square root of the last line is the triangle inequality, thereby justifying the use of $\sqrt{\langle \vec{v}, \vec{v} \rangle}$ as the norm of $\vec{v}$ and in the process validating Eq. (6.19).

$$\|\vec{v}_1 + \vec{v}_2\| = \sqrt{\langle \vec{v}_1 + \vec{v}_2, \vec{v}_1 + \vec{v}_2 \rangle} \leq \sqrt{\langle \vec{v}_1, \vec{v}_1 \rangle} + \sqrt{\langle \vec{v}_2, \vec{v}_2 \rangle} = \|\vec{v}_1\| + \|\vec{v}_2\| \tag{6.25}$$

6.10 Infinite Dimensions

Is there any real difference between the cases where the dimension of the vector space is finite and the cases where it's infinite? Yes. Most of the concepts are the same, but you have to watch out for the question of convergence. If the dimension is finite, then when you write a vector in terms of a basis $\vec{v} = \sum a_k \vec{e}_k$, the sum is finite and you don't even have to think about whether it converges or not. In the infinite-dimensional case you do.

It is even possible to have such a series converge, but not to converge to a vector. If that sounds implausible, let me take an example from a slightly different context, ordinary rational numbers. These are the number m/n where m and n are integers ($n \neq 0$). Consider the sequence

$$1, \quad 14/10, \quad 141/100, \quad 1414/1000, \quad 14142/10000, \quad 141421/100000, \;\ldots$$

These are quotients of integers, but the limit is $\sqrt{2}$ and that's *not** a rational number. Within the confines of rational numbers, this sequence doesn't converge. You have to expand the context to get a limit. That context is the real numbers. The same thing happens with vectors when the dimension of the space is infinite — in order to find a limit you sometimes have to expand the context and to expand what you're willing to call a vector.

Look at example 9 from section 6.3. These are sets of numbers $(a_1, a_2, \ldots)$ with just a finite number of non-zero entries. If you take a sequence of such vectors

$$(1,0,0,\ldots), \quad (1,1,0,0,\ldots), \quad (1,1,1,0,0,\ldots), \ldots$$

Each has a finite number of non-zero elements but the limit of the sequence does not. It isn't a vector in the original vector space. Can I expand to a larger vector space? Yes, just use example 8, allowing any number of non-zero elements.

For a more useful example of the same kind, start with the same space and take the sequence

$$(1,0,\ldots), \quad (1,\, 1/2, 0,\ldots), \quad (1,\, 1/2,\, 1/3, 0,\ldots), \ldots$$

Again the limit of such a sequence doesn't have a finite number of entries, but example 10 will hold such a limit, because $\sum_1^\infty |a_k|^2 < \infty$.

How do you know when you have a vector space without holes in it? That is, one in which these problems with limits don't occur? The answer lies in the idea of a Cauchy sequence. I'll start again with the rational numbers to demonstrate the idea. The sequence of numbers that led to the square root of two has the property that even though the elements of the sequence weren't approaching a rational number, the elements were getting close *to each other.* Let $\{r_n\}$, $n = 1, 2, \ldots$ be a sequence of rational numbers.

$$\lim_{n,m\to\infty} \bigl|r_n - r_m\bigr| = 0 \qquad \text{means}$$

$$\text{For any } \epsilon > 0 \text{ there is an } N \text{ so that if both } n \text{ and } m \text{ are } > N \text{ then } \bigl|r_n - r_m\bigr| < \epsilon. \tag{6.26}$$

---

* Proof: If it is, then express it in simplest form as $m/n = \sqrt{2} \Rightarrow m^2 = 2n^2$ where m and n have no common factor. This equation implies that m must be even: $m = 2m_1$. Substitute this value, giving $2m_1^2 = n^2$. That in turn implies that n is even, and this contradicts the assumption that the original quotient was expressed without common factors.

This property defines the sequence r_n as a Cauchy sequence. A sequence of rational numbers converges to a real number if and only if it is a Cauchy sequence; this is a theorem found in many advanced calculus texts. Still other texts will take a different approach and use the concept of a Cauchy sequence to construct the *definition* of the real numbers.

The extension of this idea to infinite dimensional vector spaces requires simply that you replace the absolute value by a norm, so that a Cauchy sequence is defined by $\lim_{n,m}\|\vec{v}_n - \vec{v}_m\| = 0$. A "complete" vector space is one in which every Cauchy sequence converges. A vector space that has a scalar product and that is also complete using the norm that this scalar product defines is called a Hilbert Space.

I don't want to imply that the differences between finite and infinite dimensional vector spaces is just a technical matter of convergence. In infinite dimensions there is far more room to move around, and the possible structures that occur are vastly more involved than in the finite dimensional case. The subject of quantum mechanics has Hilbert Spaces at the foundation of its whole structure.

Exercises

1 Determine if these are vector spaces with the usual rules for addition and multiplication by scalars. If not, which axiom(s) do they violate?
(a) Quadratic polynomials of the form $ax^2 + bx$
(b) Quadratic polynomials of the form $ax^2 + bx + 1$
(c) Quadratic polynomials $ax^2 + bx + c$ with $a + b + c = 0$
(d) Quadratic polynomials $ax^2 + bx + c$ with $a + b + c = 1$

2 What is the dimension of the vector space of (up to) 5th degree polynomials having a double root at $x = 1$?

3 Starting from three dimensional vectors (the common directed line segments) and a single fixed vector $\vec{B}$, is the set of all vectors $\vec{v}$ with $\vec{v}\cdot\vec{B} = 0$ a vector space? If so, what is it's dimension?
Is the set of all vectors $\vec{v}$ with $\vec{v}\times\vec{B} = 0$ a vector space? If so, what is it's dimension?

4 The set of all odd polynomials with the expected rules for addition and multiplication by scalars. Is it a vector space?

5 The set of all polynomials where the function "addition" is defined to be $f_3 = f_2 + f_1$ if the number $f_3(x) = f_1(-x) + f_2(-x)$. Is it a vector space?

6 Same as the preceding, but for (a) even polynomials, (b) odd polynomials

7 The set of directed line segments in the plane with the new rule for addition: add the vectors according to the usual rule then rotate the result by 10° counterclockwise. Which vector space axioms are obeyed and which not?

Problems

6.1 Fourier series represents a choice of basis for functions on an interval. For suitably smooth functions on the interval 0 to L, one basis is

$$\vec{e}_n = \sqrt{\frac{2}{L}} \sin\frac{n\pi x}{L}. \tag{6.27}$$

Use the scalar product $\langle f, g\rangle = \int_0^L f^*(x)g(x)\,dx$ and show that this is an orthogonal basis normalized to 1, *i.e.* it is orthonormal.

6.2 A function $F(x) = x(L - x)$ between zero and L. Use the basis of the preceding problem to write this vector in terms of its components:

$$F = \sum_1^\infty \alpha_n \vec{e}_n. \tag{6.28}$$

If you take the result of using this basis and write the resulting function outside the interval $0 < x < L$, graph the result.

6.3 For two dimensional real vectors with the usual parallelogram addition, interpret in pictures the first two steps of the Gram-Schmidt process, section 6.8.

6.4 For two dimensional real vectors with the usual parallelogram addition, *interpret* the vectors $\vec{u}$ and $\vec{v}$ and the parameter λ used in the proof of the Cauchy-Schwartz inequality in section 6.9. Start by considering the set of points in the plane formed by $\{\vec{u} - \lambda\vec{v}\}$ as λ ranges over the set of reals. In particular, when λ was picked to minimize the left side of the inequality (6.21), what do the vectors look like? Go through the proof and interpret it in the context of these pictures. State the idea of the whole proof geometrically.
Note: I don't mean just copy the proof. Put the geometric interpretation into words.

6.5 Start from Eq. (6.21) and show that the minimum value of the function of $\lambda = x + iy$ is given by the value stated there. Note: this derivation applies to complex vector spaces and scalar products, not just real ones. Is this a *minimum*?

6.6 For the vectors in three dimensions,

$$\vec{v}_1 = \hat{x} + \hat{y}, \qquad \vec{v}_2 = \hat{y} + \hat{z}, \qquad \vec{v}_3 = \hat{z} + \hat{x}$$

use the Gram-Schmidt procedure to construct an orthonormal basis starting from $\vec{v}_1$.
Ans: $\vec{e}_3 = (\hat{x} - \hat{y} + \hat{z})/\sqrt{3}$

6.7 For the vector space of polynomials in x, use the scalar product defined as

$$\langle f, g\rangle = \int_{-1}^{1} dx\, f(x)^* g(x)$$

(Everything is real here, so the complex conjugation won't matter.) Start from the vectors

$$\vec{v}_0 = 1, \qquad \vec{v}_1 = x, \qquad \vec{v}_2 = x^2, \qquad \vec{v}_3 = x^3$$

and use the Gram-Schmidt procedure to construct an orthonormal basis starting from $\vec{v}_0$. Compare these results to the results of section 4.11. [These polynomials appear in the study of electric potentials and in the study of angular momentum in quantum mechanics: Legendre polynomials.]

6.8 Repeat the previous problem, but use a different scalar product:

$$\langle f, g \rangle = \int_{-\infty}^{\infty} dx\, e^{-x^2} f(x)^* g(x)$$

[These polynomials appear in the study of the harmonic oscillator in quantum mechanics and in the study of certain waves in the upper atmosphere. With a conventional normalization they are called Hermite polynomials.]

6.9 Consider the set of all polynomials in x having degree $\leq N$. Show that this is a vector space and find its dimension.

6.10 Consider the set of all polynomials in x having degree $\leq N$ and only even powers. Show that this is a vector space and find its dimension. What about odd powers only?

6.11 Which of these are vector spaces?
(**a**) all polynomials of degree 3
(**b**) all polynomials of degree ≤ 3 [Is there a difference between (**a**) and (**b**)?]
(**c**) all functions such that $f(1) = 2f(2)$
(**d**) all functions such that $f(2) = f(1) + 1$
(**e**) all functions satisfying $f(x + 2\pi) = f(x)$
(**f**) all positive functions
(**g**) all polynomials of degree ≤ 4 satisfying $\int_{-1}^{1} dx\, x f(x) = 0$.
(**h**) all polynomials of degree ≤ 4 where the coefficient of x is zero.
[Is there a difference between (**g**) and (**h**)?]

6.12 (**a**) For the common picture of arrows in three dimensions, prove that the subset of vectors $\vec{v}$ that satisfy $\vec{A} \cdot \vec{v} = 0$ for fixed $\vec{A}$ forms a vector space. Sketch it.
(**b**) What if the requirement is that both $\vec{A} \cdot \vec{v} = 0$ and $\vec{B} \cdot \vec{v} = 0$ hold. Describe this and sketch it.

6.13 If a norm is defined in terms of a scalar product, $\|\vec{v}\| = \sqrt{\langle \vec{v}, \vec{v} \rangle}$, it satisfies the "parallelogram identity" (for real scalars),

$$\|\vec{u} + \vec{v}\|^2 + \|\vec{u} - \vec{v}\|^2 = 2\|\vec{u}\|^2 + 2\|\vec{v}\|^2. \tag{6.29}$$

6.14 If a norm satisfies the parallelogram identity, then it comes from a scalar product. Again, assume real scalars. Consider combinations of $\|\vec{u} + \vec{v}\|^2$, $\|\vec{u} - \vec{v}\|^2$ and construct

what ought to be the scalar product. You then have to prove the four properties of the scalar product as stated at the start of section 6.6. Numbers four and three are easy. Number one requires that you keep plugging away, using the parallelogram identity (four times by my count).
Number two is downright tricky; leave it to the end. If you can prove it for integer and rational values of the constant α, consider it a job well done. I used induction at one point in the proof. The final step, extending α to all real values, requires some arguments about limits, and is typically the sort of reasoning you will see in an advanced calculus or mathematical analysis course.

6.15 Modify the example number 2 of section 6.3 so that $f_3 = f_1 + f_2$ means $f_3(x) = f_1(x-a) + f_2(x-b)$ for fixed a and b. Is this still a vector space?

6.16 The scalar product you use depends on the problem you're solving. The fundamental equation (5.15) started from the equation $u'' = \lambda u$ and resulted in the scalar product

$$\langle u_2, u_1 \rangle = \int_a^b dx\, u_2(x)^* u_1(x)$$

Start instead from the equation $u'' = \lambda w(x) u$ and see what identity like that of Eq. (5.15) you come to. Assume w is real. What happens if it isn't? In order to have a legitimate scalar product in the sense of section 6.6, what other requirements must you make about w?

6.17 The equation describing the motion of a string that is oscillating with frequency ω about its stretched equilibrium position is

$$\frac{d}{dx}\left(T(x)\frac{dy}{dx}\right) = -\omega^2 \mu(x) y$$

Here, $y(x)$ is the sideways displacement of the string from zero; $T(x)$ is the tension in the string (not necessarily a constant); $\mu(x)$ is the linear mass density of the string (again, it need not be a constant). The time-dependent motion is really $y(x)\cos(\omega t + \phi)$, but the time dependence does not concern us here. As in the preceding problem, derive the analog of Eq. (5.15) for this equation. For the analog of Eq. (5.16) state the boundary conditions needed on y and deduce the corresponding orthogonality equation. This scalar product has the mass density for a weight.
Ans: $\left[T(x)(y_1' y_2^* - y_1 y_2^{*\prime})\right]_a^b = (\omega_2^{*2} - \omega_1^2)\int_a^b \mu(x) y_2^* y_1\, dx$

6.18 The way to define the sum in example 17 is

$$\sum_x |f(x)|^2 = \lim_{c\to 0}\{\text{the sum of } |f(x)|^2 \text{ for those } x \text{ where } |f(x)|^2 > c > 0\}. \tag{6.30}$$

This makes sense only if for each $c > 0$, $|f(x)|^2$ is greater than c for just a finite number of values of x. Show that the function

$$f(x) = \begin{cases} 1/n & \text{for } x = 1/n \\ 0 & \text{otherwise} \end{cases}$$

is in this vector space, and that the function $f(x) = x$ is not. What is a basis for this space? [Take $0 \le x \le 1$] This is an example of a vector space with non-countable dimension.

6.19 In example 10, it is assumed that $\sum_1^\infty |a_k|^2 < \infty$. Show that this implies that the sum used for the scalar product also converges: $\sum_1^\infty a_k^* b_k$. [Consider the sums $\sum |a_k + ib_k|^2$, $\sum |a_k - ib_k|^2$, $\sum |a_k + b_k|^2$, and $\sum |a_k - b_k|^2$, allowing complex scalars.]

6.20 Prove strictly from the axioms for a vector space the following four theorems. Each step in your proof must *explicitly* follow from one of the vector space axioms or from a property of scalars or from a previously proved theorem.
(**a**) The vector $\vec{O}$ is unique. [Assume that there are two, $\vec{O}_1$ and $\vec{O}_2$. Show that they're equal. First step: use axiom 4.]
(**b**) The number 0 times any vector is the zero vector: $0\vec{v} = \vec{O}$.
(**c**) The vector $\vec{v}\,'$ is unique.
(**d**) $(-1)\vec{v} = \vec{v}\,'$.

6.21 For the vector space of polynomials, are the two functions $\{1 + x^2,\ x + x^3\}$ linearly independent?

6.22 Find the dimension of the space of functions that are linear combinations of $\{1,\ \sin x,\ \cos x,\ \sin^2 x,\ \cos^2 x,\ \sin^4 x,\ \cos^4 x,\ \sin^2 x \cos^2 x\}$

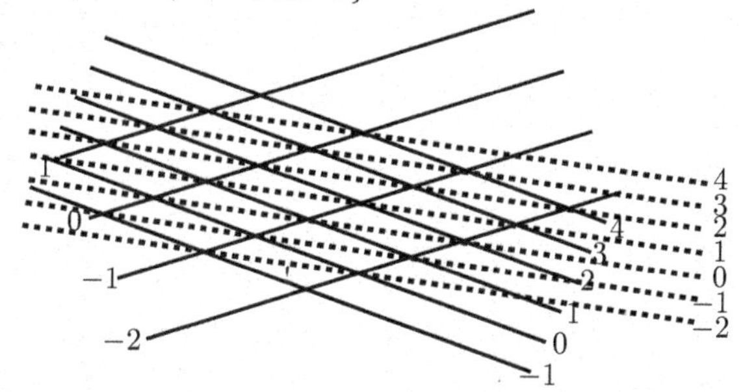

6.23 A model vector space is formed by drawing equidistant parallel lines in a plane and labelling adjacent lines by successive integers from ∞ to $+\infty$. Define multiplication by a (real) scalar so that multiplication of the vector by α means multiply the distance between the lines by $1/\alpha$. Define addition of two vectors by finding the intersections of the lines and connecting opposite corners of the parallelograms to form another set of parallel lines. The resulting lines are labelled as the sum of the two integers from the intersecting lines. (There are two choices here, if one is addition, what is the other?) Show that this construction satisfies all the requirements for a vector space. Just as a directed line segment is a good way to picture velocity, this construction is a good way to picture the gradient of a function. In the vector space of directed line segments, you pin the vectors down so that they all start from a single point. Here, you pin them down so that the lines labeled "zero" all pass through a fixed point. Did I define how to multiply by a negative scalar? If not, then you should. This picture of vectors is developed extensively in the text "Gravitation" by Misner, Wheeler, and Thorne.

6.24 In problem 6.11 (**g**), find a basis for the space. Ans: $1,\ x,\ 3x - 5x^3$.

6.25 What is the dimension of the set of polynomials of degree less than or equal to 10 and with a triple root at $x = 1$?

6.26 Verify that Eq. (6.16) does satisfy the requirements for a scalar product.

6.27 A variation on problem 6.15: $f_3 = f_1 + f_2$ means
(**a**) $f_3(x) = Af_1(x-a)+Bf_2(x-b)$ for fixed a, b, A, B. For what values of these constants is this a vector space?
(**b**) Now what about $f_3(x) = f_1(x^3) + f_2(x^3)$?

6.28 Determine if these are vector spaces:
(1) Pairs of numbers with addition defined as $(x_1, x_2) + (y_1, y_2) = (x_1 + y_2, x_2 + y_1)$ and multiplication by scalars as $c(x_1, x_2) = (cx_1, cx_2)$.
(2) Like example 2 of section 6.3, but restricted to those f such that $f(x) \geq 0$. (real scalars)
(3) Like the preceding line, but define addition as $(f + g)(x) = f(x)g(x)$ and $(cf)(x) = \big(f(x)\big)^c$.

6.29 Do the same calculation as in problem 6.7, but use the scalar product

$$\langle f, g\rangle = \int_0^1 x^2\,dx\, f^*(x)g(x)$$

6.30 Show that the following is a scalar product.

$$\langle f, g\rangle = \int_a^b dx\,\big[f^*(x)g(x) + \lambda f^{*\prime}(x)g'(x)\big]$$

where λ is a constant. What restrictions if any must you place on λ? The name Sobolev is associated with this scalar product.

6.31 (**a**) With the scalar product of problem 6.29, find the angle between the vectors 1 and x. Here the word angle appears in the sense of $\vec{A}\cdot\vec{B} = AB\cos\theta$. (**b**) What is the angle if you use the scalar product of problem 6.7? (**c**) With the first of these scalar products, what combination of 1 and x is orthogonal to 1? Ans: 14.48°

6.32 In the online text linked on the second page of this chapter, you will find that section two of chapter three has enough additional problems to keep you happy.

6.33 Show that the sequence of rational numbers $a_n = \sum_{k=1}^{n} 1/k$ is not a Cauchy sequence. What about $\sum_{k=1}^{n} 1/k^2$?

6.34 In the vector space of polynomials of the form $\alpha x + \beta x^3$, use the scalar product $\langle f, g\rangle = \int_0^1 dx\, f(x)^* g(x)$ and construct an orthogonal basis for this space. Ans: One pair is $x,\ x^3 - \frac{3}{5}x$.

6.35 You can construct the Chebyshev polynomials by starting from the successive powers, x^n, $n = 0, 1, 2, \ldots$ and applying the Gram-Schmidt process. The scalar product in this case is

$$\langle f, g \rangle = \int_{-1}^{1} dx \, \frac{f(x)^* g(x)}{\sqrt{1-x^2}}$$

The conventional normalization for these polynomials is $T_n(1) = 1$, so you should not try to make the norm of the resulting vectors one. Construct the first four of these polynomials, and show that these satisfy $T_n(\cos\theta) = \cos(n\theta)$. These polynomials are used in numerical analysis because they have the property that they oscillate uniformly between -1 and $+1$ on the domain $-1 < x < 1$. Verify that your results for the first four polynomials satisfy the recurrence relation: $T_{n+1}(x) = 2xT_n(x) - T_{n-1}(x)$. Also show that $\cos\big((n+1)\theta\big) = 2\cos\theta\cos\big(n\theta\big) - \cos\big((n-1)\theta\big)$.

6.36 In spherical coordinates (θ,ϕ), the angle θ is measured from the z-axis, and the function $f_1(\theta,\phi) = \cos\theta$ can be written in terms of rectangular coordinates as (section 8.8)

$$f_1(\theta,\phi) = \cos\theta = \frac{z}{r} = \frac{z}{\sqrt{x^2+y^2+z^2}}$$

Pick up the function f_1 and rotate it by 90° counterclockwise about the positive y-axis. Do this rotation in terms of rectangular coordinates, but express the result in terms of spherical: sines and cosines of θ and ϕ. Call it f_2. Draw a picture and figure out where the original and the rotated function are positive and negative and zero.
Now pick up the same f_1 and rotate it by 90° clockwise about the positive x-axis, again finally expressing the result in terms of spherical coordinates. Call it f_3.
If now you take the original f_1 and rotate it about some random axis by some random angle, show that the resulting function f_4 is a linear combination of the three functions f_1, f_2, and f_3. I.e., all these possible rotated functions form a three dimensional vector space. Again, calculations such as these are much easier to demonstrate in rectangular coordinates.

6.37 Take the functions f_1, f_2, and f_3 from the preceding problem and sketch the shape of the functions

$$r\,e^{-r} f_1(\theta,\phi), \qquad r\,e^{-r} f_2(\theta,\phi), \qquad r\,e^{-r} f_3(\theta,\phi)$$

To sketch these, picture them as defining some sort of density in space, ignoring the fact that they are sometimes negative. You can just take the absolute value or the square in order to visualize where they are big or small. Use dark and light shading to picture where the functions are big and small. Start by finding *where* they have the largest and smallest magnitudes. See if you can find similar pictures in an introductory chemistry text. Alternately, check out winter.group.shef.ac.uk/orbitron/

6.38 Use the results of problem 6.17 and apply it to the Legendre equation Eq. (4.55) to demonstrate that the Legendre polynomials obey $\int_{-1}^{1} dx\, P_n(x)P_m(x) = 0$ if $n \neq m$. Note: The function $T(x)$ from problem 6.17 is zero at these endpoints. That does *not* imply that there are no conditions on the functions y_1 and y_2 at those endpoints. The product of $T(x)y_1'y_2$ has to vanish there. Use the result stated just after Eq. (4.59) to show that only the Legendre polynomials and not the more general solutions of Eq. (4.58) work.

6.39 Using the result of the preceding problem that the Legendre polynomials are orthogonal, show that the equation (4.62)(a) follows from Eq. (4.62)(e). Square that equation (e) and integrate $\int_{-1}^{1} dx$. Do the integral on the left and then expand the result in an infinite series in t. On the right you have integrals of products of Legendre polynomials, and only the squared terms are non-zero. Equate like powers of t and you will have the result.

6.40 Use the scalar product of Eq. (6.16) and construct an orthogonal basis using the Gram-Schmidt process and starting from $\begin{pmatrix} 1 \\ 0 \end{pmatrix}$ and $\begin{pmatrix} 0 \\ 1 \end{pmatrix}$. Verify that your answer works in at least one special case.

6.41 For the differential equation $\ddot{x} + x = 0$, pick a set of independent solutions to the differential equation — any ones you like. Use the scalar product $\langle f, g \rangle = \int_0^1 dx\, f(x)^* g(x)$ and apply the Gram-Schmidt method to find an orthogonal basis in this space of solutions. Is there another scalar product that would make this analysis simpler? Sketch the orthogonal functions that you found.

Operators and Matrices

You've been using operators for years even if you've never heard the term. Differentiation falls into this category; so does rotation; so does wheel-alignment. In the subject of quantum mechanics, familiar ideas such as energy and momentum will be represented by operators. You probably think that pressure is simply a scalar, but no. It's an operator.

7.1 The Idea of an Operator

You can understand the subject of matrices as a set of rules that govern certain square or rectangular arrays of numbers — how to add them, how to multiply them. Approached this way the subject is remarkably opaque. Who made up these rules and why? What's the point? If you look at it as simply a way to write simultaneous linear equations in a compact way, it's perhaps convenient but certainly not the big deal that people make of it. It is a big deal.

There's a better way to understand the subject, one that relates the matrices to more fundamental ideas and that even provides some geometric insight into the subject. The technique of similarity transformations may even make a little sense. This approach is precisely parallel to one of the basic ideas in the use of vectors. You can draw pictures of vectors and manipulate the pictures of vectors and that's the right way to look at certain problems. You quickly find however that this can be cumbersome. A general method that you use to make computations tractable is to write vectors in terms of their components, then the methods for manipulating the components follow a few straight-forward rules, adding the components, multiplying them by scalars, even doing dot and cross products.

Just as you have components of vectors, which are a set of numbers that depend on your choice of basis, matrices are a set of numbers that are components of — not vectors, but functions (also called operators or transformations or tensors). I'll start with a couple of examples before going into the precise definitions.

The first example of the type of function that I'll be interested in will be a function defined on the two-dimensional vector space, arrows drawn in the plane with their starting points at the origin. The function that I'll use will rotate each vector by an angle α counterclockwise. This *is* a function, where the input is a vector and the output is a vector.

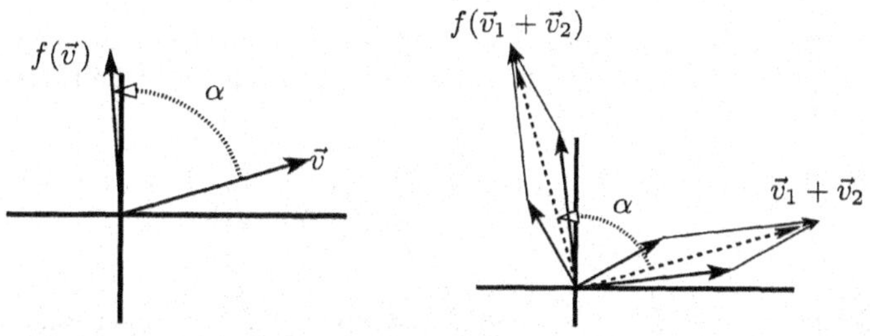

What happens if you change the argument of this function, multiplying it by a scalar? You know $f(\vec{v})$, what is $f(c\vec{v})$? Just from the picture, this is c times the vector that you got by rotating $\vec{v}$. What happens when you add two vectors and then rotate the result? The whole parallelogram defining the addition will rotate through the same angle

α, so whether you apply the function before or after adding the vectors you get the same result.

This leads to the definition of the word *linearity:*

$$\boxed{f(c\vec{v}) = cf(\vec{v}), \qquad \text{and} \qquad f(\vec{v}_1 + \vec{v}_2) = f(\vec{v}_1) + f(\vec{v}_2)} \tag{7.1}$$

Keep your eye on this pair of equations! They're central to the whole subject.

Another example of the type of function that I'll examine is from physics instead of mathematics. A rotating rigid body has some angular momentum. The greater the rotation rate, the greater the angular momentum will be. Now how do I compute the angular momentum assuming that I know the shape and the distribution of masses in the body and that I know the body's angular velocity? The body is made of a lot of point masses (atoms), but you don't need to go down to that level to make sense of the subject. As with any other integral, you start by dividing the object in to a lot of small pieces.

What is the angular momentum of a single point mass? It starts from basic Newtonian mechanics, and the equation $\vec{F} = d\vec{p}/dt$. (It's better in this context to work with this form than with the more common expressions $\vec{F} = m\vec{a}$.) Take the cross product with $\vec{r}$, the displacement vector from the origin.

$$\vec{r} \times \vec{F} = \vec{r} \times d\vec{p}/dt$$

Add and subtract the same thing on the right side of the equation (add zero) to get

$$\begin{aligned}\vec{r} \times \vec{F} &= \vec{r} \times \frac{d\vec{p}}{dt} + \frac{d\vec{r}}{dt} \times \vec{p} - \frac{d\vec{r}}{dt} \times \vec{p} \\ &= \frac{d}{dt}\big(\vec{r} \times \vec{p}\big) - \frac{d\vec{r}}{dt} \times \vec{p}\end{aligned}$$

Now recall that $\vec{p}$ is $m\vec{v}$, and $\vec{v} = d\vec{r}/dt$, so the last term in the preceding equation is zero because you are taking the cross product of a vector with itself. This means that when adding and subtracting a term from the right side above, I was really adding and subtracting zero.

$\vec{r} \times \vec{F}$ is the torque applied to the point mass m and $\vec{r} \times \vec{p}$ is the mass's angular momentum about the origin. Now if there are many masses and many forces, simply put an index on this torque equation and add the resulting equations over all the masses in the rigid body. The sums on the left and the right provide the definitions of torque and of angular momentum.

$$\vec{\tau}_{\text{total}} = \sum_k \vec{r}_k \times \vec{F}_k = \frac{d}{dt} \sum_k \big(\vec{r}_k \times \vec{p}_k\big) = \frac{d\vec{L}}{dt}$$

For a specific example, attach two masses to the ends of a light rod and attach that rod to a second, vertical one as sketched — at an angle. Now spin the vertical rod and figure out what the angular velocity and angular momentum vectors are. Since the spin is along the vertical rod, that specifies the direction of the angular velocity vector $\vec{\omega}$ to be upwards in the picture. (Viewed from above everything is rotating counter-clockwise.)

The angular momentum of one point mass is $\vec{r} \times \vec{p} = \vec{r} \times m\vec{v}$. The mass on the right has a velocity pointing *into* the page and the mass on the left has it pointing *out*. Take the origin to be where the supporting rod is attached to the axis, then $\vec{r} \times \vec{p}$ for the mass on the right is pointing up and to the left. For the other mass both $\vec{r}$ and $\vec{p}$ are reversed, so the cross product is in exactly the same direction as for the first mass. The total angular momentum the sum of these two parallel vectors, and it is *not* in the direction of the angular velocity.

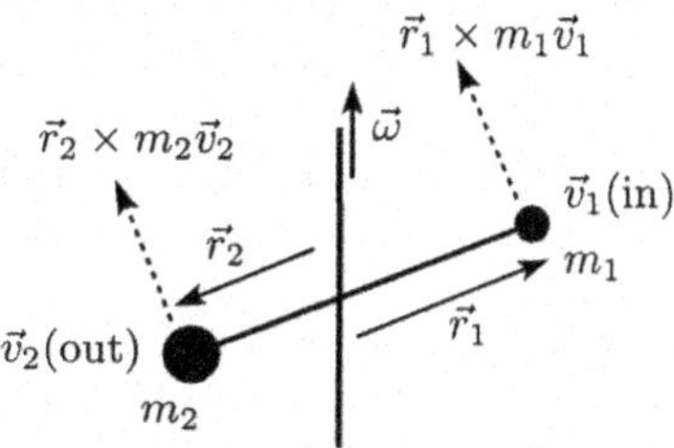

Now make this quantitative and apply it to a general rigid body. There are two basic pieces to the problem: the angular momentum of a point mass and the velocity of a point mass in terms of its angular velocity. The position of one point mass is described by its displacement vector from the origin, $\vec{r}$. Its angular momentum is then $\vec{r} \times \vec{p}$, where $\vec{p} = m\vec{v}$. If the rigid body has an angular velocity vector $\vec{\omega}$, the linear velocity of a mass at coordinate $\vec{r}$ is $\vec{\omega} \times \vec{r}$.

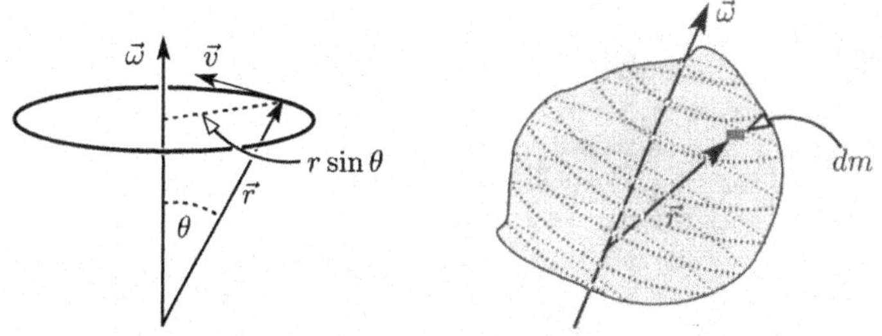

The total angular momentum of a rotating set of masses m_k at respective coordinates $\vec{r}_k$ is the sum of all the individual pieces of angular momentum

$$\vec{L} = \sum_k \vec{r}_k \times m_k\vec{v}_k, \qquad \text{and since} \qquad \vec{v}_k = \vec{\omega} \times \vec{r}_k,$$
$$\vec{L} = \sum_k \vec{r}_k \times m_k\big(\vec{\omega} \times \vec{r}_k\big) \tag{7.2}$$

If you have a continuous distribution of mass then using an integral makes more sense. For a given distribution of mass, this integral (or sum) depends on the vector $\vec{\omega}$. It defines a function having a vector as input and a vector $\vec{L}$ as output. Denote the function by I, so $\vec{L} = I(\vec{\omega})$.

$$\vec{L} = \int dm\,\vec{r} \times \big(\vec{\omega} \times \vec{r}\big) = I(\vec{\omega}) \tag{7.3}$$

This function satisfies the same linearity equations as Eq. (7.1). When you multiply $\vec{\omega}$ by a constant, the output, $\vec{L}$ is multiplied by the same constant. When you add two $\vec{\omega}$'s together as the argument, the properties of the cross product and of the integral guarantee that the corresponding $\vec{L}$'s are added.

$$I(c\vec{\omega}\,) = cI(\vec{\omega}\,), \qquad \text{and} \qquad I(\vec{\omega}_1 + \vec{\omega}_2) = I(\vec{\omega}_1) + I(\vec{\omega}_2)$$

This function I is called the "inertia operator" or more commonly the "inertia tensor." It's not simply multiplication by a scalar, so the rule that appears in an introductory course in mechanics ($\vec{L} = I\vec{\omega}$) is valid only in special cases, for example those with enough symmetry.

Note: I is not a vector and $\vec{L}$ is not a function. $\vec{L}$ is the output of the function I when you feed it the argument $\vec{\omega}$. This is the same sort of observation appearing in section 6.3 under "Function Spaces."

If an electromagnetic wave passes through a crystal, the electric field will push the electrons around, and the bigger the electric field, the greater the distance that the electrons will be pushed. They may not be pushed in the same direction as the electric field however, as the nature of the crystal can make it easier to push the electrons in one direction than in another. The relation between the applied field and the average electron displacement is a function that (for moderate size fields) obeys the same linearity relation that the two previous functions do.

$$\vec{P} = \alpha(\vec{E}\,)$$

$\vec{P}$ is the electric dipole moment density and $\vec{E}$ is the applied electric field. The function α is called the polarizability.

If you have a mass attached to six springs that are in turn attached to six walls, the mass will come to equilibrium somewhere. Now push on this mass with another (not too large) force. The mass will move, but will it move in the direction that you push it? If the six springs are all the same it will, but if they're not then the displacement will be more in the direction of the weaker springs. The displacement, $\vec{d}$, will still however depend linearly on the applied force, $\vec{F}$.

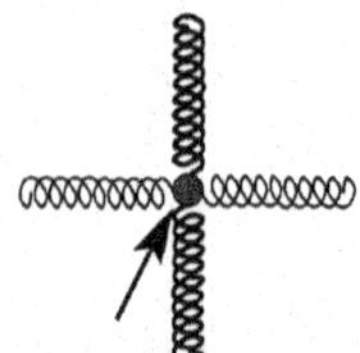

7.2 Definition of an Operator

An operator, also called a linear transformation, is a particular type of function. It is first of all, a vector valued function of a vector variable. Second, it is linear; that is, if A is such a function then $A(\vec{v})$ is a vector, and

$$A(\alpha\vec{v}_1 + \beta\vec{v}_2) = \alpha A(\vec{v}_1) + \beta A(\vec{v}_2). \tag{7.4}$$

The *domain* is the set of variables on which the operator is defined. The *range* is the set of all values put out by the function. Are there *nonlinear* operators? Yes, but not here.

7.3 Examples of Operators

The four cases that I started with, rotation in the plane, angular momentum of a rotating rigid body, polarization of a crystal by an electric field, and the mass attached to some springs all fit this definition. Other examples:

5. The simplest example of all is just multiplication by a scalar: $A(\vec{v}) \equiv c\vec{v}$ for all $\vec{v}$. This applies to any vector space and its domain is the entire space.
6. On the vector space of all real valued functions on a given interval, multiply any function f by $1+x^2$: $(Af)(x) = (1+x^2)f(x)$. The domain of A is the entire space of functions of x. This is an infinite dimensional vector space, but no matter. There's nothing special about $1+x^2$, and any other function will do to define an operator.
7. On the vector space of square integrable functions $\left[\int dx\,|f(x)|^2 < \infty\right]$ on $a < x < b$, define the operator as multiplication by x. The only distinction to make here is that if the interval is infinite, then $xf(x)$ may not itself be square integrable. The domain of this operator in this case is therefore *not* the entire space, but just those functions such that $xf(x)$ is also square-integrable. On the same vector space, differentiation is a linear operator: $(Af)(x) = f'(x)$. This too has a restriction on the domain: It is necessary that f' also exist and be square integrable.
8. On the vector space of infinitely differentiable functions, the operation of differentiation, d/dx, is itself a linear operator. It's certainly linear, and it takes a differentiable function into a differentiable function.

So where are the matrices? This chapter started by saying that I'm going to show you the inside scoop on matrices and so far I've failed to produce even one.

When you describe vectors you can use a basis as a computational tool and manipulate the vectors using their components. In the common case of three-dimensional vectors we usually denote the basis in one of several ways

$$\hat{\imath},\quad \hat{\jmath},\quad \hat{k},\qquad \text{or}\qquad \hat{x},\quad \hat{y},\quad \hat{z},\qquad \text{or}\qquad \vec{e}_1,\quad \vec{e}_2,\quad \vec{e}_3$$

and they all mean the same thing. The first form is what you see in the introductory physics texts. The second form is one that you encounter in more advanced books, and the third one is more suitable when you want to have a compact index notation. It's that third one that I'll use here; it has the advantage that it doesn't bias you to believe that you must be working in three spatial dimensions. The index could go beyond 3, and the vectors that you're dealing with may not be the usual geometric arrows. (And why does it have to start with one? Maybe I want the indices 0, 1, 2 instead.) These need not be perpendicular to each other or even to be unit vectors.

The way to write a vector $\vec{v}$ in components is

$$\vec{v} = v_x\,\hat{x} + v_y\,\hat{y} + v_z\,\hat{z}, \qquad \text{or} \qquad v_1\,\vec{e}_1 + v_2\,\vec{e}_2 + v_3\,\vec{e}_3 = \sum_k v_k\,\vec{e}_k \tag{7.5}$$

Once you've chosen a basis, you can find the three numbers that form the components of that vector. In a similar way, define the components of an operator, only that will take *nine* numbers to do it (in three dimensions). If you evaluate the effect of an operator on any one of the basis vectors, the output is a vector. That's part of the definition of the word operator. This output vector can itself be written in terms of this same basis. The defining equation for the components of an operator f is

$$\boxed{f(\vec{e}_i) = \sum_{k=1}^{3} f_{ki}\,\vec{e}_k} \tag{7.6}$$

For each input vector you have the three components of the output vector. Pay careful attention to this equation! It is the defining equation for the entire subject of matrix theory, and everything in that subject comes from this one innocuous looking equation. (And yes if you're wondering, I wrote the indices in the correct order.)

Why?

Take an arbitrary input vector for f: $\vec{u} = f(\vec{v}\,)$. Both $\vec{u}$ and $\vec{v}$ are vectors, so write them in terms of the basis chosen.

$$\vec{u} = \sum_k u_k\,\vec{e}_k = f(\vec{v}\,) = f\Big(\sum_i v_i\,\vec{e}_i\Big) = \sum_i v_i f(\vec{e}_i) \tag{7.7}$$

The last equation is the result of the linearity property, Eq. (7.1), already assumed for f. Now pull the sum and the numerical factors v_i out in front of the function, and write it out. It is then clear:

$$f(v_1\,\vec{e}_1 + v_2\,\vec{e}_2) = f(v_1\,\vec{e}_1) + f(v_2\,\vec{e}_2) = v_1 f(\vec{e}_1) + v_2 f(\vec{e}_2)$$

Now you see where the defining equation for operator components comes in. Eq. (7.7) is

$$\sum_k u_k\,\vec{e}_k = \sum_i v_i \sum_k f_{ki}\,\vec{e}_k$$

For two vectors to be equal, the corresponding coefficients of $\vec{e}_1$, $\vec{e}_2$, *etc.* must match; their respective components must be equal, and this is

$$u_k = \sum_i v_i f_{ki}, \qquad \text{usually written} \qquad u_k = \sum_i f_{ki} v_i \tag{7.8}$$

so that in the latter form it starts to resemble what you may think of as matrix manipulation. $f_{\text{row,column}}$ is the conventional way to write the indices, and multiplication is defined so that the following product *means* Eq. (7.8).

$$\begin{pmatrix} u_1 \\ u_2 \\ u_3 \end{pmatrix} = \begin{pmatrix} f_{11} & f_{12} & f_{13} \\ f_{21} & f_{22} & f_{23} \\ f_{31} & f_{32} & f_{33} \end{pmatrix} \begin{pmatrix} v_1 \\ v_2 \\ v_3 \end{pmatrix} \tag{7.9}$$

$$\begin{pmatrix} u_1 \\ u_2 \\ u_3 \end{pmatrix} = \begin{pmatrix} f_{11} & f_{12} & f_{13} \\ f_{21} & f_{22} & f_{23} \\ f_{31} & f_{32} & f_{33} \end{pmatrix} \begin{pmatrix} v_1 \\ v_2 \\ v_3 \end{pmatrix} \qquad \text{is} \qquad u_1 = f_{11}v_1 + f_{12}v_2 + f_{13}v_3 \qquad \textit{etc.}$$

And this is the reason behind the definition of how to multiply a matrix and a column matrix. The order in which the indices appear is the conventional one, and the indices appear in the matrix as they do because I chose the order of the indices in a (seemingly) backwards way in Eq. (7.6).

Components of Rotations

Apply this to the first example, rotate all vectors in the plane through the angle α. I don't want to keep using the same symbol f for *every* function, so I'll call this function

R instead, or better yet R_α. $R_\alpha(\vec{v})$ is the rotated vector. Pick two perpendicular unit vectors for a basis. You may call them $\hat{x}$ and $\hat{y}$, but again I'll call them $\vec{e}_1$ and $\vec{e}_2$. Use the definition of components to get

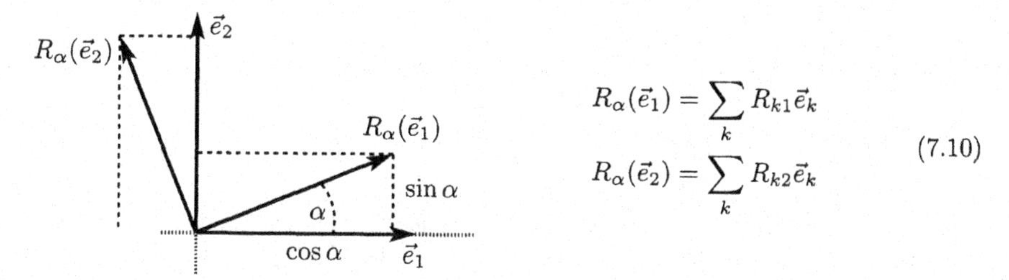

$$R_\alpha(\vec{e}_1) = \sum_k R_{k1}\vec{e}_k$$
$$R_\alpha(\vec{e}_2) = \sum_k R_{k2}\vec{e}_k \tag{7.10}$$

The rotated $\vec{e}_1$ has two components, so

$$R_\alpha(\vec{e}_1) = \vec{e}_1\cos\alpha + \vec{e}_2\sin\alpha = R_{11}\vec{e}_1 + R_{21}\vec{e}_2 \tag{7.11}$$

This determines the first column of the matrix of components,

$$R_{11} = \cos\alpha, \qquad \text{and} \qquad R_{21} = \sin\alpha$$

Similarly the effect on the other basis vector determines the second column:

$$R_\alpha(\vec{e}_2) = \vec{e}_2\cos\alpha - \vec{e}_1\sin\alpha = R_{12}\,\vec{e}_1 + R_{22}\,\vec{e}_2 \tag{7.12}$$

Check: $R_\alpha(\vec{e}_1)\cdot R_\alpha(\vec{e}_2) = 0$.

$$R_{12} = -\sin\alpha, \qquad \text{and} \qquad R_{22} = \cos\alpha$$

The component matrix is then

$$(R_\alpha) = \begin{pmatrix} \cos\alpha & -\sin\alpha \\ \sin\alpha & \cos\alpha \end{pmatrix} \tag{7.13}$$

Components of Inertia

The definition, Eq. (7.3), and the figure preceding it specify the inertia tensor as the function that relates the angular momentum of a rigid body to its angular velocity.

$$\vec{L} = \int dm\,\vec{r}\times(\vec{\omega}\times\vec{r}) = I(\vec{\omega}) \tag{7.14}$$

Use the vector identity,

$$\vec{A}\times(\vec{B}\times\vec{C}) = \vec{B}(\vec{A}\cdot\vec{C}) - \vec{C}(\vec{A}\cdot\vec{B}) \tag{7.15}$$

then the integral is

$$\vec{L} = \int dm\,\big[\vec{\omega}(\vec{r}\cdot\vec{r}) - \vec{r}(\vec{\omega}\cdot\vec{r})\big] = I(\vec{\omega}) \tag{7.16}$$

Pick the common rectangular, orthogonal basis and evaluate the components of this function. Equation (7.6) says $\vec{r} = x\,\vec{e}_1 + y\,\vec{e}_2 + z\,\vec{e}_3$ so

$$I(\vec{e}_i) = \sum_k I_{ki}\,\vec{e}_k$$

$$I(\vec{e}_1) = \int dm\,\big[\vec{e}_1(x^2+y^2+z^2) - (x\,\vec{e}_1 + y\,\vec{e}_2 + z\,\vec{e}_3)(x)\big]$$

$$= I_{11}\,\vec{e}_1 + I_{21}\,\vec{e}_2 + I_{31}\,\vec{e}_3$$

from which $\quad I_{11} = \int dm\,(y^2+z^2), \qquad I_{21} = -\int dm\,yx, \qquad I_{31} = -\int dm\,zx$

This provides the first column of the components, and you get the rest of the components the same way. The whole matrix is

$$\int dm \begin{pmatrix} y^2+z^2 & -xy & -xz \\ -xy & x^2+z^2 & -yz \\ -xz & -yz & x^2+y^2 \end{pmatrix} \tag{7.17}$$

These are the components of the tensor of inertia. The diagonal elements of the matrix may be familiar; they are the moments of inertia. x^2+y^2 is the perpendicular distance-squared to the z-axis, so the element I_{33} ($\equiv I_{zz}$) is the moment of inertia about that axis, $\int dm\,r_\perp^2$. The other components are less familiar and are called the products of inertia. This particular matrix is symmetric: $I_{ij} = I_{ji}$. That's a special property of the inertia tensor.

Components of Dumbbell

Look again at the specific case of two masses rotating about an axis. Do it quantitatively.

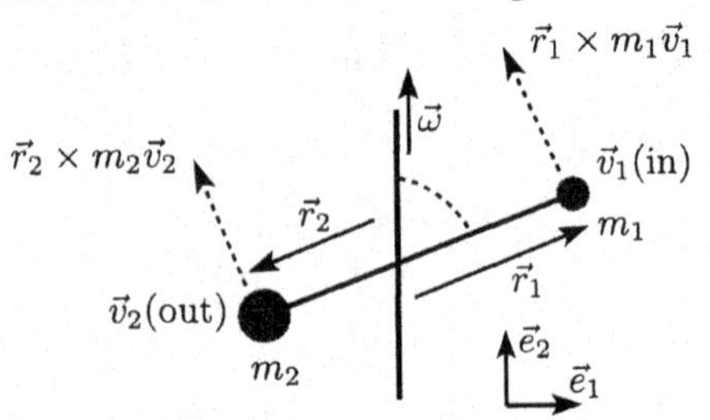

The integrals in Eq. (7.17) are simply sums this time, and the sums have just two terms. I'm making the approximation that these are point masses. Make the coordinate system match the indicated basis, with x right and y up, then z is zero for all terms in the sum, and the rest are

$$\int dm\,(y^2+z^2) = m_1r_1^2\cos^2\alpha + m_2r_2^2\cos^2\alpha$$

$$-\int dm\,xy = -m_1r_1^2\cos\alpha\sin\alpha - m_2r_2^2\cos\alpha\sin\alpha$$

$$\int dm\,(x^2+z^2) = m_1r_1^2\sin^2\alpha + m_2r_2^2\sin^2\alpha$$

$$\int dm\,(x^2+y^2) = m_1r_1^2 + m_2r_2^2$$

The matrix is then

$$(I) = (m_1 r_1^2 + m_2 r_2^2) \begin{pmatrix} \cos^2\alpha & -\cos\alpha\sin\alpha & 0 \\ -\cos\alpha\sin\alpha & \sin^2\alpha & 0 \\ 0 & 0 & 1 \end{pmatrix} \tag{7.18}$$

Don't count on all such results factoring so nicely.

In this basis, the angular velocity $\vec{\omega}$ has just one component, so what is $\vec{L}$?

$$(m_1 r_1^2 + m_2 r_2^2) \begin{pmatrix} \cos^2\alpha & -\cos\alpha\sin\alpha & 0 \\ -\cos\alpha\sin\alpha & \sin^2\alpha & 0 \\ 0 & 0 & 1 \end{pmatrix} \begin{pmatrix} 0 \\ \omega \\ 0 \end{pmatrix} =$$

$$(m_1 r_1^2 + m_2 r_2^2) \begin{pmatrix} -\omega\cos\alpha\sin\alpha \\ \omega\sin^2\alpha \\ 0 \end{pmatrix}$$

Translate this into vector form:

$$\vec{L} = (m_1 r_1^2 + m_2 r_2^2)\omega\sin\alpha\big(-\vec{e}_1\cos\alpha + \vec{e}_2\sin\alpha\big) \tag{7.19}$$

When $\alpha = 90°$, then $\cos\alpha = 0$ and the angular momentum points along the y-axis. This is the symmetric special case where everything lines up along one axis. Notice that if $\alpha = 0$ then everything vanishes, but then the masses are both *on* the axis, and they have no angular momentum. In the general case as drawn, the vector $\vec{L}$ points to the upper left, perpendicular to the line between the masses.

Parallel Axis Theorem

When you know the tensor of inertia about one origin, you can relate the result to the tensor about a different origin.

The center of mass of an object is

$$\vec{r}_{\text{cm}} = \frac{1}{M}\int \vec{r}\,dm \tag{7.20}$$

where M is the total mass. Compare the operator I using an origin at the center of mass to I about another origin.

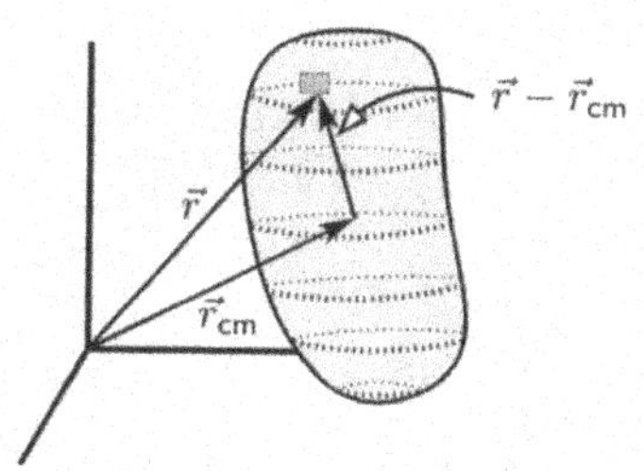

$$\begin{aligned} I(\vec{\omega}) &= \int dm\,\vec{r}\times(\vec{\omega}\times\vec{r}) = \int dm\,[\vec{r}-\vec{r}_{\text{cm}}+\vec{r}_{\text{cm}}]\times\big(\vec{\omega}\times[\vec{r}-\vec{r}_{\text{cm}}+\vec{r}_{\text{cm}}]\big) \\ &= \int dm\,[\vec{r}-\vec{r}_{\text{cm}}]\times(\vec{\omega}\times[\vec{r}-\vec{r}_{\text{cm}}]) + \int dm\,\vec{r}_{\text{cm}}\times\big(\vec{\omega}\times\vec{r}_{\text{cm}}\big) + \text{two cross terms} \end{aligned} \tag{7.21}$$

The two cross terms vanish, problem 7.17. What's left is

$$\begin{aligned} I(\vec{\omega}\,) &= \int dm\,[\vec{r}-\vec{r}_{\text{cm}}] \times \big(\vec{\omega} \times [\vec{r}-\vec{r}_{\text{cm}}]\big) + M\,r_{\text{cm}} \times \big(\vec{\omega} \times \vec{r}_{\text{cm}}\big) \\ &= I_{\text{cm}}(\vec{\omega}\,) + M\,\vec{r}_{\text{cm}} \times \big(\vec{\omega} \times \vec{r}_{\text{cm}}\big) \end{aligned} \tag{7.22}$$

Put this in words and it says that the tensor of inertia about any point is equal to the tensor of inertia about the center of mass plus the tensor of inertia of a point mass M placed *at* the center of mass.

As an example, place a disk of mass M and radius R and uniform mass density so that its center is at $(x,y,z) = (R,0,0)$ and it is lying in the x-y plane. Compute the components of the inertia tensor. First get the components about the center of mass, using Eq. (7.17).

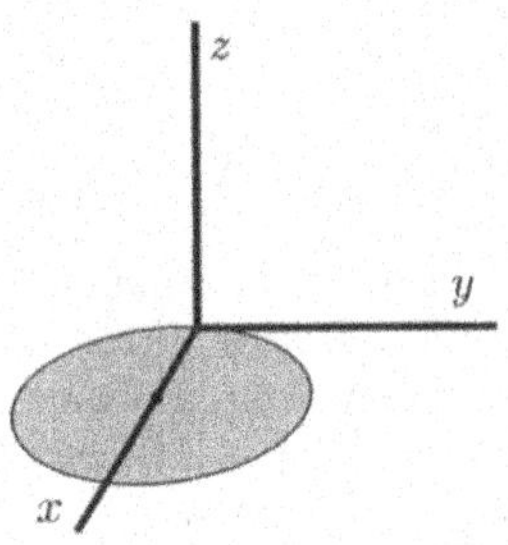

The integrals such as

$$-\int dm\,xy, \qquad -\int dm\,yz$$

are zero. For fixed y each positive value of x has a corresponding negative value to make the integral add to zero. It is odd in x (or y); remember that this is about the *center* of the disk. Next do the I_{33} integral.

$$\int dm\,(x^2+y^2) = \int dm\,r^2 = \int \frac{M}{\pi R^2} dA\,r^2$$

For the element of area, use $dA = 2\pi r\,dr$ and you have

$$I_{33} = \frac{M}{\pi R^2}\int_0^R dr\,2\pi r^3 = \frac{M}{\pi R^2}2\pi\frac{R^4}{4} = \frac{1}{2}MR^2$$

For the next two diagonal elements,

$$I_{11} = \int dm\,(y^2+z^2) = \int dm\,y^2 \qquad \text{and} \qquad I_{22} = \int dm\,(x^2+z^2) = \int dm\,x^2$$

Because of the symmetry of the disk, these two are equal, also you see that the sum is

$$I_{11} + I_{22} = \int dm\,y^2 + \int dm\,x^2 = I_{33} = \frac{1}{2}MR^2 \tag{7.23}$$

This saves integration. $I_{11} = I_{22} = MR^2/4$.

For the other term in the sum (7.22), you have a point mass at the distance R along the x-axis, $(x,y,z) = (R,0,0)$. Substitute this point mass into Eq. (7.17) and you have

$$M\begin{pmatrix} 0 & 0 & 0 \\ 0 & R^2 & 0 \\ 0 & 0 & R^2 \end{pmatrix}$$

The total about the origin is the sum of these two calculations.

$$MR^2\begin{pmatrix} 1/4 & 0 & 0 \\ 0 & 5/4 & 0 \\ 0 & 0 & 3/2 \end{pmatrix}$$

Why is this called the parallel axis theorem when you're translating a point (the origin) and not an axis? Probably because this was originally stated for the moment of inertia alone and not for the whole tensor. In that case you have only an axis to deal with.

Components of the Derivative

The set of all polynomials in x having degree ≤ 2 forms a vector space. There are three independent vectors that I can choose to be 1, x, and x^2. Differentiation is a linear operator on this space because the derivative of a sum is the sum of the derivatives and the derivative of a constant times a function is the constant times the derivative of the function. With this basis I'll compute the components of d/dx. Start the indexing for the basis from zero instead of one because it will cause less confusion between powers and subscripts.

$$\vec e_0 = 1, \qquad \vec e_1 = x, \qquad \vec e_2 = x^2$$

By the definition of the components of an operator — I'll call this one D,

$$D(\vec e_0) = \frac{d}{dx}1 = 0, \qquad D(\vec e_1) = \frac{d}{dx}x = 1 = \vec e_0, \qquad D(\vec e_2) = \frac{d}{dx}x^2 = 2x = 2\vec e_1$$

These define the three columns of the matrix.

$$(D) = \begin{pmatrix} 0 & 1 & 0 \\ 0 & 0 & 2 \\ 0 & 0 & 0 \end{pmatrix} \qquad \text{check: } \frac{dx^2}{dx} = 2x \text{ is} \qquad \begin{pmatrix} 0 & 1 & 0 \\ 0 & 0 & 2 \\ 0 & 0 & 0 \end{pmatrix}\begin{pmatrix} 0 \\ 0 \\ 1 \end{pmatrix} = \begin{pmatrix} 0 \\ 2 \\ 0 \end{pmatrix}$$

There's nothing here about the basis being orthonormal. It isn't.

7.4 Matrix Multiplication

How do you multiply two matrices? There's a rule for doing it, but where does it come from?

The composition of two functions means you first apply one function then the other, so

$$h = f \circ g \qquad \text{means} \qquad h(\vec v) = f\big(g(\vec v)\big) \tag{7.24}$$

I'm assuming that these are vector-valued functions of a vector variable, but this is the general definition of composition anyway. If f and g are linear, does it follow the h is? Yes, just check:

$$h(c\vec{v}) = f\big(g(c\vec{v})\big) = f\big(cg(\vec{v})\big) = c\,f\big(g(\vec{v})\big), \qquad \text{and}$$
$$h(\vec{v}_1 + \vec{v}_2) = f\big(g(\vec{v}_1 + \vec{v}_2)\big) = f\big(g(\vec{v}_1) + g(\vec{v}_2)\big) = f\big(g(\vec{v}_1)\big) + f\big(g(\vec{v}_2)\big)$$

What are the components of h? Again, use the definition and plug in.

$$h(\vec{e}_i) = \sum_k h_{ki}\vec{e}_k = f\big(g(\vec{e}_i)\big) = f\big(\sum_j g_{ji}\vec{e}_j\big) = \sum_j g_{ji} f(\vec{e}_j) = \sum_j g_{ji} \sum_k f_{kj}\vec{e}_k \tag{7.25}$$

and now all there is to do is to equate the corresponding coefficients of $\vec{e}_k$.

$$h_{ki} = \sum_j g_{ji} f_{kj} \qquad \text{or more conventionally} \qquad h_{ki} = \sum_j f_{kj} g_{ji} \tag{7.26}$$

This is in the standard form for matrix multiplication, recalling the subscripts are ordered as f_{rc} for row-column.

$$\begin{pmatrix} h_{11} & h_{12} & h_{13} \\ h_{21} & h_{32} & h_{23} \\ h_{31} & h_{32} & h_{33} \end{pmatrix} = \begin{pmatrix} f_{11} & f_{12} & f_{13} \\ f_{21} & f_{32} & f_{23} \\ f_{31} & f_{32} & f_{33} \end{pmatrix} \begin{pmatrix} g_{11} & g_{12} & g_{13} \\ g_{21} & g_{32} & g_{23} \\ g_{31} & g_{32} & g_{33} \end{pmatrix} \tag{7.27}$$

The computation of h_{12} from Eq. (7.26) is

$$\begin{pmatrix} h_{11} & h_{12} & h_{13} \\ h_{21} & h_{22} & h_{23} \\ h_{31} & h_{32} & h_{33} \end{pmatrix} = \begin{pmatrix} f_{11} & f_{12} & f_{13} \\ f_{21} & f_{22} & f_{23} \\ f_{31} & f_{32} & f_{33} \end{pmatrix} \begin{pmatrix} g_{11} & g_{12} & g_{13} \\ g_{21} & g_{22} & g_{23} \\ g_{31} & g_{32} & g_{33} \end{pmatrix}$$
$$\longrightarrow \quad h_{12} = f_{11}g_{12} + f_{12}g_{22} + f_{13}g_{32}$$

Matrix multiplication is just the component representation of the composition of two functions, Eq. (7.26), and there's nothing here that restricts this to three dimensions. In Eq. (7.25) I may have made it look too easy. If you try to reproduce this without looking, the odds are that you will not get the indices to match up as nicely as you see there. Remember: When an index is summed it is a *dummy*, and you are free to relabel it as anything you want. You can use this fact to make the indices come out neatly.

Composition of Rotations

In the first example, rotating vectors in the plane, the operator that rotates every vector by the angle α has components

$$(R_\alpha) = \begin{pmatrix} \cos\alpha & -\sin\alpha \\ \sin\alpha & \cos\alpha \end{pmatrix} \tag{7.28}$$

What happens if you do two such transformations, one by α and one by β? The result better be a total rotation by $\alpha + \beta$. One function, R_β is followed by the second function R_α and the composition is

$$R_{\alpha+\beta} = R_\alpha R_\beta$$

This is mirrored in the components of these operators, so the matrices must obey the same equation.

$$\begin{pmatrix} \cos(\alpha+\beta) & -\sin(\alpha+\beta) \\ \sin(\alpha+\beta) & \cos(\alpha+\beta) \end{pmatrix} = \begin{pmatrix} \cos\alpha & -\sin\alpha \\ \sin\alpha & \cos\alpha \end{pmatrix} \begin{pmatrix} \cos\beta & -\sin\beta \\ \sin\beta & \cos\beta \end{pmatrix}$$

Multiply the matrices on the right to get

$$\begin{pmatrix} \cos\alpha\cos\beta - \sin\alpha\sin\beta & -\cos\alpha\sin\beta - \sin\alpha\cos\beta \\ \sin\alpha\cos\beta + \cos\alpha\sin\beta & \cos\alpha\cos\beta - \sin\alpha\sin\beta \end{pmatrix} \tag{7.29}$$

The respective components must agree, so this gives an immediate derivation of the formulas for the sine and cosine of the sum of two angles. Cf. Eq. (3.8)

7.5 Inverses

The simplest operator is the one that does nothing. $f(\vec{v}) = \vec{v}$ for all values of the vector $\vec{v}$. This implies that $f(\vec{e}_1) = \vec{e}_1$ and similarly for all the other elements of the basis, so the matrix of its components is diagonal. The 2×2 matrix is explicitly the identity matrix

$$(I) = \begin{pmatrix} 1 & 0 \\ 0 & 1 \end{pmatrix} \qquad \text{or in index notation} \qquad \delta_{ij} = \begin{cases} 1 & (\text{if } i = j) \\ 0 & (\text{if } i \neq j) \end{cases} \tag{7.30}$$

and the index notation is completely general, not depending on whether you're dealing with two dimensions or many more. Unfortunately the words "inertia" and "identity" both start with the letter "I" and this symbol is used for both operators. Live with it. The δ symbol in this equation is the Kronecker delta — very handy.

The *inverse* of an operator is defined in terms of Eq. (7.24), the composition of functions. If the composition of two functions takes you to the identity operator, one function is said to be the inverse of the other. This is no different from the way you look at ordinary real valued functions. The exponential and the logarithm are inverse to each other because*

$$\ln(e^x) = x \qquad \text{for all } x.$$

For the rotation operator, Eq. (7.10), the inverse is obviously going to be rotation by the same angle in the opposite direction.

$$R_\alpha R_{-\alpha} = I$$

Because the matrix components of these operators mirror the the original operators, this equation must also hold for the corresponding components, as in Eqs. (7.27) and (7.29). Set $\beta = -\alpha$ in (7.29) and you get the identity matrix.

In an equation such as Eq. (7.7), or its component version Eqs. (7.8) or (7.9), if you want to solve for the vector $\vec{u}$, you are asking for the inverse of the function f.

$$\vec{u} = f(\vec{v}) \qquad \text{implies} \qquad \vec{v} = f^{-1}(\vec{u})$$

* The reverse, $e^{\ln x}$ works just for positive x, unless you recall that the logarithm of a negative number is complex. Then it works there too. This sort of question doesn't occur with finite dimensional matrices.

The translation of these equations into components is Eq. (7.9)

$$\begin{pmatrix} u_1 \\ u_2 \end{pmatrix} = \begin{pmatrix} f_{11} & f_{12} \\ f_{21} & f_{22} \end{pmatrix} \begin{pmatrix} v_1 \\ v_2 \end{pmatrix}$$

$$\text{which implies} \qquad \frac{1}{f_{11}f_{22} - f_{12}f_{21}} \begin{pmatrix} f_{22} & -f_{12} \\ -f_{21} & f_{11} \end{pmatrix} \begin{pmatrix} u_1 \\ u_2 \end{pmatrix} = \begin{pmatrix} v_1 \\ v_2 \end{pmatrix} \tag{7.31}$$

The verification that these are the components of the inverse is no more than simply multiplying the two matrices and seeing that you get the identity matrix.

7.6 Rotations, 3-d

In three dimensions there are of course more basis vectors to rotate. Start by rotating vectors about the axes and it is nothing more than the two-dimensional problem of Eq. (7.10) done three times. You do have to be careful about signs, but not much more — as long as you draw careful pictures!

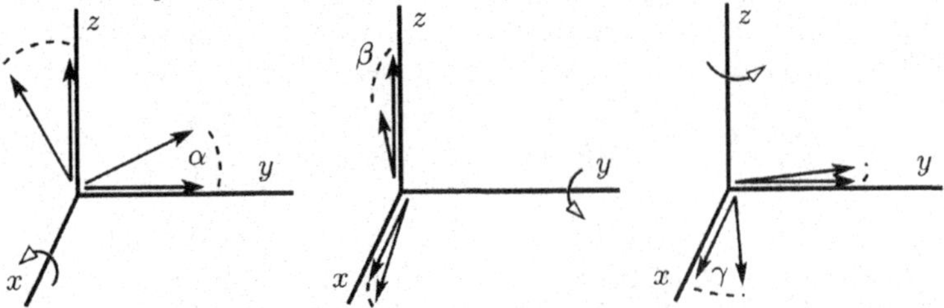

The basis vectors are drawn in the three pictures: $\vec{e}_1 = \hat{x}$, $\vec{e}_2 = \hat{y}$, $\vec{e}_3 = \hat{z}$.

In the first sketch, rotate vectors by the angle α about the x-axis. In the second case, rotate by the angle β about the y-axis, and in the third case, rotate by the angle γ about the z-axis. In the first case, the $\vec{e}_1$ is left alone. The $\vec{e}_2$ picks up a little positive $\vec{e}_3$, and the $\vec{e}_3$ picks up a little negative $\vec{e}_2$.

$$R_{\alpha\vec{e}_1}(\vec{e}_1) = \vec{e}_1, \quad R_{\alpha\vec{e}_1}(\vec{e}_2) = \vec{e}_2 \cos\alpha + \vec{e}_3 \sin\alpha, \quad R_{\alpha\vec{e}_1}(\vec{e}_3) = \vec{e}_3 \cos\alpha - \vec{e}_2 \sin\alpha \tag{7.32}$$

Here the notation $R_{\vec{\theta}}$ represents the function prescribing a rotation by θ about the axis pointing along $\hat{\theta}$. These equations are the same as Eqs. (7.11) and (7.12).

The corresponding equations for the other two rotations are now easy to write down:

$$R_{\beta\vec{e}_2}(\vec{e}_1) = \vec{e}_1 \cos\beta - \vec{e}_3 \sin\beta, \quad R_{\beta\vec{e}_2}(\vec{e}_2) = \vec{e}_2, \quad R_{\beta\vec{e}_2}(\vec{e}_3) = \vec{e}_1 \sin\beta + \vec{e}_3 \cos\beta \tag{7.33}$$

$$R_{\gamma\vec{e}_3}(\vec{e}_1) = \vec{e}_1 \cos\gamma + \vec{e}_2 \sin\gamma, \quad R_{\gamma\vec{e}_3}(\vec{e}_2) = -\vec{e}_1 \sin\gamma + \vec{e}_2 \cos\gamma, \quad R_{\gamma\vec{e}_3}(\vec{e}_3) = \vec{e}_3 \tag{7.34}$$

From these vector equations you immediate read the columns of the matrices of the components of the operators as in Eq. (7.6).

$$\underset{(R_{\alpha\vec{e}_1})}{\begin{pmatrix} 1 & 0 & 0 \\ 0 & \cos\alpha & -\sin\alpha \\ 0 & \sin\alpha & \cos\alpha \end{pmatrix}}, \quad \underset{(R_{\beta\vec{e}_2})}{\begin{pmatrix} \cos\beta & 0 & \sin\beta \\ 0 & 1 & 0 \\ -\sin\beta & 0 & \cos\beta \end{pmatrix}}, \quad \underset{(R_{\gamma\vec{e}_3})}{\begin{pmatrix} \cos\gamma & -\sin\gamma & 0 \\ \sin\gamma & \cos\gamma & 0 \\ 0 & 0 & 1 \end{pmatrix}} \tag{7.35}$$

As a check on the algebra, did you see if the rotated basis vectors from any of the three sets of equations (7.32)-(7.34) are still orthogonal sets?

Do these rotation operations commute? No. Try the case of two 90° rotations to see. Rotate by this angle about the x-axis then by the same angle about the y-axis.

$$\big(R_{\vec{e}_2\pi/2}\big)\big(R_{\vec{e}_1\pi/2}\big) = \begin{pmatrix} 0 & 0 & 1 \\ 0 & 1 & 0 \\ -1 & 0 & 0 \end{pmatrix}\begin{pmatrix} 1 & 0 & 0 \\ 0 & 0 & -1 \\ 0 & 1 & 0 \end{pmatrix} = \begin{pmatrix} 0 & 1 & 0 \\ 0 & 0 & -1 \\ -1 & 0 & 0 \end{pmatrix} \tag{7.36}$$

In the reverse order, for which the rotation about the y-axis is done first, these are

$$\big(R_{\vec{e}_1\pi/2}\big)\big(R_{\vec{e}_2\pi/2}\big) = \begin{pmatrix} 1 & 0 & 0 \\ 0 & 0 & -1 \\ 0 & 1 & 0 \end{pmatrix}\begin{pmatrix} 0 & 0 & 1 \\ 0 & 1 & 0 \\ -1 & 0 & 0 \end{pmatrix} = \begin{pmatrix} 0 & 0 & 1 \\ 1 & 0 & 0 \\ 0 & 1 & 0 \end{pmatrix} \tag{7.37}$$

Translate these operations into the movement of a physical object. Take the same x-y-z coordinate system as in this section, with x pointing toward you, y to your right and z up. Pick up a book with the cover toward you so that you can read it. Now do the operation $R_{\vec{e}_1\pi/2}$ on it so that the cover still faces you but the top is to your left. Next do $R_{\vec{e}_2\pi/2}$ and the book is face down with the top still to your left. See problem 7.57 for and algebraic version of this.

Start over with the cover toward you as before and do $R_{\vec{e}_2\pi/2}$ so that the top is toward you and the face is down. Now do the other operation $R_{\vec{e}_1\pi/2}$ and the top is toward you with the cover facing right — a different result. Do these physical results agree with the matrix products of the last two equations? For example, what happens to the vector sticking out the the cover, initially the column matrix $(\,1 \quad 0 \quad 0\,)$? This is something that you cannot simply *read*. You have to do the experiment for yourself.

7.7 Areas, Volumes, Determinants

In the two-dimensional example of arrows in the plane, look what happens to areas when an operator acts. The unit square with corners at the origin and $(0,1)$, $(1,1)$, $1,0)$ gets distorted into a parallelogram. The arrows from the origin to every point in the square become arrows that fill out the parallelogram.

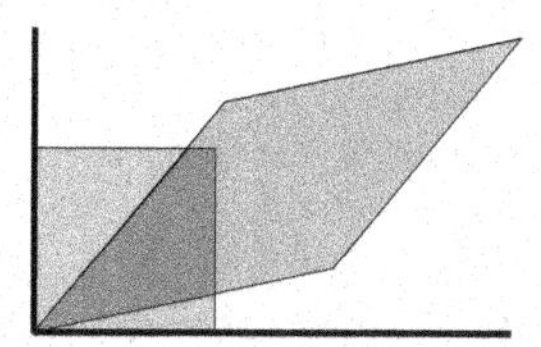

What is the area of this parallelogram?

I'll ask a more general question. (It isn't really, but it looks like it.) Start with *any* region in the plane, and say it has area A_1. The operator takes all the vectors ending in this area into some new area of a size A_2, probably different from the original. What is the ratio of the new area to the old one? A_2/A_1. How much does this transformation stretch or squeeze the area? What isn't instantly obvious is that this ratio of areas depends on the operator *alone*, and not on how you chose the initial region to be transformed. If you accept this for the moment, then you see that the question in the previous paragraph, which started with the unit square and asked for the area into which it transformed, is the same question as finding the ratio of the two more general areas. (Or the ratio of two volumes in three dimensions.) See the end of the next section for a proof.

This ratio is called the determinant of the operator.

The first example is the simplest. Rotations in the plane, R_α. Because rotations leave area unchanged, this determinant is one. For almost any other example you have to do some work. Use the component form to do the computation. The basis vector $\vec{e}_1$ is transformed into the vector $f_{11}\,\vec{e}_1 + f_{21}\,\vec{e}_2$ with a similar expression for the image of $\vec{e}_2$. You can use the cross product to compute the area of the parallelogram that these define. For another way, see problem 7.3. This is

$$\big(f_{11}\,\vec{e}_1 + f_{21}\,\vec{e}_2\big) \times \big(f_{12}\,\vec{e}_1 + f_{22}\,\vec{e}_2\big) = \big(f_{11}f_{22} - f_{21}f_{12}\big)\vec{e}_3 \tag{7.38}$$

The product in parentheses is the determinant of the transformation.

$$\det(f) = f_{11}f_{22} - f_{21}f_{12} \tag{7.39}$$

What if I had picked a different basis, maybe even one that isn't orthonormal? From the definition of the determinant it is a property of the operator and not of the particular basis and components you use to describe it, so you must get the same answer. But will the answer be the same simple formula (7.39) if I pick a different basis? Now *that's* a legitimate question. The answer is yes, and that fact will come out of the general computation of the determinant in a moment. [What is the determinant of Eq. (7.13)?]

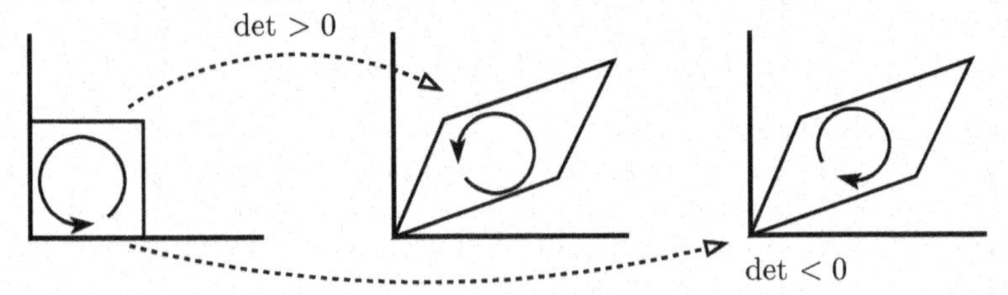

The determinant can be either positive or negative. That tells you more than simply how the transformation alters the area; it tells you whether it changes the *orientation* of the area. If you place a counterclockwise loop in the original area, does it remain counterclockwise in the image or is it reversed? In three dimensions, the corresponding plus or minus sign for the determinant says that you're changing from a right-handed set of vectors to a left-handed one. What does that mean? Make an x-y-z coordinate system out of the thumb, index finger, and middle finger of your right hand. Now do it with your left hand. You cannot move one of these and put it on top of the other (unless you have *very* unusual joints). One is a mirror image of the other.

The equation (7.39) is a special case of a rule that you've probably encountered elsewhere. You compute the determinant of a square array of numbers by some means such as expansion in minors or Gauss reduction. Here I've defined the determinant geometrically, and it has no obvious relation the traditional numeric definition. They are the same, and the reason for that comes by looking at how the area (or volume) of a parallelogram depends on the vectors that make up its sides. The derivation is slightly involved, but no one step in it is hard. Along the way you will encounter a new and important function: Λ.

Start with the basis $\vec{e}_1$, $\vec{e}_2$ and call the output of the transformation $\vec{v}_1 = f(\vec{e}_1)$ and $\vec{v}_2 = f(\vec{e}_2)$. The final area is a function of these last two vectors, call it $\Lambda(\vec{v}_1, \vec{v}_2)$, and this function has two key properties:

$$\Lambda(\vec{v}, \vec{v}) = 0, \qquad \text{and} \qquad \Lambda(\vec{v}_1, \alpha\vec{v}_2 + \beta\vec{v}_3) = \alpha\Lambda(\vec{v}_1, \vec{v}_2) + \beta\Lambda(\vec{v}_1, \vec{v}_3) \tag{7.40}$$

That the area vanishes if the two sides are the same is obvious. That the area is a linear function of the vectors forming the two sides is not so obvious. (It is linear in both arguments.) Part of the proof of linearity is easy:

$$\Lambda(\vec{v}_1, \alpha\vec{v}_2) = \alpha\Lambda(\vec{v}_1, \vec{v}_2)$$

simply says that if one side of the parallelogram remains fixed and the other changes by some factor, then the area changes by that same factor. For the other part, $\Lambda(\vec{v}_1, \vec{v}_2 + \vec{v}_3)$, start with a picture and see if the area that this function represents is the same as the sum of the two areas made from the vectors $\vec{v}_1 \& \vec{v}_2$ and $\vec{v}_1 \& \vec{v}_3$.

$\vec{v}_1 \& \vec{v}_2$ form the area OCBA. $\qquad$ $\vec{v}_1 \& \vec{v}_3$ form the area OCED.

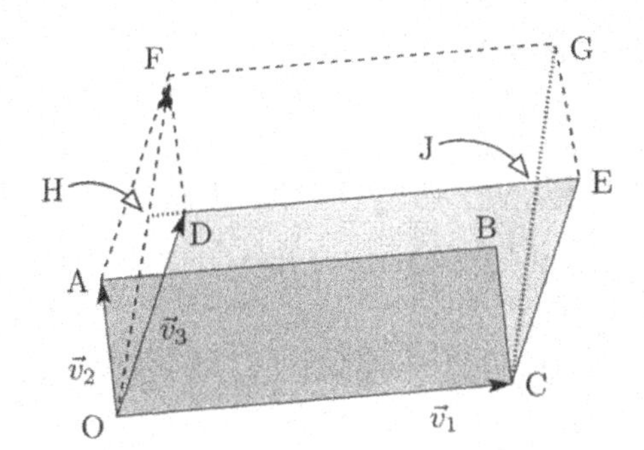

HDF $\cong$ JEG

HDO $\cong$ JEC

so

area HJGF = area DEGF = area OCBA

area OCJH = area OCED

add these equations:

area OCGF = area OCBA + area OCED

The last line is the statement that sum of the areas of the two parallelograms is the area of the parallelogram formed using the sum of the two vectors:

$$\Lambda(\vec{v}_1, \vec{v}_2 + \vec{v}_3) = \Lambda(\vec{v}_1, \vec{v}_2) + \Lambda(\vec{v}_1, \vec{v}_3)$$

This sort of function Λ, generalized to three dimensions, is characterized by

$$\begin{aligned} &(1) \qquad \Lambda(\alpha\vec{v}_1 + \beta\vec{v}_1{}', \vec{v}_2, \vec{v}_3) = \alpha\Lambda(\vec{v}_1, \vec{v}_2, \vec{v}_3) + \beta\Lambda(\vec{v}_1{}', \vec{v}_2, \vec{v}_3) \\ &(2) \qquad \Lambda(\vec{v}_1, \vec{v}_1, \vec{v}_3) = 0 \end{aligned} \tag{7.41}$$

It is linear in each variable, and it vanishes if any two arguments are equal. I've written it for three dimensions, but in N dimensions you have the same equations with N arguments, and these properties hold for all of them.

Theorem: Up to an overall constant factor, this function is unique.

An important result is that these assumptions imply the function is antisymmetric in any two arguments. Proof:

$$\Lambda(\vec{v}_1+\vec{v}_2,\vec{v}_1+\vec{v}_2,\vec{v}_3)=0=\Lambda(\vec{v}_1,\vec{v}_1,\vec{v}_3)+\Lambda(\vec{v}_1,\vec{v}_2,\vec{v}_3)+\Lambda(\vec{v}_2,\vec{v}_1,\vec{v}_3)+\Lambda(\vec{v}_2,\vec{v}_2,\vec{v}_3)$$

This is just the linearity property. Now the left side, and the 1st and 4th terms on the right, are zero because two arguments are equal. The rest is

$$\Lambda(\vec{v}_1,\vec{v}_2,\vec{v}_3)+\Lambda(\vec{v}_2,\vec{v}_1,\vec{v}_3)=0 \tag{7.42}$$

and this says that interchanging two arguments of Λ changes the sign. (The reverse is true also. Assume antisymmetry and deduce that it vanishes if two arguments are equal.)

I said that this function is unique up to a factor. Suppose that there are two of them: Λ and Λ'. Now show for some constant α, that $\Lambda-\alpha\Lambda'$ is identically zero. To do this, take three independent vectors and evaluate the number $\Lambda'(\vec{v}_a,\vec{v}_b,\vec{v}_c)$ There is some set of $\vec{v}$'s for which this is non-zero, otherwise Λ' is identically zero and that's not much fun. Now consider

$$\alpha=\frac{\Lambda(\vec{v}_a,\vec{v}_b,\vec{v}_c)}{\Lambda'(\vec{v}_a,\vec{v}_b,\vec{v}_c)}\qquad\text{and define}\qquad\Lambda_0=\Lambda-\alpha\Lambda'$$

This function Λ_0 is zero for the special argument: $(\vec{v}_a,\vec{v}_b,\vec{v}_c)$, and now I'll show why it is zero for *all* arguments. That means that it is the zero function, and says that the two functions Λ and Λ' are proportional.

The vectors $(\vec{v}_a,\vec{v}_b,\vec{v}_c)$ are independent and there are three of them (in three dimensions). They are a basis. You can write any vector as a linear combination of these. *E.g.*

$$\vec{v}_1=A\vec{v}_a+B\vec{v}_b\qquad\text{and}\qquad\vec{v}_2=C\vec{v}_a+D\vec{v}_b\qquad\text{and}$$

Put these (and let's say $\vec{v}_c$) into Λ_0.

$$\Lambda_0(\vec{v}_1,\vec{v}_2,v_c)=AC\Lambda_0(\vec{v}_a,\vec{v}_a,\vec{v}_c)+AD\Lambda_0(\vec{v}_a,\vec{v}_b,\vec{v}_c)+BC\Lambda_0(\vec{v}_b,\vec{v}_a,\vec{v}_c)+BD\Lambda_0(\vec{v}_b,\vec{v}_b,\vec{v}_c)$$

All these terms are zero. Any argument that you put into Λ_0 is a linear combination of $\vec{v}_a$, $\vec{v}_b$, and $\vec{v}_c$, and that means that this demonstration extends to any set of vectors, which in turn means that Λ_0 vanishes for any arguments. It is identically zero and that implies Λ and Λ' are, up to a constant overall factor, the same.

> In N dimensions, a scalar-valued function of N vector variables, linear in each argument and antisymmetric under interchanging any pairs of arguments, is unique up to a factor.

I've characterized this volume function Λ by two simple properties, and surprisingly enough this is all you need to *compute* it in terms of the components of the operator! With just this much information you can compute the determinant of a transformation.

Recall: $\vec{v}_1$ has for its components the first column of the matrix for the components of f, and $\vec{v}_2$ forms the second column. Adding any multiple of one vector to another leaves the volume alone. This is

$$\Lambda(\vec{v}_1, \vec{v}_2 + \alpha\vec{v}_1, \vec{v}_3) = \Lambda(\vec{v}_1, \vec{v}_2, \vec{v}_3) + \alpha\Lambda(\vec{v}_1, \vec{v}_1, \vec{v}_3) \tag{7.43}$$

and the last term is zero. Translate this into components. Use the common notation for a determinant, a square array with vertical bars, but *forget that you know how to compute this symbol!* I'm going to use it simply as a notation by keep track of vector manipulations. The numerical value will come out at the end as the computed value of a volume. $\vec{v}_i = f(\vec{e}_i) = \sum_j f_{ji}\vec{e}_j$, then $\Lambda(\vec{v}_1, \vec{v}_2, \vec{v}_3) = \Lambda(\vec{v}_1, \vec{v}_2 + \alpha\vec{v}_1, \vec{v}_3) =$

$$\begin{vmatrix} f_{11} & f_{12}+\alpha f_{11} & f_{13} \\ f_{21} & f_{22}+\alpha f_{21} & f_{23} \\ f_{31} & f_{32}+\alpha f_{31} & f_{33} \end{vmatrix} = \begin{vmatrix} f_{11} & f_{12} & f_{13} \\ f_{21} & f_{22} & f_{23} \\ f_{31} & f_{32} & f_{33} \end{vmatrix} + \alpha \begin{vmatrix} f_{11} & f_{11} & f_{13} \\ f_{21} & f_{21} & f_{23} \\ f_{31} & f_{31} & f_{33} \end{vmatrix} = \begin{vmatrix} f_{11} & f_{12} & f_{13} \\ f_{21} & f_{22} & f_{23} \\ f_{31} & f_{32} & f_{33} \end{vmatrix}$$

To evaluate this object, simply choose α to make the element $f_{12}+\alpha f_{11} = 0$. Then repeat the operation, adding a multiple of the first column to the third, making the element $f_{13}+\beta f_{11} = 0$. This operation doesn't change the original value of $\Lambda(\vec{v}_1, \vec{v}_2, \vec{v}_3)$.

$$\Lambda(\vec{v}_1, \vec{v}_2 + \alpha\vec{v}_1, \vec{v}_3 + \beta\vec{v}_1) = \begin{vmatrix} f_{11} & 0 & 0 \\ f_{21} & f_{22}+\alpha f_{21} & f_{23}+\beta f_{21} \\ f_{31} & f_{32}+\alpha f_{31} & f_{33}+\beta f_{31} \end{vmatrix} = \begin{vmatrix} f_{11} & 0 & 0 \\ f_{21} & f'_{22} & f'_{23} \\ f_{31} & f'_{32} & f'_{33} \end{vmatrix}$$

Repeat the process to eliminate f'_{23}, adding $\gamma\vec{v}'_2$ to the third argument, where $\gamma = -f'_{23}/f'_{22}$.

$$= \begin{vmatrix} f_{11} & 0 & 0 \\ f_{21} & f'_{22} & f'_{23} \\ f_{31} & f'_{32} & f'_{33} \end{vmatrix} = \begin{vmatrix} f_{11} & 0 & 0 \\ f_{21} & f'_{22} & f'_{23}+\gamma f'_{22} \\ f_{31} & f'_{32} & f'_{33}+\gamma f'_{32} \end{vmatrix} = \begin{vmatrix} f_{11} & 0 & 0 \\ f_{21} & f'_{22} & 0 \\ f_{31} & f'_{32} & f''_{33} \end{vmatrix} \tag{7.44}$$

Written in the last form, as a triangular array, the final result for the determinant *does not depend* on the elements f_{21}, f_{31}, f'_{32}. They may as well be zero. Why? Just do the same sort of column operations, but working toward the left. Eliminate f_{31} and f'_{32} by adding a constant times the third column to the first and second columns. Then eliminate f_{21} by using the second column. You don't actually have to do this, you just have to recognize that it can be done so that you can ignore the lower triangular part of the array.

Translate this back to the original vectors and Λ is unchanged:

$$\Lambda(\vec{v}_1, \vec{v}_2, \vec{v}_3) = \Lambda(f_{11}\vec{e}_1, f'_{22}\vec{e}_2, f''_{33}\vec{e}_3) = f_{11}f'_{22}f''_{33}\,\Lambda(\vec{e}_1, \vec{e}_2, \vec{e}_3)$$

The volume of the original box is $\Lambda(\vec{e}_1, \vec{e}_2, \vec{e}_3)$, so the quotient of the new volume to the old one is

$$\det = f_{11}f'_{22}f''_{33} \tag{7.45}$$

The fact that Λ is unique up to a constant factor doesn't matter. Do you want to measure volume in cubic feet, cubic centimeters, or cubic light-years? This algorithm is called

Gauss elimination. It's development started with the geometry and used vector manipulations to recover what you may recognize from elsewhere as the traditional computed value of the determinant.

Did I leave anything out in this computation of the determinant? Yes, one point. What if in Eq. (7.44) the number $f'_{22} = 0$? You can't divide by it then. You can however interchange any two arguments of Λ, causing simply a sign change. If this contingency occurs then you need only interchange the two columns to get a component of zero where you want it. Just keep count of such switches whenever they occur.

Trace

There's a property closely related to the determinant of an operator. It's called the trace. If you have an operator f, then consider the determinant of $M = I + \epsilon f$, where I is the identity. This combination is very close to the identity if ϵ is small enough, so its determinant is very close to one. How close? The first order in ϵ is called the trace of f, or more formally

$$\operatorname{Tr}(f) = \frac{d}{d\epsilon} \det\big(I + \epsilon f\big)\Big|_{\epsilon=0} \tag{7.46}$$

Express this in components for a two dimensional case, and

$$(f) = \begin{pmatrix} a & b \\ c & d \end{pmatrix} \Rightarrow \det\big(I + \epsilon f\big) = \det\begin{pmatrix} 1+\epsilon a & \epsilon b \\ \epsilon c & 1+\epsilon d \end{pmatrix} = (1+\epsilon a)(1+\epsilon d) - \epsilon^2 bc \tag{7.47}$$

The first order coefficient of ϵ is $a + d$, the sum of the diagonal elements of the matrix. This is the form of the result in any dimension, and the proof involves carefully looking at the method of Gauss elimination for the determinant, remembering at every step that you're looking for *only* the first order term in ϵ. See problem 7.53.

7.8 Matrices as Operators

There's an important example of a vector space that I've avoided mentioning up to now. Example 5 in section 6.3 is the set of n-tuples of numbers: $(a_1, a_2, \ldots, a_n)$. I can turn this on its side, call it a column matrix, and it forms a perfectly good vector space. The functions (operators) on this vector space are the matrices themselves.

When you have a system of linear equations, you can translate this into the language of vectors.

$$ax + by = e \qquad \text{and} \qquad cx + dy = f \qquad \longrightarrow \qquad \begin{pmatrix} a & b \\ c & d \end{pmatrix}\begin{pmatrix} x \\ y \end{pmatrix} = \begin{pmatrix} e \\ f \end{pmatrix}$$

Solving for x and y is inverting a matrix.

There's an aspect of this that may strike you as odd. This matrix is an operator on the vector space of column matrices. What are the components of this operator? What? Isn't the matrix a set of components already? That depends on your choice of basis. Take an example

$$M = \begin{pmatrix} 1 & 2 \\ 3 & 4 \end{pmatrix} \qquad \text{with basis} \qquad \vec e_1 = \begin{pmatrix} 1 \\ 0 \end{pmatrix}, \qquad \vec e_2 = \begin{pmatrix} 0 \\ 1 \end{pmatrix}$$

Compute the components as usual.

$$M\vec e_1 = \begin{pmatrix} 1 & 2 \\ 3 & 4 \end{pmatrix}\begin{pmatrix} 1 \\ 0 \end{pmatrix} = \begin{pmatrix} 1 \\ 3 \end{pmatrix} = 1\vec e_1 + 3\vec e_2$$

This says that the first column of the components of M in this basis are $\begin{pmatrix} 1 \\ 3 \end{pmatrix}$. What else would you expect? Now select a different basis.

$$\vec e_1 = \begin{pmatrix} 1 \\ 1 \end{pmatrix}, \qquad \vec e_2 = \begin{pmatrix} 1 \\ -1 \end{pmatrix}$$

Again compute the component.

$$M\vec e_1 = \begin{pmatrix} 1 & 2 \\ 3 & 4 \end{pmatrix}\begin{pmatrix} 1 \\ 1 \end{pmatrix} = \begin{pmatrix} 3 \\ 7 \end{pmatrix} = 5\begin{pmatrix} 1 \\ 1 \end{pmatrix} - 2\begin{pmatrix} 1 \\ -1 \end{pmatrix} = 5\vec e_1 - 2\vec e_2$$

$$M\vec e_2 = \begin{pmatrix} 1 & 2 \\ 3 & 4 \end{pmatrix}\begin{pmatrix} 1 \\ -1 \end{pmatrix} = \begin{pmatrix} -1 \\ -1 \end{pmatrix} = -1\vec e_1$$

$$\text{The components of } M \text{ in this basis are } \begin{pmatrix} 5 & -1 \\ -2 & 0 \end{pmatrix}$$

It doesn't look at all the same, but it represents the same operator. Does this matrix have the same determinant, using Eq. (7.39)?

Determinant of Composition

If you do one linear transformation followed by another one, that is the composition of the two functions, each operator will then have its own determinant. What is the determinant of the composition? Let the operators be F and G. One of them changes areas by a scale factor $\det(F)$ and the other ratio of areas is $\det(G)$. If you use the composition of the two functions, FG or GF, the overall ratio of areas from the start to the finish will be the same:

$$\det(FG) = \det(F)\cdot\det(G) = \det(G)\cdot\det(F) = \det(GF) \tag{7.48}$$

Recall that the the determinant measures the ratio of areas for any input area, not just a square; it can be a parallelogram. The overall ratio of the product of the individual ratios, $\det(F)\det(G)$. The product of these two numbers is the total ratio of a new area to the original area and it is independent of the order of F and G, so the determinant of the composition of the functions is also independent of order.

Now what about the statement that the definition of the determinant doesn't depend on the original area that you start with. To show this takes a couple of steps. First, start with a square that's not at the origin. You can always picture it as a piece of a square that *is* at the origin. The shaded square that is 1/16 the area of the big square goes over to a parallelogram that's 1/16 the area of the big parallelogram. Same ratio.

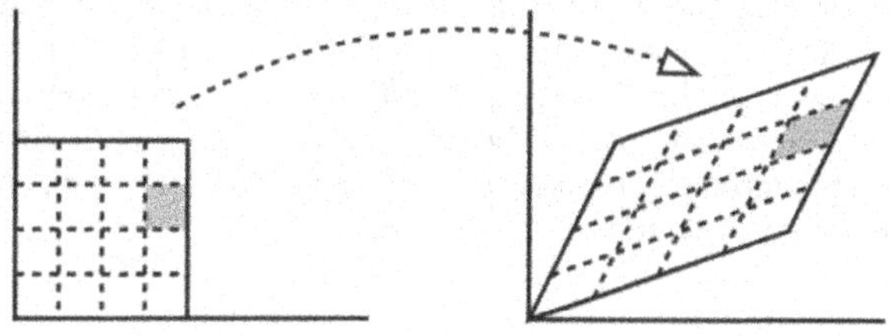

An arbitrary shape can be divided into a lot of squares. That's how you do an integral. The image of the whole area is distorted, but it retains the fact that a square that was inside the original area will become a parallelogram that is inside the new area. In the limit as the number of squares goes to infinity you still maintain the same ratio of areas as for the single original square.

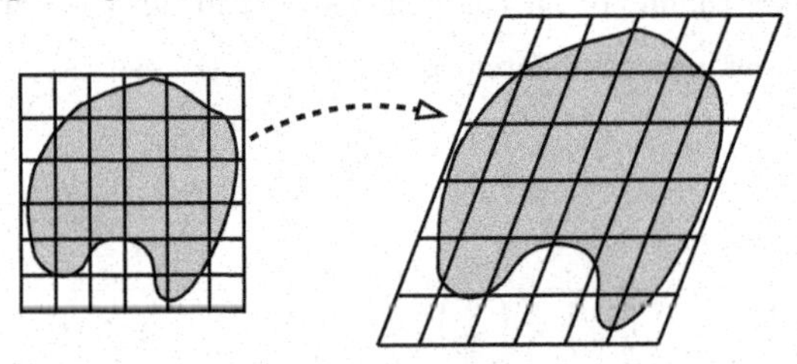

7.9 Eigenvalues and Eigenvectors

There is a particularly important basis for an operator, the basis in which the components form a diagonal matrix. Such a basis almost always* exists, and it's easy to see *from the definition* as usual just what this basis must be.

$$f(\vec{e}_i) = \sum_{k=1}^{N} f_{ki}\vec{e}_k$$

To be diagonal simply means that $f_{ki} = 0$ for all $i \neq k$, and that in turn means that all but one term in the sum disappears. This defining equation reduces to

$$f(\vec{e}_i) = f_{ii}\,\vec{e}_i \qquad \text{(with no sum this time)} \tag{7.49}$$

This is called an *eigenvalue* equation. It says that for any one of these special vectors, the operator f on it returns a scalar multiple of that same vector. These multiples are called the eigenvalues, and the corresponding vectors are called the eigenvectors. The eigenvalues are then the diagonal elements of the matrix in this basis.

The inertia tensor is the function that relates the angular momentum of a rigid body to its angular velocity. The axis of rotation is defined by those points in the rotating body that aren't moving, and the vector $\vec{\omega}$ lies along that line. The angular momentum is computed from Eq. (7.3) and when you've done all those vector products and integrals you can't really expect the angular momentum to line up with $\vec{\omega}$ unless there is some exceptional reason for it. As the body rotates around the $\vec{\omega}$ axis, $\vec{L}$ will be carried with it, making $\vec{L}$ rotate about the direction of $\vec{\omega}$. The vector $\vec{L}$ is time-dependent and that implies there will be a torque necessary to keep it going, $\vec{\tau} = d\vec{L}/dt$. Because $\vec{L}$ is rotating with frequency ω, this rotating torque will be felt as a vibration at this rotation frequency. If however the angular momentum happens to be parallel to the angular velocity, the angular momentum will not be changing; $d\vec{L}/dt = 0$ and the torque $\vec{\tau} = d\vec{L}/dt$ will be zero, implying that the vibrations will be absent. Have you ever taken your car in for servicing and asked the mechanic to make the angular momentum and the angular velocity vectors of the wheels parallel? It's called wheel-alignment.

---

* See section 7.12.

How do you compute these eigenvectors? Just move everything to the left side of the preceding equation.

$$f(\vec{e}_i) - f_{ii}\,\vec{e}_i = 0, \qquad \text{or} \qquad (f - f_{ii}I)\vec{e}_i = 0$$

I is the identity operator, output equals input. This notation is cumbersome. I'll change it.

$$f(\vec{v}) = \lambda\vec{v} \;\leftrightarrow\; (f - \lambda I)\vec{v} = 0 \tag{7.50}$$

λ is the eigenvalue and $\vec{v}$ is the eigenvector. This operator $(f - \lambda I)$ takes some non-zero vector into the zero vector. In two dimensions then it will squeeze an area down to a line or a point. In three dimensions it will squeeze a volume down to an area (or a line or a point). In any case the ratio of the final area (or volume) to the intial area (or volume) is zero. That says the determinant is zero, and that's the key to computing the eigenvectors. Figure out which λ's will make this determinant vanish.

Look back at section 4.9 and you'll see that the analysis there closely parallels what I'm doing here. In that case I didn't use the language of matrices or operators, but was asking about the possible solutions of two simultaneous linear equations.

$$ax + by = 0 \qquad \text{and} \qquad cx + dy = 0, \qquad \text{or} \qquad \begin{pmatrix} a & b \\ c & d \end{pmatrix}\begin{pmatrix} x \\ y \end{pmatrix} = \begin{pmatrix} 0 \\ 0 \end{pmatrix}$$

The explicit algebra there led to the conclusion that there can be a non-zero solution (x, y) to the two equations only if the determinant of the coefficients vanishes, $ad - bc = 0$, and that's the same thing that I'm looking for here: a non-zero vector solution to Eq. (7.50).

Write the problem in terms of components, and of course you aren't yet in the basis where the matrix is diagonal. If you were, you're already done. The defining equation is $f(\vec{v}) = \lambda\vec{v}$, and in components this reads

$$\sum_i f_{ki}v_i = \lambda v_k, \qquad \text{or} \qquad \begin{pmatrix} f_{11} & f_{12} & f_{13} \\ f_{21} & f_{22} & f_{23} \\ f_{31} & f_{32} & f_{33} \end{pmatrix}\begin{pmatrix} v_1 \\ v_2 \\ v_3 \end{pmatrix} = \lambda\begin{pmatrix} v_1 \\ v_2 \\ v_3 \end{pmatrix}$$

Here I arbitrarily wrote the equation for three dimensions. That will change with the problem. Put everything on the left side and insert the components of the identity, the unit matrix.

$$\left[\begin{pmatrix} f_{11} & f_{12} & f_{13} \\ f_{21} & f_{22} & f_{23} \\ f_{31} & f_{32} & f_{33} \end{pmatrix} - \lambda\begin{pmatrix} 1 & 0 & 0 \\ 0 & 1 & 0 \\ 0 & 0 & 1 \end{pmatrix}\right]\begin{pmatrix} v_1 \\ v_2 \\ v_3 \end{pmatrix} = \begin{pmatrix} 0 \\ 0 \\ 0 \end{pmatrix} \tag{7.51}$$

The one way that this has a non-zero solution for the vector $\vec{v}$ is for the determinant of the whole matrix on the left-hand side to be zero. This equation is called the *characteristic equation* of the matrix, and in the example here that's a cubic equation in λ. If it has all distinct roots, no double roots, then you're guaranteed that this procedure will work, and you will be able to find a basis in which the components form a diagonal matrix. If this equation has a multiple root then there is no guarantee. It may work, but it may not; you have to look closer. See section 7.12. If the operator has certain symmetry properties then it's guaranteed to work. For example, the symmetry property found in problem 7.16

is enough to insure that you can find a basis in which the matrix for the inertia tensor is diagonal. It is even an orthogonal basis in that case.

Example of Eigenvectors

To keep the algebra to a minimum, I'll work in two dimensions and will specify an arbitrary but simple example:

$$f(\vec{e}_1) = 2\vec{e}_1 + \vec{e}_2, \quad f(\vec{e}_2) = 2\vec{e}_2 + \vec{e}_1 \qquad \text{with components} \qquad M = \begin{pmatrix} 2 & 1 \\ 1 & 2 \end{pmatrix} \tag{7.52}$$

The eigenvalue equation is, in component form

$$\begin{pmatrix} 2 & 1 \\ 1 & 2 \end{pmatrix} \begin{pmatrix} v_1 \\ v_2 \end{pmatrix} = \lambda \begin{pmatrix} v_1 \\ v_2 \end{pmatrix} \qquad \text{or} \qquad \left[\begin{pmatrix} 2 & 1 \\ 1 & 2 \end{pmatrix} - \lambda \begin{pmatrix} 1 & 0 \\ 0 & 1 \end{pmatrix} \right] \begin{pmatrix} v_1 \\ v_2 \end{pmatrix} = 0 \tag{7.53}$$

The condition that there be a non-zero solution to this is

$$\det \left[\begin{pmatrix} 2 & 1 \\ 1 & 2 \end{pmatrix} - \lambda \begin{pmatrix} 1 & 0 \\ 0 & 1 \end{pmatrix} \right] = 0 = (2-\lambda)^2 - 1$$

The solutions to this quadratic are $\lambda = 1,\ 3$. For these values then, the apparently two equation for the two unknowns v_1 and v_2 are really one equation. The other is not independent. Solve this single equation in each case. Take the first of the two linear equations for v_1 and v_2 as defined by Eq. (7.53).

$$2v_1 + v_2 = \lambda v_1$$
$$\lambda = 1 \text{ implies} \qquad v_2 = -v_1, \qquad \lambda = 3 \text{ implies} \qquad v_2 = v_1$$

The two new basis vectors are then

$$\vec{e}\,'_1 = (\vec{e}_1 - \vec{e}_2) \qquad \text{and} \qquad \vec{e}\,'_2 = (\vec{e}_1 + \vec{e}_2) \tag{7.54}$$

and in this basis the matrix of components is the diagonal matrix of eigenvalues.

$$\begin{pmatrix} 1 & 0 \\ 0 & 3 \end{pmatrix}$$

If you like to keep your basis vectors normalized, you may prefer to say that the new basis is $(\vec{e}_1 - \vec{e}_2)/\sqrt{2}$ and $(\vec{e}_1 + \vec{e}_2)/\sqrt{2}$. The eigenvalues are the same, so the new matrix is the same.

Example: Coupled Oscillators

Another example drawn from physics: Two masses are connected to a set of springs and fastened between two rigid walls. This is a problem that appeared in chapter 4, Eq. (4.45).

$$m_1 d^2x_1/dt^2 = -k_1x_1 - k_3(x_1 - x_2), \qquad \text{and} \qquad m_2 d^2x_2/dt^2 = -k_2x_2 - k_3(x_2 - x_1)$$

The exponential form of the solution was

$$x_1(t) = Ae^{i\omega t}, \qquad x_2(t) = Be^{i\omega t}$$

The algebraic equations that you get by substituting these into the differential equations are a pair of linear equations for A and B, Eq. (4.47). In matrix form these equations are, after rearranging some minus signs,

$$\begin{pmatrix} k_1 + k_3 & -k_3 \\ -k_3 & k_2 + k_3 \end{pmatrix} \begin{pmatrix} A \\ B \end{pmatrix} = \omega^2 \begin{pmatrix} m_1 & 0 \\ 0 & m_2 \end{pmatrix} \begin{pmatrix} A \\ B \end{pmatrix}$$

You can make it look more like the previous example with some further arrangement

$$\left[\begin{pmatrix} k_1 + k_3 & -k_3 \\ -k_3 & k_2 + k_3 \end{pmatrix} - \omega^2 \begin{pmatrix} m_1 & 0 \\ 0 & m_2 \end{pmatrix} \right] \begin{pmatrix} A \\ B \end{pmatrix} = \begin{pmatrix} 0 \\ 0 \end{pmatrix}$$

The matrix on the left side maps the column matrix to zero. That can happen only if the matrix has zero determinant (or the column matrix is zero). If you write out the determinant of this 2×2 matrix you have a quadratic equation in ω^2. It's simple but messy, so rather than looking first at the general case, look at a special case with more symmetry. Take $m_1 = m_2 = m$ and $k_1 = k_2$.

$$\det \left[\begin{pmatrix} k_1 + k_3 & -k_3 \\ -k_3 & k_1 + k_3 \end{pmatrix} - \omega^2 m \begin{pmatrix} 1 & 0 \\ 0 & 1 \end{pmatrix} \right] = 0 = \left(k_1 + k_3 - m\omega^2\right)^2 - k_3^2$$

This is now so simple that you don't even need the quadratic formula; it factors directly.

$$\left(k_1 + k_3 - m\omega^2 - k_3\right)\left(k_1 + k_3 - m\omega^2 + k_3\right) = 0$$

The only way that the product of two numbers is zero is if one of the numbers is zero, so either

$$k_1 - m\omega^2 = 0 \qquad \text{or} \qquad k_1 + 2k_3 - m\omega^2 = 0$$

This determines two possible frequencies of oscillation.

$$\omega_1 = \sqrt{\frac{k_1}{m}} \qquad \text{and} \qquad \omega_2 = \sqrt{\frac{k_1 + 2k_3}{m}}$$

You're not done yet; these are just the eigenvalues. You still have to find the eigenvectors and *then* go back to apply them to the original problem. This is $\vec{F} = m\vec{a}$ after all. Look back to section 4.10 for the development of the solutions.

7.10 Change of Basis

In many problems in physics and mathematics, the correct choice of basis can enormously simplify a problem. Sometimes the obvious choice of a basis turns out in the end not to be the best choice, and you then face the question: Do you start over with a new basis, or can you use the work that you've already done to transform everything into the new basis?

For linear transformations, this becomes the problem of computing the components of an operator in a new basis in terms of its components in the old basis.

First: Review how to do this for vector components, something that ought to be easy to do. The equation (7.5) defines the components with respect to a basis, *any* basis.

If I have a second proposed basis, then by the definition of the word basis, every vector in that second basis can be written as a linear combination of the vectors in the first basis. I'll call the vectors in the first basis, $\vec e_i$ and those in the second basis $\vec e\,'_i$, for example in the plane you could have

$$\vec e_1 = \hat x, \quad \vec e_2 = \hat y, \qquad \text{and} \qquad \vec e\,'_1 = 2\hat x + 0.5\hat y, \quad \vec e\,'_2 = 0.5\hat x + 2\hat y \tag{7.55}$$

Each vector $\vec e\,'_i$ is a linear combination* of the original basis vectors:

$$\vec e\,'_i = S(\vec e_i) = \sum_j S_{ji}\vec e_j \tag{7.56}$$

This follows the standard notation of Eq. (7.6); you have to put the indices in this order in order to make the notation come out right in the end. One vector expressed in two different bases is still one vector, so

$$\vec v = \sum_i v'_i \vec e\,'_i = \sum_i v_i \vec e_i$$

and I'm using the fairly standard notation of v'_i for the ith component of the vector $\vec v$ with respect to the second basis. Now insert the relation between the bases from the preceding equation (7.56).

$$\vec v = \sum_i v'_i \sum_j S_{ji}\vec e_j = \sum_j v_j \vec e_j$$

and this used the standard trick of changing the last dummy label of summation from i to j so that it is easy to compare the components.

$$\sum_i S_{ji}v'_i = v_j \qquad \text{or in matrix notation} \qquad (S)(v') = (v), \implies (v') = (S)^{-1}(v)$$

Similarity Transformations

Now use the definition of the components of an operator to get the components in the new basis.

$$\begin{aligned} f\big(\vec e\,'_i\big) = & \qquad\qquad\qquad\qquad\qquad\qquad\qquad\qquad = \sum_j f'_{ji}\vec e\,'_j \\ f\Big(\sum_j S_{ji}\vec e_j\Big) &= \sum_j S_{ji} f\big(\vec e_j\big) = \sum_j S_{ji}\sum_k f_{kj}\vec e_k = \sum_j f'_{ji}\sum_k S_{kj}\vec e_k \end{aligned}$$

The final equation comes from the preceding line. The coefficients of $\vec e_k$ must agree on the two sides of the equation.

$$\sum_j S_{ji}f_{kj} = \sum_j f'_{ji}S_{kj}$$

---

* There are two possible conventions here. You can write $\vec e\,'_i$ in terms of the $\vec e_i$, calling the coefficients S_{ji}, or you can do the reverse and call *those* components S_{ji}. $[\vec e_i = S(\vec e\,'_i)]$ Naturally, both conventions are in common use. The reverse convention will interchange the roles of the matrices S and S^{-1} in what follows.

Now rearrange this in order to place the indices in their conventional $_{\text{row,column}}$ order.

$$\sum_j S_{kj} f'_{ji} = \sum_j f_{kj} S_{ji}$$
$$\begin{pmatrix} S_{11} & S_{12} \\ S_{21} & S_{22} \end{pmatrix} \begin{pmatrix} f'_{11} & f'_{12} \\ f'_{21} & f'_{22} \end{pmatrix} = \begin{pmatrix} f_{11} & f_{12} \\ f_{21} & f_{22} \end{pmatrix} \begin{pmatrix} S_{11} & S_{12} \\ S_{21} & S_{22} \end{pmatrix} \tag{7.57}$$

In turn, this matrix equation is usually written in terms of the inverse matrix of S,

$$(S)(f') = (f)(S) \qquad \text{is} \qquad (f') = (S)^{-1}(f)(S) \tag{7.58}$$

and this is called a similarity transformation. For the example Eq. (7.55) this is

$$\vec{e}\,'_1 = 2\hat{x} + 0.5\hat{y} = S_{11}\vec{e}_1 + S_{21}\vec{e}_2$$

which determines the first column of the matrix (S), then $\vec{e}\,'_2$ determines the second column.

$$(S) = \begin{pmatrix} 2 & 0.5 \\ 0.5 & 2 \end{pmatrix} \qquad \text{then} \qquad (S)^{-1} = \frac{1}{3.75}\begin{pmatrix} 2 & -0.5 \\ -0.5 & 2 \end{pmatrix}$$

Eigenvectors

In defining eigenvalues and eigenvectors I pointed out the utility of having a basis in which the components of an operator form a diagonal matrix. Finding the non-zero solutions to Eq. (7.50) is then the way to find the basis in which this holds. Now I've spent time showing that you can find a matrix in a new basis by using a similarity transformation. Is there a relationship between these two subjects? Another way to ask the question: I've solved the problem to find all the eigenvectors and eigenvalues, so what is the similarity transformation that accomplishes the change of basis (and why is it necessary to know it if I already know that the transformed, diagonal matrix is just the set of eigenvalues, and I already know them)?

For the last question, the simplest answer is that you *don't* need to know the explicit transformation once you already know the answer. It is however useful to know that it exists and how to construct it. *If* it exists — I'll come back to that presently. Certain manipulations are more easily done in terms of similarity transformations, so you ought to know how they are constructed, especially because almost all the work in constructing them is done when you've found the eigenvectors.

The equation (7.57) tells you the answer. Suppose that you want the transformed matrix to be diagonal. That means that $f'_{12} = 0$ and $f'_{21} = 0$. Write out the first column of the product on the right.

$$\begin{pmatrix} f_{11} & f_{12} \\ f_{21} & f_{22} \end{pmatrix} \begin{pmatrix} S_{11} & S_{12} \\ S_{21} & S_{22} \end{pmatrix} \longrightarrow \begin{pmatrix} f_{11} & f_{12} \\ f_{21} & f_{22} \end{pmatrix} \begin{pmatrix} S_{11} \\ S_{21} \end{pmatrix}$$

This equals the first column on the left of the same equation

$$f'_{11} \begin{pmatrix} S_{11} \\ S_{21} \end{pmatrix}$$

This is the eigenvector equation that you've supposedly already solved. The first column of the component matrix of the similarity transformation is simply the set of components of the first eigenvector. When you write out the second column of Eq. (7.57) you'll see that it's the defining equation for the second eigenvector. You already know these, so you can immediately write down the matrix for the similarity transformation.

For the example Eq. (7.52) the eigenvectors are given in Eq. (7.54). In components these are

$$\vec{e}\,'_1 \to \begin{pmatrix} 1 \\ -1 \end{pmatrix}, \qquad \text{and} \qquad \vec{e}\,'_2 \to \begin{pmatrix} 1 \\ 1 \end{pmatrix}, \qquad \text{implying} \qquad S = \begin{pmatrix} 1 & 1 \\ -1 & 1 \end{pmatrix}$$

The inverse to this matrix is

$$S^{-1} = \frac{1}{2}\begin{pmatrix} 1 & -1 \\ 1 & 1 \end{pmatrix}$$

You should verify that $S^{-1}MS$ is diagonal.

7.11 Summation Convention

In all the manipulation of components of vectors and components of operators you have to do a lot of sums. There are so many sums over indices that a convention* was invented (by Einstein) to simplify the notation.

A repeated index in a term is summed.

Eq. (7.6) becomes $f(\vec{e}_i) = f_{ki}\vec{e}_k$.
Eq. (7.8) becomes $u_k = f_{ki}v_i$.
Eq. (7.26) becomes $h_{ki} = f_{kj}g_{ji}$.
$IM = M$ becomes $\delta_{ij}M_{jk} = M_{ik}$.

What if there are *three* identical indices in the same term? Then you made a mistake; that can't happen. What about Eq. (7.49)? That has three indices. Yes, and there I explicitly said that there is no sum. This sort of rare case you have to handle as an exception.

7.12 Can you Diagonalize a Matrix?

At the beginning of section 7.9 I said that the basis in which the components of an operator form a diagonal matrix "almost always exists." There's a technical sense in which this is precisely true, but that's not what you need to know in order to manipulate matrices; the theorem that you need to have is that every matrix is the limit of a sequence of diagonalizable matrices. If you encounter a matrix that cannot be diagonalized, then you can approximate it as closely as you want by a matrix that can be diagonalized, do your calculations, and finally take a limit. You already did this if you did problem 4.11, but in that chapter it didn't look anything like a problem involving matrices, much less diagonalization of matrices. Yet it is the same.

Take the matrix

$$\begin{pmatrix} 1 & 2 \\ 0 & 1 \end{pmatrix}$$

---

* There is a modification of this convention that appears in chapter 12, section 12.5

You can't diagonalize this. If you try the standard procedure, here is what happens:

$$\begin{pmatrix} 1 & 2 \\ 0 & 1 \end{pmatrix} \begin{pmatrix} v_1 \\ v_2 \end{pmatrix} = \lambda \begin{pmatrix} v_1 \\ v_2 \end{pmatrix} \qquad \text{then} \qquad \det \begin{pmatrix} 1-\lambda & 2 \\ 0 & 1-\lambda \end{pmatrix} = 0 = (1-\lambda)^2$$

The resulting equations you get for $\lambda = 1$ are

$$0v_1 + 2v_2 = 0 \qquad \text{and} \qquad 0 = 0$$

This provides only one eigenvector, a multiple of $\begin{pmatrix} 1 \\ 0 \end{pmatrix}$. You need two for a basis.

Change this matrix in any convenient way to make the two roots of the characteristic equation different from each other. For example,

$$M_\epsilon = \begin{pmatrix} 1+\epsilon & 2 \\ 0 & 1 \end{pmatrix}$$

The eigenvalue equation is now

$$(1+\epsilon-\lambda)(1-\lambda) = 0$$

and the resulting equations for the eigenvectors are

$$\lambda = 1: \ \epsilon v_1 + 2v_2 = 0, \quad 0 = 0 \qquad \lambda = 1+\epsilon: \ 0v_1 + 2v_2 = 0, \quad \epsilon v_2 = 0$$

Now you have two distinct eigenvectors,

$$\lambda = 1: \ \begin{pmatrix} 1 \\ -\epsilon/2 \end{pmatrix}, \qquad \text{and} \qquad \lambda = 1+\epsilon: \ \begin{pmatrix} 1 \\ 0 \end{pmatrix}$$

You see what happens to these vectors as $\epsilon \to 0$.

Differential Equations at Critical

Problem 4.11 was to solve the damped harmonic oscillator for the critical case that $b^2 - 4km = 0$.

$$m\frac{d^2x}{dt^2} = -kx - b\frac{dx}{dt} \tag{7.59}$$

Write this as a pair of equations, using the velocity as an independent variable.

$$\frac{dx}{dt} = v_x \qquad \text{and} \qquad \frac{dv_x}{dt} = -\frac{k}{m}x - \frac{b}{m}v_x$$

In matrix form, this is a matrix differential equation.

$$\frac{d}{dt}\begin{pmatrix} x \\ v_x \end{pmatrix} = \begin{pmatrix} 0 & 1 \\ -k/m & -b/m \end{pmatrix} \begin{pmatrix} x \\ v_x \end{pmatrix}$$

This is a linear, constant-coefficient differential equation, but now the constant coefficients are matrices. Don't let that slow you down. The reason that an exponential form of solution works is that the derivative of an exponential is an exponential. Assume such a solution here.

$$\begin{pmatrix} x \\ v_x \end{pmatrix} = \begin{pmatrix} A \\ B \end{pmatrix} e^{\alpha t}, \qquad \text{giving} \qquad \alpha \begin{pmatrix} A \\ B \end{pmatrix} e^{\alpha t} = \begin{pmatrix} 0 & 1 \\ -k/m & -b/m \end{pmatrix} \begin{pmatrix} A \\ B \end{pmatrix} e^{\alpha t} \tag{7.60}$$

When you divide the equation by $e^{\alpha t}$, you're left with an eigenvector equation where the eigenvalue is α. As usual, to get a non-zero solution set the determinant of the coefficients to zero and the characteristic equation is

$$\det \begin{pmatrix} 0-\alpha & 1 \\ -k/m & -b/m-\alpha \end{pmatrix} = \alpha(\alpha + b/m) + k/m = 0$$

with familiar roots

$$\alpha = \big(-b \pm \sqrt{b^2 - 4km}\big)/2m$$

If the two roots are equal you *may* not have distinct eigenvectors, and in this case you *do* not. No matter, you can solve any such problem for the case that $b^2 - 4km \neq 0$ and then take the limit as this approaches zero.

The eigenvectors come from the either one of the two equations represented by Eq. (7.60). Pick the simpler one, $\alpha A = B$. The column matrix $\begin{pmatrix} A \\ B \end{pmatrix}$ is then $A\begin{pmatrix} 1 \\ \alpha \end{pmatrix}$.

$$\begin{pmatrix} x \\ v_x \end{pmatrix}(t) = A_+ \begin{pmatrix} 1 \\ \alpha_+ \end{pmatrix} e^{\alpha_+ t} + A_- \begin{pmatrix} 1 \\ \alpha_- \end{pmatrix} e^{\alpha_- t}$$

Pick the initial conditions that $x(0) = 0$ and $v_x(0) = v_0$. You must choose *some* initial conditions in order to apply this technique. In matrix terminology this is

$$\begin{pmatrix} 0 \\ v_0 \end{pmatrix} = A_+ \begin{pmatrix} 1 \\ \alpha_+ \end{pmatrix} + A_- \begin{pmatrix} 1 \\ \alpha_- \end{pmatrix}$$

These are two equations for the two unknowns

$$A_+ + A_- = 0, \qquad \alpha_+ A_+ + \alpha_- A_- = v_0, \qquad \text{so} \qquad A_+ = \frac{v_0}{\alpha_+ - \alpha_-}, \qquad A_- = -A_+$$

$$\begin{pmatrix} x \\ v_x \end{pmatrix}(t) = \frac{v_0}{\alpha_+ - \alpha_-}\left[\begin{pmatrix} 1 \\ \alpha_+ \end{pmatrix} e^{\alpha_+ t} - \begin{pmatrix} 1 \\ \alpha_- \end{pmatrix} e^{\alpha_- t} \right]$$

If you now take the limit as $b^2 \to 4km$, or equivalently as $\alpha_- \to \alpha_+$, this expression is just the definition of a derivative.

$$\begin{pmatrix} x \\ v_x \end{pmatrix}(t) \longrightarrow v_0 \frac{d}{d\alpha} \begin{pmatrix} 1 \\ \alpha \end{pmatrix} e^{\alpha t} = v_0 \begin{pmatrix} t e^{\alpha t} \\ (1 + \alpha t) e^{\alpha t} \end{pmatrix} \qquad \alpha = -\frac{b}{2m} \tag{7.61}$$

7.13 Eigenvalues and Google

The motivating idea behind the search engine Google is that you want the first items returned by a search to be the most important items. How do you do this? How do you program a computer to decide which web sites are the most important?

A simple idea is to count the number of sites that contain a link to a given site, and the site that is linked to the most is then the most important site. This has the drawback that all links are treated as equal. If your site is referenced from the home page of Al Einstein, it counts no more than if it's referenced by Joe Blow. This shouldn't be.

A better idea is to assign each web page a numerical importance rating. If your site, #1, is linked from sites #11, #59, and #182, then your rating, x_1, is determined by adding those ratings (and multiplying by a suitable scaling constant).

$$x_1 = C\big(x_{11} + x_{59} + x_{182}\big)$$

Similarly the second site's rating is determined by what links to it, as

$$x_2 = C\big(x_{137} + x_{157983} + x_1 + x_{876}\big)$$

But this assumes that you already know the ratings of the sites, and that's what you're trying to find!

Write this in matrix language. Each site is an element in a huge column matrix $\{x_i\}$.

$$x_i = C\sum_{j=1}^{N} \alpha_{ij}x_j \qquad \text{or} \qquad \begin{pmatrix} x_1 \\ x_2 \\ \vdots \end{pmatrix} = C\begin{pmatrix} 0 & 0 & 1 & 0 & 1 & \ldots \\ 1 & 0 & 0 & 0 & 0 & \ldots \\ 0 & 1 & 0 & 1 & 1 & \ldots \\ \ldots \end{pmatrix}\begin{pmatrix} x_1 \\ x_2 \\ \vdots \end{pmatrix}$$

An entry of 1 indicates a link and a 0 is no link. This is an eigenvector problem with the eigenvalue $\lambda = 1/C$, and though there are many eigenvectors, there is a constraint that lets you pick the right one. All the x_is must be non-negative, and there's a theorem (Perron-Frobenius) guaranteeing that you can find such an eigenvector. This algorithm is a key idea behind Google's ranking methods. They have gone well beyond this basic technique of course, but the spirit of the method remains.

See www-db.stanford.edu/~backrub/google.html for more on this.

7.14 Special Operators

Symmetric
Antisymmetric
Hermitian
Antihermitian
Orthogonal
Unitary
Idempotent
Nilpotent
Self-adjoint

In no particular order of importance, these are names for special classes of operators. It is often the case that an operator defined in terms of a physical problem will be in some way special, and it's then worth knowing the consequent simplifications. The first ones involve a scalar product.

$$\textbf{Symmetric:} \quad \langle \vec u, S(\vec v\,)\rangle = \langle S(\vec u\,), \vec v\,\rangle$$
$$\textbf{Antisymmetric:} \quad \langle \vec u, A(\vec v\,)\rangle = -\langle A(\vec u\,), \vec v\,\rangle$$

The inertia operator of Eq. (7.3) is symmetric.

$$I(\vec\omega\,) = \int dm\, \vec r \times \left(\vec\omega \times \vec r\,\right) \quad \text{satisfies} \quad \langle \vec\omega_1, I(\vec\omega_2)\rangle = \vec\omega_1 \cdot I(\vec\omega_2) = \langle I(\vec\omega_1), \vec\omega_2\,\rangle = I(\vec\omega_1)\cdot\vec\omega_2$$

Proof: Plug in.

$$\begin{aligned}\vec\omega_1 \cdot I(\vec\omega_2) &= \vec\omega_1 \cdot \int dm\, \vec r \times \left(\vec\omega_2 \times \vec r\,\right) = \vec\omega_1 \cdot \int dm\, \left[\vec\omega_2\, r^2 - \vec r(\vec\omega_2 \cdot \vec r\,)\right] \\ &= \int dm\, \left[\vec\omega_1 \cdot \vec\omega_2\, r^2 - (\vec\omega_1 \cdot \vec r\,)(\vec\omega_2 \cdot \vec r\,)\right] = I(\vec\omega_1)\cdot\vec\omega_2\end{aligned}$$

What good does this do? You will be guaranteed that all eigenvalues are real, all eigenvectors are orthogonal, and the eigenvectors form an orthogonal basis. In this example, the eigenvalues are moments of inertia about the axes defined by the eigenvectors, so these moments better be real. The magnetic field operator (problem 7.28) is antisymmetric.

Hermitian operators obey the same identity as symmetric: $\langle \vec u, H(\vec v\,)\rangle = \langle H(\vec u\,), \vec v\,\rangle$. The difference is that in this case you allow the scalars to be complex numbers. That means that the scalar product has a complex conjugation implied in the first factor. You saw this sort of operator in the chapter on Fourier series, section 5.3, but it didn't appear under this name. You will become familiar with this class of operators when you hit quantum mechanics. Then they are ubiquitous. The same theorem as for symmetric operators applies here, that the eigenvalues are real and that the eigenvectors are orthogonal.

Orthogonal operators satisfy $\langle O(\vec u\,), O(\vec v\,)\rangle = \langle \vec u, \vec v\,\rangle$. The most familiar example is rotation. When you rotate two vectors, their magnitudes and the angle between them do not change. That's all that this equation says — scalar products are preserved by the transformation.

Unitary operators are the complex analog of orthogonal ones: $\langle U(\vec u\,), U(\vec v\,)\rangle = \langle \vec u, \vec v\,\rangle$, but all the scalars are complex and the scalar product is modified accordingly.

The next couple you don't see as often. **Idempotent** means that if you take the square of the operator, it equals the original operator.

Nilpotent means that if you take successive powers of the operator you eventually reach the zero operator.

Self-adjoint in a finite dimensional vector space is exactly the same thing as Hermitian. In infinite dimensions it is not, and in quantum mechanics the important operators are the self-adjoint ones. The issues involved are a bit technical. As an aside, in infinite dimensions you need one extra hypothesis for unitary and orthogonal: that they are invertible.

Problems

7.1 Draw a picture of the effect of these linear transformations on the unit square with vertices at $(0,0)$, $(1,0)$, $(1,1)$, $(0,1)$. The matrices representing the operators are

$$(\text{a})\ \begin{pmatrix} 1 & 2 \\ 3 & 4 \end{pmatrix}, \qquad (\text{b})\ \begin{pmatrix} 1 & -2 \\ 2 & -4 \end{pmatrix}, \qquad (\text{c})\ \begin{pmatrix} -1 & 2 \\ 1 & 2 \end{pmatrix}$$

Is the orientation preserved or not in each case? See the figure at the end of section 7.7

7.2 Using the same matrices as the preceding question, what is the picture resulting from doing (a) followed by (c)? What is the picture resulting from doing (c) followed by (a)? The results of section 7.4 may prove helpful.

7.3 Look again at the parallelogram that is the image of the unit square in the calculation of the determinant. In Eq. (7.39) I used the cross product to get its area, but sometimes a brute-force method is more persuasive. If the transformation has components $\begin{pmatrix} a & b \\ c & d \end{pmatrix}$ The corners of the parallelogram that is the image of the unit square are at $(0,0)$, (a,c), $(a+b,c+d)$, (b,d). You can compute its area as sums and differences of rectangles and triangles. Do so; it should give the same result as the method that used a cross product.

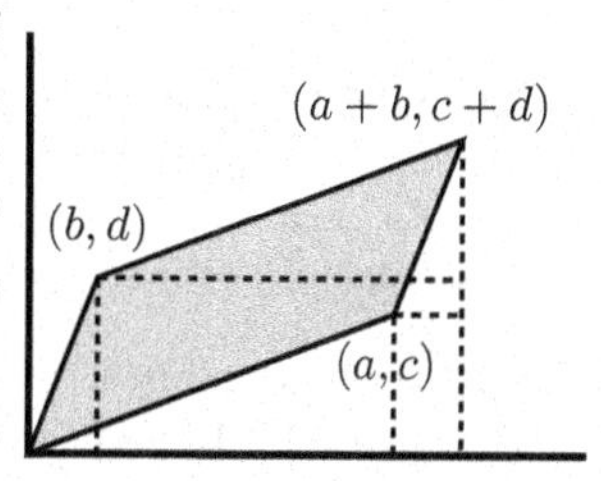

7.4 In three dimensions, there is an analogy to the geometric interpretation of the cross product as the area of a parallelogram. The triple scalar product $\vec{A} \cdot \vec{B} \times \vec{C}$ is the volume of the parallelepiped having these three vectors as edges. Prove both of these statements starting from the geometric definitions of the two products. That is, from the $AB\cos\theta$ and $AB\sin\theta$ definitions of the dot product and the magnitude of the cross product (and its direction).

7.5 Derive the relation $\vec{v} = \vec{\omega} \times \vec{r}$ for a point mass rotating about an axis. Refer to the figure before Eq. (7.2).

7.6 You have a mass attached to four springs in a plane and that are in turn attached to four walls as on page 160; the mass is at equilibrium. Two opposing spring have spring constant k_1 and the other two are k_2. Push on the mass with a (small) force $\vec{F}$ and the resulting displacement of m is $\vec{d} = f(\vec{F})$, defining a linear operator. Compute the components of f in an obvious basis and check a couple of special cases to see if the displacement is in a plausible direction, especially if the two k's are quite different.

7.7 On the vector space of quadratic polynomials, degree ≤ 2, the operator d/dx is defined: the derivative of such a polynomial is a polynomial. (**a**) Use the basis $\vec{e}_0 = 1$, $\vec{e}_1 = x$, and $\vec{e}_2 = x^2$ and compute the components of this operator. (**b**) Compute the components of the operator d^2/dx^2. (**c**) Compute the square of the first matrix and compare it to the result for (**b**). Ans: (**a**)^2 = (**b**)

7.8 Repeat the preceding problem, but look at the case of cubic polynomials, a four-dimensional space.

7.9 In the preceding problem the basis 1, x, x^2, x^3 is too obvious. Take another basis, the Legendre polynomials:

$$P_0(x) = 1, \qquad P_1(x) = x, \qquad P_2(x) = \frac{3}{2}x^2 - \frac{1}{2}, \qquad P_3(x) = \frac{5}{2}x^3 - \frac{3}{2}x$$

and repeat the problem, finding components of the first and second derivative operators. Verify an example explicitly to check that your matrix reproduces the effect of differentiation on a polynomial of your choice. Pick one that will let you test your results.

7.10 What is the determinant of the inverse of an operator, explaining why?
Ans: $1/\det(\text{original operator})$

7.11 Eight identical point masses m are placed at the corners of a cube that has one corner at the origin of the coordinates and has its sides along the axes. The side of the cube is length $= a$. In the basis that is placed along the axes as usual, compute the components of the inertia tensor. Ans: $I_{11} = 8ma^2$

7.12 For the dumbbell rotating about the off-axis axis in Eq. (7.19), what is the time-derivative of $\vec{L}$? In very short time dt, what new direction does $\vec{L}$ take and what then is $d\vec{L}$? That will tell you $d\vec{L}/dt$. Prove that this is $\vec{\omega} \times \vec{L}$.

7.13 A cube of uniform volume mass density, mass m, and side a has one corner at the origin of the coordinate system and the adjacent edges are placed along the coordinate axes. Compute the components of the tensor of inertia. Do it (**a**) directly and (**b**) by using the parallel axis theorem to check your result.

Ans: $ma^2 \begin{pmatrix} 2/3 & -1/4 & -1/4 \\ -1/4 & 2/3 & -1/4 \\ -1/4 & -1/4 & 2/3 \end{pmatrix}$

7.14 Compute the cube of Eq. (7.13) to find the trigonometric identities for the cosine and sine of triple angles in terms of single angle sines and cosines. Compare the results of problem 3.9.

7.15 On the vectors of column matrices, the operators are matrices. For the two dimensional case take $M = \begin{pmatrix} a & b \\ c & d \end{pmatrix}$ and find its components in the basis $\begin{pmatrix} 1 \\ 1 \end{pmatrix}$ and $\begin{pmatrix} 1 \\ -1 \end{pmatrix}$.
What is the determinant of the resulting matrix? Ans: $M_{11} = (a+b+c+d)/2$, and the determinant is *still* $ad - bc$.

7.16 Show that the tensor of inertia, Eq. (7.3), satisfies $\vec{\omega}_1 \cdot I(\vec{\omega}_2) = I(\vec{\omega}_1) \cdot \vec{\omega}_2$. What does this identity tell you about the components of the operator when you use the ordinary orthonormal basis? First determine in such a basis what $\vec{e}_1 \cdot I(\vec{e}_2)$ is.

7.17 Use the definition of the center of mass to show that the two cross terms in Eq. (7.21) are zero.

7.18 Prove the Perpendicular Axis Theorem. This says that for a mass that lies flat in a plane, the moment of inertia about an axis perpendicular to the plane equals the sum of the two moments of inertia about the two perpendicular axes that lie in the plane and that intersect the third axis.

7.19 Verify in the conventional, non-matrix way that Eq. (7.61) really does provide a solution to the original second order differential equation (7.59).

7.20 The Pauli spin matrices are

$$\sigma_x = \begin{pmatrix} 0 & 1 \\ 1 & 0 \end{pmatrix}, \qquad \sigma_y = \begin{pmatrix} 0 & -i \\ i & 0 \end{pmatrix}, \qquad \sigma_z = \begin{pmatrix} 1 & 0 \\ 0 & -1 \end{pmatrix}$$

Show that $\sigma_x \sigma_y = i\sigma_z$ and the same for cyclic permutations of the indices x, y, z. Compare the products $\sigma_x \sigma_y$ and $\sigma_y \sigma_x$ and the other pairings of these matrices.

7.21 Interpret $\vec{\sigma} \cdot \vec{A}$ as $\sigma_x A_x + \sigma_y A_y + \sigma_z A_z$ and prove that

$$\vec{\sigma} \cdot \vec{A}\, \vec{\sigma} \cdot \vec{B} = \vec{A} \cdot \vec{B} + i\vec{\sigma} \cdot \vec{A} \times \vec{B}$$

where the first term on the right has to include the identity matrix for this to make sense.

7.22 Evaluate the matrix

$$\frac{I}{I - \vec{\sigma} \cdot \vec{A}} = \left(I - \vec{\sigma} \cdot \vec{A}\right)^{-1}$$

Evaluate this by two methods: (**a**) You may assume that $\vec{A}$ is in some sense small enough for you to manipulate by infinite series methods. This then becomes a geometric series that you can sum. Use the results of the preceding problem.
(**b**) You can manipulate the algebra directly without series. I suggest that you recall the sort of manipulation that allows you to write the complex number $1/(1-i)$ without any i's in the denominator.
I suppose you could do it a third way, writing out the 2×2 matrix and explicitly inverting it, but I definitely don't recommend this.

7.23 Evaluate the sum of the infinite series defined by $e^{-i\sigma_y \theta}$. Where have you seen this result before? The first term in the series must be interpreted as the identity matrix. Ans: $I\cos\theta - i\sigma_y \sin\theta$

7.24 For the moment of inertia about an axis, the integral is $\int r_\perp^2 \, dm$. State precisely what this m function must be for this to make sense as a Riemann-Stieltjes integral, Eq. (1.28). For the case that you have eight masses, all m_0 at the 8 corners of a cube of side a, write explicitly what this function is and evaluate the moment of inertia about an axis along one edge of the cube.

7.25 The summation convention allows you to write some compact formulas. Evaluate these, assuming that you're dealing with three dimensions. Note Eq. (7.30). Define the alternating symbol ϵ_{ijk} to be
1: It is totally anti-symmetric. That is, interchange any two indices and you change the

sign of the value.
2: $\epsilon_{123} = 1$. [*E.g.* $\epsilon_{132} = -1$, $\epsilon_{312} = +1$]

$$\delta_{ii}, \qquad \epsilon_{ijk}A_jB_k, \qquad \delta_{ij}\epsilon_{ijk}, \qquad \delta_{mn}A_mB_n, \qquad S_{mn}u_mv_n, \qquad u_nv_n,$$
$$\epsilon_{ijk}\epsilon_{mnk} = \delta_{im}\delta_{jn} - \delta_{in}\delta_{jm}$$

Multiply the last identity by $A_jB_mC_n$ and interpret.

7.26 The set of Hermite polynomials starts out as

$$H_0 = 1, \qquad H_1 = 2x, \qquad H_2 = 4x^2 - 2, \qquad H_3 = 8x^3 - 12x, \qquad H_4 = 16x^4 - 48x^2 + 12,$$

(**a**) For the vector space of cubic polynomials in x, choose a basis of Hermite polynomials and compute the matrix of components of the differentiation operator, d/dx.
(**b**) Compute the components of the operator d^2/dx^2 and show the relation between this matrix and the preceding one.

7.27 On the vector space of functions of x, define the translation operator

$$T_af = g \qquad \text{means} \qquad g(x) = f(x-a)$$

This picks up a function and moves it by a to the right.
(**a**) Pick a simple example function f and test this definition graphically to verify that it does what I said.
(**b**) On the space of cubic polynomials and using a basis of your choice, find the components of this operator.
(**c**) Square the resulting matrix and verify that the result is as it should be.
(**d**) What is the inverse of the matrix? (You should be able to guess the answer and then verify it. Or you can work out the inverse the traditional way.)
(**e**) What if the parameter a is *huge*? Interpret some of the components of this first matrix and show why they are clearly correct. (If they are.)
(**f**) What is the determinant of this operator?
(**g**) What are the eigenvectors and eigenvalues of this operator?

7.28 The force by a magnetic field on a moving charge is $\vec{F} = q\vec{v} \times \vec{B}$. The operation $\vec{v} \times \vec{B}$ defines a linear operator on $\vec{v}$, stated as $f(\vec{v}) = \vec{v} \times \vec{B}$. What are the components of this operator expressed in terms of the three components of the vector $\vec{B}$? What are the eigenvectors and eigenvalues of this operator? For this last part, pick the basis in which you want to do the computations. If you're not careful about this choice, you are asking for a lot of algebra. Ans: eigenvalues: 0, $\pm iB$

7.29 In section 7.8 you have an operator M expressed in two different bases. What is its determinant computed in each basis?

7.30 In a given basis, an operator has the values

$$A(\vec{e}_1) = \vec{e}_1 + 3\vec{e}_2 \qquad \text{and} \qquad A(\vec{e}_2) = 2\vec{e}_1 + 4\vec{e}_4$$

(**a**) Draw a picture of what this does. (**b**) Find the eigenvalues and eigenvectors and determinant of A and see how this corresponds to the picture you just drew.

7.31 The characteristic polynomial of a matrix M is $\det(M - \lambda I)$. I is the identity matrix and λ is the variable in the polynomial. Write the characteristic polynomial for the general 2×2 matrix. Then in place of λ in this polynomial, put the matrix M itself. The constant term will have to include the factor I as usual. For this 2×2 case verify the Cayley-Hamilton Theorem, that the matrix satisfies its own characteristic equation, making this polynomial in M the zero matrix.

7.32 (**a**) For the magnetic field operator defined in problem 7.28, place $\hat{z} = \vec{e}_3$ along the direction of $\vec{B}$. Then take $\vec{e}_1 = (\hat{x} - i\hat{y})/\sqrt{2}$, $\vec{e}_2 = (\hat{x} + i\hat{y})/\sqrt{2}$ and find the components of the linear operator representing the magnetic field. (**b**) A charged particle is placed in this field and the equations of motion are $m\vec{a} = \vec{F} = q\vec{v} \times \vec{B}$. Translate this into the operator language with a matrix like that of problem 7.28, and write $\vec{F} = m\vec{a}$ in this language and this basis. Ans: (part) $m\ddot{r}_1 = -iqBr_1$, where $r_1 = (x + iy)/\sqrt{2}$. $m\ddot{r}_2 = +iqBr_2$, where $r_2 = (x - iy)/\sqrt{2}$.

7.33 For the operator in problem 7.27 part (**b**), what are the eigenvectors and eigenvalues?

7.34 A *nilpotent* operator was defined in section 7.14. For the operator defined in problem 7.8, show that it is nilpotent. How does this translate into the successive powers of its matrix components?

7.35 A cube of uniform mass density has side a and mass m. Evaluate its moment of inertia about an axis along a longest diagonal of the cube. Note: If you find yourself entangled in a calculation having multiple integrals with hopeless limits of integration, toss it out and start over. You may even find problem 7.18 useful. Ans: $ma^2/6$

7.36 Show that the set of all 2×2 matrices forms a vector space. Produce a basis for it, and so what is its dimension?

7.37 In the vector space of the preceding problem, the following transformation defines an operator. $f(M) = SMS^{-1}$. For S, use the rotation matrix of Eq. (7.13) and compute the components of this operator f. The obvious choice of basis would be matrices with a single non-zero element 1. Instead, try the basis I, σ_x, σ_y, σ_z. Ans: A rotation by 2α about the y-axis, *e.g.* $f(\vec{e}_1) = \vec{e}_1 \cos 2\alpha - \vec{e}_3 \sin 2\alpha$.

7.38 What are the eigenvectors and eigenvalues of the operator in the preceding problem? Now you'll be happy I suggested the basis that I did.

7.39 (**a**) The commutator of two matrices is defined to be $[A, B] = AB - BA$. Show that this commutator satisfies the Jacobi identity.

$$[A, [B, C]] + [B, [C, A]] + [C, [A, B]] = 0$$

(**b**) The anti-commutator of two matrices is $\{A, B\} = AB + BA$. Show that there is an identity like the Jacobi identity, but with one of the two commutators (the inner one or the outer one) replaced by an anti-commutator. I'll leave it to you to figure out which.

7.40 Diagonalize each of the Pauli spin matrices of problem 7.20. That is, find their eigenvalues and specify the respective eigenvectors as the basis in which they are diagonal.

7.41 What are the eigenvalues and eigenvectors of the rotation matrix Eq. (7.13)? Translate the answer back into a statement about rotating vectors, not just their components.

7.42 Same as the preceding problem, but replace the circular trigonometric functions in Eq. (7.13) with hyperbolic ones. Also change the sole minus sign in the matrix to a plus sign. Draw pictures of what this matrix does to the basis vectors. What is its determinant?

7.43 Compute the eigenvalues and eigenvectors of the matrix Eq. (7.18). Interpret each.

7.44 Look again at the vector space of problem 6.36 and use the basis f_1, f_2, f_3 that you constructed there. (**a**) In this basis, what are the components of the two operators described in that problem?
(**b**) What is the product of these two matrices? Do it in the order so that it represents the composition of the first rotation followed by the second rotation.
(**c**) Find the eigenvectors of this product and from the result show that the combination of the two rotations is a third rotation about an axis that you can now specify. Can you anticipate before solving it, what one of the eigenvalues will be?
(**d**) Does a sketch of this rotation axis agree with what you should get by doing the two original rotations in order?

7.45 Verify that the Gauss elimination method of Eq. (7.44) agrees with (7.38).

7.46 What is the determinant of a nilpotent operator? See problem 7.34.

7.47 (**a**) Recall (or look up) the method for evaluating a determinant using cofactors (or minors). For $2{\times}2$, $3{\times}3$, and in fact, for $N{\times}N$ arrays, how many multiplication operations are required for this. Ignore the time to do any additions and assume that a computer can do a product in 10^{-10} seconds. How much time does it take by this method to do the determinant for $10{\times}10$, $20{\times}20$, and $30{\times}30$ arrays? Express the times in convenient units.
(**b**) Repeat this for the Gauss elimination algorithm at Eq. (7.44). How much time for the above three matrices and for $100{\times}100$ and $1000{\times}1000$? Count division as taking the same time as multiplication. Ans: For the first method, $30{\times}30$ requires $10\,000{\times}$age of universe. For Gauss it is $3\,\mu$s.

7.48 On the vector space of functions on $0 < x < L$, (**a**) use the basis of complex exponentials, Eq. (5.20), and compute the matrix components of d/dx.
(**b**) Use the basis of Eq. (5.17) to do the same thing.

7.49 Repeat the preceding problem, but for d^2/dx^2. Compare the result here to the squares of the matrices from that problem.

7.50 Repeat problem 7.27 but using a different vector space of functions with basis
(**a**) $e^{n\pi ix/L}$, $(n = 0, \pm1, \pm2, \ldots)$
(**b**) $\cos(n\pi x/L)$, and $\sin(m\pi x/L)$.

These functions will be a basis in the set of periodic functions of x, and these will be *very* big matrices.

7.51 (**a**) What is the determinant of the translation operator of problem 7.27?
(**b**) What is the determinant of d/dx on the vector space of problem 7.26?

7.52 (**a**) Write out the construction of the trace in the case of a three dimensional operator, analogous to Eq. (7.47). What are the coefficients of ϵ^2 and ϵ^3? (**b**) Back in the two dimensional case, draw a picture of what $(I + \epsilon f)$ does to the basis vectors to first order in ϵ.

7.53 Evaluate the trace for arbitrary dimension. Use the procedure of Gauss elimination to compute the determinant, and note at every step that you are keeping terms only through ϵ^0 and ϵ^1. Any higher orders can be dropped as soon as they appear.
Ans: $\sum_{i=1}^{N} f_{ii}$

7.54 The set of all operators on a given vector space forms a vector space.* (Show this.) Consider whether you can or should restrict yourself to real numbers or if you ought to be dealing with complex scalars.
Now what about the list of operators in section 7.14. Which of them form vector spaces?
Ans: Yes(real), Yes(real), No, No, No, No, No, No, No

7.55 In the vector space of cubic polynomials, choose the basis

$$\vec{e}_0 = 1, \qquad \vec{e}_1 = 1 + x, \qquad \vec{e}_2 = 1 + x + x^2, \qquad \vec{e}_3 = 1 + x + x^2 + x^3.$$

In this basis, compute the matrix of components of the operator P, where this is the parity operator, defined as the operator that takes the variable x and changes it to $-x$. For example $P(\vec{e}_1) = 1 - x$. Compute the square of the resulting matrix. What is the determinant of P? If you had only the quadratic polynomials with basis $\vec{e}_0$, $\vec{e}_1$, $\vec{e}_2$, what is the determinant? What about linear polynomials, with basis $\vec{e}_0$, $\vec{e}_1$? Maybe even constant polynomials?

7.56 On the space of quadratic polynomials define an operator that permutes the coefficients: $f(x) = ax^2 + bx + c$, then $Of = g$ has $g(x) = bx^2 + cx + a$. Find the eigenvalues and eigenvectors of this operator.

7.57 The results in Eq. (7.36) is a rotation about *some* axis. Where is it? Notice that a rotation about an axis leaves the axis itself alone, so this is an eigenvector problem. If it leaves the vector alone, you even know what the eigenvalue is, so you can easily find the vector. Repeat for the other rotation, found in Eq. (7.37) Ans: $\vec{e}_1 + \vec{e}_2 - \vec{e}_3$

7.58 Find the eigenvectors and eigenvalues of the matrices in problem 7.1.

---

* If you're knowledgeable enough to recognize the difficulty caused by the question of domains, you'll recognize that this is false in infinite dimensions. But if you know that much then you don't need to be reading this chapter.

Multivariable Calculus

The world is not one-dimensional, and calculus doesn't stop with a single independent variable. The ideas of partial derivatives and multiple integrals are not too different from their single-variable counterparts, but some of the details about manipulating them are not so obvious. Some are downright tricky.

8.1 Partial Derivatives

The basic idea of derivatives and of integrals in two, three, or more dimensions follows the same pattern as for one dimension. They're just more complicated.

The derivative of a function of one variable is defined as

$$\frac{df(x)}{dx} = \lim_{\Delta x \to 0} \frac{f(x + \Delta x) - f(x)}{\Delta x} \tag{8.1}$$

You would think that the definition of a derivative of a function of x and y would then be defined as

$$\frac{\partial f(x,y)}{\partial x} = \lim_{\Delta x \to 0} \frac{f(x + \Delta x, y) - f(x,y)}{\Delta x} \tag{8.2}$$

and more-or-less it is. The ∂ notation instead of d is a reminder that there are other coordinates floating around that are temporarily being treated as constants.

In order to see why I used the phrase "more-or-less," take a very simple example: $f(x,y) = y$. Use the preceding definition, and because y is being held constant, the derivative $\partial f/\partial x = 0$. What could be easier?

I don't like these variables so I'll switch to a different set of coordinates, x' and y':

$$y' = x + y \qquad \text{and} \qquad x' = x$$

What is $\partial f/\partial x'$ now?

$$f(x,y) = y = y' - x = y' - x'$$

Now the derivative of f with respect to x' is -1, because I'm keeping the other coordinate fixed. Or is the derivative still zero because $x' = x$ and I'm taking $\partial f/\partial x$ and why should that change just because I'm using a different coordinate system?

The problem is that the *notation* is ambiguous. When you see $\partial f/\partial x$ it doesn't tell you what to hold constant. Is it to be y or y' or yet something else? In some contexts the answer is clear and you won't have any difficulty deciding, but you've already encountered cases for which the distinction is crucial. In thermodynamics, when you add heat to a gas to raise its temperature does this happen at constant pressure or at constant volume or with some other constraint? The specific heat at constant pressure is not the same as the specific heat at constant volume; it is necessarily bigger because during an expansion some of the energy has to go into the work of changing the volume. This sort of derivative depends on type of process that you're using, and for a classical ideal gas the difference between the two molar specific heats obeys the equation

$$c_p - c_v = R$$

If the gas isn't ideal, this equation is replaced by a more complicated and general one, but the same observation applies, that the two derivatives dQ/dT aren't the same.

In thermodynamics there are so many variables in use that there is a standard notation for a partial derivative, indicating exactly which other variables are to be held constant.

$$\left(\frac{\partial U}{\partial V}\right)_T \qquad \text{and} \qquad \left(\frac{\partial U}{\partial V}\right)_P$$

represent the change in the internal energy of an object per change in volume during processes in which respectively the temperature and the pressure are held constant. In the previous example with the function $f = y$, this says

$$\left(\frac{\partial f}{\partial x}\right)_y = 0 \qquad \text{and} \qquad \left(\frac{\partial f}{\partial x}\right)_{y'} = -1$$

This notation is a way to specify the *direction* in the x-y plane along which you're taking the derivative.

8.2 Chain Rule

For functions of one variable, the chain rule allows you to differentiate with respect to still another variable: y a function of x and x a function of t allows

$$\frac{dy}{dt} = \frac{dy}{dx}\frac{dx}{dt} \tag{8.3}$$

You can derive this simply from the definition of a derivative.

$$\begin{aligned}\frac{\Delta y}{\Delta t} &= \frac{y\big(x(t+\Delta t)\big) - y\big(x(t)\big)}{\Delta t} \\ &= \frac{y\big(x(t+\Delta t)\big) - y\big(x(t)\big)}{x(t+\Delta t) - x(t)} \cdot \frac{x(t+\Delta t) - x(t)}{\Delta t} = \frac{\Delta y}{\Delta x} \cdot \frac{\Delta x}{\Delta t}\end{aligned}$$

Take the limit of this product as $\Delta t \to 0$. Necessarily then you have that $\Delta x \to 0$ too (unless the derivative doesn't exist anyway). The second factor is then the definition of the derivative dx/dt, and the first factor is the definition of dy/dx. The Leibnitz *notation* as written in Eq. (8.3) leads you to the required proof.

What happens with more variables? Roughly the same thing but with more manipulation, the same sort of manipulation that you use to derive the rule for differentiating more complicated functions of one variable (as in section 1.5).

$$\text{Compute} \quad \frac{d}{dt} f\big(x(t), y(t)\big)$$

Back to the Δ's. The manipulation is much like the preceding except that you have to add and subtract a term in the second line.

$$\begin{aligned}
\frac{\Delta f}{\Delta t} &= \frac{f\big(x(t+\Delta t), y(t+\Delta t)\big) - f\big(x(t), y(t)\big)}{\Delta t} \\
&= \frac{f\big(x(t+\Delta t), y(t+\Delta t)\big) - f\big(x(t), y(t+\Delta t)\big) + f\big(x(t), y(t+\Delta t)\big) - f\big(x(t), y(t)\big)}{\Delta t} \\
&= \frac{f\big(x(t+\Delta t), y(t+\Delta t)\big) - f\big(x(t), y(t+\Delta t)\big)}{x(t+\Delta t) - x(t)} \cdot \frac{x(t+\Delta t) - x(t)}{\Delta t} \\
&\qquad + \frac{f\big(x(t), y(t+\Delta t)\big) - f\big(x(t), y(t)\big)}{y(t+\Delta t) - y(t)} \cdot \frac{y(t+\Delta t) - y(t)}{\Delta t} \\
&= \frac{\Delta f}{\Delta x} \cdot \frac{\Delta x}{\Delta t} + \frac{\Delta f}{\Delta y} \cdot \frac{\Delta y}{\Delta t}
\end{aligned}$$

In the first factor of the first term, $\Delta f/\Delta x$, the variable x is changed but y is not. In the first factor of the second term, the reverse holds true. The limit of this expression is then

$$\lim_{\Delta t \to 0} \frac{\Delta f}{\Delta t} = \frac{df}{dt} = \left(\frac{\partial f}{\partial x}\right)_y \frac{dx}{dt} + \left(\frac{\partial f}{\partial y}\right)_x \frac{dy}{dt} \tag{8.4}$$

If these manipulations look familiar, it's probably because they mimic the procedures of section 1.5. That case is like this one, with the special values $x \equiv y \equiv t$.

Example: (When you want to check out an equation, you should construct an example so that it reveals a lot of structure without requiring a lot of calculation.)

$$f(x, y) = Axy^2, \qquad \text{and} \qquad x(t) = Ct^3, \qquad y(t) = Dt^2$$

First do it using the chain rule.

$$\begin{aligned}
\frac{df}{dt} &= \left(\frac{\partial f}{\partial x}\right)_y \frac{dx}{dt} + \left(\frac{\partial f}{\partial y}\right)_x \frac{dy}{dt} \\
&= \big(Ay^2\big)\big(3Ct^2\big) + \big(2Axy\big)\big(2Dt\big) \\
&= \big(A(Dt^2)^2\big)\big(3Ct^2\big) + \big(2A(Ct^3)(Dt^2)\big)\big(2Dt\big) \\
&= 7ACD^2t^6
\end{aligned}$$

Now repeat the calculation by first substituting the values of x and y and then differentiating.

$$\begin{aligned}
\frac{df}{dt} &= \frac{d}{dt}\left[A(Ct^3)(Dt^2)^2\right] \\
&= \frac{d}{dt}\left[ACD^2t^7\right] \\
&= 7ACD^2t^6
\end{aligned}$$

What if f also has an explicit t in it: $f\big(t, x(t), y(t)\big)$? That simply adds another term. Remember, $dt/dt = 1$.

$$\frac{df}{dt} = \left(\frac{\partial f}{\partial t}\right)_{x,y} + \left(\frac{\partial f}{\partial x}\right)_{y,t} \frac{dx}{dt} + \left(\frac{\partial f}{\partial y}\right)_{x,t} \frac{dy}{dt} \tag{8.5}$$

Sometimes you see the chain rule written in a slightly different form. You can change coordinates from (x, y) to (r, ϕ), switching from rectangular to polar. You can switch from (x, y) to a system such as $(x', y') = (x + y, x - y)$. The function can be expressed in the new coordinates explicitly. Solve for x, y in terms of r, ϕ or x', y' and then differentiate with respect to the new coordinate. OR you can use the chain rule to differentiate with respect to the new variable.

$$\boxed{\left(\frac{\partial f}{\partial x'}\right)_{y'} = \left(\frac{\partial f}{\partial x}\right)_y \left(\frac{\partial x}{\partial x'}\right)_{y'} + \left(\frac{\partial f}{\partial y}\right)_x \left(\frac{\partial y}{\partial x'}\right)_{y'}} \tag{8.6}$$

This is actually not a different equation from Eq. (8.4). It only looks different because in addition to t there's another variable that you have to keep constant: $t \to x'$, and y' is constant.

Example: When you switch from rectangular to plane polar coordinates what is $\partial f/\partial\phi$ in terms of the x and y derivatives?

$$x = r\cos\phi, \qquad y = r\sin\phi, \qquad \text{so}$$

$$\begin{aligned}\left(\frac{\partial f}{\partial \phi}\right)_r &= \left(\frac{\partial f}{\partial x}\right)_y \left(\frac{\partial x}{\partial \phi}\right)_r + \left(\frac{\partial f}{\partial y}\right)_x \left(\frac{\partial y}{\partial \phi}\right)_r \\ &= \left(\frac{\partial f}{\partial x}\right)_y (-r\sin\phi) + \left(\frac{\partial f}{\partial y}\right)_x (r\cos\phi)\end{aligned}$$

If $f(x, y) = x^2 + y^2$ this better be zero, because I'm finding how f changes when r is held fixed. Check it out; it is. The equation (8.6) presents the form that is most important in many applications.

Example: What is the derivative of y with respect to ϕ at constant x?

$$\begin{aligned}\left(\frac{\partial y}{\partial \phi}\right)_x &= \left(\frac{\partial y}{\partial r}\right)_\phi \left(\frac{\partial r}{\partial \phi}\right)_x + \left(\frac{\partial y}{\partial \phi}\right)_r \left(\frac{\partial \phi}{\partial \phi}\right)_x \\ &= [\sin\phi] \cdot \left[r\frac{\sin\phi}{\cos\phi}\right] + [r\cos\phi] \cdot 1 = r\frac{1}{\cos\phi}\end{aligned} \tag{8.7}$$

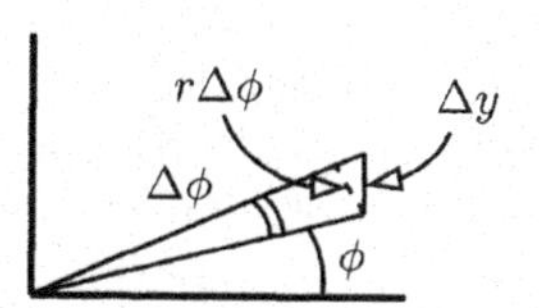

You see a graphical interpretation of the calculation in this diagram: ϕ changes by $\Delta\phi$, so the coordinate moves up by Δy (x is constant). The angle between the lines Δy and $r\Delta\phi$ is ϕ itself. This means that $\Delta y \div r\Delta\phi = 1/\cos\phi$, and that is precisely the preceding equation for $(\partial y/\partial\phi)_x$.

In doing the calculation leading to Eq. (8.7), do you see how to do the calculation for $(\partial r/\partial\phi)_x$? Differentiate the equation $x = r\cos\phi$ with respect to ϕ.

$$x = r\cos\phi \;\to\; \left(\frac{\partial x}{\partial \phi}\right)_x = 0 = \left(\frac{\partial r}{\partial \phi}\right)_x \cos\phi + r\left(\frac{\partial \cos\phi}{\partial \phi}\right)_x = \left(\frac{\partial r}{\partial \phi}\right)_x \cos\phi - r\sin\phi$$

Solve for the unknown derivative and you have the result.

Another example: $f(x,y) = x^2 - 2xy$. The transformation between rectangular and polar coordinates is $x = r\cos\phi$, $y = r\sin\phi$. What is $(\partial f/\partial x)_r$?

$$\left(\frac{\partial f}{\partial x}\right)_r = \left(\frac{\partial f}{\partial x}\right)_y \left(\frac{\partial x}{\partial x}\right)_r + \left(\frac{\partial f}{\partial y}\right)_x \left(\frac{\partial y}{\partial x}\right)_r = (2x - 2y) + (-2x)\left(\frac{\partial y}{\partial x}\right)_r$$

$$\left(\frac{\partial y}{\partial x}\right)_r = \frac{(\partial y/\partial\phi)_r}{(\partial x/\partial\phi)_r} = \frac{r\cos\phi}{-r\sin\phi} = -\cot\phi \tag{8.8}$$

(Remember problem 1.49?) Put these together and

$$\left(\frac{\partial f}{\partial x}\right)_r = (2x - 2y) + (-2x)(-\cot\phi) = 2x - 2y + 2x\cot\phi \tag{8.9}$$

The brute-force way to do this is to express the function f explicitly in terms of the variables x and r, eliminating y and ϕ.

$$y = r\sin\phi = \sqrt{r^2 - x^2}, \qquad \text{then}$$

$$\left(\frac{\partial f}{\partial x}\right)_r = \frac{\partial}{\partial x}\left[x^2 - 2x\sqrt{r^2 - x^2}\right]_r$$

$$= 2x - 2\sqrt{r^2 - x^2} - 2x\frac{1}{\sqrt{r^2 - x^2}}(-x) = 2x + \frac{-2(r^2 - x^2) + 2x^2}{\sqrt{r^2 - x^2}} \tag{8.10}$$

You can see that this is the same as the the equation (8.9) if you look at the next-to-last form of equation (8.10).

$$\frac{x}{\sqrt{r^2 - x^2}} = \frac{r\cos\phi}{\sqrt{r^2 - r^2\cos^2\phi}} = \cot\phi$$

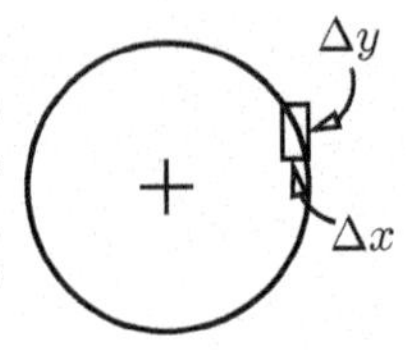

Is this result reasonable? Look at what happens to y when you change x by a little bit. Constant r is a circle, and if ϕ puts the position over near the right side (ten or twenty degrees), a little change in x causes a big change in y as shown by the rectangle. As drawn, $\Delta y/\Delta x$ is big and negative, sort of like the (negative) cotangent of ϕ as in Eq. (8.8).

8.3 Differentials

For a function of a single variable you can write

$$df = \frac{df}{dx}dx \tag{8.11}$$

and read (sort of) that the infinitesimal change in the function f is the slope times the infinitesimal change in x. Does this really make any sense? What is an infinitesimal change? Is it zero? Is dx a number or isn't it? What's going on?

It *is* possible to translate this intuitive idea into something fairly simple and that makes perfectly good sense. Once you understand what it really means you'll be able to use the intuitive idea and its notation with more security.

Let g be a function of *two* variables, x and h.

$$g(x,h) = \frac{df(x)}{dx}h \quad \text{has the property that} \quad \frac{1}{h}\big|f(x+h) - f(x) - g(x,h)\big| \longrightarrow 0 \quad \text{as } h \to 0$$

That is, the function $g(x,h)$ approximates *very well* the change in f as you go from x to $x+h$. The difference between g and $\Delta f = f(x+h) - f(x)$ goes to zero so fast that even after you've divided by h the difference goes to zero.

The usual notation is to use the symbol dx instead of h and to call the function df instead* of g.

$$df(x,dx) = f'(x)\,dx \quad \text{has the property that}$$
$$\frac{1}{dx}\big|f(x+dx) - f(x) - df(x,dx)\big| \longrightarrow 0 \quad \text{as } dx \to 0 \tag{8.12}$$

In this language dx is just another variable that can go from $-\infty$ to $+\infty$ and df is just a specified function of two variables. The point is that this function is useful because when the variable dx is small df provides a very good approximation to the increment Δf in f.

What is the volume of the peel on an orange? The volume of a sphere is $V = 4\pi r^3/3$, so its differential is $dV = 4\pi r^2\,dr$. If the radius of the orange is 3 cm and the thickness of the peel is 2 mm, the volume of the peel is

$$dV = 4\pi r^2\,dr = 4\pi(3\,\text{cm})^2(0.2\,\text{cm}) = 23\,\text{cm}^3$$

The whole volume of the orange is $\frac{4}{3}\pi(3\,\text{cm})^3 = 113\,\text{cm}^3$, so this peel is about 20% of the volume.

Differentials in Several Variables

The analog of Eq. (8.11) for several variables is

$$df = df(x,y,dx,dy) = \left(\frac{\partial f}{\partial x}\right)_y dx + \left(\frac{\partial f}{\partial y}\right)_x dy \tag{8.13}$$

Roughly speaking, near a point in the x-y plane, the value of the function f changes as a linear function of the coordinates as you move a (little) distance away. This function df describes this change to high accuracy. It bears the same relation to Eq. (8.4) that (8.11) bears to Eq. (8.3).

For example, take the function $f(x,y) = x^2 + y^2$. At the point $(x,y) = (1,2)$, the differential is

$$df(1,2,dx,dy) = (2x)\Big|_{(1,2)} dx + (2y)\Big|_{(1,2)} dy = 2dx + 4dy$$

---

* Who says that a variable in algebra must be a single letter? You would never write a computer program that way. $d\,\mathit{Fred}^2/d\,\mathit{Fred} = 2\,\mathit{Fred}$ is perfectly sensible.

so that

$$f(1.01, 1.99) \approx f(1,2) + df(1,2,.01,-.01) = 1^2 + 2^2 + 2(.01) + 4(-.01) = 4.98$$

compared to the exact answer, 4.9802.

The equation analogous to (8.12) is

$$df(x,y,dx,dy) \qquad \text{has the property that}$$
$$\frac{1}{dr}\big|f(x+dx,y+dy) - f(x,y) - df(x,y,dx,dy)\big| \longrightarrow 0 \qquad \text{as } dr \to 0 \tag{8.14}$$

where $dr = \sqrt{dx^2 + dy^2}$ is the distance to (x, y). It's not that you will be able to do a lot more with this precise definition than you could with the intuitive idea. You will however be able to work with a better understanding of you're actions. When you say that "dx is an infinitesimal" you can understand that this means simply that dx is *any* number but that the equations using it are useful only for very small values of that number.

You can't use this notation for everything as the notation for the derivative demonstrates. The symbol "df/dx" does not mean to divide a function by a length; it refers to a well-defined limiting process. This notation is however constructed so that it provides an intuitive guide, and even if you *do* think of it as the function df divided by the variable dx, you get the right answer.

Why should such a thing as a differential exist? It's essentially the first terms after the constant in the power series representation of the original function: section 2.5. But how to tell if such a series works anyway? I've been notably cavalier about proofs. The answer is that there is a proper theorem guaranteeing Eq. (8.14) works. It is that if both partial derivatives exist in the neighborhood of the expansion point and if these derivatives are continuous there, then the differential exists and has the value that I stated in Eq. (8.13). It has the properties stated in Eq. (8.14). For all this refer to one of many advanced calculus texts, such as Apostol's.*

8.4 Geometric Interpretation

For one variable, the picture of the differential is simple. Start with a graph of the function and at a point $(x, y) = (x, f(x))$, find the straight line that best approximates the function in the immediate neighborhood of that point. Now set up a new coordinate system with origin at this (x, y) and call the new coordinates dx and dy. In this coordinate system the straight line passes through the origin and the slope is the derivative $df(x)/dx$. The equation for the straight line is then Eq. (8.11), describing the differential.

$$dy = \frac{df(x)}{dx}dx$$

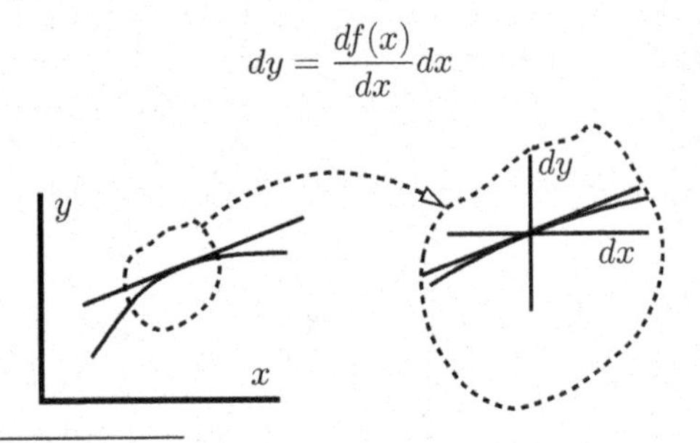

* Mathematical Analysis, Addison-Wesley

For two variables, the picture parallels this one. At a point $(x, y, z) = (x, y, f(x, y))$ find the *plane* that best approximates the function in the immediate neighborhood of that point. Set up a new coordinate system with origin at this (x, y, z) and call the new coordinates dx, dy, and dz. The equation for a plane that passes through this origin is $\alpha\, dx + \beta\, dy + \gamma\, dz = 0$, and for this best approximating plane, the equation is nothing more than the equation for the differential, Eq. (8.13).

$$dz = \left(\frac{\partial f(x,y)}{\partial x}\right)_y dx + \left(\frac{\partial f(x,y)}{\partial y}\right)_x dy$$

The picture is a bit harder to draw, but with a little practice you can do it.

For the case of three independent variables, I'll leave the sketch to you.

Examples

The temperature on the surface of a heated disk is given to be $T(r,\phi) = T_0 + T_1(1 - r^2/a^2)$, where a is the radius of the disk and T_0 and T_1 are constants. If you start at position $x = c < a$, $y = 0$ and move parallel to the y-axis at speed v_0 what is the rate of change of temperature that you feel?

Use Eq. (8.4), and the relation $r = \sqrt{x^2 + y^2}$.

$$\begin{aligned}\frac{dT}{dt} &= \left(\frac{\partial T}{\partial r}\right)_\phi \frac{dr}{dt} + \left(\frac{\partial T}{\partial \phi}\right)_r \frac{d\phi}{dt} = \left(\frac{\partial T}{\partial r}\right)_\phi \left[\left(\frac{\partial r}{\partial x}\right)_y \frac{dx}{dt} + \left(\frac{\partial r}{\partial y}\right)_x \frac{dy}{dt}\right] \\ &= \left(-2T_1\frac{r}{a^2}\right)\left[\frac{y}{\sqrt{x^2+y^2}}v_0\right] = -2T_1\frac{\sqrt{c^2+v_0^2t^2}}{a^2}\cdot\frac{v_0^2t}{\sqrt{c^2+v_0^2t^2}} \\ &= -2T_1\frac{v_0^2t}{a^2}\end{aligned}$$

As a check, the dimensions are correct (are they?). At time zero, this vanishes, and that's what you should expect because at the beginning of the motion you're starting to move in the direction *perpendicular* to the direction in which the temperature is changing. The farther you go, the more nearly parallel to the direction of the radius you're moving. If you are moving exactly parallel to the radius, this time-derivative is easier to calculate; it's then almost a problem in a single variable.

$$\frac{dT}{dt} \approx \frac{dT}{dr}\frac{dr}{dt} \approx -2T_1\frac{r}{a^2}v_0 \approx -2T_1\frac{V_0t}{a^2}v_0$$

So the approximate and the exact calculation agree. In fact they agree so well that you should try to find out if this is a lucky coincidence or if there some special aspect of the problem that you might have seen from the beginning and that would have made the whole thing much simpler.

8.5 Gradient

The equation (8.13) for the differential has another geometric interpretation. For a function such as $f(x,y) = x^2 + 4y^2$, the equations representing constant values of f describe curves in the x-y plane. In this example, they are ellipses. If you start from any fixed point in the plane and start to move away from it, the rate at which the value of f changes will depend on the direction in which you move. If you move along the curve defined by $f =$constant then f won't change at all. If you move perpendicular to that direction then f may change a lot.

> The gradient of f at a point is the vector pointing in the direction in which f is increasing most rapidly, and the component of the gradient along that direction is the derivative of f with respect to the distance in that direction.

To relate this to the partial derivatives that we've been using, and to understand how to compute and to use the gradient, return to Eq. (8.13) and write it in vector form. Use the common notation for the basis: $\hat{x}$ and $\hat{y}$. Then let

$$d\vec{r} = dx\,\hat{x} + dy\,\hat{y} \qquad \text{and} \qquad \vec{G} = \left(\frac{\partial f}{\partial x}\right)_y \hat{x} + \left(\frac{\partial f}{\partial y}\right)_x \hat{y} \tag{8.15}$$

The equation for the differential is now

$$df = df(x, y, dx, dy) = \vec{G} \cdot d\vec{r} \tag{8.16}$$

Because you know the properties of the dot product, you know that this is $G\,dr\cos\theta$ and it is largest when the directions of $d\vec{r}$ and of $\vec{G}$ are the same. It's zero when they are perpendicular. You also know that df is zero when $d\vec{r}$ is in the direction along the curve where f is constant. The vector $\vec{G}$ is therefore perpendicular to this curve. It is in the direction in which f is changing most rapidly. Also because $df = G\,dr\cos 0$, you see that G is the derivative of f with respect to distance along that direction. $\vec{G}$ is the gradient.

For the example $f(x,y) = x^2 + 4y^2$, $\vec{G} = 2x\hat{x} + 8y\hat{y}$. At each point in the x-y plane it provides a vector showing the steepness of f at that point and the direction in which f is changing most rapidly.

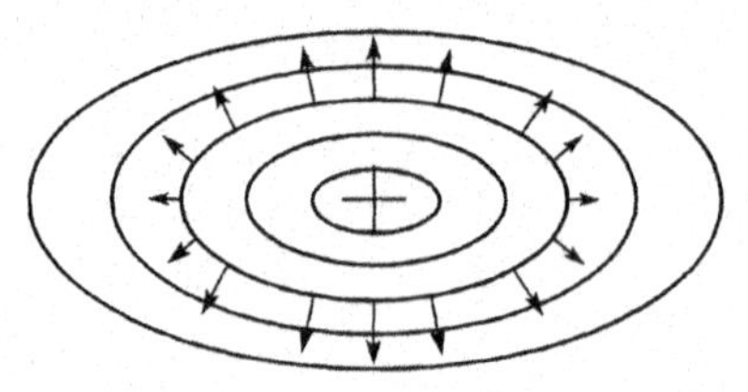

Notice that the gradient vectors are twice as long where the ellipses are closest together as they are at the ends where the ellipses are farthest apart. The function changes more rapidly in the y-direction.

The U.S.Coast and Geodetic Survey makes a large number of maps, and hikers are particularly interested in the contour maps. They show curves indicating the lines of constant altitude. When Apollo 16 went to the Moon in 1972, NASA prepared a similar map for the astronauts, and this is a small segment of that map. The contour lines represent 10 meter increments in altitude.*

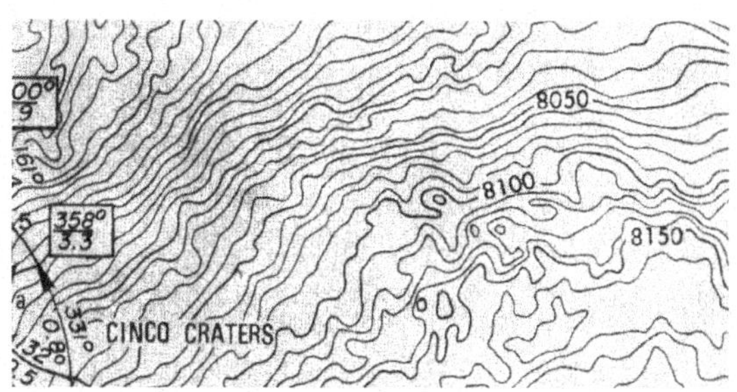

The gravitational potential energy of a mass m near the Earth's (or Moon's) surface is mgh. This divided by m is the gravitational potential, gh. These lines of constant altitude are then lines of constant potential, equipotentials of the gravitational field. Walk along an equipotential and you are doing no work against gravity, just walking on the level.

8.6 Electrostatics

The electric field can be described in terms of a gradient. For a single point charge at the origin the electric field is

$$\vec{E}(x, y, z) = \frac{kq}{r^2}\hat{r}$$

where $\hat{r}$ is the unit vector pointing away from the origin and r is the distance to the origin. This vector can be written as a gradient. Because this $\vec{E}$ is everywhere pointing away from the origin, it's everywhere perpendicular to the sphere centered at the origin.

$$\vec{E} = -\text{grad}\,\frac{kq}{r}$$

You can verify this a several ways. The first is to go straight to the definition of a gradient. (There's a blizzard of minus signs in this approach, so have a little patience. It will get better.) This function is increasing most rapidly in the direction moving toward the origin. $(1/r)$ The derivative with respect to distance in this direction is $-d/dr$, so $-d/dr(1/r) = +1/r^2$. The direction of greatest increase is along $-\hat{r}$, so $\text{grad}\,(1/r) = -\hat{r}(1/r^2)$. But the relation to the electric field has another -1 in it, so

$$-\text{grad}\frac{kq}{r} = +\hat{r}\frac{kq}{r^2}$$

There's got to be a better way.

* history.nasa.gov/alsj/a16/ scan by Robin Wheeler

Yes, instead of insisting that you move in the direction in which the function is *increasing* most rapidly, simply move in the direction in which it is changing most rapidly. The derivative with respect to distance in that direction is the component in that direction and the plus or minus signs take care of themselves. The derivative with respect to r of $(1/r)$ is $-1/r^2$. That is the component in the direction $\hat{r}$, the direction in which you took the derivative. This says $\operatorname{grad}(1/r) = -\hat{r}(1/r^2)$. You get the same result as before but without so much fussing. This also makes it look more like the familiar ordinary derivative in one dimension.

Still another way is from the Stallone-Schwarzenegger brute force school of computing. Put everything in rectangular coordinates and do the partial derivatives using Eqs. (8.15) and (8.6).

$$\left(\frac{\partial(1/r)}{\partial x}\right)_{y,z} = \left(\frac{\partial(1/r)}{\partial r}\right)_{\theta,\phi}\left(\frac{\partial r}{\partial x}\right)_{y,z} = -\frac{1}{r^2}\frac{\partial}{\partial x}\sqrt{x^2+y^2+z^2} = -\frac{1}{r^2}\frac{x}{\sqrt{x^2+y^2+z^2}}$$

Repeat this for y and z with similar results and assemble the output.

$$-\operatorname{grad}\frac{kq}{r} = \frac{kq}{r^2}\frac{x\hat{x}+y\hat{y}+z\hat{z}}{\sqrt{x^2+y^2+z^2}} = \frac{kq}{r^2}\frac{\vec{r}}{r} = \frac{kq}{r^2}\hat{r}$$

The symbol ∇ is commonly used for the gradient operator. This vector operator will appear in several other places, the curl of a vector field will be the one you see most often.

$$\nabla = \hat{x}\frac{\partial}{\partial x} + \hat{y}\frac{\partial}{\partial y} + \hat{z}\frac{\partial}{\partial z} \tag{8.17}$$

From Eq. (8.15) you have

$$\operatorname{grad} f = \nabla f \tag{8.18}$$

8.7 Plane Polar Coordinates

When doing integrals in the plane there are many coordinate systems to choose from, but rectangular and polar coordinates are the most common. You can find the element of area with a simple sketch: The lines (or curves) of constant coordinate enclose an area that is, for small enough increments in the coordinates, a rectangle. Then you just multiply the sides. In one case $\Delta x \cdot \Delta y$ and in the other case $\Delta r \cdot r\Delta\phi$.

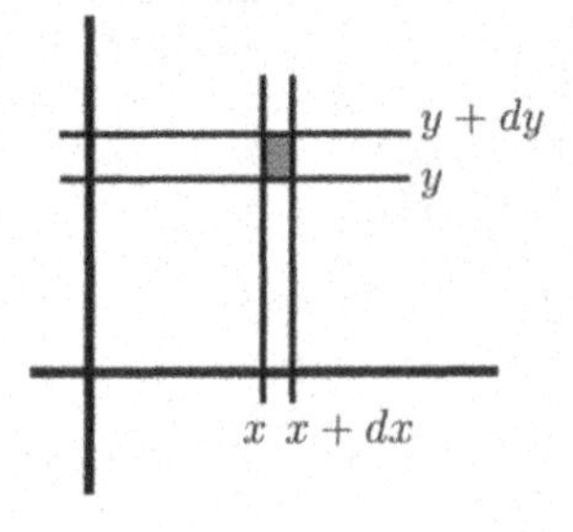

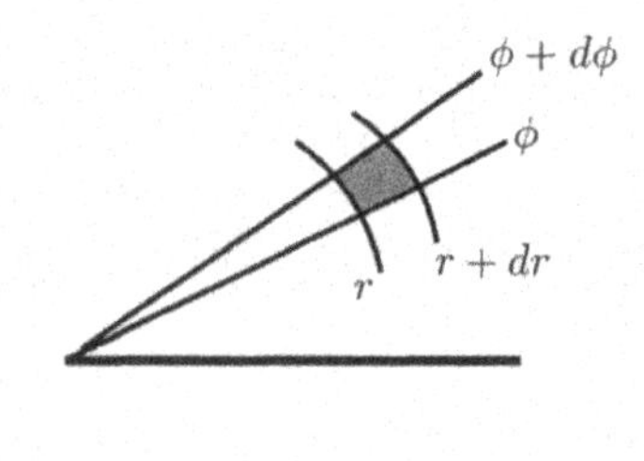

Vibrating Drumhead

A circular drumhead can vibrate in many complicated ways. The simplest and lowest

frequency mode is approximately

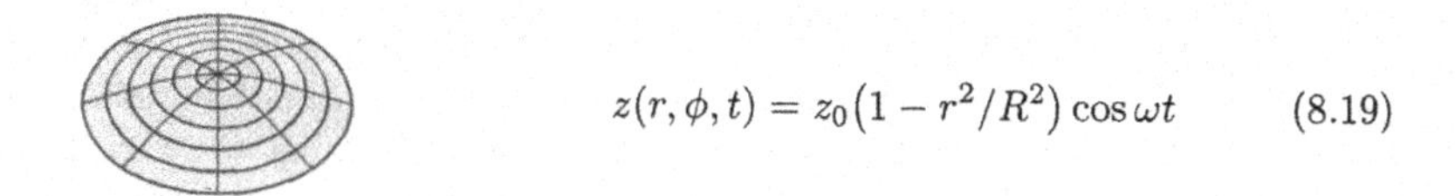

$$z(r,\phi,t) = z_0\big(1 - r^2/R^2\big)\cos\omega t \tag{8.19}$$

where R is the radius of the drum and ω is the frequency of oscillation. (The shape is more accurately described by Eq. (4.22) but this approximation is pretty good for a start.) The kinetic energy density of the moving drumhead is $u = \frac{1}{2}\sigma\big(\partial z/\partial t\big)^2$. That is, in a small area ΔA, the kinetic energy is $\Delta K = u\Delta A$ and the limit as $\Delta A \to 0$ of $\Delta K/\Delta A$ is the area-energy-density. In the same way, σ is the area mass density, dm/dA.

What is the total kinetic energy because of this oscillation? It is $\int u\,dA = \int u\,d^2r$. To evaluate it, use polar coordinates and integrate over the area of the drumhead. The notation d^2r is another notation for dA just as d^3r is used for a piece of volume.

$$\begin{aligned}
\int u\,dA &= \int_0^R r\,dr \int_0^{2\pi} d\phi\, \frac{\sigma}{2} z_0^2\big((1 - r^2/R^2)\omega\sin\omega t\big)^2 \\
&= \frac{\sigma}{2} 2\pi z_0^2\omega^2 \sin^2\omega t \int_0^R dr\, r\big(1 - r^2/R^2\big)^2 \\
&= \sigma\pi z_0^2\omega^2\sin^2\omega t \frac{1}{2}\int_{r=0}^{r=R} d(r^2)\,\big(1 - r^2/R^2\big)^2 \\
&= \sigma\pi z_0^2\omega^2\sin^2\omega t \frac{1}{2} R^2 \frac{1}{3}\big(1 - r^2/R^2\big)^3(-1)\Big|_0^{r=R} \\
&= \frac{1}{6}\sigma R^2\pi z_0^2\omega^2\sin^2\omega t
\end{aligned} \tag{8.20}$$

See problem 8.10 and following for more on this.*

8.8 Cylindrical, Spherical Coordinates

The three common coordinate systems used in three dimensions are rectangular, cylindrical, and spherical coordinates, and these are the ones you have to master. When you need to use prolate spheroidal coordinates you can look them up.

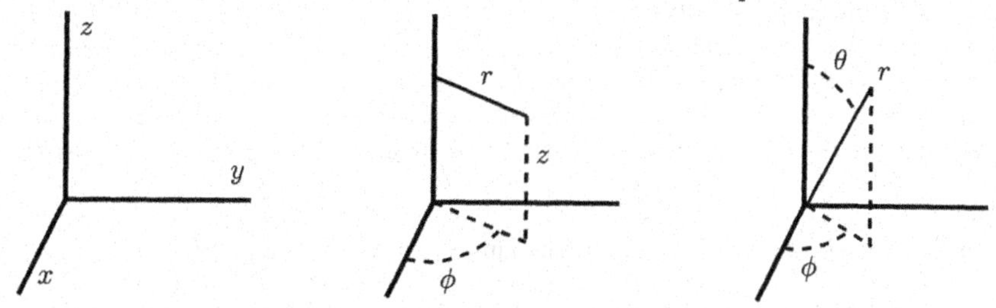

* For some animations showing the these oscillations and others, check out www.physics.miami.edu/nearing/mathmethods/drumhead-animations.html

rectangular	cylindrical	spherical
$-\infty < x < \infty$	$0 < r < \infty$	$0 < r < \infty$
$-\infty < y < \infty$	$0 < \phi < 2\pi$	$0 < \theta < \pi$
$-\infty < z < \infty$	$-\infty < z < \infty$	$0 < \phi < 2\pi$

The surfaces that have constant values of these coordinates are planes in rectangular coordinates; planes and cylinders in cylindrical; planes, spheres, and cones in spherical. In every one of these cases the constant-coordinate surfaces intersect each other at right angles, hence the name "orthogonal coordinate" systems. In spherical coordinates I used the coordinate θ as the angle from the z-axis and ϕ as the angle around the axis. In mathematics books these are typically reversed, so watch out for the notation. On the globe of the Earth, ϕ is like the longitude and θ like the latitude except that longitude goes 0 to 180° East and 0 to 180° West from the Greenwich meridian instead of zero to 2π. Latitude is 0 to 90° North or South from the equator instead of zero to π from the pole. Except for the North-South terminology, latitude is $90° - \theta$.

The volume elements for these systems come straight from the drawings, just as the area elements do in plane coordinates. In every case you can draw six surfaces, bounded by constant coordinates, and surrounding a small box. Because these are orthogonal coordinates you can compute the volume of the box easily as the product of its three edges.

In the spherical case, one side is Δr. Another side is $r\Delta\theta$. The third side is not $r\Delta\phi$; it is $r\sin\theta\Delta\phi$. The reason for the factor $\sin\theta$ is that the arc of the circle made at constant r and constant θ is not in a plane passing through the origin. It is in a plane parallel to the x-y plane, so it has a radius $r\sin\theta$.

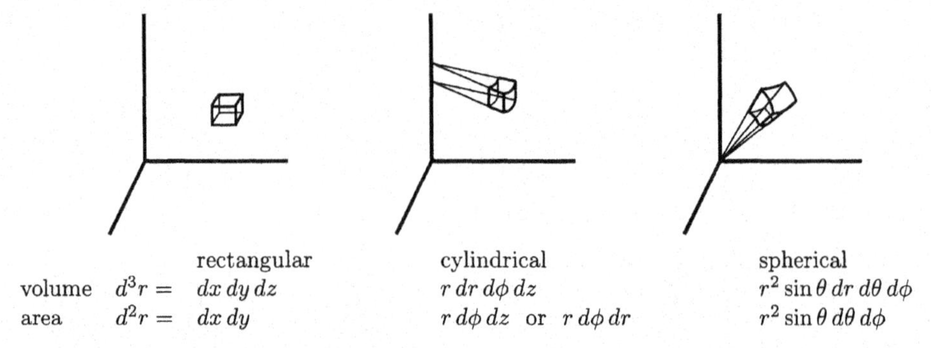

		rectangular	cylindrical	spherical
volume	$d^3r =$	$dx\,dy\,dz$	$r\,dr\,d\phi\,dz$	$r^2\sin\theta\,dr\,d\theta\,d\phi$
area	$d^2r =$	$dx\,dy$	$r\,d\phi\,dz$ or $r\,d\phi\,dr$	$r^2\sin\theta\,d\theta\,d\phi$

Examples of Multiple Integrals

Even in rectangular coordinates integration can be tricky. That's because you have to pay attention to the limits of integration far more closely than you do for simple one dimensional integrals. I'll illustrate this with two dimensional rectangular coordinates first, and will choose a problem that is easy but still shows what you have to look for.

An Area

Find the area in the x-y plane between the curves $y = x^2/a$ and $y = x$.

$$\text{(A)} \quad \int_0^a dx \int_{x^2/a}^{x} dy\, 1 \qquad \text{and} \qquad \text{(B)} \quad \int_0^a dy \int_y^{\sqrt{ay}} dx\, 1$$

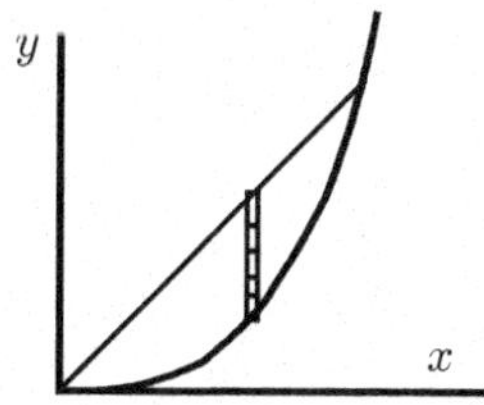

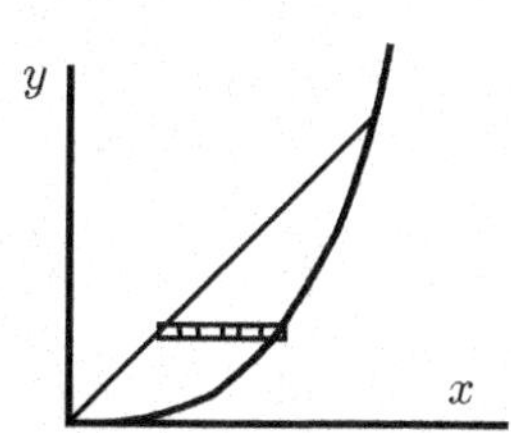

In the first instance I fix x and add the pieces of dy in the strip indicated. The lower limit of the dy integral comes from the specified equation of the lower curve. The upper limit is the value of y for the given x at the upper curve. After that the limits on the sum over dx comes from the intersection of the two curves: $y = x = x^2/a$ gives $x = a$ for that limit.

In the second instance I fix y and sum over dx first. The left limit is easy, $x = y$, and the upper limit comes from solving $y = x^2/a$ for x in terms of y. When that integral is done, the remaining dy integral starts at zero and goes up to the intersection at $y = x = a$.

Now do the integrals.

$$\text{(A)} \quad \int_0^a dx\big[x - x^2/a\big] = \frac{a^2}{2} - \frac{a^3}{3a} = \frac{a^2}{6}$$

$$\text{(B)} \quad \int_0^a dy\big[\sqrt{ay} - y\big] = a^{1/2}\frac{a^{3/2}}{3/2} - \frac{a^2}{2} = \frac{a^2}{6}$$

If you would care to try starting this calculation from the beginning, *without* drawing any pictures, be my guest.

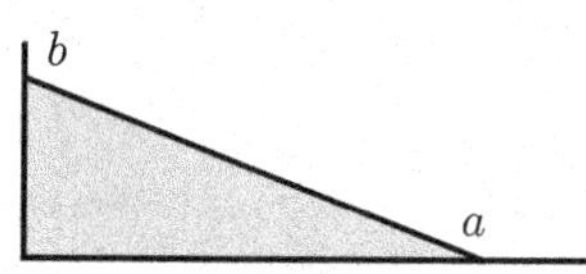

A Moment of Inertia

The moment of inertia about an axis is $\int r_\perp^2\, dm$. Here, $r_\perp$ is the perpendicular distance to the axis. What is the moment of inertia of a uniform sheet of mass M in the shape of a right triangle of sides a and b? Take the moment about the right angled vertex. The area mass density, $\sigma = dm/dA$ is $2M/ab$. The moment of inertia is then

$$\begin{aligned}
\int (x^2+y^2)\sigma\, dA &= \int_0^a dx \int_0^{b(a-x)/a} dy\sigma\,(x^2+y^2) = \int_0^a dx\sigma\left[x^2y + y^3/3\right]_0^{b(a-x)/a} \\
&= \int_0^a dx\sigma\left[x^2\frac{b}{a}(a-x) + \frac{1}{3}\left(\frac{b}{a}\right)^3(a-x)^3\right] \\
&= \sigma\left[\frac{b}{a}\left(\frac{a^4}{3} - \frac{a^4}{4}\right) + \frac{1}{3}\left(\frac{b^3}{a^3}\frac{a^4}{4}\right)\right] \\
&= \frac{1}{12}\sigma\left(ba^3 + ab^3\right) = \frac{M}{6}\left(a^2+b^2\right)
\end{aligned}$$

The dimensions are correct. For another check take the case where $a = 0$, reducing this to $Mb^2/6$. But wait, this now looks like a thin rod, and I remember that the moment of inertia of a thin rod about its end is $Mb^2/3$. What went wrong? Nothing. Look again more closely. Show why this limiting answer ought to be less than $Mb^2/3$.

Volume of a Sphere

What is the volume of a sphere of radius R? The most obvious approach would be to use spherical coordinates. See problem 8.16 for that. I'll use cylindrical coordinates instead. The element of volume is $dV = r\,dr d\phi dz$, and the integrals can be done a couple of ways.

$$\int d^3r = \int_0^R r\,dr \int_0^{2\pi} d\phi \int_{-\sqrt{R^2-r^2}}^{+\sqrt{R^2-r^2}} dz = \int_{-R}^{+R} dz \int_0^{2\pi} d\phi \int_0^{\sqrt{R^2-z^2}} r\,dr \tag{8.21}$$

You can finish these now, see problem 8.17.

A Surface Charge Density

An example that appears in electrostatics: The surface charge density, dq/dA, on a sphere of radius R is $\sigma(\theta,\phi) = \sigma_0 \sin^2\theta \cos^2\phi$. What is the total charge on the sphere?

The element of area is $R^2 \sin\theta\, d\theta d\phi$, so the total charge is $\int \sigma\, dA$,

$$Q = \int_0^{\pi} \sin\theta\, d\theta R^2 \int_0^{2\pi} d\phi\, \sigma_0 \sin^2\theta \cos^2\phi = R^2 \int_{-1}^{+1} d\cos\theta\, \sigma_0 \left(1 - \cos^2\theta\right) \int_0^{2\pi} d\phi\, \cos^2\phi$$

The mean value of $\cos^2$ is $1/2$. so the ϕ integral gives π. For the rest, it is

$$\sigma_0 \pi R^2 \left[\cos\theta - \frac{1}{3}\cos^3\theta\right]_{-1}^{+1} = \frac{4}{3}\sigma_0 \pi R^2$$

Limits of Integration

Sometimes the trickiest part of multiple integrals is determining the limits of integration. Especially when you have to change the order of integration, the new limits may not be obvious. Are there any special techniques or tricks to doing this? Yes, there is one, perhaps obscure, method that you may not be accustomed to.

Draw Pictures.

If you have an integral such as the first one, you have to draw a picture of the integration domain to switch limits.

$$\int_0^1 dy \int_y^{\sqrt{2-y^2}} dx\, f(x,y) \qquad \left[\int_0^1 dx \int_0^x dy + \int_1^{\sqrt{2}} dx \int_0^{\sqrt{2-x^2}} dy\right] f(x,y) \tag{8.22}$$

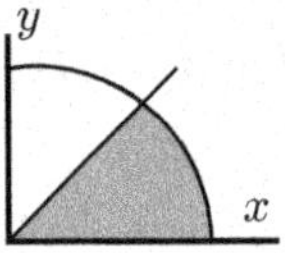

Of course, once you've drawn the picture you may realize that simply interchanging the order of integration won't help, but that polar coordinates may.

$$\int_0^{\sqrt{2}} r\,dr \int_0^{\pi/4} d\phi$$

8.9 Vectors: Cylindrical, Spherical Bases

When you describe vectors in three dimensions are you restricted to the basis $\hat{x}$, $\hat{y}$, $\hat{z}$? In a different coordinate system you should use basis vectors that are adapted to that system. In rectangular coordinates these vectors have the convenient property that they point along the direction perpendicular to the plane where the corresponding coordinate is constant. They also point in the direction in which the other two coordinates are constant. *E.g.* the unit vector $\hat{x}$ points perpendicular to the plane of constant x (the y-z plane); it also point along the line where y and z are constant.

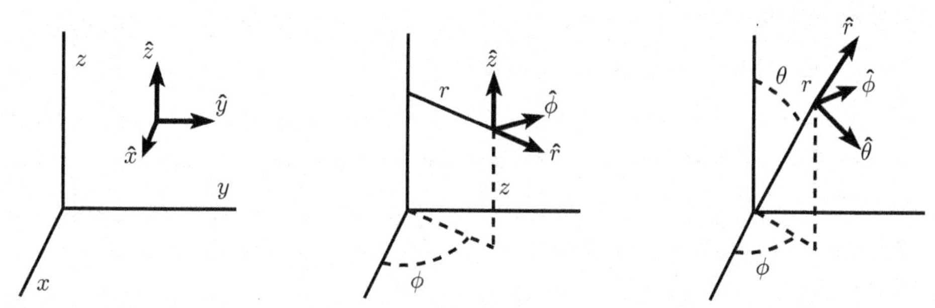

Do the same thing for cylindrical coordinates. The unit vector $\hat{z}$ points perpendicular to the x-y plane. The unit vector $\hat{r}$ points perpendicular to the cylinder $r =$constant. The unit vector $\hat{\phi}$ points perpendicular to the plane $\phi =$constant and along the direction for which r and z are constant. The conventional right-hand rule specifies $\hat{z} = \hat{r} \times \hat{\phi}$.

For spherical coordinates $\hat{r}$ points perpendicular to the sphere $r =$constant. The $\hat{\phi}$ vector is perpendicular to the plane $\phi =$constant and points along the direction where $r =$ constant and $\theta =$constant and toward increasing coordinate ϕ. Finally $\hat{\theta}$ is perpendicular to the cone $\theta =$constant and again, points toward increasing θ. Then $\hat{\phi} = \hat{r} \times \hat{\theta}$, and on the Earth, these vectors $\hat{r}$, $\hat{\theta}$, and $\hat{\phi}$ are up̂, Sout̂h, and Eâst.

Solenoid

A standard solenoid is cylindrical coil of wire, so that when the wire carries a current it produces a magnetic field. To describe this field, it seems that cylindrical coordinates are advised. Until you know something about the field the most general thing that you can write is

$$\vec{B}(r,\phi,z) = \hat{r}\,B_r(r,\phi,z) + \hat{\phi}\,B_\phi(r,\phi,z) + \hat{z}\,B_z(r,\phi,z)$$

In a real solenoid that's it; all three of these components are present. If you have an ideal, infinitely long solenoid, with the current going strictly around in the $\hat{\phi}$ direction, (found only in textbooks) the use of Maxwell's equations and appropriately applied symmetry arguments will simplify this to $\hat{z}\,B_z(r)$.

Gravitational Field

The gravitational field of the Earth is simple, $\vec{g} = -\hat{r}\,GM/r^2$, pointing straight toward the center of the Earth. Well no, not really. The Earth has a bulge at the equator; its equatorial diameter is about 43 km larger than its polar diameter. This changes the $\vec{g}$-field so that it has a noticeable $\hat{\theta}$ component. At least it's noticeable if you're trying to place a satellite in orbit or to send a craft to another planet.

A better approximation to the gravitational field of the Earth is

$$\vec{g} = -\hat{r}\frac{GM}{r^2} - G\frac{3Q}{r^4}\left[\hat{r}\left(3\cos^2\theta - 1\right)/2 + \hat{\theta}\cos\theta\sin\theta\right] \tag{8.23}$$

The letter Q stands for the quadrupole moment. $|Q| \ll MR^2$, and it's a measure of the bulge. By convention a football (American football) has a positive Q; the Earth's Q is negative. (What about a European football?)

Nuclear Magnetic Field

The magnetic field from the nucleus of many atoms (even as simple an atom as hydrogen) is proportional to

$$\frac{1}{r^3}\left[2\hat{r}\cos\theta + \hat{\theta}\sin\theta\right] \tag{8.24}$$

As with the preceding example these are in spherical coordinates, and the component along the $\hat{\phi}$ direction is zero. This field's effect on the electrons in the atom is small but detectable. The magnetic properties of the nucleus are central to the subject of nuclear magnetic resonance (NMR), and that has its applications in magnetic resonance imaging* (MRI).

8.10 Gradient in other Coordinates

The equation for the gradient computed in rectangular coordinates is Eq. (8.15) or (8.18). How do you compute it in cylindrical or spherical coordinates? You do it the same way that you got Eq. (8.15) from Eq. (8.13). The coordinates r, ϕ, and z are just more variables, so Eq. (8.13) is simply

$$df = df(r,\phi,z,dr,d\phi,dz) = \left(\frac{\partial f}{\partial r}\right)_{\phi,z} dr + \left(\frac{\partial f}{\partial \phi}\right)_{r,z} d\phi + \left(\frac{\partial f}{\partial z}\right)_{r,\phi} dz \tag{8.25}$$

All that's left is to write $d\vec{r}$ in these coordinates, just as in Eq. (8.15).

$$d\vec{r} = \hat{r}\,dr + \hat{\phi}r\,d\phi + \hat{z}\,dz \tag{8.26}$$

The part in the $\hat{\phi}$ direction is the *displacement* of $d\vec{r}$ in that direction. As ϕ changes by a small amount the distance moved is not $d\phi$; it is $r\,d\phi$. The equation

$$df = df(r,\phi,z,dr,d\phi,dz) = \operatorname{grad} f \cdot d\vec{r}$$

combined with the two equations (8.25) and (8.26) gives $\operatorname{grad} f$ as

$$\operatorname{grad} f = \hat{r}\frac{\partial f}{\partial r} + \hat{\phi}\frac{1}{r}\frac{\partial f}{\partial \phi} + \hat{z}\frac{\partial f}{\partial z} = \nabla f \tag{8.27}$$

Notice that the units work out right too.

---

* In medicine MRI was originally called NMR, but someone decided that this would disconcert the patients.

In spherical coordinates the procedure is identical. All that you have to do is to identify what $d\vec r$ is.

$$d\vec r = \hat r\, dr + \hat\theta\, r\, d\theta + \hat\phi\, r \sin\theta\, d\phi$$

Again with this case you have to look at the distance moved when the coordinates changes by a small amount. Just as with cylindrical coordinates this determines the gradient in spherical coordinates.

$$\operatorname{grad} f = \hat r \frac{\partial f}{\partial r} + \hat\theta \frac{1}{r}\frac{\partial f}{\partial\theta} + \hat\phi \frac{1}{r\sin\theta}\frac{\partial f}{\partial\phi} = \nabla f \tag{8.28}$$

The equations (8.15), (8.27), and (8.28) define the gradient (and correspondingly ∇) in three coordinate systems.

8.11 Maxima, Minima, Saddles

With one variable you can look for a maximum or a minimum by taking a derivative and setting it to zero. For several variables you do it several times so that you will get as many equations as you have unknown coordinates.

Put this in the language of gradients: $\nabla f = 0$. The derivative of f vanishes in every direction as you move from such a point. As examples,

$$f(x,y) = x^2 + y^2, \qquad \text{or} \qquad = -x^2 - y^2, \qquad \text{or} \qquad = x^2 - y^2$$

For all three of these the gradient is zero at $(x,y) = (0,0)$; the first has a minimum there, the second a maximum, and the third neither — it is a "saddle point." Draw a picture to see the reason for the name. The generic term for all three of these is "critical point."

An important example of finding a minimum is "least square fitting" of functions. How close are two functions to each other? The most commonly used, and in every way the simplest, definition of the distance (squared) between f and g on the interval $a < x < b$ is

$$\int_a^b dx\, \big|f(x) - g(x)\big|^2 \tag{8.29}$$

This means that a large deviation of one function from the other in a small region counts more than smaller deviations spread over a larger domain. The square sees to that. As a specific example, take a function f on the interval $0 < x < L$ and try to fit it to the sum of a couple of trigonometric functions. The best fit will be the one that minimizes the distance between f and the sum. (Take f to be a real-valued function for now.)

$$D^2(\alpha,\beta) = \int_0^L dx\,\left(f(x) - \alpha\sin\frac{\pi x}{L} - \beta\sin\frac{2\pi x}{L}\right)^2 \tag{8.30}$$

D is the distance between the given function and the sines used to fit it. To minimize the distance, take derivatives with respect to the parameters α and β.

$$\begin{aligned}
\frac{\partial D^2}{\partial\alpha} &= 2\int_0^L dx\,\left(f(x) - \alpha\sin\frac{\pi x}{L} - \beta\sin\frac{2\pi x}{L}\right)\left(-\sin\frac{\pi x}{L}\right) = 0 \\
\frac{\partial D^2}{\partial\beta} &= 2\int_0^L dx\,\left(f(x) - \alpha\sin\frac{\pi x}{L} - \beta\sin\frac{2\pi x}{L}\right)\left(-\sin\frac{2\pi x}{L}\right) = 0
\end{aligned}$$

These two equations determine the parameters α and β.

$$\alpha \int_0^L dx\, \sin^2 \frac{\pi x}{L} = \int_0^L dx\, f(x) \sin \frac{\pi x}{L}$$
$$\beta \int_0^L dx\, \sin^2 \frac{2\pi x}{L} = \int_0^L dx\, f(x) \sin \frac{2\pi x}{L}$$

The other integrals vanish because of the orthogonality of $\sin \pi x/L$ and $\sin 2\pi x/L$ on this interval. What you get is exactly the coefficients of the Fourier series expansion of f. The Fourier series is the best fit (in the least square sense) of a sum of orthogonal functions to f. See section 11.6 for more on this

Is it a minimum? Yes. Look at the coefficients of α^2 and β^2 in Eq. (8.30). They are positive; $+\alpha^2 + \beta^2$ has a minimum, not a maximum or saddle point, and there is no cross term in $\alpha\beta$ to mess it up.

The distance function Eq. (8.29) is simply (the square of) the norm in the vector space sense of the difference of the two vectors f and g. Equations(6.12) and (6.7) here become

$$\|f - g\|^2 = \langle f - g, f - g \rangle = \int_a^b dx\, \big|f(x) - g(x)\big|^2$$

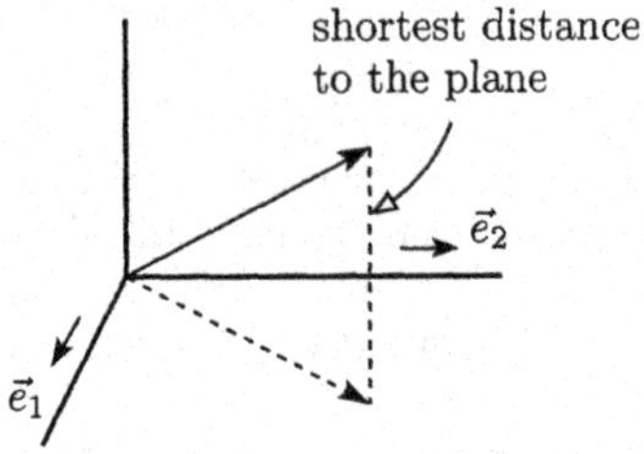

shortest distance
to the plane
$\vec e_2$
$\vec e_1$

The geometric meaning of Eq. (8.30) is that $\vec e_1$ and $\vec e_2$ provide a basis for the two dimensional space

$$\alpha \vec e_1 + \beta \vec e_2 = \alpha \sin \frac{\pi x}{L} + \beta \sin \frac{2\pi x}{L}$$

The plane is the set of all linear combinations of the two vectors, and for a general vector not in this plane, the shortest distance to the plane defines the vector *in* the plane that is the best fit to the given vector. It's the one that's closest. Because the vectors $\vec e_1$ and $\vec e_2$ are orthogonal it makes it easy to find the closest vector. You require that the difference, $\vec v - \alpha \vec e_1 - \beta \vec e_2$ has only an $\vec e_3$ component. That is Fourier series.

Hessian

In this example leading to Fourier components, it's pretty easy to see that you are dealing with a minimum and not anything else. In other situations it may not be so easy. You may have a lot of variables. You may have complicated cross terms. Is $x^2 + xy + y^2$ a minimum at the origin? Is $x^2 + 3xy + y^2$? (Yes and No respectively.)

When there's just one variable there is a simple rule that lets you decide. Check the second derivative. If it's positive you have a minimum; if it's negative you have a maximum. If it's zero you have more work to do. Is there a similar method for several variables? Yes, and I'll show it explicitly for two variables. Once you see how to do it in two dimensions, the generalization to N is just a matter of how much work you're willing to do (or how much computer time you can use).

The Taylor series in two variables, Eq. (2.16), is to second order

$$f(x+dx, y+dy) = f(x,y) + \frac{\partial f}{\partial x}dx + \frac{\partial f}{\partial y}dy + \frac{\partial^2 f}{\partial x^2}dx^2 + 2\frac{\partial^2 f}{\partial x \partial y}dx\,dy + \frac{\partial^2 f}{\partial y^2}dy^2 + \cdots$$

Write this in a more compact notation in order to emphasize the important parts.

$$f(\vec{r}+d\vec{r}) - f(\vec{r}) = \nabla f \cdot d\vec{r} + \langle d\vec{r}, H\,d\vec{r}\rangle + \cdots$$

The part with the gradient is familiar, and to have either a minimum or a maximum, that will have to be zero. The next term introduces a new idea, the *Hessian*, constructed from all the second derivative terms. Write these second order terms as a matrix to see what they are, and in order to avoid a lot of clumsy notation use subscripts as an abbreviation for the partial derivatives.

$$\langle d\vec{r}, H\,d\vec{r}\rangle = \begin{pmatrix} dx & dy \end{pmatrix}\begin{pmatrix} f_{xx} & f_{xy} \\ f_{yx} & f_{yy}\end{pmatrix}\begin{pmatrix} dx \\ dy \end{pmatrix} \qquad \text{where} \qquad d\vec{r} = \hat{x}\,dx + \hat{y}\,dy \tag{8.31}$$

This matrix is symmetric because of the properties of mixed partials. How do I tell from this whether the function f has a minimum or a maximum (or neither) at a point where the gradient of f is zero? Eq. (8.31) describes a function of two variables even *after* I've fixed the values of x and y by saying that $\nabla f = 0$. It is a quadratic function of dx and dy. Expressed in the language of vectors this says that f has a minimum if (8.31) is positive no matter what the direction of $d\vec{r}$ is — H is *positive definite.*

Pull back from the problem a step. This is a 2×2 symmetric matrix sandwiched inside a scalar product.

$$h(x,y) = \begin{pmatrix} x & y \end{pmatrix}\begin{pmatrix} a & b \\ b & c\end{pmatrix}\begin{pmatrix} x \\ y \end{pmatrix} \tag{8.32}$$

Is h positive definite? That is, positive for all x, y? If this matrix is diagonal it's much easier to see what is happening, so diagonalize it. Find the eigenvectors and use those for a basis.

$$\begin{pmatrix} a & b \\ b & c\end{pmatrix}\begin{pmatrix} x \\ y \end{pmatrix} = \lambda\begin{pmatrix} x \\ y \end{pmatrix} \qquad \text{requires} \qquad \det\begin{pmatrix} a-\lambda & b \\ b & c-\lambda\end{pmatrix} = 0$$

$$\lambda^2 - \lambda(a+c) + ac - b^2 = 0 \implies \lambda = \left[(a+c) \pm \sqrt{(a-c)^2 + b^2}\,\right]/2 \tag{8.33}$$

For the applications here all the a, b, c are the real partial derivatives, so the eigenvalues are real and the only question is whether the λs are positive or negative, because they will be the (diagonal) components of the Hessian matrix in the new basis. If this is a double root, the matrix was already diagonal. You can verify that the eigenvalues are positive if $a > 0$, $c > 0$, and $4ac > b^2$, and that will indicate a minimum point.

Geometrically the equation $z = h(x,y)$ from Eq. (8.32) defines a surface. If it is positive definite the surface is a paraboloid opening upward. If negative definite it is a paraboloid opening down. The mixed case is a hyperboloid — a saddle.

In this 2×2 case you have a quadratic formula to fall back on, and with more variables there are standard algorithms for determining eigenvalues of matrices, but I'll leave those to some other book.

8.12 Lagrange Multipliers

This is an incredibly clever method to handle problems of maxima and minima in several variables when there are constraints.

An example: "What is the largest rectangle?" obviously has no solution, but "What is the largest rectangle contained in an ellipse?" does.

Another: Particles are to be placed into states of specified energies. You know the total number of particles; you know the total energy. All else being equal, what is the most probable distribution of the number of particles in each state?

I'll describe this procedure for two variables; it's the same for more. The problem stated is that I want to find the maximum (or minimum) of a function $f(x,y)$ given the fact that the coordinates x and y must lie on the curve $\phi(x,y) = 0$. If you can solve the ϕ equation for y in terms of x explicitly, then you can substitute it into f and turn it into a problem in ordinary one variable calculus. What if you can't?

Analyze this graphically. The equation $\phi(x,y) = 0$ represents one curve in the plane. The succession of equations $f(x,y) =$ constant represent many curves in the plane, one for each constant. Think of equipotentials.

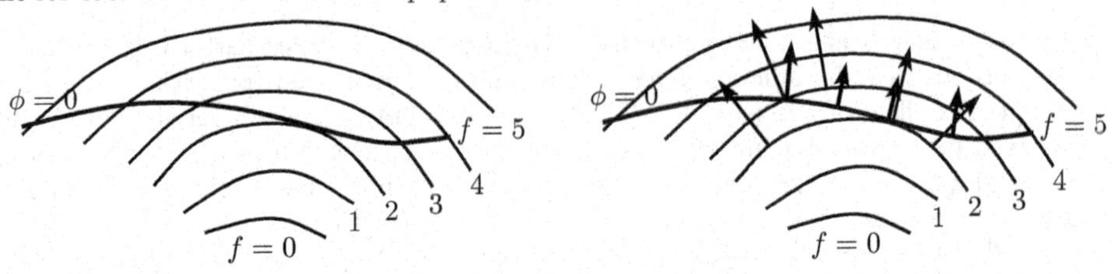

Look at the intersections of the ϕ-curve and the f-curves. Where they intersect, they will usually cross each other. Ask if such a crossing could possibly be a point where f is a maximum. Clearly the answer is no, because as you move along the ϕ-curve you're then moving from a point where f has one value to where it has another.

The one way to have f be a maximum at a point on the ϕ-curve is for the two curves to touch and not to cross. When that happens the values of f will increase as you approach the point from one side and decrease on the other. That makes it a maximum. In this sketch, the values of f decrease from 4 to 3 to 2 and then back to to 3, 4, and 5. This point where the curve $f = 2$ touches the $\phi = 0$ curve is then a minimum of f along $\phi = 0$.

To implement this picture so that you can compute with it, look at the gradient of f and the gradient of ϕ. The gradient vectors are perpendicular to the curves $f =$constant and $\phi =$constant respectively, and at the point where the curves are tangent to each other these gradients are in the same direction (or opposite, no matter). Either way one vector is a scalar times the other.

$$\nabla f = \lambda \nabla \phi \tag{8.34}$$

In the second picture, the arrows are the gradient vectors for f and for ϕ. Break this into components and you have

$$\frac{\partial f}{\partial x} - \lambda \frac{\partial \phi}{\partial x} = 0, \qquad \frac{\partial f}{\partial y} - \lambda \frac{\partial \phi}{\partial y} = 0, \qquad \phi(x,y) = 0$$

There are three equations in three unknowns (x, y, λ), and these are the equations to solve for the position of the maximum or minimum value of f. You are looking for x and y, so you'll be tempted to ignore the third variable λ and to eliminate it. Look again. This parameter, the Lagrange multiplier, has a habit of being significant.

Examples of Lagrange Multipliers

The first example that I mentioned: What is the largest rectangle that you can inscribe in an ellipse? Let the ellipse and the rectangle be centered at the origin. The upper right corner of the rectangle is at (x, y), then the area of the rectangle is

$$\begin{aligned} \text{Area } &= f(x, y) = 4xy, \\ \text{with constraint } \phi(x, y) &= \frac{x^2}{a^2} + \frac{y^2}{b^2} - 1 = 0 \end{aligned}$$

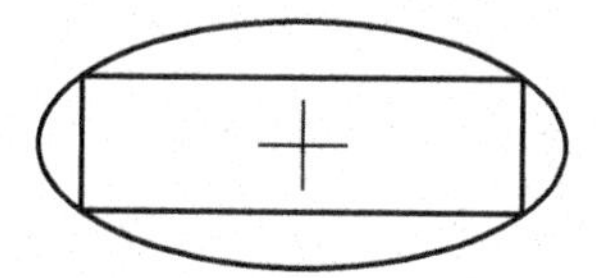

The equations to solve are now

$$\begin{aligned} &\nabla(f - \lambda\phi) = 0, \qquad \text{and} \qquad \phi = 0, \qquad \text{which become} \\ &4y - \lambda\frac{2x}{a^2} = 0, \qquad 4x - \lambda\frac{2y}{b^2} = 0, \qquad \frac{x^2}{a^2} + \frac{y^2}{b^2} - 1 = 0 \end{aligned} \tag{8.35}$$

The solutions to these three equations are straight-forward. They are $x = a/\sqrt{2}$, $y = b/\sqrt{2}$, $\lambda = 2ab$. The maximum area is then $4xy = 2ab$. The Lagrange multiplier turns out to be the required area. Does this reduce to the correct result for a circle?

The second example said that you have several different allowed energies, typical of what happens in quantum mechanics. If the total number of particles and the total energy are given, how are the particles distributed among the different energies?

If there are N particles and exactly two energy levels, E_1 and E_2,

$$N = n_1 + n_2, \qquad \text{and} \qquad E = n_1 E_1 + n_2 E_2$$

you have two equations in two unknowns and all you have to do is solve them for the numbers n_1 and n_2, the number of particles in each state. If there are three or more possible energies the answer isn't uniquely determined by just two equations, and there can be many ways that you can put particles into different energy states and still have the same number of particles and the same total energy.

If you're dealing with four particles and three energies, you can perhaps count the possibilities by hand. How many ways can you put four particles in three states? (400), (310), (301), (220), 211), *etc.* There's only one way to get the (400) configuration: All four particles go into state 1. For (310) there are four ways to do it; any one of the four particles can be in the second state and the rest in the first. Keep going. If you have 10^{20} particles you have to find a better way.

If you have a total of N particles and you place n_1 of them in the first state, the number of ways that you can do that is N for the first particle, $(N-1)$ for the second particle, *etc.* $= N(N-1)(N-2)\cdots(N-n_1+1) = N!/(N-n_1)!$. This is over-counting because you don't care which one went into the first state first, just that it's there. There are $n_1!$ rearrangements of these n_1 particles, so you have to divide by that to get the number of ways that you can get this number of particles into state 1: $N!/n_1!(N-n_1)!$ For example, $N=4$, $n_1=4$ as in the (400) configuration in the preceding paragraph is $4!/0!4!=1$, or $4!/3!1!=4$ as in the (310) configuration.

Once you've got n_1 particles into the first state you want to put n_2 into the second state (out of the remaining $N-n_1$). Then on to state 3.

The total number of ways that you can do this is the product of all of these numbers. For three allowed energies it is

$$\frac{N!}{n_1!(N-n_1)!}\cdot\frac{(N-n_1)!}{n_2!(N-n_1-n_2)!}\cdot\frac{(N-n_1-n_2)!}{n_3!(N-n_1-n_2-n_3)!}=\frac{N!}{n_1!n_2!n_3!} \tag{8.36}$$

There's a lot of cancellation and the final factor in the denominator is one because of the constraint $n_1+n_2+n_3=N$.

Lacking any other information about the particles, the most probable configuration is the one for which Eq. (8.36) is a maximum. This calls for Lagrange multipliers because you want to maximize a complicated function of several variables subject to constraints on N and on E. Now all you have to do is to figure out out to differentiate with respect to integers. Answer: If N is large you will be able to treat these variables as continuous and to use standard calculus to manipulate them.

For large n, recall Stirling's formula, Eq. (2.20),

$$n!\sim\sqrt{2\pi n}\,n^n e^{-n} \qquad \text{or its log:} \qquad \ln(n!)\sim\ln\sqrt{2\pi n}+n\ln n-n \tag{8.37}$$

This, I can differentiate. Maximizing (8.36) is the same as maximizing its logarithm, and that's easier to work with.

$$\begin{aligned}&\text{maximize } f=\ln(N!)-\ln(n_1!)-\ln(n_2!)-\ln(n_3!)\\ &\text{subject to } n_1+n_2+n_3=N \qquad \text{and} \qquad n_1E_1+n_2E_2+n_3E_3=E\end{aligned}$$

There are two constraints here, so there are two Lagrange multipliers.

$$\nabla\big(f-\lambda_1(n_1+n_2+n_3-N)-\lambda_2(n_1E_1+n_2E_2+n_3E_3-E)\big)=0$$

For f, use Stirling's approximation, but not quite. The term $\ln\sqrt{2\pi n}$ is negligible. For n as small as 10^6, it is about 6×10^{-7} of the whole. Logarithms are much smaller than powers. That means that I can use

$$\nabla\left(\sum_{\ell=1}^{3}\big(-n_\ell\ln(n_\ell)+n_\ell\big)-\lambda_1 n_\ell-\lambda_2 n_\ell E_\ell\right)=0$$

This is easier than it looks because each derivative involves only one coordinate.

$$\frac{\partial}{\partial n_1}\to-\ln n_1-1+1-\lambda_1-\lambda_2E_1=0,\ \textit{etc.}$$

This is

$$n_\ell=e^{-\lambda_1-\lambda_2E_\ell},\quad \ell=1,\,2,\,3$$

There are two unknowns here, λ_1 and λ_2. There are two equations, for N and E, and the parameter λ_1 simply determines an overall constant, $e^{-\lambda_1} = C$.

$$C\sum_{\ell=1}^{3} e^{-\lambda_2 E_\ell} = N, \qquad \text{and} \qquad C\sum_{\ell=1}^{3} E_\ell\, e^{-\lambda_2 E_\ell} = E$$

The quantity λ_2 is usually denoted β in this type of problem, and it is related to temperature by $\beta = 1/kT$ where as usual the Lagrange multiplier is important on its own. It is usual to manipulate these results by defining the "partition function"

$$Z(\beta) = \sum_{\ell=1}^{3} e^{-\beta E_\ell} \tag{8.38}$$

In terms of this function Z you have

$$C = N/Z, \qquad \text{and} \qquad E = -\frac{N}{Z}\frac{dZ}{d\beta} \tag{8.39}$$

For a lot more on this subject, you can refer to any one of many books on thermodynamics or statistical physics. There for example you can find the reason that β is related to the temperature and how the partition function can form the basis for computing everything there is to compute in thermodynamics. Especially there you will find that more powerful versions of the same ideas will arise when you allow the total energy and the total number of particles to be variables too.

8.13 Solid Angle

The extension of the concept of angle to three dimensions is called "solid angle." To explain what this is, I'll first show a definition of ordinary angle that's different from what you're accustomed to. When you see that, the extension to one more dimension is easy.

Place an object in the plane somewhere not at the origin. You are at the origin and look at it. I want a definition that describes what fraction of the region around you is spanned by this object. For this, draw a circle of radius R centered at the origin and draw all the lines from everywhere on the object to the origin. These lines will intersect the circle on an arc (or even a set of arcs) of length s. *Define* the angle subtended by the object to be $\theta = s/R$.

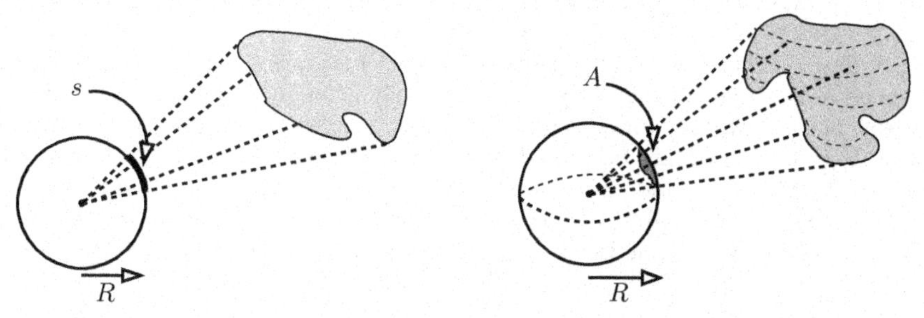

Now step up to three dimensions and again place yourself at the origin. This time place a sphere of radius R around the origin and draw all the lines from the three dimensional object to the origin. This time the lines intersect the sphere on an area of size A. *Define* the solid angle subtended by the object to be $\Omega = A/R^2$. (If you want four or more dimensions, see problem 8.52.)

For the circle, the circumference is $2\pi R$, so if you're surrounded, the angle subtended is $2\pi R/R = 2\pi$ radians. For the sphere, the area is $4\pi R^2$, so this time if you're surrounded, the solid angle subtended is $4\pi R^2/R^2 = 4\pi$ sterradians. That is the name for this unit.

All very pretty. Is it useful? Only if you want to describe radiative transfer, nuclear scattering, illumination, the structure of the atom, or rainbows. Except for illumination, these subjects center around one idea, that of a "cross section."

Cross Section, Absorption

Before showing how to use solid angle to describe scattering, I'll take a simpler example: absorption. There is a hole in a wall and I propose to measure its area. Instead of taking a ruler to it I blindly fire bullets at the wall and see how many go in. The bigger the area, the larger the fraction that will go into the hole of course, but I have to make this quantitative to make it useful.

Define the flux of bullets: $f = dN/(dt\, dA)$. That is, suppose that I'm firing all the bullets in the same direction, but not starting from the same place. Pick an area ΔA perpendicular to the stream of bullets and pick a time interval Δt. How many bullets pass through this area in this time? ΔN, and that's proportional to both ΔA and Δt. The limit of this quotient is the flux.

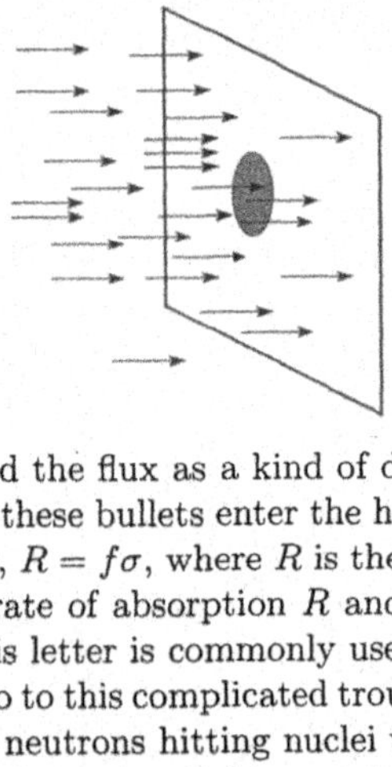

$$\lim_{\substack{\Delta t \to 0 \\ \Delta A \to 0}} \frac{\Delta N}{\Delta t \Delta A} = f \tag{8.40}$$

Having defined the flux as a kind of density, call the (unknown) area of the hole σ. The rate at which these bullets enter the hole is proportional to the size of the hole and to the flux of bullets, $R = f\sigma$, where R is the rate of entry and σ is the area of the hole. If I can measure the rate of absorption R and the flux f, I have measured the area of the hole, $\sigma = R/f$. This letter is commonly used for cross sections.

Why go to this complicated trouble for a hole? I probably shouldn't, but to measure absorption of neutrons hitting nuclei this is precisely what you do. I can't use a ruler on a nucleus, but I can throw things at it. In this example, neutron absorption by nuclei, the value of the measured absorption cross section can vary from millibarns to kilobarns, where a barn is $10^{-24}\,\text{cm}^2$. The radii of nuclei vary by a factor of only about six from hydrogen through uranium ($\sqrt[3]{238} = 6.2$), so the cross section measured by bombarding the nucleus has little to do with the geometric area πr^2. It is instead a measure of interaction strength

Cross Section, Scattering

There are many types of cross sections besides absorption, and the next simplest is the scattering cross section, especially the differential scattering cross section.

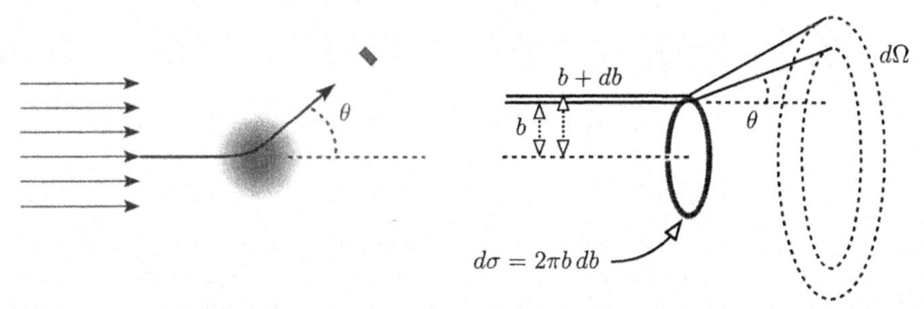

The same flux of particles that you throw at an object may not be absorbed, but may scatter instead. You detect the scattering by using a detector. (You were expecting a catcher's mitt?) The detector will have an area ΔA facing the particles and be at a distance r from the center of scattering. The detection rate will be proportional the the area of the detector, but if I double r for the same ΔA, the detection rate will go down by a factor of four. The detection rate is proportional to $\Delta A/r^2$, but this is just the solid angle of the detector from the center:

$$\Delta\Omega = \Delta A/r^2 \tag{8.41}$$

The detection rate is proportional to the incoming flux and to the solid angle of the detector. The proportionality is an effective scattering area, $\Delta\sigma$.

$$\Delta R = f\Delta\sigma, \qquad \text{so} \qquad \frac{d\sigma}{d\Omega} = \frac{dR}{f d\Omega} \tag{8.42}$$

This is the differential scattering cross section.

You can compute this if you know something about the interactions involved. The one thing that you need is the relationship between where the particle comes in and the direction in which it leaves. That is, the incoming particle is aimed to hit at a distance b (called the impact parameter) from the center and it scatters at an angle θ, called of course the scattering angle, from its original direction. Particles that come in at distance between b and $b + db$ from the axis through the center will scatter into directions between θ and $\theta + d\theta$.

The cross section for being sent in a direction between these two angles is the area of the ring: $d\sigma = 2\pi b\, db$. Anything that hits in there will scatter into the outgoing angles shown. How much solid angle is this? Put the z-axis of spherical coordinates to the right, so that θ is the usual spherical coordinate angle from z. The element of area on the surface of a sphere is $dA = r^2 \sin\theta d\theta d\phi$, so the integral over all the azimuthal angles ϕ around the ring just gives a factor 2π. The element of solid angle is then

$$d\Omega = \frac{dA}{r^2} = 2\pi \sin\theta d\theta$$

As a check on this, do the integral over all theta to get the total solid angle around a point, verifying that it is 4π.

Divide the effective area for this scattering by the solid angle, and the result is the differential scattering cross section.

$$\frac{d\sigma}{d\Omega} = \frac{2\pi b\, db}{2\pi \sin\theta d\theta} = \frac{b}{\sin\theta}\frac{db}{d\theta}$$

If you have θ as a function of b, you can compute this. There are a couple of very minor modifications that you need in order to complete this development. The first is that the derivative $db/d\theta$ can easily be negative, but both the area and the solid angle are positive. That means that you need an absolute value here. One other complication is that one value of θ can come from several values of b. It may sound unlikely, but it happens routinely. It even happens in the example that comes up in the next section.

$$\frac{d\sigma}{d\Omega} = \sum_i \frac{b_i}{\sin\theta}\left|\frac{db_i}{d\theta}\right| \tag{8.43}$$

The differential cross section often becomes much more involved than this, especially the when it involves nuclei breaking up in a collision, resulting in a range of possible energies of each part of the debris. In such collisions particles can even be created, and the probabilities and energy ranges of the results are described by their own differential cross sections. You will wind up with differential cross sections that look like $d\sigma/d\Omega_1\, d\Omega_2 \ldots dE_1\, dE_2 \ldots$. These rapidly become so complex that it takes some elaborate computer programming to handle the information.

8.14 Rainbow

An interesting, if slightly complicated example is the rainbow. Sunlight scatters from small drops of water in the air and the detector is your eye. The water drops are small enough that I'll assume them to be spheres, where surface tension is enough to hold them in this shape for the ordinary small sizes of water droplets in the air. The first and simplest model uses geometric optics and Snell's law to figure out where the scattered light goes. This model ignores the wave nature of light and it does not take into account the fraction of the light that is transmitted and reflected at each surface.

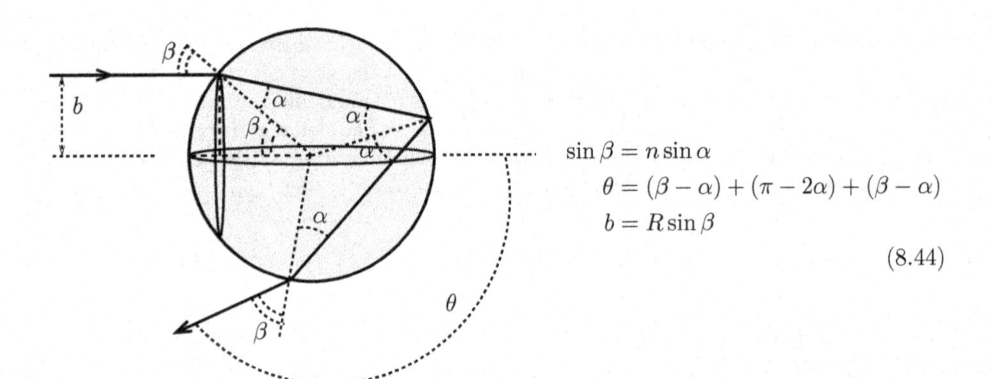

$$\begin{aligned} \sin\beta &= n\sin\alpha \\ \theta &= (\beta-\alpha) + (\pi - 2\alpha) + (\beta - \alpha) \\ b &= R\sin\beta \end{aligned} \tag{8.44}$$

The light comes in at the indicated distance b from the axis through the center of the sphere. It is then refracted, reflected, and refracted. Snell's law describes the first and third of these, and the middle one has equal angles of incidence and reflection. The dashed lines are from the center of the sphere. The three terms in Eq. (8.44) for the evaluation of θ come from the three places at which the light changes direction, and they are the amount of deflection at each place. The third equation simply relates b to the radius of the sphere.

From these three equations, eliminate the two variables α and β to get the single relation between b and θ that I'm looking for. When you do this, you find that the resulting equations are a bit awkward. It's sometimes easier to use one of the two intermediate angles as a parameter, and in this case you will want to use β. From the picture you know that it varies from zero to $\pi/2$. The third equation gives b in terms of β. The first equation gives α in terms of β. The second equation determines θ in terms of β and the α that you've just found.

The parametrized relation between b and θ is then

$$b = R\sin\beta, \theta \qquad = \pi + 2\beta - 4\sin^{-1}\left(\frac{1}{n}\sin\beta\right), \qquad (0 < \beta < \pi/2) \tag{8.45}$$

or you can carry it through and eliminate β.

$$\theta = \pi + 2\sin^{-1}\left(\frac{b}{R}\right) - 4\sin^{-1}\left(\frac{1}{n}\frac{b}{R}\right) \tag{8.46}$$

The derivative $db/d\theta = 1/[d\theta/db]$. Compute this.

$$\frac{d\theta}{db} = \frac{2}{\sqrt{R^2 - b^2}} - \frac{4}{\sqrt{n^2R^2 - b^2}} \tag{8.47}$$

In the parametrized form this is

$$\frac{db}{d\theta} = \frac{db/d\beta}{d\theta/d\beta} = \frac{R\cos\beta}{2 - 4\cos\beta/\sqrt{n^2 - \sin^2\beta}}$$

In analyzing this, it's convenient to have both forms, as you never know which one will be easier to interpret. (Have you checked to see if they agree with each other in any special cases?)

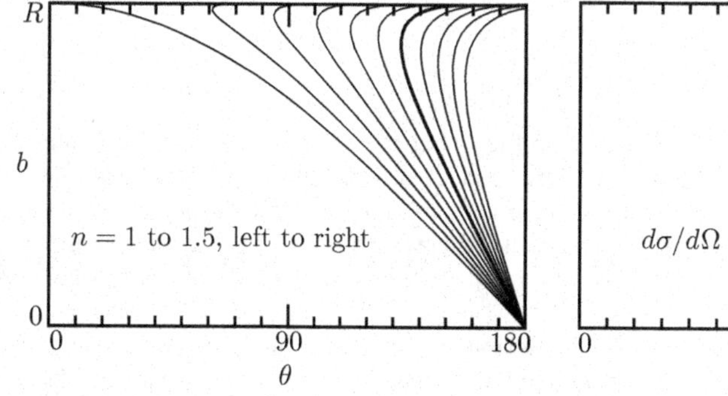

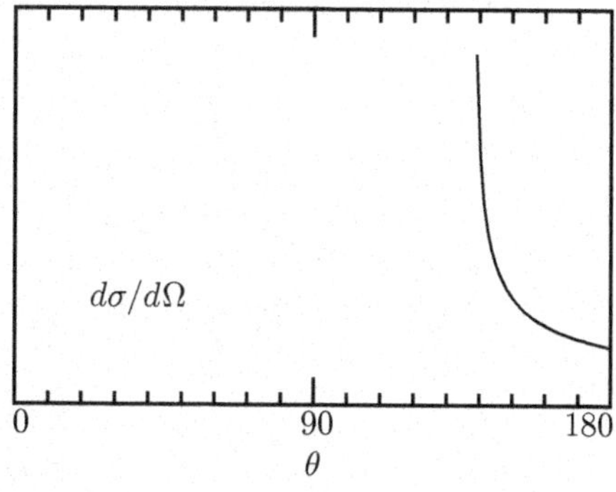

These graphs are generated from Eq. (8.45) for eleven values of the index of refraction equally spaced from 1 to 1.5, and the darker curve corresponds to $n = 1.3$. The key factor that enters the cross-section calculation, Eq. (8.43), is $db/d\theta$, because it goes to infinity when the curve has a vertical tangent. For water, with $n = 1.33$, the b-θ curve has a vertical slope that occurs for θ a little less than 140°. *That* is the rainbow.

To complete this I should finish with $d\sigma/d\Omega$. The interesting part of the problem is near the vertical part of the curve. To see what happens near such a point use a power series expansion near there. Not $b(\theta)$ but $\theta(b)$. This has zero derivative here, so near the vertical point

$$\theta(b) = \theta_0 + \gamma(b - b_0)^2$$

At (b_0, θ_0), Eq. (8.47) gives zero and Eq. (8.46) tells you θ_0. The coefficient γ comes from the second derivative of Eq. (8.46) at b_0. What is the differential scattering cross section in this neighborhood?

$$b = b_0 \pm \sqrt{(\theta - \theta_0)/\gamma}, \qquad \text{so} \qquad db/d\theta = \pm\frac{1}{2\sqrt{\gamma(\theta - \theta_0)}}$$

$$\begin{aligned}\frac{d\sigma}{d\Omega} &= \sum_i \frac{b_i}{\sin\theta}\left|\frac{db_i}{d\theta}\right| \\ &= \frac{b_0 + \sqrt{(\theta - \theta_0)/\gamma}}{\sin\theta}\,\frac{1}{2\sqrt{\gamma(\theta - \theta_0)}} + \frac{b_0 - \sqrt{(\theta - \theta_0)/\gamma}}{\sin\theta}\,\frac{1}{2\sqrt{\gamma(\theta - \theta_0)}} \\ &= \frac{b_0}{\sin\theta\sqrt{\gamma(\theta - \theta_0)}} \approx \frac{b_0}{\sin\theta_0\sqrt{\gamma(\theta - \theta_0)}}\end{aligned} \tag{8.48}$$

In the final expression, because this is near $\theta - \theta_0$ and because I'm doing a power series expansion of the exact solution anyway, I dropped all the θ-dependence except the dominant factors. This is the only consistent thing to do because I've previously dropped higher order terms in the expansion of $\theta(b)$.

Why is this a rainbow? (1) With the sun at your back you see a bright arc of a circle in the direction for which the scattering cross-section is very large. The angular radius of this circle is $\pi - \theta_0 \approx 42°$. (2) The value of θ_0 depends on the index of refraction, n, and that varies slightly with wavelength. The variation of this angle of peak intensity is

$$\frac{d\theta_0}{d\lambda} = \frac{d\theta_0}{db_0}\frac{db_0}{dn}\frac{dn}{d\lambda} \tag{8.49}$$

When you graph Eq. (8.48) note carefully that it is zero on the left of θ_0 (smaller θ) and large on the right. Large scattering angles correspond to the region of the sky underneath the rainbow, toward the center of the circular arc. This implies that there is much more light scattered toward your eye underneath the arc of the rainbow than there is above it. Look at your next rainbow and compare the area of sky below and above the rainbow.

There's a final point about this calculation. I didn't take into account the fact that when light hits a surface, some is transmitted and some is reflected. The largest effect is at the point of internal reflection, because typically only about two percent of the light is reflected and the rest goes through. The cross section should be multiplied by this factor to be complete. The detailed equations for this are called the Fresnel formulas and they tell you the fraction of the light transmitted and reflected at a surface as a function of angle and polarization.

This is far from the whole story about rainbows. Light is a wave, and the geometric optics approximation that I've used doesn't account for everything. In fact Eq. (8.43) doesn't apply to waves, so the whole development has to be redone. To get an idea of some of the other phenomena associated with the rainbow, see for example
www.usna.edu/Users/oceano/raylee/RainbowBridge/Chapter_8.html
www.philiplaven.com/links.html

Exercises

1 For the functions $f(x,y) = Axy^2\sin(xy)$, $x(t) = Ct^3$, $y(t) = Dt^2$, compute df/dt two ways. First use the chain rule, then do explicit substitution and compute it directly.

2 Compute $(\partial f/\partial x)_y$ and $(\partial f/\partial y)_x$ for

$$(a)\ f(x,y) = x^2 - 2xy + y^2, \qquad (b)\ f(x,y) = \ln(y/x), \qquad (c)\ f(x,y) = (y+x)/(y-x)$$

3 Compute df/dx using the chain rule for

$$(a)\ f(x,y) = \ln(y/x), \quad y = x^2, \qquad (b)\ f(x,y) = (y+x)/(y-x), \quad y = \alpha x,$$
$$(c)\ f(x,y) = \sin(xy), \quad y = 1/x$$

Also calculate the results by substituting y explicitly and then differentiating, comparing the results.

4 Let $f(x,y) = x^2 - 2xy$, and the polar coordinates are $x = r\cos\phi$, $y = r\sin\phi$. Compute

$$\left(\frac{\partial f}{\partial x}\right)_y, \quad \left(\frac{\partial f}{\partial y}\right)_x, \quad \left(\frac{\partial f}{\partial x}\right)_r, \quad \left(\frac{\partial f}{\partial y}\right)_r, \quad \left(\frac{\partial f}{\partial x}\right)_\phi, \quad \left(\frac{\partial f}{\partial y}\right)_\phi$$

5 Let $f(x,y) = x^2 - 2xy$, and the polar coordinates are $x = r\cos\phi$, $y = r\sin\phi$. Compute

$$\left(\frac{\partial f}{\partial r}\right)_\phi, \quad \left(\frac{\partial f}{\partial \phi}\right)_r, \quad \left(\frac{\partial f}{\partial r}\right)_x, \quad \left(\frac{\partial f}{\partial \phi}\right)_x, \quad \left(\frac{\partial f}{\partial r}\right)_y, \quad \left(\frac{\partial f}{\partial \phi}\right)_y$$

6 For the function $f(u,v) = u^3 - v^3$, what is the value at $(u,v) = (2,1)$? Approximately what is its value at $(u,v) = (2.01, 1.01)$? Approximately what is its value at $(u,v) = (2.01, 0.99)$?

7 Assume the Earth's atmosphere is uniform density and 10 km high, what is its volume? What is the ratio of this volume to the Earth's volume?

8 For a cube 1 m on a side, what volume of paint will you need in order to paint it to a thickness of 0.2 mm? Don't forget to paint all the sides.

9 What is grad r^2? Do it in both rectangular and polar coordinates. Two dimensions will do. Are your results *really* the same?

10 What is grad $\left(\alpha x^2 + \beta y^2\right)$. Do this in both rectangular and polar coordinates. For the polar form, put x and y in terms of r and ϕ, then refer to Eq. (8.27) for the polar form of the gradient. Finally, compare the two results.

11 The Moon has a radius about 1740 km and its distance from Earth averages about 384 000 km from Earth. What solid angle does the Moon subtend from Earth? What solid angle does Earth (radius 6400 km) subtend from the Moon?

12 Express the cylindrical unit vectors $\hat{r}$, $\hat{\phi}$, $\hat{z}$ in terms of the rectangular ones. And vice versa.

13 Evaluate the volume of a sphere by integration in spherical coordinates.

Problems

8.1 Let $r = \sqrt{x^2 + y^2}$, $x = A\sin\omega t$, $y = B\cos\omega t$. Use the chain rule to compute the derivative with respect to t of e^{kr}. Notice the various checks you can do on the result, verifying (or disproving) your result.

8.2 Sketch these functions* in plane polar coordinates:
(a) $r = a\cos\phi$ (b) $r = a\sec\phi$ (c) $r = a\phi$ (d) $r = a/\phi$ (e) $r^2 = a^2\sin 2\phi$

8.3 The two coordinates x and y are related by $f(x, y) = 0$. What is the derivative of y with respect to x under these conditions? [What is df *along* this curve? And have you drawn a sketch?] Make up a test function (with enough structure to be a test but still simple enough to verify your answer independently) and see if your answer is correct.
Ans: $-(\partial f/\partial x)/(\partial f/\partial y)$

8.4 If $x = u + v$ and $y = u - v$, show that

$$\left(\frac{\partial y}{\partial x}\right)_u = -\left(\frac{\partial y}{\partial x}\right)_v$$

Do this by application of the chain rule, Eq. (8.6). Then as a check do the calculation by explicit elimination of the respective variables v and u.

8.5 If $x = r\cos\phi$ and $y = r\sin\phi$, compute

$$\left(\frac{\partial x}{\partial r}\right)_\phi \qquad \text{and} \qquad \left(\frac{\partial x}{\partial r}\right)_y$$

8.6 What is the differential of $f(x, y, z) = \ln(xyz)$.

8.7 If $f(x, y) = x^3 + y^3$ and you switch to plane polar coordinates, use the chain rule to evaluate

$$\left(\frac{\partial f}{\partial r}\right)_\phi, \quad \left(\frac{\partial f}{\partial \phi}\right)_r, \quad \left(\frac{\partial^2 f}{\partial r^2}\right)_\phi, \quad \left(\frac{\partial^2 f}{\partial \phi^2}\right)_r, \quad \left(\frac{\partial^2 f}{\partial r\,\partial \phi}\right)$$

Check one or more of these by substituting r and ϕ explicitly and doing the derivatives.

8.8 When current I flows through a resistance R the heat produced is I^2R. Two terminals are connected in parallel by two resistors having resistance R_1 and R_2. Given that the total current is divided as $I = I_1 + I_2$, show that the condition that the total heat generated is a minimum leads to the relation $I_1R_1 = I_2R_2$. You don't need Lagrange multipliers to solve this problem, but try them anyway.

8.9 Sketch the magnetic field represented by Eq. (8.24). I suggest that you start by fixing r and drawing the $\vec{B}$-vectors at various values of θ. It will probably help your sketch if you first compute the magnitude of B to see how it varies around the circle. Recall, this

---

* See www-groups.dcs.st-and.ac.uk/~history/Curves/Curves.html for more.

field is expressed in spherical coordinates, though you can take advantage of its symmetry about the z-axis to make the drawing simpler. Don't stop with just the field at fixed r as I suggested you begin. The field fills space, so try to describe it.

8.10 A drumhead can vibrate in more complex modes. One such mode that vibrates at a frequency higher than that of Eq. (8.19) looks approximately like

$$z(r,\phi,t) = A\,r\big(1 - r^2/R^2\big)\sin\phi\cos\omega_2 t$$

(**a**) Find the total kinetic energy of this oscillating drumhead.
(**b**) Sketch the shape of the drumhead at $t = 0$. Compare it to the shape of Eq. (8.19). At the instant that the total kinetic energy is a maximum, what is the shape of the drumhead?
Ans: $\frac{\pi}{48}\sigma A^2\omega_2^2 R^4\sin^2\omega_2 t$

8.11 Just at there is kinetic energy in a vibrating drumhead, there is potential energy, and as the drumhead moves its total potential energy will change because of the slight stretching of the material. The potential energy density (dP.E.$/dA$) in a drumhead is

$$u_\text{p} = \frac{1}{2}T\big(\nabla z\big)^2$$

T is the tension in the drumhead. It has units of Newtons/meter and it is the force per length you would need if you cut a small slit in the surface and had to hold the two sides of the slit together. This potential energy arises from the slight stretching of the drumhead as it moves away from the plane of equilibrium.
(**a**) For the motion described by Eq. (8.19) compute the total potential energy. (Naturally, you will have checked the dimensions first to see if the claimed expression for u_p is sensible.)
(**b**) Energy is conserved, so the sum of the total potential energy and the total kinetic energy from Eq. (8.20) must be a constant. What must the frequency ω be for this to hold? Is this a plausible result? A more accurate result, from solving a differential equation, is $2.405\sqrt{T/\sigma R^2}$. Ans: $\sqrt{6T/\sigma R^2} = 2.45\sqrt{T/\sigma R^2}$

8.12 Repeat the preceding problem for the drumhead mode of problem 8.10. The exact result, calculated in terms of roots of Bessel functions is $3.832\sqrt{T/\sigma R^2}$. Ans: $4\sqrt{T/\sigma R^2}$

8.13 Sketch the gravitational field of the Earth from Eq. (8.23). Is the direction of the field plausible? Draw lots of arrows.

8.14 Prove that the unit vectors in polar coordinates are related to those in rectangular coordinates by

$$\hat r = \hat x\cos\phi + \hat y\sin\phi, \qquad \hat\phi = -\hat x\sin\phi + \hat y\cos\phi$$

What are $\hat x$ and $\hat y$ in terms of $\hat r$ and $\hat\phi$?

8.15 Prove that the unit vectors in spherical coordinates are related to those in rectangular coordinates by

$$\begin{aligned}
\hat r &= \hat x\sin\theta\cos\phi + \hat y\sin\theta\sin\phi + \hat z\cos\theta \\
\hat\theta &= \hat x\cos\theta\cos\phi + \hat y\cos\theta\sin\phi - \hat z\sin\theta \\
\hat\phi &= -\hat x\sin\phi + \hat y\cos\phi
\end{aligned}$$

8.16 Compute the volume of a sphere using spherical coordinates. Also do it using rectangular coordinates. Also do it in cylindrical coordinates.

8.17 Finish both integrals Eq. (8.21). Draw sketches to demonstrate that the limits stated there are correct.

8.18 Find the volume under the plane $2x + 2y + z = 8a$ and over the triangle bounded by the lines $x = 0$, $y = 2a$, and $x = y$ in the x-y plane. Ans: $8a^3$

8.19 Find the volume enclosed by the doughnut-shaped surface (spherical coordinates) $r = a\sin\theta$. Ans: $\pi^2 a^3/4$

8.20 In plane polar coordinates, compute $\partial\hat{r}/\partial\phi$, also $\partial\hat{\phi}/\partial\phi$. This means that r is fixed and you're finding the change in these vectors as you move around a circle. In both cases express the answer in terms of the $\hat{r}$-$\hat{\phi}$ vectors. Draw pictures that will demonstrate that your answers are at least in the right direction. Ans: $\partial\hat{\phi}/\partial\phi = -\hat{r}$

8.21 Compute the gradient of the distance from the origin (in three dimensions) in three coordinate systems and verify that they agree.

8.22 Taylor's power series expansion of a function of several variables was discussed in section 2.5. The Taylor series in one variable was expressed in terms of an exponential in problem 2.30. Show that the series in three variables can be written as

$$e^{\vec{h}\cdot\nabla} f(x, y, z)$$

8.23 The wave equation is (a) below. Change variables to $z = x - vt$ and $w = x + vt$ and show that in these coordinates this equation is (b) (except for a constant factor). Did you *explicitly* note which variables are kept fixed at each stage of the calculation? See also problem 8.53.

$$(a)\ \frac{\partial^2 u}{\partial x^2} - \frac{1}{v^2}\frac{\partial^2 u}{\partial t^2} = 0 \qquad\qquad (b)\ \frac{\partial^2 u}{\partial z\,\partial w} = 0$$

8.24 The equation (8.23) comes from taking the gradient of the Earth's gravitational potential in an expansion to terms in $1/r^3$.

$$V = -\frac{GM}{r} - \frac{GQ}{r^3}P_2(\cos\theta)$$

where $P_2(\cos\theta) = \frac{3}{2}\cos^2\theta - \frac{1}{2}$ is the second order Legendre polynomial. Compute $\vec{g} = -\nabla V$.

8.25 In problem 2.25 you computed the electric potential at large distances from a pair of charges, $-q$ at the origin and $+q$ at $z = a$ $(r \gg a)$. The result was

$$V = \frac{kqa}{r^2}P_1(\cos\theta)$$

where $P_1(\cos\theta) = \cos\theta$ is the first order Legendre polynomial. Compute the electric field from this potential, $\vec{E} = -\nabla V$. And sketch it of course.

8.26 In problem 2.26 you computed the electric potential at large distances from a set of three charges, $-2q$ at the origin and $+q$ at $z = \pm a$ $(r \gg a)$. The result was

$$V = \frac{kqa^2}{r^3} P_2(\cos\theta)$$

where $P_2(\cos\theta)$ is the second order Legendre polynomial. Compute the electric field from this potential, $\vec{E} = -\nabla V$. And sketch it of course.

8.27 Compute the area of an ellipse having semi major and semi-minor axes a and b. Compare your result to that of Eq. (8.35). Ans: πab

8.28 Two equal point charges q are placed at $z = \pm a$. The origin is a point of equilibrium; $\vec{E} = 0$ there. (**a**) Compute the potential near the origin, writing V in terms of powers of x, y, and z near there, carrying the powers high enough to describe the nature of the equilibrium point. Is V maximum, minimum, or saddle point there? It will be easier if you carry the calculation as far as possible using vector notation, such as $|\vec{r} - a\hat{z}| = \sqrt{(\vec{r} - a\hat{z})^2}$, and $r \ll a$.
(**b**) Write your result for V near the origin in spherical coordinates also.
Ans: $\frac{2q}{4\pi\epsilon_0 a}\left[1 + \frac{r^2}{a^2}\left(\frac{3}{2}\cos^2\theta - \frac{1}{2}\right)\right]$

8.29 When current I flows through a resistance R the heat produced is I^2R. Two terminals are connected in parallel by three resistors having resistance R_1, R_2, and R_3. Given that the total current is divided as $I = I_1 + I_2 + I_3$, show that the condition that the total heat generated is a minimum leads to the relation $I_1R_1 = I_2R_2 = I_3R_3$. You can easily do problem 8.8 by eliminating a coordinate the doing a derivative. Here it's starting to get sufficiently complex that you should use Lagrange multipliers. Does λ have any significance this time?

8.30 Given a right circular cylinder of volume V, what radius and height will provide the minimum total area for the cylinder. Ans: $r = (V/2\pi)^{1/3}$, $h = 2r$

8.31 Sometimes the derivative isn't zero at a maximum or a minimum. Also, there are two types of maxima and minima; local and global. The former is one that is max or min in the immediate neighborhood of a point and the latter is biggest or smallest over the entire domain of the function. Examine these functions for maxima and minima both inside the domains and on the boundary.

$$|x|, \; (-1 \le x \le +2)$$
$$T_0(x^2 - y^2)/a^2, \; (-a \le x \le a, \; -a \le y \le a)$$
$$V_0(r^2/R^2)P_2(\cos\theta), \; (r \le R, \text{ 3 dimensions})$$

8.32 In Eq. (8.39) it is more common to specify N and $\beta = 1/kT$, the Lagrange multiplier, than it is to specify N and E, the total energy. Pick three energies, E_ℓ, to be 1, 2, and

3 electron volts. (**a**) What is the average energy, E/N, as $\beta \to \infty$ ($T \to 0$)?
(**b**) What is the average energy as $\beta \to 0$?
(**c**) What are n_1, n_2, and n_3 in these two cases?

8.33 (**a**) Find the gradient of V, where $V = V_0(x^2+y^2+z^2)a^{-2}e^{-\sqrt{x^2+y^2+z^2}\,/a}$. (**b**) Find the gradient of V, where $V = V_0(x+y+z)a^{-1}e^{-(x+y+z)/a}$.

8.34 A billiard ball of radius R is suspended in space and is held rigidly in position. Very small pellets are thrown at it and the scattering from the surface is completely elastic, with no friction. Compute the relation between the impact parameter b and the scattering angle θ. Then compute the differential scattering cross section $d\sigma/d\Omega$.
Finally compute the total scattering cross section, the integral of this over $d\Omega$.

8.35 Modify the preceding problem so that the incoming object is a ball of radius R_1 and the fixed billiard ball has radius R_2.

8.36 Find the differential scattering cross section from a spherical drop of water, but instead of Snell's law, use a pre-Snell law: $\beta = n\alpha$, without the sines. Is there a rainbow in this case? Sketch $d\sigma/d\Omega$ versus θ.
Ans: $R^2 \sin 2\beta/\big[4\sin\theta|1-2/n|\big]$, where $\theta = \pi + 2(1-2/n)\beta$

8.37 From the equation (8.43), assuming just a single b for a given θ, what is the integral over all $d\Omega$ of $d\sigma/d\Omega$? Ans: πb_{max}^2

8.38 Solve Eq. (8.47) for b when $d\theta/db = 0$. For $n = 1.33$ what value of θ does this give?

8.39 If the scattering angle $\theta = \frac{\pi}{2}\sin(\pi b/R)$ for $0 < b < R$, what is the resulting differential scattering cross section (with graph). What is the total scattering cross section? Start by sketching a graph of θ versus b. Ans: $2R^2/\big[\pi^2\sin\theta\sqrt{1-(2\theta/\pi)^2}\big]$

8.40 Find the signs of all the factors in Eq. (8.49), and determine from that whether red or blue is on the outside of the rainbow. Ans: Look

8.41 If it suddenly starts to rain small, spherical diamonds instead of water, what happens to the rainbow? $n = 2.4$ for diamond.

8.42 What would the rainbow look like for $n = 2$? You'll have to look closely at the expansions in this case. For small b, where does the ray hit the inside surface of the drop?

8.43 (**a**) The secondary rainbow occurs because there can be two internal reflections before the light leave the drop. What is the analog of Eqs. (8.44) for this case? (**b**) Repeat problems 8.38 and 8.40 for this case.

8.44 What is the shortest distance from the origin to the plane defined by $\vec{A}\cdot(\vec{r}-\vec{r}_0) = 0$? Do this using Lagrange multipliers, and then explain why of course the answer is correct.

8.45 The U.S. Post Office has decided to use a norm like Eq. (6.11)(2) to measure boxes. The size is defined to be the sum of the height and the circumference of the box, and the circumference is around the thickest part of the package: "length plus girth." What is the maximum volume you can ship if this size is constrained to be less than 130 inches?

For this purpose, assume the box is rectangular, not cylindrical, though you may expect the cylinder to improve the result. Assume that the box's dimensions are a, a, b, with volume a^2b.
(**a**) Show that if you assume that the girth is $4a$, then you will conclude that $b > a$ and that you *didn't* measure the girth at the thickest part of the package.
(**b**) Do it again with the opposite assumption, that you assume b is big so that the girth is $2b + 2a$. Again show that it is a contradiction.
(**c**) You have two inequalities that you must satisfy: girth plus length measured one way is less than $L = 130$ inches and girth plus length measured the other way is too. That is, $4a+b < L$ and $3a+2b < L$. Plot these regions in the a-b plane, showing the allowed region in a-b space. Also plot some curves of constant volume, $V = a^2b$. Show that the point of maximum volume subject to these constraints is on the edge of this allowed region, and that it is at the corner of intersection of the two inequalities. This is the beginning of the subject called "linear programming."
Ans: a cube

8.46 Plot θ versus b in equation (8.45) or (8.46).

8.47 A disk of radius R is at a distance c above the x-y plane and parallel to that plane. What is the solid angle that this disk subtends from the origin? Ans: $2\pi\left[1-c/\sqrt{c^2+R^2}\right]$

8.48 Within a sphere of radius R, what is the volume contained between the planes defined by $z = a$ and $z = b$? Ans: $\pi(b-a)\left(R^2 - \frac{1}{3}(b^2+ab+a^2)\right)$

8.49 Find the mean-square distance, $\frac{1}{V}\int r^2\,dV$, from a point on the surface of a sphere to points inside the sphere. Note: Plan ahead and try to make this problem as easy as possible. Ans: $8R^2/5$

8.50 Find the mean distance, $\frac{1}{V}\int r\,dV$, from a point on the surface of a sphere to points inside the sphere. Unlike the preceding problem, this requires some brute force. Ans: $6R/5$

8.51 A volume mass density is specified in spherical coordinates to be

$$\rho(r,\theta,\phi) = \rho_0\left(1 + r^2/R^2\right)\left[1 + \tfrac{1}{2}\cos\theta\sin^2\phi + \tfrac{1}{4}\cos^2\theta\sin^3\phi\right]$$

Compute the total mass in the volume $0 < r < R$. Ans: $32\pi\rho_0R^3/15$

8.52 The circumference of a circle is some constant times its radius (C_1r). For the two-dimensional surface that is a sphere in three dimensions the area is of the form C_2r^2. Start from the fact that you know the integral $\int_{-\infty}^{\infty} dx\, e^{-x^2} = \pi^{1/2}$ and write out the following two dimensional integral twice. It is over the entire plane.

$$\int dA\, e^{-r^2} \qquad \text{using} \qquad dA = dx\,dy \qquad \text{and using} \qquad dA = C_1r\,dr$$

From this, evaluate C_1. Repeat this for dV and C_2r^2 in three dimensions, evaluating C_2. Now repeat this in arbitrary dimensions to evaluate C_n. Do you need to reread chapter one? In particular, what is C_3? It tells you about the three dimensional hypersphere in four dimensions. From this, what is the total "hypersolid angle" in four dimensions (like 4π in three)? Ans: $2\pi^2$

8.53 Do the reverse of problem 8.23. Start with the second equation there and change variables to see that it reverts to a constant times the first equation.

8.54 Carry out the interchange of limits in Eq. (8.22). Does the drawing really represent the integral?

8.55 Is $x^2 + xy + y^2$ a minimum or maximum or something else at $(0,0)$? Do the same question for $x^2 + 2xy + y^2$ and for $x^2 + 3xy + y^2$. Sketch the surface $z = f(x,y)$ in each case.

8.56 Derive the conditions stated after Eq. (8.33), expressing the circumstances under which the Hessian matrix is positive definite.

8.57 In the spirit of problems 8.10 *et seq.* what happens if you have a rectangular drumhead instead of a circular one? Let $0 < x < a$ and $0 < y < b$. The drumhead is tied down at its edges, so an appropriate function that satisfies these conditions is

$$z(x,y) = A\sin(n\pi x/a)\sin(m\pi y/b)\cos\omega t$$

Compute the total kinetic and the total potential energy for this oscillation, a function of time. For energy to be conserved the total energy must be a constant, so compute the frequency ω for which this is true. As compared to the previous problems about a circular drumhead, this turns out to give the exact results instead of only approximate ones. Ans: $\omega^2 = \pi^2 \frac{\mu}{T}\left[\frac{n^2}{a^2} + \frac{m^2}{b^2}\right]$

8.58 Repeat problem 8.45 by another method. Instead of assuming that the box has a square end, allow it to be any rectangular box, so that its volume is $V = abc$. Now you have three independent variables to use, maximizing the volume subject to the post office's constraint on length plus girth. This looks like it will have to be harder. Instead, it's much easier. Draw pictures! Ans: still a cube

8.59 An asteroid is headed in the general direction of Earth, and its speed when far away is v_0 relative to the Earth. What is the total cross section for it's hitting Earth? It is not necessary to compute the complete orbit; all you have to do is use a couple of conservation laws. Express the result in terms of the escape speed from Earth.
Ans: $\sigma = \pi R^2\left(1 + (v_{\text{esc}}/v_0)^2\right)$

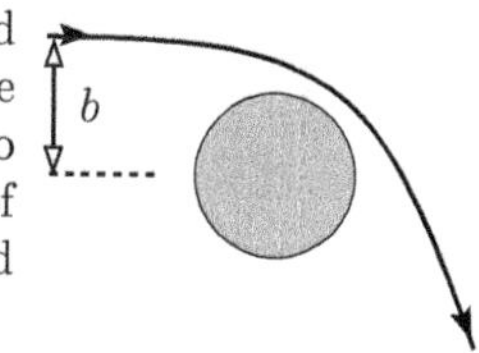

8.60 In three dimensions the differential scattering cross section appeared in Eqs. (8.42) and (8.43). If the world were two dimensional this area would be a length instead. What are the two corresponding equations in that case, giving you an expression for $d\ell/d\theta$. Apply this to the light scattering from a (two dimensional) drop of water and describe the scattering results. For simplicity this time, assume the pre-Snell law as in problem 8.36.

8.61 As in the preceding problem, but use the regular Snell law instead.

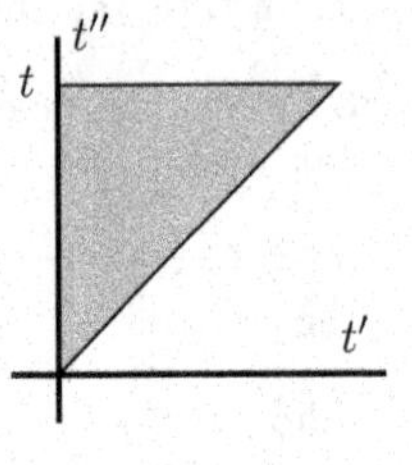

8.62 This double integral is over the isosceles right triangle in the figure. The function to be integrated is $f(t') = \alpha t'^3$, BUT FIRST, set it up for an arbitrary $f(t')$ and then set it up again but with the order of integration reversed. In one of the two cases you should be able to do one integral without knowing f. Having done this, apply your two results to this particular f as a test case that your work was correct. In the figure, t' and t'' are the two coordinates and t is the coordinate of the top of the triangle.

Vector Calculus 1

The first rule in understanding vector calculus is *draw lots of pictures.* This subject can become rather abstract if you let it, but try to visualize all the manipulations. Try a lot of special cases and explore them. Keep relating the manipulations to the underlying pictures and don't get lost in the forest of infinite series. Along with the pictures, there are three types of derivatives, a couple of types of integrals, and some theorems relating them.

9.1 Fluid Flow

When water or any fluid moves through a pipe, what is the relationship between the motion of the fluid and the total rate of flow through the pipe (volume per time)? Take a rectangular pipe of sides a and b with fluid moving at constant speed through it and with the velocity of the fluid being the same throughout the pipe. It's a simple calculation: In time Δt the fluid moves a distance $v\Delta t$ down the pipe. The cross-section of the pipe has area $A = ab$, so the volume that move past a given flat surface is $\Delta V = Av\Delta t$. The flow rate is the volume per time, $\Delta V/\Delta t = Av$. (The usual limit as $\Delta t \to 0$ isn't needed here.)

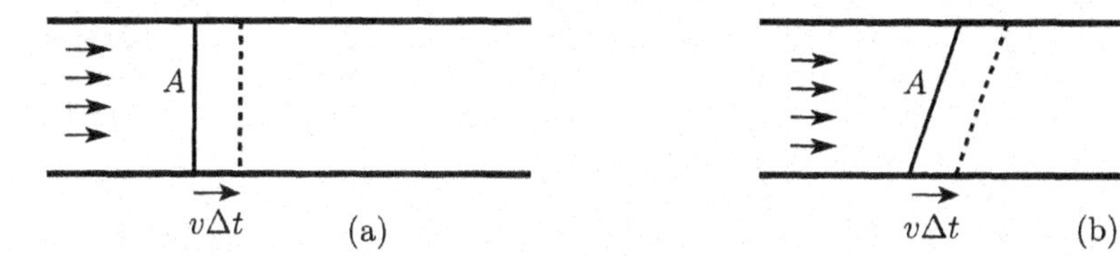

Just to make the problem look a little more involved, what happens to the result if I ask for the flow through a surface that is tilted at an angle to the velocity. Do the calculation the same way as before, but use the drawing (b) instead of (a). The fluid still moves a distance $v\Delta t$, but the volume that moves past this flat but tilted surface is not its new (bigger) area A times $v\Delta t$. The area of a parallelogram is not the product of its sides and the volume of a parallelepiped is not the area of a base times the length of another side.

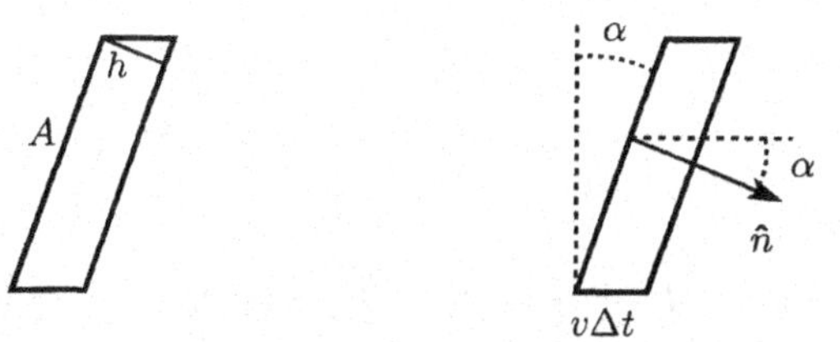

The area of a parallelogram is the length of one side times the perpendicular distance from that side to its opposite side. Similarly the volume of a parallelepiped is the area of one side times the perpendicular distance from that side to the side opposite. The perpendicular distance is not the distance that the fluid moved ($v\Delta t$). This perpendicular distance is smaller by a factor $\cos\alpha$, where α is the angle that the plane is tilted. It is most easily described by the angle that the *normal* to the plane makes with the direction of the fluid velocity.

$$\Delta V = Ah = A(v\Delta t)\cos\alpha$$

The flow rate is then $\Delta V/\Delta t = Av\cos\alpha$. Introduce the unit normal vector $\hat{n}$, then this expression can be rewritten in terms of a dot product,

$$Av\cos\alpha = A\vec{v}\cdot\hat{n} = \vec{A}\cdot\vec{v} \tag{9.1}$$

where α is the angle between the direction of the fluid velocity and the normal to the area.

This invites the definition of the area itself as a vector, and that's what I wrote in the final expression. The vector $\vec{A}$ is a notation for $A\hat{n}$, and defines the area vector. If it looks a little odd to have an area be a vector, do you recall the geometric interpretation of a cross product? That's the vector perpendicular to two given vectors and it has a magnitude equal to the area of the parallelogram between the two vectors. It's the same thing.

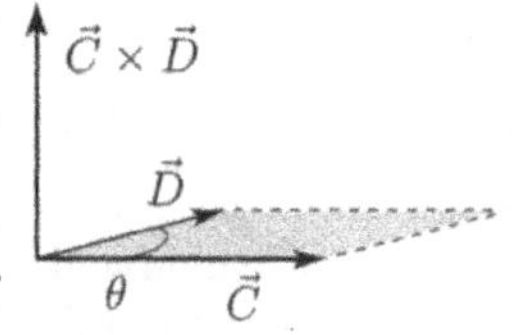

General Flow, Curved Surfaces

The fluid velocity will not usually be a constant in space. It will be some function of position. The surface doesn't have to be flat; it can be cylindrical or spherical or something more complicated. How do you handle this? That's why integrals were invented.

The idea behind an integral is that you will divide a complicated thing into small pieces and add the results of the small pieces to *estimate* the whole. Improve the estimation by making more, smaller pieces, and in the limit as the size of the pieces goes to zero get an exact answer. That's the procedure to use here.

The concept of the surface integral is that you systematically divide a surface into a number (N) of pieces ($k = 1, 2, \ldots N$). The pieces have area ΔA_k and each piece has a unit normal vector $\hat{n}_k$. Within the middle of each of these areas the fluid has a velocity $\vec{v}_k$. This may not be a constant, but as usual with integrals, you pick a point somewhere in the little area and pick the $\vec{v}$ there; in the limit as all the pieces of area shrink to zero it won't matter exactly where you picked it. The flow rate through one of these pieces is Eq. (9.1), $\vec{v}_k\cdot\hat{n}_k\Delta A_k$, and the corresponding estimate of the total flow through the surface is, using the notation $\Delta\vec{A}_k = \hat{n}_k\Delta A_k$,

$$\sum_{k=1}^{N}\vec{v}_k\cdot\Delta\vec{A}_k$$

This limit as the size of each piece is shrunk to zero and correspondingly the number of pieces goes to infinity is the definition of the integral

$$\int\vec{v}\cdot d\vec{A} = \lim_{\Delta A_k\to 0}\sum_{k=1}^{N}\vec{v}_k\cdot\Delta\vec{A}_k \tag{9.2}$$

Example of Flow Calculation

In the rectangular pipe above, suppose that the flow exhibits shear, rising from zero at the bottom to v_0 at the top. The velocity field is

$$\vec{v}(x,y,z) = v_x(y)\hat{x} = v_0\frac{y}{b}\hat{x} \tag{9.3}$$

The origin is at the bottom of the pipe and the y-coordinate is measured upward from the origin. What is the flow rate through the area indicated, tilted at an angle ϕ from the vertical? The distance in and out of the plane of the picture (the z-axis) is the length a. Can such a fluid flow really happen? Yes, real fluids such as water have viscosity, and if you construct a very wide pipe but not too high, and leave the top open to the wind, the horizontal wind will drag the fluid at the top with it (even if not as fast). The fluid at the bottom is kept at rest by the friction with the bottom surface. In between you get a gradual transition in the flow that is represented by Eq. (9.3).

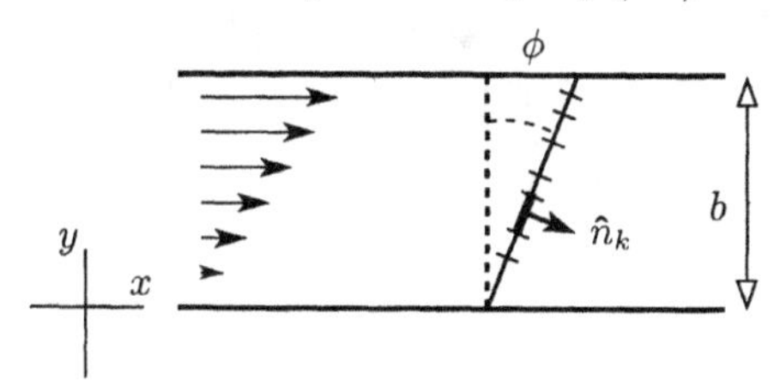

Now to implement the calculation of the flow rate:
Divide the area into N pieces of length $\Delta\ell_k$ along the slant.
The length in and out is a so the piece of area is $\Delta A_k = a\Delta\ell_k$.
The unit normal is $\hat n_k = \hat x\cos\phi - \hat y\sin\phi$. (It happens to be independent of the index k, but that's special to this example.)
The velocity vector at the position of this area is $\vec v = v_0\,\hat x\,y_k/b$.
Put these together and you have the piece of the flow rate contributed by this area.

$$\begin{aligned}\Delta\text{flow}_k = \vec v\cdot\Delta\vec A_k &= v_0\frac{y_k}{b}\hat x\cdot a\Delta\ell_k\big(\hat x\cos\phi - \hat y\sin\phi\big)\\ &= v_0\frac{y_k}{b}a\Delta\ell_k\cos\phi = v_0\frac{\ell_k\cos\phi}{b}a\Delta\ell_k\cos\phi\end{aligned}$$

In the last line I put all the variables in terms of ℓ, using $y = \ell\cos\phi$.

Now sum over all these areas and take a limit.

$$\begin{aligned}\lim_{\Delta\ell_k\to 0}\sum_{k=1}^{N} v_0\frac{\ell_k\cos\phi}{b}a\Delta\ell_k\cos\phi &= \int_0^{b/\cos\phi} d\ell\, v_0\frac{a}{b}\ell\cos^2\phi = v_0\frac{a}{b}\frac{\ell^2}{2}\cos^2\phi\bigg|_0^{b/\cos\phi}\\ &= v_0\frac{a}{2b}\left(\frac{b}{\cos\phi}\right)^2\cos^2\phi = v_0\frac{ab}{2}\end{aligned}$$

This turns out to be independent of the orientation of the plane; the parameter ϕ is absent from the result. If you think of two planes, at angles ϕ_1 and ϕ_2, what flows into one flows out of the other. Nothing is lost in between.

Another Flow Calculation

Take the same sort of fluid flow in a pipe, but make it a little more complicated. Instead of a flat surface, make it a cylinder. The axis of the cylinder is in and out in the picture and its radius is half the width of the pipe. Describe the coordinates on the surface by the angle θ as measured from the midline. That means that $-\pi/2 < \theta < \pi/2$. Divide the surface into pieces that are rectangular strips of length a (in and out in the picture) and width $b\Delta\theta_k/2$. (The radius of the cylinder is $b/2$.)

$$\Delta A_k = a\frac{b}{2}\Delta\theta_k, \qquad \text{and} \qquad \hat{n}_k = \hat{x}\cos\theta_k + \hat{y}\sin\theta_k \tag{9.4}$$

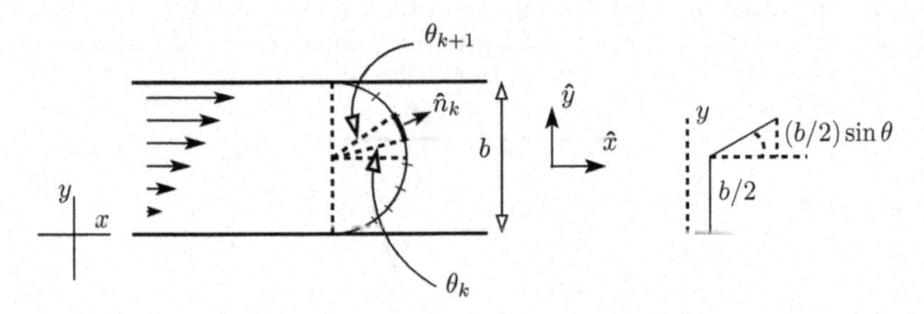

The velocity field is the same as before, $\vec{v}(x,y,z) = v_0\,\hat{x}\,y/b$, so the contribution to the flow rate through this piece of the surface is

$$\vec{v}_k \cdot \Delta\vec{A}_k = v_0\,\frac{y_k}{b}\hat{x}\cdot a\,\frac{b}{2}\,\Delta\theta_k\,\hat{n}_k$$

The value of y_k at the angle θ_k is

$$y_k = \frac{b}{2} + \frac{b}{2}\sin\theta_k, \qquad \text{so} \qquad \frac{y_k}{b} = \frac{1}{2}[1+\sin\theta_k]$$

Put the pieces together and you have

$$v_0\frac{1}{2}[1+\sin\theta_k]\hat{x}\cdot a\frac{b}{2}\Delta\theta_k\big[\hat{x}\cos\theta_k + \hat{y}\sin\theta_k\big] = v_0\frac{1}{2}\big[1+\sin\theta_k\big]a\frac{b}{2}\Delta\theta_k\cos\theta_k$$

The total flow is the sum of these over k and then the limit as$\Delta\ \theta_k \to 0$.

$$\lim_{\Delta\theta_k\to 0}\sum_k v_0\frac{1}{2}[1+\sin\theta_k]a\frac{b}{2}\Delta\theta_k\cos\theta_k = \int_{-\pi/2}^{\pi/2} v_0\frac{1}{2}\big[1+\sin\theta\big]a\frac{b}{2}\,d\theta\,\cos\theta$$

Finally you can do the two terms of the integral: Look at the second term first. You can of course start grinding away and find the right trigonometric formula to do the integral, OR, you can sketch a graph of the integrand, $\sin\theta\cos\theta$, on the interval $-\pi/2 < \theta < \pi/2$ and write the answer down by inspection. The first part of the integral is

$$v_0\frac{ab}{4}\int_{-\pi/2}^{\pi/2}\cos\theta = v_0\frac{ab}{4}\sin\theta\bigg|_{-\pi/2}^{\pi/2} = v_0\frac{ab}{2}$$

And this is the same result as for the flat surface calculation. I set it up so that the two results are the same; it's easier to check that way. Gauss's theorem of vector calculus will guarantee that you get the same result for any surface spanning this pipe *and* for this particular velocity function.

9.2 Vector Derivatives

I want to show the underlying ideas of the vector derivatives, divergence and curl, and as the names themselves come from the study of fluid flow, that's where I'll start. You can describe the flow of a fluid, either gas or liquid or anything else, by specifying its velocity field, $\vec{v}(x, y, z) = \vec{v}(\vec{r})$.

For a single real-valued function of a real variable, it is often too complex to capture all the properties of a function at one glance, so it's going to be even harder here. One of the uses of ordinary calculus is to provide information about the *local* properties of a function without attacking the whole function at once. That is what derivatives do. If you know that the derivative of a function is positive at a point then you know that it is increasing there. This is such an ordinary use of calculus that you hardly give it a second thought (until you hit some advanced calculus and discover that some continuous functions don't even *have* derivatives). The geometric concept of derivative is the slope of the curve at a point — the tangent of the angle between the x-axis and the straight line that best approximates the curve at that point. Going from this geometric idea to calculating the derivative takes some effort.

How can you do this for fluid flow? If you inject a small amount of dye into the fluid at some point it will spread into a volume that depends on how much you inject. As time goes on this region will move and distort and possibly become very complicated, too complicated to grasp in one picture.

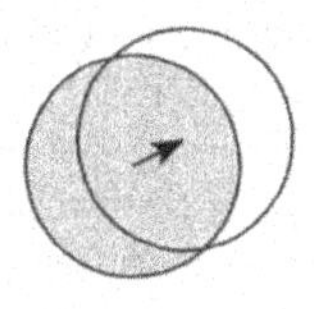 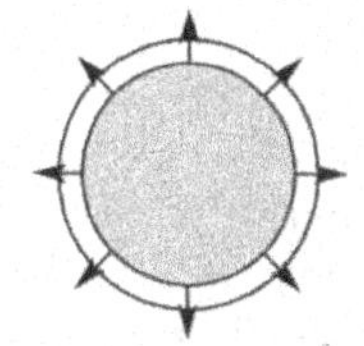 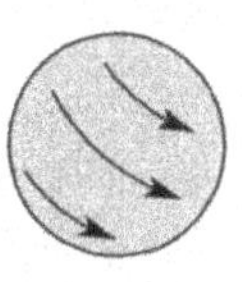 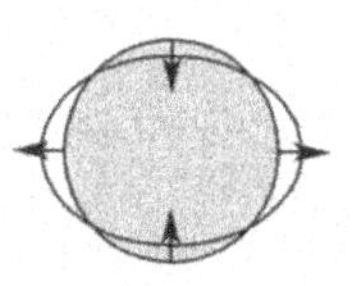

There must be a way to get a simpler picture. There is. Do it in the same spirit that you introduce the derivative, and concentrate on a little piece of the picture. Inject just a little bit of dye and wait only a little time. To make it explicit, assume that the initial volume of dye forms a sphere of (small) volume V and let the fluid move for a little time.

1. In a small time Δt the center of the sphere will move.
2. The sphere can expand or contract, changing its volume.
3. The sphere can rotate.
4. The sphere can distort.

Div, Curl, Strain

The first one, the motion of the center, tells you about the velocity at the center of the sphere. It's like knowing the value of a function at a point, and that tells you nothing about the behavior of the function in the neighborhood of the point.

The second one, the volume, gives new information. You can simply take the time derivative dV/dt to see if the fluid is expanding or contracting; just check the sign and determine if it's positive or negative. But how big is it? That's not yet in a useful form because the size of this derivative will depend on how much the original volume is. If you put in twice as much dye, each part of the volume will change and there will be twice as much rate of change in the total volume. Divide the time derivative by the volume itself and this effect will cancel. Finally, to get the effect at one point take the limit as the volume approaches a point. This defines a kind of derivative of the velocity field called the divergence.

$$\lim_{V\to 0}\frac{1}{V}\frac{dV}{dt} = \text{divergence of } \vec v \tag{9.5}$$

This doesn't tell you how to compute it any more than saying that the derivative is the slope tells you how to compute an ordinary* derivative. I'll have to work that out.

But first look at the third way that the sphere can move: rotation. Again, if you take a large object it will distort a lot and it will be hard to define a single rotation for it. Take a very small sphere instead. The time derivative of this rotation is its angular velocity, the vector $\vec\omega$. In the limit as the sphere approaches a point, this tells me about the rotation of the fluid in the immediate neighborhood of that point. If I place a tiny paddlewheel in the fluid, how will it rotate?

$$2\vec\omega = \text{curl of } \vec v \tag{9.6}$$

The factor of 2 is for later convenience.

After considering expansion and rotation, the final way that the sphere can change is that it can alter its shape. In a very small time interval, the sphere can slightly distort into an ellipsoid. This will lead to the mathematical concept of the *strain.* This is important in the subject of elasticity and viscosity, but I'll put it aside for now save for one observation: how much information is needed to describe whatever it is? The sphere changes to an ellipsoid, and the first question is: what is the longest axis and how much stretch occurs along it — that's the three components of a vector. After that what is the shortest axis and how much contraction occurs along *it*? That's one more vector, but you need only two new components to define its direction because it's perpendicular to the long axis. After this there's nothing left. The direction of the third axis is determined and so is its length if you assume that the total volume hasn't changed. You can assume that is so because the question of volume change is already handled by the divergence; you don't need it here too. The total number of components needed for this object is $2 + 3 = 5$. It comes under the heading of tensors.

9.3 Computing the divergence

Now how do you calculate these? I'll start with the simplest, the divergence, and compute the time derivative of a volume from the velocity field. To do this, go back to the definition of derivative:

$$\frac{dV}{dt} = \lim_{\Delta t\to 0}\frac{V(t+\Delta t) - V(t)}{\Delta t} \tag{9.7}$$

---

* Can you start from the definition of the derivative as a slope, use it directly with no limits, and compute the derivative of x^2 with respect to x, getting $2x$? It *can* be done.

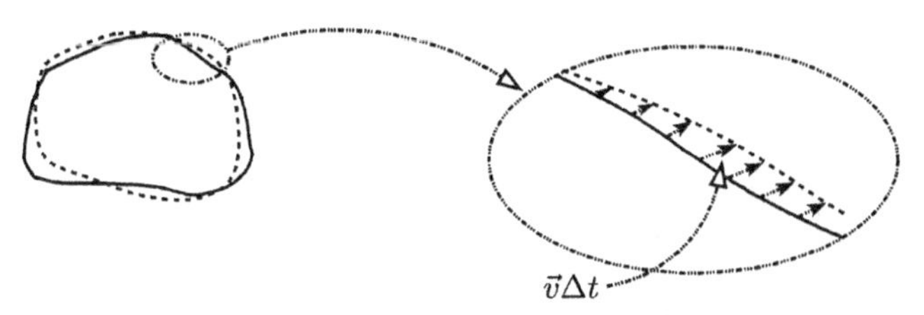

Pick an arbitrary surface to start with and see how the volume changes as the fluid moves, carrying the surface with it. In time Δt a point on the surface will move by a distance $\vec{v}\Delta t$ and it will carry with it a piece of neighboring area ΔA. This area sweeps out a volume. This piece of volume is not ΔA times $v\Delta t$ because the motion isn't likely to be perpendicular to the surface. It's only the *component* of the velocity normal to the surface that contributes to the volume swept out. Use $\hat{n}$ to denote the unit vector perpendicular to ΔA, then this volume is $\Delta A\,\hat{n}\cdot\vec{v}\,\Delta t$. This is the same as the the calculation for fluid flow except that I'm interpreting the picture differently.

If at a particular point on the surface the normal $\hat{n}$ is more or less in the direction of the velocity then this dot product is positive and the change in volume is positive. If it's opposite the velocity then the change is negative. The total change in volume of the whole initial volume is the sum over the entire surface of all these changes. Divide the surface into a lot of pieces ΔA_i with accompanying unit normals $\hat{n}_i$, then

$$\Delta V_{\text{total}} = \sum_i \Delta A_i\,\hat{n}_i \cdot \vec{v}_i\,\Delta t$$

Not really. I have to take a limit before this becomes an equality. The limit of this as all the $\Delta A_i \to 0$ defines an integral

$$\Delta V_{\text{total}} = \oint dA\,\hat{n}\cdot\vec{v}\,\Delta t$$

and this integral notation is special; the circle through the integral designates an integral over the whole closed surface and the direction of $\hat{n}$ is always taken to be outward. Finally, divide by Δt and take the limit as Δt approaches zero.

$$\frac{dV}{dt} = \oint dA\,\hat{n}\cdot\vec{v} \tag{9.8}$$

The $\vec{v}\cdot\hat{n}\,dA$ is the rate at which the area dA sweeps out volume as it is carried with the fluid. Note: There's nothing in this calculation saying that I have to take the limit as $V \to 0$; it's a perfectly general expression for the rate of change of volume in a surface being carried with the fluid. It's also a completely general expression for the rate of flow of fluid through a fixed surface as the fluid moves past it. I'm interested in the first interpretation for now, but the second is just as valid in other contexts.

Again, use the standard notation in which the area vector combines the unit normal and the area: $d\vec{A} = \hat{n}\,dA$.

$$\text{divergence of } \vec{v} = \lim_{V\to 0}\frac{1}{V}\frac{dV}{dt} = \lim_{V\to 0}\frac{1}{V}\oint \vec{v}\cdot d\vec{A} \tag{9.9}$$

If the fluid is on average moving away from a point then the divergence there is positive. It's diverging.

The Divergence as Derivatives

This is still a long way from something that you can easily compute. I'll first go through a detailed analysis of how you turn this into a simple result, and I'll then go back to try to capture the essence of the derivation so you can see how it applies in a wide variety of coordinate systems. At that point I'll also show how to get to the result with a lot less algebra. You will see that a lot of the terms that appear in this first calculation will vanish in the end. It's important then to go back and see what was really essential to the calculation and what was not. As you go through this derivation then, try to anticipate which terms are going to be important and which terms are going to disappear.

Express the velocity in rectangular components, $v_x\,\hat{x} + v_y\,\hat{y} + v_z\,\hat{z}$. For the small volume, choose a rectangular box with sides parallel to the axes. One corner is at point (x_0, y_0, z_0) and the opposite corner has coordinates that differ from these by $(\Delta x, \Delta y, \Delta z)$. Expand everything in a power series about the first corner as in section 2.5. Instead of writing out (x_0, y_0, z_0) every time, I'll abbreviate it by $({}_0)$.

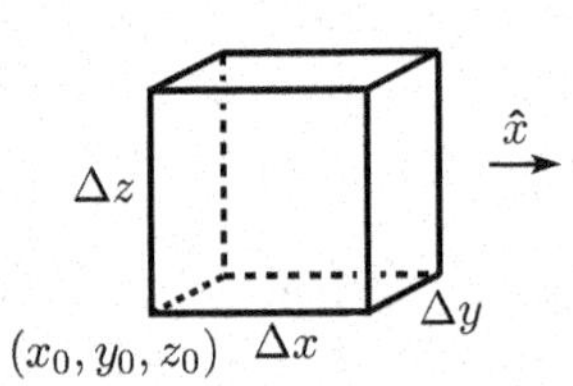

Δz $\quad (x_0, y_0, z_0)$ $\quad \Delta x$ $\quad \Delta y$ $\quad \hat{x}$

$$\begin{aligned} v_x(x,y,z) = v_x({}_0) &+ (x-x_0)\frac{\partial v_x}{\partial x}({}_0) + (y-y_0)\frac{\partial v_x}{\partial y}({}_0) \\ &+ (z-z_0)\frac{\partial v_x}{\partial z}({}_0) + \frac{1}{2}(x-x_0)^2\frac{\partial^2 v_x}{\partial x^2}({}_0) \\ &+ (x-x_0)(y-y_0)\frac{\partial^2 v_x}{\partial x\,\partial y}({}_0) + \cdots \end{aligned} \tag{9.10}$$

There are six integrals to do, one for each face of the box, and there are three functions, v_x, v_y, and v_z to expand in three variables x, y, and z. *Don't Panic.* A lot of these are zero. If you look at the face on the right in the sketch you see that it is parallel to the y-z plane and has normal $\hat{n} = \hat{x}$. When you evaluate $\vec{v}\cdot\hat{n}$, only the v_x term survives; flow parallel to the surface (v_y, v_z) contributes nothing to volume change along this part of the surface. Already then, many terms have simply gone away.

Write the two integrals over the two surfaces parallel to the y-z plane, one at x_0 and one at $x_0 + \Delta x$.

$$\begin{aligned} &\int_{\text{right}} \vec{v}\cdot d\vec{A} + \int_{\text{left}} \vec{v}\cdot d\vec{A} \\ &\qquad = \int_{y_0}^{y_0+\Delta y} dy \int_{z_0}^{z_0+\Delta z} dz\, v_x(x_0+\Delta x, y, z) - \int_{y_0}^{y_0+\Delta y} dy \int_{z_0}^{z_0+\Delta z} dz\, v_x(x_0, y, z) \end{aligned}$$

The minus sign comes from the dot product because $\hat{n}$ points left on the left side. You can evaluate these integrals by using their power series representations, and though you may have an infinite number of terms to integrate, at least they're all easy. Take the first of them:

$$\int_{y_0}^{y_0+\Delta y} dy \int_{z_0}^{z_0+\Delta z} dz \Bigg[v_x(0)+$$
$$(\Delta x)\frac{\partial v_x}{\partial x}(0) + (y-y_0)\frac{\partial v_x}{\partial y}(0) + (z-z_0)\frac{\partial v_x}{\partial z}(0) + \frac{1}{2}(\Delta x)^2\frac{\partial^2 v_x}{\partial x^2}(0) + \cdots \Bigg]$$
$$= v_x(0)\Delta y\Delta z + (\Delta x)\frac{\partial v_x}{\partial x}(0)\Delta y\Delta z +$$
$$\frac{1}{2}(\Delta y)^2\Delta z\frac{\partial v_x}{\partial y}(0) + \frac{1}{2}(\Delta z)^2\Delta y\frac{\partial v_x}{\partial z}(0) + \frac{1}{2}(\Delta x)^2\frac{\partial^2 v_x}{\partial x^2}(0)\Delta y\Delta z + \cdots$$

Now look at the second integral, the one that you have to subtract from this one. Before plunging in to the calculation, stop and look around. What will cancel; what will contribute; what will not contribute? The only difference is that this is now evaluated at x_0 instead of at $x_0 + \Delta x$. The terms that have Δx in them simply won't appear this time. All the rest are exactly the same as before. That means that all the terms in the above expression that do *not* have a Δx in them will be canceled when you subtract the second integral. All the terms that *do* have a Δx will be untouched. The combination of the two integrals is then

$$(\Delta x)\frac{\partial v_x}{\partial x}(0)\Delta y\Delta z + \frac{1}{2}(\Delta x)^2\frac{\partial^2 v_x}{\partial x^2}(0)\Delta y\Delta z + \frac{1}{2}(\Delta x)\frac{\partial^2 v_x}{\partial x\partial y}(0)(\Delta y)^2\Delta z + \cdots$$

Two down four to go, but not really. The other integrals are the same except that x becomes y and y becomes z and z becomes x. The integral over the two faces with y constant are then

$$(\Delta y)\frac{\partial v_y}{\partial y}(0)\Delta z\Delta x + \frac{1}{2}(\Delta y)^2\frac{\partial^2 v_y}{\partial y^2}(0)\Delta z\Delta x + \cdots$$

and a similar expression for the final two faces. The definition of Eq. (9.9) says to add all three of these expressions, divide by the volume, and take the limit as the volume goes to zero. $V = \Delta x\Delta y\Delta z$, and you see that this is a common factor in all of the terms above. Cancel what you can and you have

$$\frac{\partial v_x}{\partial x}(0) + \frac{\partial v_y}{\partial y}(0) + \frac{\partial v_x}{\partial x}(0) + \frac{1}{2}(\Delta x)\frac{\partial^2 v_x}{\partial x^2}(0) + \frac{1}{2}(\Delta y)\frac{\partial^2 v_y}{\partial y^2}(0) + \frac{1}{2}(\Delta z)\frac{\partial^2 v_z}{\partial z^2}(0) + \cdots$$

In the limit that the all the Δx, Δy, and Δz shrink to zero the terms with a second derivative vanish, as do all the other higher order terms. You are left then with a rather simple expression for the divergence.

$$\text{divergence of } \vec{v} = \text{div}\,\vec{v} = \frac{\partial v_x}{\partial x} + \frac{\partial v_y}{\partial y} + \frac{\partial v_x}{\partial x} \tag{9.11}$$

This is abbreviated by using the differential operator ∇, "del."

$$\nabla = \hat{x}\frac{\partial}{\partial x} + \hat{y}\frac{\partial}{\partial y} + \hat{z}\frac{\partial}{\partial z} \tag{9.12}$$

Then you can write the preceding equation as

$$\text{divergence of } \vec{v} = \operatorname{div} \vec{v} = \nabla \cdot \vec{v} \tag{9.13}$$

The symbol ∇ will take other forms in other coordinate systems.

Now that you've waded through this rather technical set of manipulations, is there an easier way? Yes *but*, without having gone through the preceding algebra you won't be able to see and to understand which terms are important and which terms are going to cancel or otherwise disappear. When you need to apply these ideas to something besides rectangular coordinates you have to know what to keep and what to ignore. Once you know this, you can go straight to the end of the calculation and write down those terms that you know are going to survive, dropping the others. *This takes practice.*

Simplifying the derivation

In the long derivation of the divergence, the essence is that you find $\vec{v} \cdot \hat{n}$ on one side of the box (maybe take it in the center of the face), and multiply it by the area of that side. Do this on the other side, remembering that $\hat{n}$ isn't in the same direction there, and combine the results. Do this for each side and divide by the volume of the box.

$$\begin{aligned} \big[v_x(x_0 + \Delta x, &y_0 + \Delta y/2, z_0 + \Delta z/2)\Delta y \Delta z \\ &- v_x(x_0, y_0 + \Delta y/2, z_0 + \Delta z/2)\Delta y \Delta z\big] \div (\Delta x \Delta y \Delta z) \end{aligned} \tag{9.14}$$

the Δy and Δz factors cancel, and what's left is, in the limit $\Delta x \to 0$, the derivative $\partial v_x/\partial x$.

I was careful to evaluate the values of v_x in the center of the sides, but you see that it didn't matter. In the limit as all the sides go to zero I could just as easily taken the coordinates at one corner and simplified the steps still more. Do this for the other sides, add, and you get the result. It all looks very simple when you do it this way, but what if you need to do it in cylindrical coordinates?

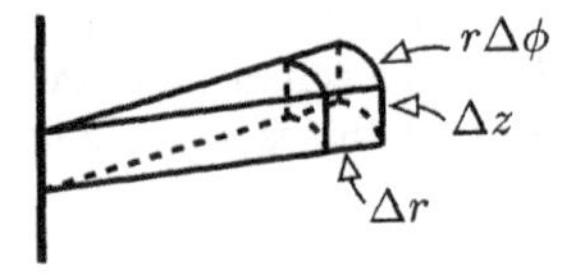

When everything is small, the volume is close to a rectangular box, so its volume is $V = (\Delta r)(\Delta z)(r\Delta\phi)$. Go through the simple version for the calculation of the surface integral. The top and bottom present nothing significantly different from the rectangular case.

$$\big[v_z(r_0, \phi_0, z_0 + \Delta z) - v_z(r_0, \phi_0, z_0)\big](\Delta r)(r_0\Delta\phi) \div r_0 \Delta r_0 \Delta\phi \Delta z \longrightarrow \frac{\partial v_z}{\partial z}$$

The curved faces of constant r are a bit different, because the areas of the two opposing faces aren't the same.

$$\big[v_r(r_0 + \Delta r, \phi_0, z_0)(r_0 + \Delta r)\Delta\phi\Delta z - v_r(r_0, \phi_0, z_0) r_0 \Delta\phi\Delta z\big] \div r_0 \Delta r \Delta\phi \Delta z \longrightarrow \frac{1}{r}\frac{\partial (r v_r)}{\partial r}$$

A bit more complex than the rectangular case, but not too bad.

Now for the constant ϕ sides. Here the areas of the two faces are the same, so even though they are not precisely parallel to each other this doesn't cause any difficulties.

$$\big[v_\phi(r_0, \phi_0 + \Delta\phi, z_0) - v_z(r_0, \phi_0, z_0)\big](\Delta r)(\Delta z) \div r_0 \Delta r \Delta\phi \Delta z \longrightarrow \frac{1}{r}\frac{\partial v_\phi}{\partial \phi}$$

The sum of all these terms is the divergence expressed in cylindrical coordinates.

$$\text{div}\, \vec{v} = \frac{1}{r}\frac{\partial(r v_r)}{\partial r} + \frac{1}{r}\frac{\partial v_\phi}{\partial \phi} + \frac{\partial v_z}{\partial z} \tag{9.15}$$

The corresponding expression in spherical coordinates is found in exactly the same way, problem 9.4.

$$\text{div}\, \vec{v} = \frac{1}{r^2}\frac{\partial(r^2 v_r)}{\partial r} + \frac{1}{r\sin\theta}\frac{\partial(\sin\theta v_\theta)}{\partial\theta} + \frac{1}{r\sin\theta}\frac{\partial v_\phi}{\partial \phi} \tag{9.16}$$

These are the three commonly occurring coordinates system, though the same simplified method will work in any other orthogonal coordinate system. The coordinate system is orthogonal if the surfaces made by setting the value of the respective coordinates to a constant intersect at right angles. In the spherical example this means that a surface of constant r is a sphere. A surface of constant θ is a half-plane starting from the z-axis. These intersect perpendicular to each other. If you set the third coordinate, ϕ, to a constant you have a cone that intersects the other two at right angles. Look back to section 8.8.

9.4 Integral Representation of Curl

The calculation of the divergence was facilitated by the fact the the equation (9.5) could be manipulated into the form of an integral, Eq. (9.9). Is there a similar expression for the curl? Yes.

$$\text{curl}\, \vec{v} = \lim_{V\to 0} \frac{1}{V} \oint d\vec{A} \times \vec{v} \tag{9.17}$$

For the divergence there was a logical and orderly development to derive Eq. (9.9) from (9.5). Is there a similar intuitively clear path here? I don't know of one. The best that I can do is to show that it gives the right answer.

And what's that surface integral doing with a $\times$ instead of a $\cdot$? No mistake. Just replace the dot product by a cross product in the definition of the integral. This time however you have to watch the order of the factors.

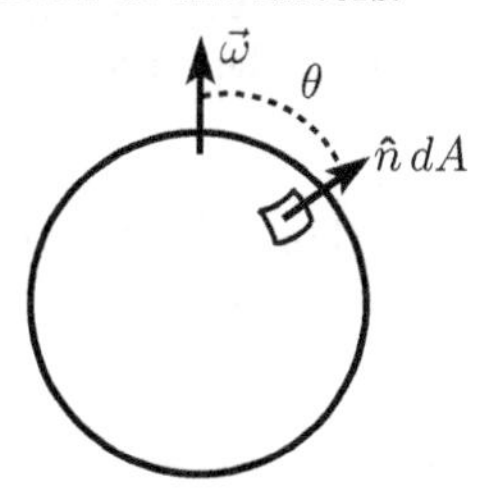

To verify that this does give the correct answer, use a vector field that represents pure rigid body rotation. You're going to take the limit as $\Delta V \to 0$, so it may as well be uniform. The velocity field for this is the same as from problem 7.5.

$$\vec{v} = \vec{\omega} \times \vec{r} \tag{9.18}$$

To evaluate the integral use a sphere of radius R centered at the origin, making $\hat{n} = \hat{r}$. You also need the identity $\vec{A} \times (\vec{B} \times \vec{C}) = \vec{B}(\vec{A} \cdot \vec{C}) - \vec{C}(\vec{A} \cdot B)$.

$$d\vec{A} \times (\vec{\omega} \times \vec{r}) = \vec{\omega}(d\vec{A} \cdot \vec{r}) - \vec{r}(\vec{\omega} \cdot d\vec{A}) \tag{9.19}$$

Choose a spherical coordinate system with the z-axis along $\vec{\omega}$.

$$d\vec{A} = \hat{n}\, dA = \hat{r}\, dA, \qquad \text{and} \qquad \vec{\omega} \cdot d\vec{A} = \omega\, dA \cos\theta$$

$$\begin{aligned}
\oint d\vec{A} \times \vec{v} &= \oint \vec{\omega} R\, dA - \vec{r}\omega\, dA \cos\theta \\
&= \vec{\omega} R\, 4\pi R^2 - \omega \int_0^\pi R^2 \sin\theta\, d\theta \int_0^{2\pi} d\phi\, \hat{z}\, R\cos\theta\cos\theta \\
&= \vec{\omega} R\, 4\pi R^2 - \omega\hat{z}\, 2\pi R^3 \int_0^\pi \sin\theta d\theta \cos^2\theta \\
&= \vec{\omega} R\, 4\pi R^2 - \omega\hat{z}\, 2\pi R^3 \int_{-1}^1 \cos^2\theta\, d\cos\theta = \vec{\omega} R\, 4\pi R^2 - \omega\hat{z}\, 2\pi R^3 \cdot \frac{2}{3} \\
&= \vec{\omega} \frac{8}{3} \pi R^3
\end{aligned}$$

Divide by the volume of the sphere and you have $2\vec{\omega}$ as promised. In the first term on the first line of the calculation, $\vec{\omega} R$ is a constant over the surface so you can pull it out of the integral. In the second term, $\vec{r}$ has components in the $\hat{x}$, $\hat{y}$, and $\hat{z}$ directions; the first two of these integrate to zero because for every vector with a positive $\hat{x}$-component there is one that has a negative component. Same for $\hat{y}$. All that is left of $\vec{r}$ is $\hat{z}\, R\cos\theta$.

The Curl in Components

With the integral representation, Eq. (9.17), available for the curl, the process is much like that for computing the divergence. Start with rectangular of course. Use the same equation, Eq. (9.10) and the same picture that accompanied that equation. With the experience gained from computing the divergence however, you don't have to go through all the complications of the first calculation. Use the simpler form that followed.

In Eq. (9.14) you have $\vec{v} \cdot \Delta A = v_x \Delta y \Delta z$ on the right face and on the left face. This time replace the dot with a cross (in the right order).

On the right,

$$\Delta\vec{A} \times \vec{v} = \Delta y \Delta z\, \hat{x} \times \vec{v}(x_0 + \Delta x, y_0 + \Delta y/2, z_0 + \Delta z/2) \tag{9.20}$$

On the left it is

$$\Delta\vec{A} \times \vec{v} = \Delta y \Delta z\, \hat{x} \times \vec{v}(x_0, y_0 + \Delta y/2, z_0 + \Delta z/2) \tag{9.21}$$

When you subtract the second from the first and divide by the volume, $\Delta x \Delta y \Delta z$, what is left is (in the limit $\Delta x \to 0$) a derivative.

$$\begin{aligned}
\hat{x} \times \frac{\vec{v}(x_0 + \Delta x, y_0, z_0) - \vec{v}(x_0, y_0, z_0)}{\Delta x} &\longrightarrow \hat{x} \times \frac{\partial \vec{v}}{\partial x} \\
&= \hat{x} \times \left(\hat{x}\frac{\partial v_x}{\partial x} + \hat{y}\frac{\partial v_y}{\partial x} + \hat{z}\frac{\partial v_z}{\partial x} \right) \\
&= \hat{z}\frac{\partial v_y}{\partial x} - \hat{y}\frac{\partial v_z}{\partial x}
\end{aligned}$$

Similar calculations for the other four faces of the box give results that you can get simply by changing the labels: $x \to y \to z \to x$, a cyclic permutation of the indices. The result can be expressed most succinctly in terms of ∇.

$$\operatorname{curl} v = \nabla \times \vec{v} \tag{9.22}$$

In the process of this calculation the normal vector $\hat{x}$ was parallel on the opposite faces of the box (except for a reversal of direction). Watch out in other coordinate systems and you'll see that this isn't always true. Just draw the picture in cylindrical coordinates and this will be clear.

9.5 The Gradient

The gradient is the closest thing to an ordinary derivative here, taking a scalar-valued function into a vector field. The simplest geometric definition is "the derivative of a function with respect to distance along the direction in which the function changes most rapidly," and the direction of the gradient vector is along that most-rapidly-changing direction. If you're dealing with one dimension, ordinary real-valued functions of real variables, the gradient *is* the ordinary derivative. Section 8.5 has some discussion and examples of this, including its use in various coordinate systems. It is most conveniently expressed in terms of ∇.

$$\operatorname{grad} f = \nabla f \tag{9.23}$$

The equations (8.15), (8.27), and (8.28) show the gradient (and correspondingly ∇) in three coordinate systems.

$$\begin{aligned}
\text{rectangular:} \quad \nabla &= \hat{x}\frac{\partial}{\partial x} + \hat{y}\frac{\partial}{\partial y} + \hat{z}\frac{\partial}{\partial z} \\
\text{cylindrical:} \quad \nabla &= \hat{r}\frac{\partial}{\partial r} + \hat{\phi}\frac{1}{r}\frac{\partial}{\partial \phi} + \hat{z}\frac{\partial}{\partial z} \\
\text{spherical:} \quad \nabla &= \hat{r}\frac{\partial}{\partial r} + \hat{\theta}\frac{1}{r}\frac{\partial}{\partial \theta} + \hat{\phi}\frac{1}{r\sin\theta}\frac{\partial}{\partial \phi}
\end{aligned} \tag{9.24}$$

In all nine of these components, the denominator (*e.g.* $r \sin\theta\, d\phi$) is the element of displacement along the direction indicated.

9.6 Shorter Cut for div and curl

There is another way to compute the divergence and curl in cylindrical and rectangular coordinates. A direct application of Eqs. (9.13), (9.22), and (9.24) gets the the result quickly. The major caution is that you have to be careful that the unit vectors are *inside* the derivative, so you have to differentiate them too.

$\nabla\cdot\vec v$ is the divergence of $\vec v$, and in cylindrical coordinates

$$\nabla\cdot\vec v = \left(\hat r\frac{\partial}{\partial r} + \hat\phi\frac{1}{r}\frac{\partial}{\partial\phi} + \hat z\frac{\partial}{\partial z}\right)\cdot\left(\hat r\, v_r + \hat\phi v_\phi + \hat z\, v_z\right) \tag{9.25}$$

The unit vectors $\hat r$, $\hat\phi$, and $\hat z$ don't change as you alter r or z. They do change as you alter ϕ. (except for $\hat z$).

$$\frac{\partial\hat r}{\partial r} = \frac{\partial\hat\phi}{\partial r} = \frac{\partial\hat z}{\partial r} = \frac{\partial\hat r}{\partial z} = \frac{\partial\hat\phi}{\partial z} = \frac{\partial\hat z}{\partial z} = \frac{\partial\hat z}{\partial\phi} = 0 \tag{9.26}$$

Next come $\partial\hat r/\partial\phi$ and $\partial\hat\phi/\partial\phi$. This is problem 8.20. You can do this by first showing that

$$\hat r = \hat x\cos\phi + \hat y\sin\phi \qquad \text{and} \qquad \hat\phi = -\hat x\sin\phi + \hat y\cos\phi \tag{9.27}$$

and differentiating with respect to ϕ. This gives

$$\partial\hat r/\partial\phi = \hat\phi, \qquad \text{and} \qquad \partial\hat\phi/\partial\phi = -\hat r \tag{9.28}$$

Put these together and you have

$$\begin{aligned}\nabla\cdot\vec v &= \frac{\partial v_r}{\partial r} + \hat\phi\cdot\frac{1}{r}\frac{\partial}{\partial\phi}\left(\hat r\, v_r + \hat\phi v_\phi\right) + \frac{\partial v_z}{\partial z}\\ &= \frac{\partial v_r}{\partial r} + \hat\phi\cdot\frac{1}{r}\left(v_r\frac{d\hat r}{d\phi} + \hat\phi\frac{\partial v_\phi}{\partial\phi}\right) + \frac{\partial v_z}{\partial z}\\ &= \frac{\partial v_r}{\partial r} + \frac{1}{r}v_r + \frac{1}{r}\frac{\partial v_\phi}{\partial\phi} + \frac{\partial v_z}{\partial z}\end{aligned} \tag{9.29}$$

This agrees with equation (9.15).

Similarly you can use the results of problem 8.15 to find the derivatives of the corresponding vectors in spherical coordinates. The non-zero values are

$$\begin{aligned}&\frac{d\hat r}{d\phi} = \hat\phi\sin\theta \qquad \frac{d\hat\theta}{d\phi} = \hat\phi\cos\theta \qquad \frac{d\hat\phi}{d\phi} = -\hat r\sin\theta - \hat\theta\cos\theta\\ &\frac{d\hat r}{d\theta} = \hat\theta \qquad \frac{d\hat\theta}{d\theta} = -\hat r\end{aligned} \tag{9.30}$$

The result is for spherical coordinates

$$\nabla\cdot\vec v = \frac{1}{r^2}\frac{\partial(r^2 v_r)}{\partial r} + \frac{1}{r\sin\theta}\frac{\partial(\sin\theta v_\theta)}{\partial\theta} + \frac{1}{r\sin\theta}\frac{\partial v_\phi}{\partial\phi} \tag{9.31}$$

The expressions for the curl are, cylindrical:

$$\nabla\times\vec v=\hat r\left(\frac{1}{r}\frac{\partial v_z}{\partial\phi}-\frac{\partial v_\phi}{\partial z}\right)+\hat\phi\left(\frac{\partial v_r}{\partial z}-\frac{\partial v_z}{\partial r}\right)+\hat z\left(\frac{1}{r}\frac{\partial(rv_\phi)}{\partial r}-\frac{1}{r}\frac{\partial v_r}{\partial\phi}\right)\tag{9.32}$$

and spherical:

$$\begin{aligned}\nabla\times\vec v=\hat r\frac{1}{r\sin\theta}&\left(\frac{\partial(\sin\theta v_\phi)}{\partial\theta}-\frac{\partial v_\theta}{\partial\phi}\right)\\ &+\hat\theta\left(\frac{1}{r\sin\theta}\frac{\partial v_r}{\partial\phi}-\frac{1}{r}\frac{\partial(rv_\phi)}{\partial r}\right)+\hat\phi\frac{1}{r}\left(\frac{\partial(rv_\theta)}{\partial r}-\frac{\partial v_r}{\partial\theta}\right)\end{aligned}\tag{9.33}$$

9.7 Identities for Vector Operators

Some of the common identities can be proved simply by computing them in rectangular components. These are vectors, and if you show that one vector equals another vector it doesn't matter that you used a simple coordinate system to demonstrate the fact. Of course there are some people who quite properly complain about the inelegance of such a procedure. They're called mathematicians.

$$\nabla\cdot\nabla\times\vec v=0\qquad\nabla\times\nabla f=0\qquad\nabla\times\nabla\times\vec v=\nabla(\nabla\cdot\vec v)-(\nabla\cdot\nabla)\vec v\tag{9.34}$$

There are many other identities, but these are the big three.

$$\oint\vec v\cdot d\vec A=\int d^3r\,\nabla\cdot\vec v\qquad\oint\vec v\cdot d\vec r=\int\nabla\times\vec v\cdot d\vec A\tag{9.35}$$

are the two fundamental integral relationships, going under the names of Gauss and Stokes. See chapter 13 for the proofs of these integral relations.

9.8 Applications to Gravity

The basic equations to describe the gravitational field in Newton's theory are

$$\nabla\cdot\vec g=-4\pi G\rho,\qquad\text{and}\qquad\nabla\times\vec g=0\tag{9.36}$$

In these equations, the vector field $\vec g$ is defined by placing a (very small) test mass m at a point and measuring the gravitational force on it. This force is proportional to m itself, and the proportionality factor is called the gravitational field $\vec g$. The other symbol used here is ρ, and that is the volume mass density, dm/dV of the matter that is generating the gravitational field. G is Newton's gravitational constant: $G=6.67\times10^{-11}\,\text{N}\cdot\text{m}^2/\text{kg}^2$.

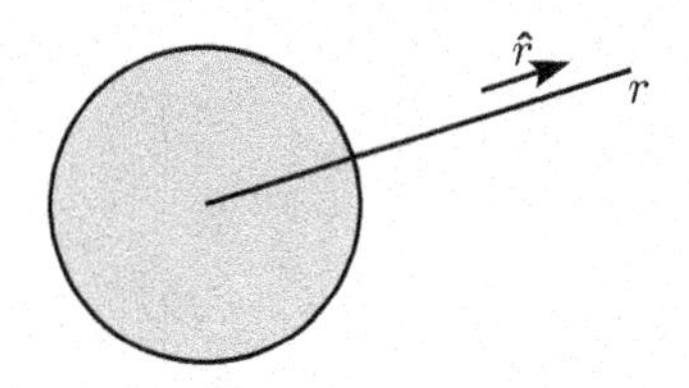

For the first example of solutions to these equations, take the case of a spherically symmetric mass that is the source of a gravitational field. Assume also that its density is constant inside; the total mass is M and it occupies a sphere of radius R. Whatever $\vec{g}$ is, it has only a radial component, $\vec{g} = g_r\,\hat{r}$. Proof: Suppose it has a sideways component at some point. Rotate the whole system by 180° about an axis that passes through this point and through the center of the sphere. The system doesn't change because of this, but the sideways component of $\vec{g}$ would reverse. That can't happen.

The component g_r can't depend on either θ or ϕ because the source doesn't change if you rotate it about any axis; it's spherically symmetric.

$$\vec{g} = g_r(r)\hat{r} \tag{9.37}$$

Now compute the divergence and the curl of this field. Use Eqs. (9.16) and (9.33) to get

$$\nabla\cdot g_r(r)\hat{r} = \frac{1}{r^2}\frac{d(r^2 g_r)}{dr} \qquad \text{and} \qquad \nabla\times g_r(r)\hat{r} = 0$$

The first equation says that the divergence of $\vec{g}$ is proportional to ρ.

$$\frac{1}{r^2}\frac{d(r^2 g_r)}{dr} = -4\pi G\rho \tag{9.38}$$

Outside the surface $r = R$, the mass density is zero, so this is

$$\frac{1}{r^2}\frac{d(r^2 g_r)}{dr} = 0, \qquad \text{implying} \qquad r^2 g_r = C, \qquad \text{and} \qquad g_r = \frac{C}{r^2}$$

where C is some as yet undetermined constant. Now do this inside.

$$\frac{1}{r^2}\frac{d(r^2 g_r)}{dr} = -4\pi G\rho_0, \qquad \text{where} \qquad \rho_0 = 3M/4\pi R^3$$

This is

$$\frac{d(r^2 g_r)}{dr} = -4\pi G\rho_0\, r^2, \qquad \text{so} \qquad r^2 g_r = -\frac{4}{3}\pi G\rho_0\, r^3 + C',$$

$$\text{or} \qquad g_r(r) = -\frac{4}{3}\pi G\rho_0\, r + \frac{C'}{r^2}$$

There are two constants that you have to evaluate: C and C'. The latter has to be zero, because $C'/r^2 \to \infty$ as $r \to 0$, and there's nothing in the mass distribution that will cause this. As for the other, note that g_r must be continuous at the surface of the mass. If it isn't, then when you try to differentiate it in Eq. (9.38) you'll be differentiating a step function and you get an infinite derivative there (and the mass density isn't infinite there).

$$g_r(R-) = -\frac{4}{3}\pi G\rho_0\, R = g_r(R+) = \frac{C}{R^2}$$

Solve for C and you have

$$C = -\frac{4}{3}\pi G\rho_0\, R^3 = -\frac{4}{3}\pi G\frac{3M}{4\pi R^3}R^3 = -GM$$

Put this all together and express the density ρ_0 in terms of M and R to get

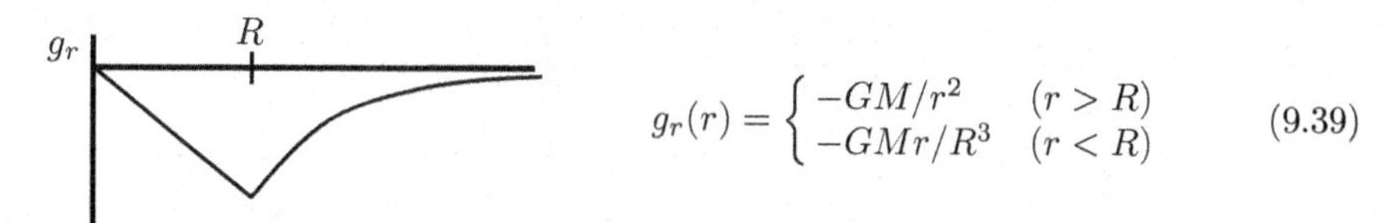

$$g_r(r) = \begin{cases} -GM/r^2 & (r > R) \\ -GMr/R^3 & (r < R) \end{cases} \tag{9.39}$$

This says that outside the spherical mass distribution you can't tell what its radius R is. It creates the same gravitational field as a point mass. Inside the uniform sphere, the field drops to zero linearly toward the center. For a slight variation on how to do this calculation see problem 9.14.

Non-uniform density

The density of the Earth is not uniform; it's bigger in the center. The gravitational field even increases as you go down below the Earth's surface. What does this tell you about the density function? $\nabla\cdot\vec g = -4\pi G\rho$ remains true, and I'll continue to make the approximation of spherical symmetry, so this is

$$\frac{1}{r^2}\frac{d(r^2 g_r)}{dr} = \frac{dg_r}{dr} + \frac{2}{r}g_r = -4\pi G\rho(r) \tag{9.40}$$

That gravity increases with depth (for a little while) says

$$\frac{dg_r}{dr} = -4\pi G\rho(r) - \frac{2}{r}g_r > 0$$

Why > 0? Remember: g_r is itself negative and r is measured outward. Sort out the signs. I can solve for the density to get

$$\rho(r) < -\frac{1}{2\pi Gr}g_r$$

At the surface, $g_r(R) = -GM/R^2$, so this is

$$\rho(R) < \frac{M}{2\pi R^3} = \frac{2}{3}\cdot\frac{3M}{4\pi R^3} = \frac{2}{3}\rho_{\text{average}}$$

The mean density of the Earth is 5.5 gram/cm^3, so this bound is 3.7 gram/cm^3. Pick up a random rock. What is its density?

9.9 Gravitational Potential

The gravitational potential is that function V for which

$$\vec g = -\nabla V \tag{9.41}$$

That such a function even *exists* is not instantly obvious, but it is a consequence of the second of the two defining equations (9.36). If you grant that, then you can get an immediate equation for V by substituting it into the first of (9.36).

$$\nabla\cdot\vec g = -\nabla\cdot\nabla V = -4\pi G\rho, \qquad \text{or} \qquad \nabla^2 V = 4\pi G\rho \tag{9.42}$$

This is a scalar equation instead of a vector equation, so it will often be easier to handle. Apply it to the same example as above, the uniform spherical mass.

The Laplacian, ∇^2 is the divergence of the gradient, so to express it in spherical coordinates, combine Eqs. (9.24) and (9.31).

$$\nabla^2 V = \frac{1}{r^2}\frac{\partial}{\partial r}\left(r^2\frac{\partial V}{\partial r}\right) + \frac{1}{r^2\sin\theta}\frac{\partial}{\partial\theta}\left(\sin\theta\frac{\partial V}{\partial\theta}\right) + \frac{1}{r^2\sin^2\theta}\frac{\partial^2 V}{\partial\phi^2} \tag{9.43}$$

Because the mass is spherical it doesn't change no matter how you rotate it so the same thing holds for the solution, $V(r)$. Use this spherical coordinate representation of ∇^2 and for this case the θ and ϕ derivatives vanish.

$$\frac{1}{r^2}\frac{d}{dr}\left(r^2\frac{dV}{dr}\right) = 4\pi G\rho(r) \tag{9.44}$$

I changed from ∂ to d because there's now one independent variable, not several. Just as with Eq. (9.38) I'll divide thisinto two cases, inside and outside.

$$\text{Outside:}\quad \frac{1}{r^2}\frac{d}{dr}\left(r^2\frac{dV}{dr}\right) = 0, \qquad \text{so} \qquad r^2\frac{dV}{dr} = C$$

Continue solving this and you have

$$\frac{dV}{dr} = \frac{C}{r^2} \longrightarrow V(r) = -\frac{C}{r} + D \qquad (r > R) \tag{9.45}$$

$$\text{Inside:}\quad \frac{1}{r^2}\frac{d}{dr}\left(r^2\frac{dV}{dr}\right) = 4\pi G\rho_0 \qquad \text{so} \qquad r^2\frac{dV}{dr} = 4\pi G\rho_0\frac{r^3}{3} + C'$$

Continue, dividing by r^2 and integrating,

$$V(r) = 4\pi G\rho_0\frac{r^2}{6} - \frac{C'}{r} + D' \qquad (r < R) \tag{9.46}$$

There are now four arbitrary constants to examine. Start with C'. It's the coefficient of $1/r$ in the domain where $r < R$. That means that it blows up as $r \to 0$, but there's nothing at the origin to cause this. $C' = 0$. Notice that the same argument does *not* eliminate C because (9.45) applies only for $r > R$.

Boundary Conditions

Now for the boundary conditions at $r = R$. There are a couple of ways to determine this. I find the simplest and the most general approach is to recognize the the equations (9.42) and (9.44) must be satisfied *everywhere*. That means not just outside, not just inside, but *at* the surface too. The consequence of this statement is the result*

$$V \text{ is continuous at } r = R \qquad\qquad dV/dr \text{ is continuous at } r = R \tag{9.47}$$

---

* Watch out for the similar looking equations that appear in electrostatics. Only the first of these equations holds there; the second must be modified by a dielectric constant.

Where do these continuity conditions come from? Assume for a moment that the first one is false, that V is discontinuous at $r = R$, and look at the proposition graphically. If V changes value in a very small interval the graphs of V, of dV/dr, and of d^2V/dr^2 look like

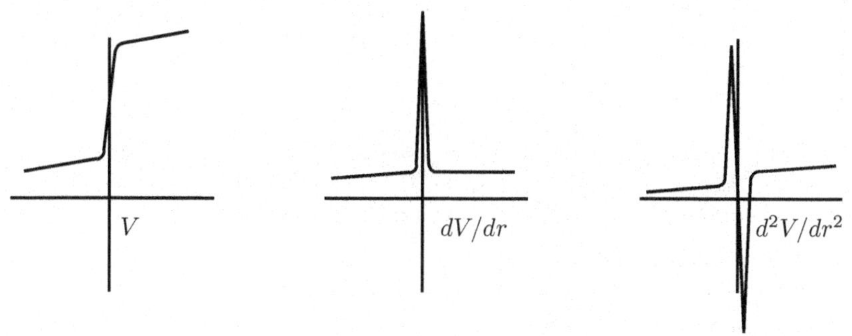

The second derivative on the left side of Eq. (9.44) has a double spike that does not appear on the right side. It can't be there, so my assumption that V is discontinuous is false and V must be continuous.

Assume next that V is continuous but its derivative is not. The graphs of V, of dV/dr, and of d^2V/dr^2 then look like

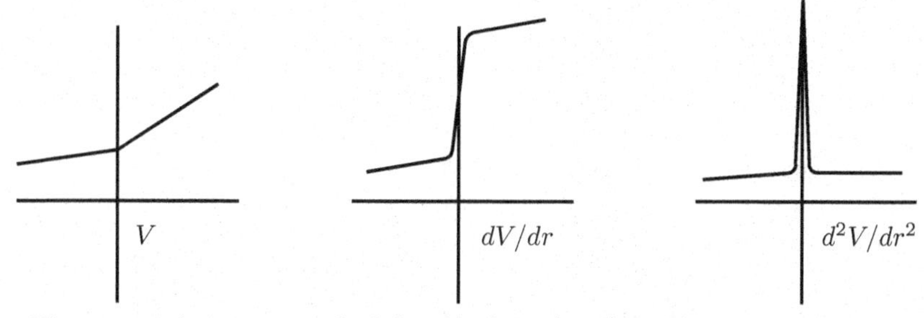

The second derivative on the left side of Eq. (9.44) still has a spike in it and there is no such spike in the ρ on the right side. This is impossible, so dV/dr too must be continuous.

Back to the Problem

Of the four constants that appear in Eqs. (9.45) and (9.46), one is already known, C'. For the rest,

$$V(R-) = V(R+) \qquad \text{is} \qquad 4\pi G\rho_0 \frac{R^2}{6} + D' = -\frac{C}{R} + D$$

$$\frac{dV}{dr}(R-) = \frac{dV}{dr}(R+) \qquad \text{is} \qquad 8\pi G\rho_0 \frac{R}{6} = +\frac{C}{R^2}$$

These two equations determine two of the constants.

$$C = 4\pi G\rho_0 \frac{R^3}{3}, \qquad \text{then} \qquad D - D' = 4\pi G\rho_0 \frac{R^2}{6} + 4\pi G\rho_0 \frac{R^2}{3} = 2\pi G\rho_0 R^2$$

Put this together and you have

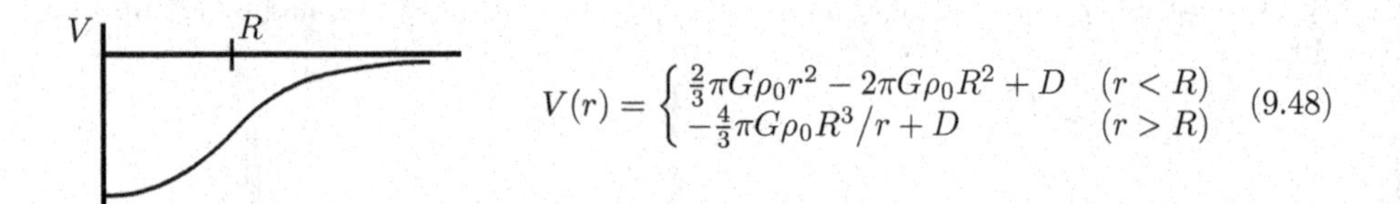

$$V(r) = \begin{cases} \frac{2}{3}\pi G\rho_0 r^2 - 2\pi G\rho_0 R^2 + D & (r < R) \\ -\frac{4}{3}\pi G\rho_0 R^3/r + D & (r > R) \end{cases} \tag{9.48}$$

Did I say that the use of potentials is supposed to simplify the problems? Yes, but only the harder problems. The negative gradient of Eq. (9.48) should be $\vec{g}$. Is it? The constant D can't be determined and is arbitrary. You may choose it to be zero.

Magnetic Boundary Conditions
The equations for (time independent) magnetic fields are

$$\nabla \times \vec{B} = \mu_0 \vec{J} \qquad \text{and} \qquad \nabla \cdot \vec{B} = 0 \tag{9.49}$$

The vector $\vec{J}$ is the current density, the current per area, defined so that across a tiny area $d\vec{A}$ the current that flows through the area is $dI = \vec{J} \cdot d\vec{A}$. (This is precisely parallel to Eq. (9.1) for fluid flow rate.) In a wire of radius R, carrying a uniform current I, the magnitude of J is $I/\pi R^2$. These equations are sort of the reverse of Eq. (9.36).

If there *is* a discontinuity in the current density at a surface such as the edge of a wire, will there be some sort of corresponding discontinuity in the magnetic field? Use the same type of analysis as followed Eq. (9.47) for the possible discontinuities in the potential function. Take the surface of discontinuity of the current density to be the x-y plane, $z = 0$ and write the divergence equation

$$\frac{\partial B_x}{\partial x} + \frac{\partial B_y}{\partial y} + \frac{\partial B_z}{\partial z} = 0$$

If there is a discontinuity, it will be in the z variable. Perhaps B_x or B_z is discontinuous at the x-y plane. The divergence equation has a derivative with respect to z only on B_z. If one of the other components changes abruptly at the surface, this equation causes no problem — nothing special happens in the x or y direction. If B_z changes at the surface then the derivative $\partial B_z/\partial z$ has a spike. Nothing else in the equation has a spike, so there's no way that you can satisfy the equation. Conclusion: The normal component of $\vec{B}$ is continuous at the surface.

What does the curl say?

$$\hat{x}\left(\frac{\partial B_z}{\partial y} - \frac{\partial B_y}{\partial x}\right) + \hat{y}\left(\frac{\partial B_x}{\partial z} - \frac{\partial B_z}{\partial y}\right) + \hat{z}\left(\frac{\partial B_y}{\partial x} - \frac{\partial B_x}{\partial y}\right) = \mu_0\left(\hat{x}\, J_x + \hat{y}\, J_y + \hat{z}\, J_z\right)$$

Derivatives with respect to x or y don't introduce a problem at the surface, all the action is again along z. Only the terms with a $\partial/\partial z$ will raise a question. If B_x is discontinuous at the surface, then its derivative with respect to z will have a spike in the $\hat{y}$ direction that has no other term to balance it. (J_y has a *step* here but not a spike.) Similarly for

B_y. This can't happen, so the conclusion: The tangential component of $\vec{B}$ is continuous at the surface.

What if the surface isn't a plane. Maybe it is a cylinder or a sphere. In a small enough region, both of these look like planes. That's why there is still* a Flat Earth Society.

9.10 Index Notation

In section 7.11 I introduced the summation convention for repeated indices. I'm now going to go over it again and emphasize its utility in practical calculations.

When you want to work in a rectangular coordinate system, with basis vectors $\hat{x}$, $\hat{y}$, and $\hat{z}$, it is convenient to use a more orderly notation for the basis vectors instead of just a sequence of letters of the alphabet. Instead, call them $\hat{e}_1$, $\hat{e}_2$, and $\hat{e}_3$. (More indices if you have more dimensions.) I'll keep the assumption that these are orthogonal unit vectors so that

$$\hat{e}_1 \cdot \hat{e}_2 = 0, \qquad \hat{e}_3 \cdot \hat{e}_3 = 1, \ etc.$$

More generally, write this in the compact notation

$$\hat{e}_i \cdot \hat{e}_j = \delta_{ij} = \begin{cases} 1 & (i = j) \\ 0 & (i \neq j) \end{cases} \tag{9.50}$$

The Kronecker delta is either one or zero depending on whether $i = j$ or $i \neq j$, and this equation sums up the properties of the basis in a single, compact notation. You can now write a vector in this basis as

$$\vec{A} = A_1 \hat{e}_1 + A_2 \hat{e}_2 + A_3 \hat{e}_3 = A_i \hat{e}_i$$

The last expression uses the summation convention that a repeated index is summed over its range. When an index is repeated in a term, you will invariably have exactly two instances of the index; if you have three it's a mistake.

When you add or subtract vectors, the index notation is

$$\vec{A} + \vec{B} = \vec{C} = A_i \hat{e}_i + B_i \hat{e}_i = (A_i + B_i)\hat{e}_i = C_i \hat{e}_i \qquad \textbf{or} \qquad A_i + B_i = C_i$$

Does it make sense to write $\vec{A} + \vec{B} = A_i \hat{e}_i + B_k \hat{e}_k$? Yes, but it's sort of pointless and confusing. You can change any summed index to any label you find convenient — they're just dummy variables.

$$A_i \hat{e}_i = A_\ell e_\ell = A_m \hat{e}_m = A_1 \hat{e}_1 + A_2 \hat{e}_2 + A_3 \hat{e}_3$$

You can sometime use this freedom to help do manipulations, but in the example $A_i \hat{e}_i + B_k \hat{e}_k$ it's no help at all.

Combinations such as

$$E_i + F_i \qquad \text{or} \qquad E_k F_k G_i = H_i \qquad \text{or} \qquad M_{k\ell} D_\ell = F_k$$

* en.wikipedia.org/wiki/Flat_Earth_Society

are valid. The last is simply Eq. (7.8) for a matrix times a column matrix.

$$A_i + B_j = C_k \qquad \text{or} \qquad E_m F_m G_m \qquad \text{or} \qquad C_k = A_{ij} B_j$$

have no meaning.

You can manipulate the indices for your convenience as long as you follow the rules.

$$A_i = B_{ij} C_j \qquad \text{is the same as} \qquad A_k = B_{kn} C_n \qquad \text{or} \qquad A_\ell = B_{\ell p} C_p$$

The scalar product has a simple form in index notation:

$$\vec{A} \cdot \vec{B} = A_i \hat{e}_i \cdot B_j \hat{e}_j = A_i B_j \hat{e}_i \cdot \hat{e}_j = A_i B_j \delta_{ij} = A_i B_i \tag{9.51}$$

The final equation comes by doing one of the two sums (say j), and only the term with $j = i$ survives, producing the final expression. The result shows that the sum over the repeated index is the scalar product in disguise.

Just as the dot product of the basis vectors is the delta-symbol, the cross product provides another important index function, the alternating symbol.

$$\hat{e}_i \cdot \hat{e}_j \times \hat{e}_k = \epsilon_{ijk} \tag{9.52}$$

$$\begin{aligned}
\hat{e}_2 \times \hat{e}_3 \equiv \hat{y} \times \hat{z} = \hat{e}_1, &\qquad \text{so} \qquad \epsilon_{123} = \hat{e}_1 \cdot \hat{e}_2 \times \hat{e}_3 = 1 \\
\hat{e}_2 \times \hat{e}_3 = -\hat{e}_3 \times \hat{e}_2, &\qquad \text{so} \qquad \epsilon_{132} = -\epsilon_{123} = -1 \\
\hat{e}_3 \cdot \hat{e}_1 \times \hat{e}_2 = \hat{e}_3 \cdot \hat{e}_3 = 1, &\qquad \text{so} \qquad \epsilon_{312} = \epsilon_{123} = 1 \\
\hat{e}_1 \cdot \hat{e}_3 \times \hat{e}_3 = \hat{e}_1 \cdot 0 = 0 &\qquad \text{so} \qquad \epsilon_{133} = 0
\end{aligned}$$

If the indices are a cyclic permutation of 123, (231 or 312), the alternating symbol is 1.
If the indices are an odd permutation of 123, (132 or 321 or 213), the symbol is -1.
If any two of the indices are equal the alternating symbol is zero, and that finishes all the cases. The last property is easy to see because if you interchange any two indices the sign changes. If the indices are the same, the sign can't change so it must be zero.

Use the alternating symbol to write the cross product itself.

$$\vec{A} \times \vec{B} = A_i \hat{e}_i \times B_j \hat{e}_j \qquad \text{and the } k\text{-component is}$$
$$\hat{e}_k \cdot \vec{A} \times \vec{B} = \hat{e}_k \cdot A_i B_j \hat{e}_i \times \hat{e}_j = \epsilon_{kij} A_i B_j$$

You can use the summation convention to advantage in calculus too. The ∇ vector operator has components

$$\nabla_i \qquad \text{or some people prefer} \qquad \partial_i$$

For unity of notation, use $x_1 = x$ and $x_2 = y$ and $x_3 = z$. In this language,

$$\partial_1 \equiv \nabla_1 \qquad \text{is} \qquad \frac{\partial}{\partial x} \equiv \frac{\partial}{\partial x_1} \tag{9.53}$$

Note: This notation applies to rectangular component calculations only! ($\hat{e}_i \cdot \hat{e}_j = \delta_{ij}$.) The generalization to curved coordinate systems will wait until chapter 12.

$$\operatorname{div} \vec{v} = \nabla \cdot \vec{v} = \partial_i v_i = \frac{\partial v_1}{\partial x_1} + \frac{\partial v_2}{\partial x_2} + \frac{\partial v_3}{\partial x_3} \tag{9.54}$$

You should verify that $\partial_i x_j = \delta_{ij}$.

Similarly the curl is expressed using the alternating symbol.

$$\operatorname{curl} \vec{v} = \nabla \times \vec{v} \qquad \text{becomes} \qquad \epsilon_{ijk} \partial_j v_k = \left(\operatorname{curl} \vec{v}\right)_i \tag{9.55}$$

the ith components of the curl.

An example of manipulating these object: Prove that $\operatorname{curl} \operatorname{grad} \phi = 0$.

$$\operatorname{curl} \operatorname{grad} \phi = \nabla \times \nabla \phi \longrightarrow \epsilon_{ijk} \partial_j \partial_k \phi \tag{9.56}$$

You can interchange the order of the differentiation as always, and the trick here is to relabel the indices — that is a standard technique in this business.

$$\epsilon_{ijk} \partial_j \partial_k \phi = \epsilon_{ijk} \partial_k \partial_j \phi = \epsilon_{ikj} \partial_j \partial_k \phi$$

The first equation interchanges the order of differentiation. In the second equation, I call "j" "k" and I call "k" "j". These are dummy indices, summed over, so this can't affect the result, but now this expression looks like that of Eq. (9.56) except that two (dummy) indices are reversed. The ϵ symbol is antisymmetric under interchange of any of its indices, so the final expression is the negative of the expression in Eq. (9.56). The only number equal to minus itself is zero, so the identity is proved.

9.11 More Complicated Potentials

The gravitational field from a point mass is $\vec{g} = -Gm\hat{r}/r^2$, so the potential for this point mass is $\phi = -Gm/r$. This satisfies

$$\vec{g} = -\nabla \phi = -\nabla \frac{-Gm}{r} = \hat{r} \frac{\partial}{\partial r} \frac{Gm}{r} = -\frac{Gm\hat{r}}{r^2}$$

For several point masses, the gravitational field is the vector sum of the contributions from each mass. In the same way the gravitational potential is the (scalar) sum of the potentials contributed by each mass. This is almost always easier to calculate than the vector sum. If the distribution is continuous, you have an integral.

$$\phi_{\text{total}} = \sum -\frac{Gm_k}{r_k} \qquad \text{or} \qquad -\int \frac{G\,dm}{r}$$

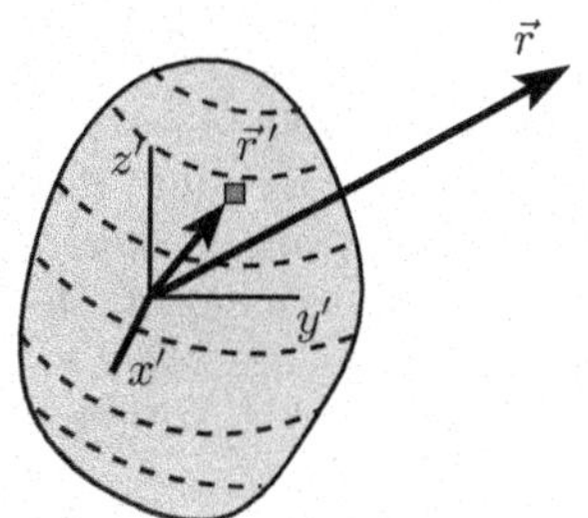

This sort of very abbreviated notation for the sums and integrals is normal once you have done a lot of them, but when you're just getting started it is useful to go back and forth between this terse notation and a more verbose form. Expand the notation and you have

$$\phi_{\text{total}}(\vec{r}\,) = -G\int \frac{dm}{\left|\vec{r} - \vec{r}\,'\right|} \tag{9.57}$$

This is still not very explicit, so expand it some more. Let

$$\vec{r}\,' = \hat{x}\,x' + \hat{y}\,y' + \hat{z}\,z' \qquad \text{and} \qquad \vec{r} = \hat{x}\,x + \hat{y}\,y + \hat{z}\,z$$

$$\text{then} \qquad \phi(x,y,z) = -G\int dx'dy'dz'\,\rho(x',y',z')\,\frac{1}{\sqrt{(x-x')^2 + (y-y')^2 + (z-z')^2}}$$

where ρ is the volume mass density so that $dm = \rho\,dV = \rho\,d^3r'$, and the limits of integration are such that this extends over the whole volume of the mass that is the source of the potential. The primed coordinates represent the positions of the masses, and the non-primed ones are the position of the point where you are evaluating the potential, the field point. The combination d^3r is a common notation for a volume element in three dimensions.

For a simple example, what is the gravitational potential from a uniform thin rod? Place its center at the origin and its length $= 2L$ along the z-axis. The potential is

$$\phi(\vec{r}\,) = -\int \frac{G\,dm}{r} = -G\int \frac{\lambda\,dz'}{\sqrt{x^2 + y^2 + (z-z')^2}}$$

where $\lambda = M/2L$ is its linear mass density. This is an elementary integral. Let $u = z' - z$, and $a = \sqrt{x^2 + y^2}$.

$$\phi = -G\lambda\int_{-L-z}^{L-z} \frac{du}{\sqrt{a^2 + u^2}} = -G\lambda\int d\theta = -G\lambda\,\theta\,\Big|_{u=-L-z}^{u=L-z}$$

where $u = a\sinh\theta$. Put this back into the original variables and you have

$$\phi = -G\lambda\left[\sinh^{-1}\left(\frac{L-z}{\sqrt{x^2+y^2}}\right) + \sinh^{-1}\left(\frac{L+z}{\sqrt{x^2+y^2}}\right)\right] \tag{9.58}$$

The inverse hyperbolic function is a logarithm as in Eq. (1.4), so this can be rearranged and the terms combined into the logarithm of a function of x, y, and z, but the $\sinh^{-1}$s are easier to work with so there's not much point. This is not too complicated a result, and it is far easier to handle than the vector field you get if you take its gradient. It's still necessary to analyze it in order to understand it and to check for errors. See problem 9.48.

Exercises

1 Prove that the geometric interpretation of a cross product is as an area.

2 Start from a picture of $\vec{C} = \vec{A} - \vec{B}$ and use the definition and properties of the dot product to derive the law of cosines. (If this takes you more than about three lines, start over, and no components, just vectors.)

3 Start from a picture of $\vec{C} = \vec{A} - \vec{B}$ and use the definition and properties of the cross product to derive the law of sines. (If this takes you more than a few lines, start over.)

4 Show that $\vec{A}\cdot\vec{B}\times\vec{C} = \vec{A}\times\vec{B}\cdot\vec{C}$. Do this by drawing pictures of the three vectors and find the geometric meaning of each side of the equation, showing that they are the same (including sign).

5 (a) If the dot product of a given vector $\vec{F}$ with every vector results in zero, show that $\vec{F} = 0$. (b) Same for the cross product.

6 From the definition of the dot product, and in two dimensions, draw pictures to interpret $\vec{A}\cdot(\vec{B}+\vec{C})$ and from there prove the distributive law: $\vec{A}\cdot = \vec{A}\cdot\vec{B} + \vec{A}\cdot\vec{C}$. (Draw $\vec{B}+\vec{C}$ tip-to-tail.)

7 For a sphere, from the definition of the integral, what is $\oint d\vec{A}$? What is $\oint dA$?

8 What is the divergence of $\hat{x}\,xy + \hat{y}\,yz + \hat{z}\,zx$?

9 What is the divergence of $\hat{r}\,r\sin\theta + \hat{\theta}\,r_0\sin\theta\cos\phi + \hat{\phi}\,r\cos\phi$? (spherical)

10 What is the divergence of $\hat{r}\,r\sin\phi + \hat{\phi}\,z\sin\phi + \hat{z}\,zr$? (cylindrical)

11 In cylindrical coordinates draw a picture of the vector field $\vec{v} = \hat{\phi}r^2(2+\cos\phi)$ (for $z=0$). Compute the divergence of $\vec{v}$ and indicate in a second sketch what it looks like. Draw field lines for $\vec{v}$.

12 What is the curl of the vector field in the preceding exercise (and indicate in a sketch what it's like).

13 Calculate $\nabla\cdot\hat{r}$. (a) in cylindrical and (b) in spherical coordinates.

14 Compute $\partial_i x_j$.

15 Compute div curl $\vec{v}$ using index notation.

16 Show that $\epsilon_{ijk} = \frac{1}{2}(i-j)(j-k)(k-i)$.

17 Use index notation to derive $\nabla\cdot(f\vec{v}) = (\nabla f)\cdot\vec{v} + f\nabla\cdot\vec{v}$.

18 Write Eq. (8.6) in index notation, where $(x,y)\to x_i$ and $(x',y')\to x'_j$.

Problems

9.1 Use the same geometry as that following Eq. (9.3), and take the velocity function to be $\vec{v} = \hat{x}\, v_0 xy/b^2$. Take the bottom edge of the plane to be at $(x, y) = (0, 0)$ and calculate the flow rate. Should the result be independent of the angle ϕ? Sketch the flow to understand this point. Does the result check for any special, simple value of ϕ? Ans: $(v_0 ab \tan\phi)/3$

9.2 Repeat the preceding problem using the cylindrical surface of Eq. (9.4), but place the bottom point of the cylinder at coordinate $(x, y) = (x_0, 0)$. Ans: $(v_0 a/4)(2x_0 + \pi b/4)$

9.3 Use the same velocity function $\vec{v} = \hat{x}\, v_0 xy/b^2$ and evaluate the flow integral outward from the *closed* surface of the rectangular box, $(c < x < d)$, $(0 < y < b)$, $(0 < z < a)$. The convention is that the unit normal vector points outward from the six faces of the box. Ans: $v_0 a(d - c)/2$

9.4 Work out the details, deriving the divergence of a vector field in spherical coordinates, Eq. (9.16).

9.5 (**a**) For the vector field $\vec{v} = A\vec{r}$, that is pointing away from the origin with a magnitude proportional to the distance from the origin, express this in rectangular components and compute its divergence.
(**b**) Repeat this in cylindrical coordinates (still pointing away from the origin though).
(**c**) Repeat this in spherical coordinates, Eq. (9.16).

9.6 Gauss's law for electromagnetism says $\oint \vec{E} \cdot d\vec{A} = q_{\text{encl}}/\epsilon_0$. If the electric field is given to be $\vec{E} = A\vec{r}$, what is the surface integral of $\vec{E}$ over the whole closed surface of the cube that spans the region from the origin to $(x, y, z) = (a, a, a)$?
(**a**) What is the charge enclosed in the cube?
(**b**) Compute the volume integral, $\int d^3r\, \nabla \cdot \vec{E}$ inside the same cube?

9.7 Evaluate the surface integral, $\oint \vec{v} \cdot d\vec{A}$, of $\vec{v} = \hat{r}\, Ar^2 \sin^2\theta + \hat{\theta}\, Br\cos\theta\sin\phi$ over the surface of the sphere centered at the origin and of radius R. Recall section 8.8.

9.8 (**a**) What is the area of the spherical cap on the surface of a sphere of radius R: $0 \le \theta \le \theta_0$?
(**b**) Does the result have the correct behavior for both small and large θ_0?
(**c**) What are the surface integrals over this cap of the vector field $\vec{v} = \hat{r} v_0 \cos\theta \sin^2\phi$? Consider both $\int \vec{v} \cdot d\vec{A}$ and $\int \vec{v} \times d\vec{A}$. Ans: $v_0 \pi R^2(1 - \cos^2\theta_0)/2$

9.9 A rectangular area is specified parallel to the x-y plane at $z = d$ and $0 < x < a$, $a < y < b$. A vector field is $\vec{v} = \left(\hat{x}\, Axyz + \hat{y}\, Byx^2 + \hat{z}\, Cx^2yz^2\right)$ Evaluate the two integrals over this surface

$$\int \vec{v} \cdot d\vec{A}, \qquad \text{and} \qquad \int d\vec{A} \times \vec{v}$$

9.10 For the vector field $\vec{v} = Ar^n\vec{r}$, compute the integral over the surface of a sphere of radius R centered at the origin: $\oint \vec{v}\cdot d\vec{A}$. $(n \ge 0)$
Compute the integral over the volume of this same sphere $\int d^3r\, \nabla\cdot\vec{v}$.

9.11 The velocity of a point in a rotating rigid body is $\vec{v} = \omega \times \vec{r}$. See problem 7.5. Compute its divergence and curl. Do this in rectangular, cylindrical, and spherical coordinates.

9.12 Fill in the missing steps in the calculation of Eq. (9.29).

9.13 Mimic the calculation in section 9.6 for the divergence in cylindrical coordinates, computing the curl in cylindrical coordinates, $\nabla \times \vec{v}$. Ans: Eq. (9.32).

9.14 Another way to get to Eq. (9.39) is to work with Eq. (9.38) directly and to write the function $\rho(r)$ explicitly as two cases: $r < R$ and $r > R$. Multiply Eq. (9.38) by r^2 and integrate it from zero to r, being careful to handle the integral differently when the upper limit is $< R$ and when it is $> R$.

$$r^2 g_r(r) = -4\pi G \int_0^r dr'\, r'^2 \rho(r')$$

Note: This is not simply reproducing that calculation that I've already done. This is doing it a different way.

9.15 If you have a very large (assume it's infinite) slab of mass of thickness d the gravitational field will be perpendicular to its plane. To be specific, say that there is a uniform mass density ρ_0 between $z = \pm d/2$ and that $\vec{g} = g_z(z)\,\hat{z}$. Use Eqs. (9.36) to find $g_z(z)$. Be precise in your reasoning when you evaluate any constants. (What happens when you rotate the system about the x-axis?) Does your graph of the result make sense?
Ans: in part, $g_z = +2\pi G\rho_0 d$, $(z < -d/2)$

9.16 Use Eqs. (9.36) to find the gravitational field of a very long solid cylinder of uniform mass density ρ_0 and radius R. (Assume it's infinitely long.) Start from the assumption that in cylindrical coordinates the field is $\vec{g} = g_r(r,\phi,z)\,\hat{r}$, and apply both equations.
Ans: in part $g_r = -2\pi G r$, $(0 < r < R)$

9.17 The gravitational field in a spherical region $r < R$ is stated to be $\vec{g}(r) = -\hat{r}\,C/r$, where C is a constant. What mass density does this imply?
If there is no mass for $r > R$, what is $\vec{g}$ there?

9.18 In Eq. (8.23) you have an approximate expression for the gravitational field of Earth, including the effect of the equatorial bulge. Does it satisfy Eqs. (9.36)? $(r > R_{\text{Earth}})$

9.19 Compute the divergence of the velocity function in problem 9.3 and integrate this divergence over the volume of the box specified there. Ans: $(d-c)av_0$

9.20 The gravitational potential, equation (9.42), for the case that the mass density is zero says to set the Laplacian Eq. (9.43) equal to zero. Assume a solution to $\nabla^2 V = 0$ to be a function of the spherical coordinates r and θ alone and that

$$V(r,\theta) = Ar^{-(\ell+1)} f(x), \qquad \text{where} \qquad x = \cos\theta$$

Show that this works provided that f satisfies a certain differential equation and show that it is the Legendre equation of Eq. (4.18) and section 4.11.

9.21 The volume energy density, dU/dV in the electric field is $\epsilon_0 E^2/2$. The electrostatic field equations are the same as the gravitational field equations, Eq. (9.36).

$$\nabla\cdot\vec{E} = \rho/\epsilon_0, \qquad \text{and} \qquad \nabla\times\vec{E} = 0$$

A uniformly charged ball of radius R has charge density ρ_0 for $r < R$, $Q = 4\pi\rho_0 R^3/3$.
(**a**) What is the electric field everywhere due to this charge distribution?
(**b**) The total energy of this electric field is the integral over all space of the energy density. What is it?
(**c**) If you want to account for the mass of the electron by saying that all this energy that you just computed *is* the electron's mass via $E_0 = mc^2$, then what must the electron's radius be? What is its numerical value? Ans: $r_e = {}^3\!/\!_5\,(e^2/4\pi\epsilon_0 mc^2) = 1.69\,\text{fm}$

9.22 The equations relating a magnetic field, $\vec{B}$, to the current producing it are, for the stationary case,

$$\nabla\times\vec{B} = \mu_0\vec{J} \qquad \text{and} \qquad \nabla\cdot\vec{B} = 0$$

Here $\vec{J}$ is the current density, current per area, defined so that across a tiny area $d\vec{A}$ the current that flows through the area is $dI = \vec{J}\cdot d\vec{A}$. A cylindrical wire of radius R carries a total current I distributed uniformly across the cross section of the wire. Put the z-axis of a cylindrical coordinate system along the central axis of the wire with positive z in the direction of the current flow. Write the function $\vec{J}$ explicitly in these coordinates (for all values of $r < R$, $r > R$). Use the curl and divergence expressed in cylindrical coordinates and assume a solution in the form $\vec{B} = \hat{\phi}\,B_\phi(r,\phi,z)$. Write out the divergence and curl equations and show that you can satisfy these equations relating $\vec{J}$ and $\vec{B}$ with such a form, solving for B_ϕ. Sketch a graph of the result. At a certain point in the calculation you will have to match the boundary conditions at $r = R$. Recall that the tangential component of $\vec{B}$ (here B_ϕ) is continuous at the boundary.
Ans: in part, $\mu_0 Ir/2\pi R^2 \;\; (r < R)$

9.23 A long cylinder of radius R has a uniform charge density inside it, ρ_0 and it is rotating about its long axis with angular speed ω. This provides an azimuthal current density $\vec{J} = \rho_0 r\omega\hat{\phi}$ in cylindrical coordinates. Assume the form of the magnetic field that this creates has only a z-component: $\vec{B} = B_z(r,\phi,z)\hat{z}$ and apply the equations of the preceding problem to determine this field both inside and outside. The continuity condition at $r = R$ is that the tangential component of $\vec{B}$ (here it is B_z) is continuous there. The divergence and the curl equations will (almost) determine the rest. Ans: in part, $-\rho r^2\omega/2 + C \; (r < R)$

9.24 By analogy to Eqs. (9.9) and (9.17) the expression

$$\lim_{V\to 0}\frac{1}{V}\oint \phi\, d\vec{A}$$

is the gradient of the scalar function ϕ. Compute this in rectangular coordinates by mimicking the derivation that led to Eq. (9.11) or (9.15), showing that it has the correct components.

9.25 (**a**) A fluid of possibly non-uniform mass density is in equilibrium in a possibly non-uniform gravitational field. Pick a volume and write down the total force vector on the fluid in that volume; the things acting on it are gravity and the surrounding fluid. Take the limit as the volume shrinks to zero, and use the result of the preceding problem in order to get the equation for equilibrium.
(**b**) Now apply the result to the special case of a uniform gravitational field and a constant mass density to find the pressure variation with height. Starting from an atmospheric pressure of $1.01\times10^5\,\mathrm{N/m^2}$, how far must you go under water to reach double this pressure?
Ans: $\nabla p = -\rho\vec{g}$; about 10 meters

9.26 The volume energy density, $u = dU/dV$, in the gravitational field is $g^2/8\pi G$. [Check the units to see if it makes sense.] Use the results found in Eq. (9.39) for the gravitational field of a spherical mass and get the energy density. An extension of Newton's theory of gravity is that the source of gravity is *energy* not just mass! This energy that you just computed from the gravitational field is then the source of more gravity, and this energy density contributes as a mass density $\rho = u/c^2$ would.
(**a**) Find the additional gravitational field $g_r(r)$ that this provides and add it to the previous result for $g_r(r)$.
(**b**) For our sun, its mass is 2×10^{30} kg and its radius is 700,000 km. Assume its density is constant throughout so that you can apply the results of this problem. At the sun's surface, what is the ratio of this correction to the original value?
(**c**) What radius would the sun have to be so that this correction is equal to the original $g_r(R)$, resulting in double gravity? Ans: (**a**) $g_{\mathrm{correction}}/g_{\mathrm{original}} = GM/10Rc^2$

9.27 Continuing the ideas of the preceding problem, the energy density, $u = dU/dV$, in the gravitational field is $g^2/8\pi G$, and the source of gravity is energy not just mass. In the region of space that is empty of matter, show that the divergence equation for gravity, (9.36), then becomes

$$\nabla\cdot\vec{g} = -4\pi Gu/c^2 = -g^2/2c^2$$

Assume that you have a spherically symmetric system, $\vec{g} = g_r(r)\hat{r}$, and write the differential equation for g_r.
(**a**) Solve it and apply the boundary condition that as $r\to\infty$, the gravitational field should go to $g_r(r)\to -GM/r^2$. How does this solution behave as $r\to0$ and compare its behavior to that of the usual gravitational field of a point mass.
(**b**) Can you explain why the behavior is different? Note that in this problem it is the gravitational field itself that is the source of the gravitational field; mass as such isn't present.
(**c**) A characteristic length appears in this calculation. Evaluate it for the sun. It is 1/4 the Schwarzchild radius that appears in the theory of general relativity.
Ans: (**a**) $g_r = -GM/\big[r(r+R)\big]$, where $R = GM/2c^2$

9.28 In the preceding problem, what is the total energy in the gravitational field, $\int u\,dV$? How does this ($\div c^2$) compare to the mass M that you used in setting the value of g_r as $r\to\infty$?

9.29 Verify that the solution Eq. (9.48) does satisfy the continuity conditions on V and V'.

9.30 The r-derivatives in Eq. (9.43), spherical coordinates, can be written in a different and more convenient form. Show that they are equivalent to

$$\frac{1}{r}\frac{\partial^2(rV)}{\partial r^2}$$

9.31 The gravitational potential from a point mass M is $-GM/r$ where r is the distance to the mass. Place a single point mass at coordinates $(x, y, z) = (0, 0, d)$ and write its potential V. Write this expression in terms of spherical coordinates about the origin, (r, θ), and then expand it for the case $r > d$ in a power series in d/r, putting together the like powers of d/r. Do this through order $(d/r)^3$. Express the result in the language of Eq. (4.61).
Ans: $-\frac{GM}{r} - \frac{GMd}{r^2}\big[\cos\theta\big] - \frac{GMd^2}{r^3}\big[\frac{3}{2}\cos^2\theta - \frac{1}{2}\big] - \frac{GMd^3}{r^4}\big[\frac{5}{2}\cos^3\theta - \frac{3}{2}\cos\theta\big]$

9.32 As in the preceding problem a point mass M has potential $-GM/r$ where r is the distance to the mass. The mass is at coordinates $(x, y, z) = (0, 0, d)$. Write its potential V in terms of spherical coordinates about the origin, (r, θ), but this time take $r < d$ and expand it in a power series in r/d. Do this through order $(r/d)^3$.
Ans: $(-GM/d)[1 + (r/d)P_1(\cos\theta) + (r^2/d^2)P_2(\cos\theta) + (r^3/d^3)P_3(\cos\theta) + \cdots]$

9.33 Theorem: Given that a vector field satisfies $\nabla \times \vec{v} = 0$ everywhere, then it follows that you can write $\vec{v}$ as the gradient of a scalar function, $\vec{v} = -\nabla\psi$. For each of the following vector fields find if possible, and probably by trail and error if so, a function ψ that does this. *First* determine is the curl is zero, because if it isn't then your hunt for a ψ will be futile. You should try however — the hunt will be instructive.

$$\text{(a) } \hat{x}\,y^3 + 3\hat{y}\,xy^2, \qquad \text{(c) } \hat{x}\,y\cos(xy) + \hat{y}\,x\cos(xy),$$
$$\text{(b) } \hat{x}\,x^2y + \hat{y}\,xy^2, \qquad \text{(d) } \hat{x}\,y^2\sinh(2xy^2) + 2\hat{y}\,xy\sinh(2xy^2)$$

9.34 A hollow sphere has inner radius a, outer radius b, and mass M, with uniform mass density in this region.
(a) Find (and sketch) its gravitational field $g_r(r)$ everywhere.
(b) What happens in the limit that $a \to b$? In this limiting case, graph g_r. Use $g_r(r) = -dV/dr$ and compute and graph the potential function $V(r)$ for this limiting case. This violates Eq. (9.47). Why?
(c) Compute the area mass density, $\sigma = dM/dA$, in this limiting case and find the relationship between the discontinuity in dV/dr and the value of σ.

9.35 Evaluate

$$\delta_{ij}\epsilon_{ijk}, \qquad \epsilon_{mjk}\epsilon_{njk}, \qquad \partial_i x_j, \qquad \partial_i x_i, \qquad \epsilon_{ijk}\epsilon_{ijk}, \qquad \delta_{ij}v_j$$
$$\text{and show that} \qquad \epsilon_{ijk}\epsilon_{mnk} = \delta_{im}\delta_{jn} - \delta_{in}\delta_{jm}$$

9.36 Verify the identities for arbitrary $\vec{A}$,

$$(\vec{A}\cdot\nabla)\vec{r} = \vec{A} \qquad \text{or} \qquad A_i\partial_i x_j = A_j$$
$$\nabla\cdot\nabla\times\vec{v} = 0 \qquad \text{or} \qquad \partial_i\epsilon_{ijk}\partial_j v_k = 0$$
$$\nabla\cdot(f\vec{A}) = (\nabla f)\cdot\vec{A} + f(\nabla\cdot\vec{A}) \qquad \text{or} \qquad \partial_i(fA_i) = (\partial_i f)A_i + f\partial_i A_i$$

You can try proving all these in the standard vector notation, but use the index notation instead. It's a lot easier.

9.37 Use index notation to prove $\nabla\times\nabla\times\vec{v} = \nabla(\nabla\cdot\vec{v}) - \nabla^2\vec{v}$. First, what identity you have to prove about ϵ's.

9.38 Is $\nabla\times\vec{v}$ perpendicular to $\vec{v}$? Either prove it's true or give an explicit example for which it is false.

9.39 If for arbitrary A_i and arbitrary B_j it is known that $a_{ij}A_iB_j = 0$, prove then that all the a_{ij} are zero.

9.40 Compute the divergences of

$Ax\,\hat{x} + By^2\,\hat{y} + C\,\hat{z}$ in rectangular coordinates.
$Ar\,\hat{r} + B\theta^2\,\hat{\theta} + C\,\hat{\phi}$ in spherical coordinates.

How do the pictures of these vector fields correspond to the results of these calculations?

9.41 Compute the divergence and the curl of

$$\frac{y\,\hat{x} - x\,\hat{y}}{x^2+y^2}, \qquad \text{and of} \qquad \frac{y\,\hat{x} - x\,\hat{y}}{(x^2+y^2)^2}$$

9.42 Translate the preceding vector fields into polar coordinates, then take their divergence and curl. And of course draw some pictures.

9.43 As a review of ordinary vector algebra, and perhaps some practice in using index notation, translate the triple scalar product into index notation and prove first that it is invariant under cyclic permutations of the vectors.
(**a**) $\vec{A}\cdot\vec{B}\times\vec{C} = \vec{B}\cdot\vec{C}\times\vec{A} = \vec{C}\cdot\vec{A}\times\vec{B}$. Then that
(**b**) $\vec{A}\cdot\vec{B}\times\vec{C} = \vec{A}\times\vec{B}\cdot\vec{C}$.
(**c**) What is the result of interchanging any pair of the vectors in the product?
(**d**) Show why the geometric interpretation of this product is as the volume of a parallelepiped.

9.44 What is the total flux, $\oint\vec{E}\cdot d\vec{A}$, out of the cube of side a with one corner at the origin?
(**a**) $\vec{E} = \alpha\,\hat{x} + \beta\,\hat{y} + \gamma\,\hat{z}$
(**b**) $\vec{E} = \alpha x\,\hat{x} + \beta y\,\hat{y} + \gamma z\,\hat{z}$.

9.45 The electric potential from a single point charge q is kq/r. Two charges are on the z-axis: $-q$ at position $z = z_0$ and $+q$ at position $z_0 + a$.

(**a**) Write the total potential at the point (r, θ, ϕ) in spherical coordinates.
(**b**) Assume that $r \gg a$ and $r \gg z_0$, and use the binomial expansion to find the series expansion for the total potential out to terms of order $1/r^3$.
(**c**) how does the coefficient of the $1/r^2$ term depend on z_0? The coefficient of the $1/r^3$ term? These tell you the total electric dipole moment and the total quadrupole moment.
(**d**) What is the curl of the gradient of each of these two terms?
The polynomials of section 4.11 will appear here, with argument $\cos\theta$.

9.46 charges q_1 and q_2, the electric field very far away will look like that of a single point charge $q_1 + q_2$. Go the next step beyond this and show that the electric field at large distances will approach a direction such that it points along a line that passes through the "center of charge" (like the center of mass): $(q_1\vec{r}_1 + q_2\vec{r}_2)/(q_1 + q_2)$. What happens to this calculation if $q_2 = -q_1$? You will find the results of problem 9.31 useful. Sketch various cases of course. At a certain point in the calculation, you will probably want to pick a particular coordinate system and to place the charges conveniently, probably one at the origin and the other on the z-axis. You should keep terms in the expansion for the potential up through $1/r^2$ and then take $-\nabla V$. Unless of course you find a better way.

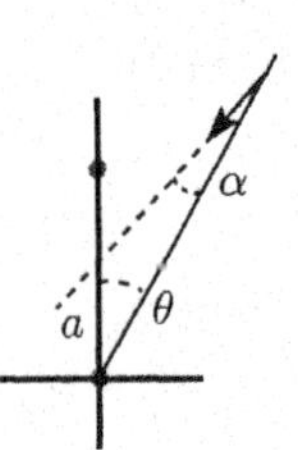

9.47 Fill in the missing steps in deriving Eq. (9.58).

9.48 Analyze the behavior of Eq. (9.58). The first thing you will have to do is to derive the behavior of $\sinh^{-1}$ in various domains and maybe to do some power series expansions. In every case seek an explanation of why the result comes out as it does.
(**a**) If $z = 0$ and $r = \sqrt{x^2 + y^2} \gg L$ what is it and what should it be? (And no, zero won't do.)
(**b**) If $z = 0$ and $r \ll L$ what is it and what should it be?
(**c**) If $z > L$ and $r \to 0$ what is this and what should it be? Be careful with your square roots here.
(**d**) What is the result of (**c**) for $z \gg L$ and for $z - L \ll L$?

9.49 Use the magnetic field equations as in problem 9.22, and assume a current density that is purely azimuthal and dependent on r alone. That is, in cylindrical coordinates it is

$$\vec{J} = J_0 \hat{\phi} f(r)$$

Look for a solution for the magnetic field in the form $\vec{B} = \hat{z} B_z(r)$. What is the total current per length in this case? That is, for a length Δz how much current is going around the axis and how is this related to B_z along $r = 0$?
Examine also the special case for which $f(r) = 0$ except in a narrow range $a < r < b$ with $b - a \ll b$ (thin). Compare this result to what you find in introductory texts about the solenoid.

9.50 For a spherical mass distribution such as the Earth, what would the mass density function have to be so that the gravitational field has constant strength as a function of depth?
Ans: $\rho \propto 1/r$.

9.51 Use index notation to derive

$$(\vec{A}\times\vec{B})\cdot(\vec{C}\times\vec{D}) = (\vec{A}\cdot\vec{C})(\vec{B}\cdot\vec{D}) - (\vec{A}\cdot\vec{D})(\vec{B}\cdot\vec{C})$$

9.52 Show that $\nabla\cdot(\vec{A}\times\vec{B}) = \vec{B}\cdot\nabla\times\vec{A} - \vec{A}\cdot\nabla\times\vec{B}$. Use index notation to derive this.

9.53 Use index notation to compute $\nabla e^{i\vec{k}\cdot\vec{r}}$. Also compute the Laplacian of the same exponential, $\nabla^2 = \text{div grad}$.

9.54 Derive the force of one charged ring on another, as shown in equation (2.35).

9.55 A point mass m is placed at a distance $d > R$ from the center of a spherical shell of radius R and mass M. Starting from Newton's gravitational law for point masses, derive the force on m from M. Place m on the z-axis and use spherical coordinates to express the piece of dM within $d\theta$ and $d\phi$. (This problem slowed Newton down when he first tried to solve it, as he had to stop and invent integral calculus first.)

9.56 The volume charge density is measured near the surface of a (not quite perfect) conductor to be $\rho(x) = \rho_0 e^{x/a}$ for $x < 0$. The conductor occupies the region $x < 0$, so the electric field in the conductor is zero once you're past the thin surface charge density. Find the electric field everywhere and graph it. Assume that everything is independent of y and z (and t).

Partial Differential Equations

If the subject of ordinary differential equations is large, this is enormous. I am going to examine only one corner of it, and will develop only one tool to handle it: Separation of Variables. Another major tool is the method of characteristics and I'll not go beyond mentioning the word. When I develop a technique to handle the heat equation or the potential equation, don't think that it stops there. The same set of tools will work on the Schroedinger equation in quantum mechanics and on the wave equation in its many incarnations.

10.1 The Heat Equation

The flow of heat in one dimension is described by the heat conduction equation

$$P = -\kappa A \frac{\partial T}{\partial x} \tag{10.1}$$

where P is the power in the form of heat energy flowing toward positive x through a wall and A is the area of the wall. κ is the wall's thermal conductivity. Put this equation into words and it says that if a thin slab of material has a temperature on one side different from that on the other, then heat energy will flow through the slab. If the temperature difference is big or the wall is thin ($\partial T/\partial x$ is big) then there's a big flow. The minus sign says that the energy flows from hot toward cold.

When more heat comes into a region than leaves it, the temperature there will rise. This is described by the specific heat, c.

$$dQ = mc\,dT, \qquad \text{or} \qquad \frac{dQ}{dt} = mc\frac{dT}{dt} \tag{10.2}$$

Again in words, the temperature rise in a chunk of material is proportional to the amount of heat added to it and inversely proportional to its mass.

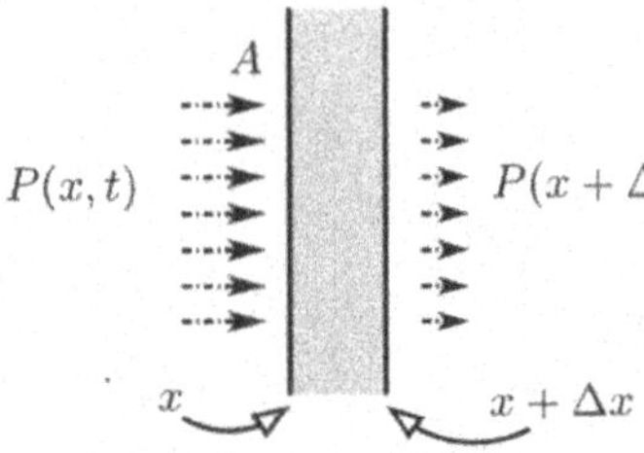

For a slab of area A, thickness Δx, and mass density ρ, let the coordinates of the two sides be x and $x + \Delta x$.

$$m = \rho A \Delta x, \qquad \text{and} \qquad \frac{dQ}{dt} = P(x,t) - P(x+\Delta x, t)$$

The net power into this volume is the power in from one side minus the power out from the other. Put these three equations together.

$$\frac{dQ}{dt} = mc\frac{dT}{dt} = \rho A \Delta x\, c\frac{dT}{dt} = -\kappa A \frac{\partial T(x,t)}{\partial x} + \kappa A \frac{\partial T(x+\Delta x,t)}{\partial x}$$

If you let $\Delta x \to 0$ here, all you get is $0 = 0$, not very helpful. Instead divide by Δx first and then take the limit.

$$\frac{\partial T}{\partial t} = +\frac{\kappa A}{\rho c A}\left(\frac{\partial T(x+\Delta x,t)}{\partial x} - \frac{\partial T(x,t)}{\partial x}\right)\frac{1}{\Delta x}$$

and in the limit this is

$$\frac{\partial T}{\partial t} = \frac{\kappa}{c\rho}\frac{\partial^2 T}{\partial x^2} \tag{10.3}$$

I was a little cavalier with the notation in that I didn't specify the argument of T on the left side. You could say that it was $(x + \Delta x/2, t)$, but in the limit everything is evaluated at (x,t) anyway. I also assumed that κ, the thermal conductivity, is independent of x. If not, then it stays inside the derivative,

$$\frac{\partial T}{\partial t} = \frac{1}{c\rho}\frac{\partial}{\partial x}\left(\kappa\frac{\partial T}{\partial x}\right) \tag{10.4}$$

In Three Dimensions

In three dimensions, this becomes

$$\frac{\partial T}{\partial t} = \frac{\kappa}{c\rho}\nabla^2 T \tag{10.5}$$

Roughly speaking, the temperature in a box can change because of heat flow in any of three directions. More precisely, the correct three dimensional equation that replaces Eq. (10.1) is

$$\vec{H} = -\kappa\nabla T \tag{10.6}$$

where $\vec{H}$ is the heat flow vector. That is the power per area in the direction of the energy transport. $\vec{H}\cdot d\vec{A} = dP$, the power going across the area $d\vec{A}$. The total heat flowing into a volume is

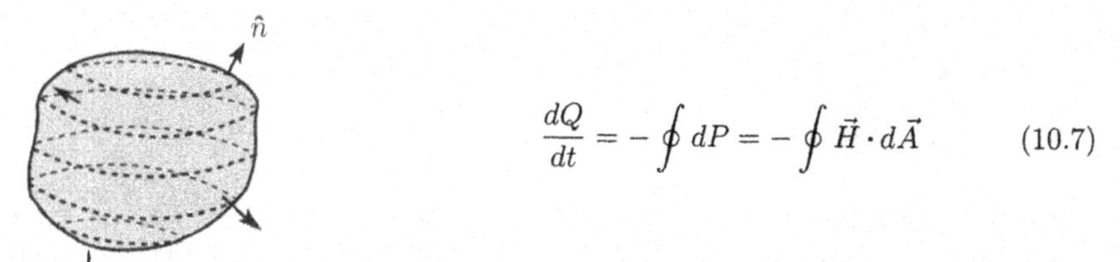

$$\frac{dQ}{dt} = -\oint dP = -\oint \vec{H}\cdot d\vec{A} \tag{10.7}$$

where the minus sign occurs because this is the heat flow *in*. For a small volume ΔV, you now have $m = \rho\Delta V$ and

$$mc\frac{\partial T}{\partial t} = \rho\Delta V\, c\frac{\partial T}{\partial t} = -\oint \vec{H}\cdot d\vec{A}$$

Divide by ΔV and take the limit as $\Delta V \to 0$. The right hand side is the divergence, Eq. (9.9).

$$\rho c\frac{\partial T}{\partial t} = -\lim_{\Delta V\to 0}\frac{1}{\Delta V}\oint \vec{H}\cdot d\vec{A} = -\nabla\cdot\vec{H} = +\nabla\cdot\kappa\nabla T = +\kappa\nabla^2 T$$

The last step again assumes that the thermal conductivity, κ, is independent of position.

10.2 Separation of Variables

How do you solve these equations? I'll start with the one-dimensional case and use the method of *separation of variables*. The trick is to start by looking for a solution to the equation in the form of a product of a function of x and a function of t. $T(x,t) = f(t)g(x)$. *I do not* assume that every solution to the equation will look like this — that's just not true. What will happen is that I'll be able to express every solution as a sum of such factored forms. That this is so is a theorem that I don't plan to prove here. For that you should go to a purely mathematical text on PDEs.

If you want to find out if you have a solution, plug in:

$$\frac{\partial T}{\partial t} = \frac{\kappa}{c\rho}\frac{\partial^2 T}{\partial x^2} \qquad \text{is} \qquad \frac{df}{dt}g = \frac{\kappa}{c\rho}f\frac{d^2 g}{dx^2}$$

Denote the constant by $\kappa/c\rho = D$ and divide by the product fg.

$$\frac{1}{f}\frac{df}{dt} = D\frac{1}{g}\frac{d^2 g}{dx^2} \tag{10.8}$$

The left side of this equation is a function of t alone, no x. The right side is a function of x alone with no t, hence the name separation of variables. Because x and t can vary quite independently of each other, the only way that this can happen is if the two side are constant (the same constant).

$$\frac{1}{f}\frac{df}{dt} = \alpha \qquad \text{and} \qquad D\frac{1}{g}\frac{d^2 g}{dx^2} = \alpha \tag{10.9}$$

At this point, the constant α can be anything, even complex. For a particular specified problem there will be boundary conditions placed on the functions, and those will constrain the α's. If α is real and positive then

$$g(x) = A\sinh\sqrt{\alpha/D}\,x + B\cosh\sqrt{\alpha/D}\,x \qquad \text{and} \qquad f(t) = e^{\alpha t} \tag{10.10}$$

For negative real α, the hyperbolic functions become circular functions.

$$g(x) = A\sin\sqrt{-\alpha/D}\,x + B\cos\sqrt{-\alpha/D}\,x \qquad \text{and} \qquad f(t) = e^{\alpha t} \tag{10.11}$$

If $\alpha = 0$ then

$$g(x) = Ax + B, \qquad \text{and} \qquad f(t) = \text{constant} \tag{10.12}$$

For imaginary α the $f(t)$ is oscillating and the $g(x)$ has both exponential and oscillatory behavior in space. This can really happen in very ordinary physical situations; see section 10.3.

This analysis provides a solution to the original equation (10.3) valid for any α. A sum of such solutions for different α's is also a solution, for example

$$T(x,t) = A_1 e^{\alpha_1 t}\sin\sqrt{-\alpha_1/D}\,x + A_2 e^{\alpha_2 t}\sin\sqrt{-\alpha_2/D}\,x$$

or any other linear combination with various α's

$$T(x,t) = \sum_{\{\alpha' s\}} f_\alpha(t) g_\alpha(x)$$

It is the combined product that forms a solution to the original partial differential equation, not the separate factors. Determining the details of the sum is a job for Fourier series.

Example

A specific problem: You have a slab of material of thickness L and at a uniform temperature T_0. Plunge it into ice water at temperature $T = 0$ and find the temperature inside at later times. The boundary condition here is that the surface temperature is zero, $T(0,t) = T(L,t) = 0$. This constrains the separated solutions, requiring that $g(0) = g(L) = 0$. For this to happen you can't use the hyperbolic functions of x that occur when $\alpha > 0$, you will need the circular functions of x, sines and cosines, implying that $\alpha < 0$. That is also compatible with your expectation that the temperature should approach zero eventually, and that needs a negative exponential in time,, Eq. (10.11).

$$g(x) = A\sin kx + B\cos kx, \qquad \text{with} \qquad k^2 = -\alpha/D \qquad \text{and} \qquad f(t) = e^{-Dk^2 t}$$

$g(0) = 0$ implies $B = 0$. $g(L) = 0$ implies $\sin kL = 0$.
The sine vanishes for the values $n\pi$ where n is any integer, positive, negative, or zero. This implies $kL = n\pi$, or $k = n\pi/L$. The corresponding values of α are $\alpha_n = -Dn^2\pi^2/L^2$, and the separated solution is

$$\sin\left(n\pi x/L\right) e^{-n^2\pi^2 Dt/L^2} \tag{10.13}$$

If $n = 0$ this whole thing vanishes, so it's not much of a solution. (Not so fast there! See problem 10.2.) Notice that the sine is an odd function so when $n < 0$ this expression just reproduces the positive n solution except for an overall factor of (-1), and that factor was arbitrary anyway. The negative n solutions are redundant, so ignore them.

The general solution is a sum of separated solutions, see problem 10.3.

$$T(x,t) = \sum_1^\infty a_n \sin\frac{n\pi x}{L}\, e^{-n^2\pi^2 Dt/L^2} \tag{10.14}$$

The problem now is to determine the coefficients a_n. *This* is why Fourier series were invented. (Yes, literally, the problem of heat conduction is where Fourier series started.) At time $t = 0$ you know the temperature distribution is $T = T_0$, a constant on $0 < x < L$. This general sum must equal T_0 at time $t = 0$.

$$T(x,0) = \sum_1^\infty a_n \sin\frac{n\pi x}{L} \qquad (0 < x < L)$$

Multiply by $\sin\left(m\pi x/L\right)$ and integrate over the domain to isolate the single term, $n = m$.

$$\int_0^L dx\, T_0 \sin\frac{m\pi x}{L} = a_m \int_0^L dx\, \sin^2\frac{m\pi x}{L}$$
$$T_0[1 - \cos m\pi]\frac{L}{m\pi} = a_m \frac{L}{2}$$

This expression for a_m vanishes for even m, and when you assemble the whole series for the temperature you have

$$T(x,t) = \frac{4}{\pi}T_0 \sum_{m\ \mathrm{odd}} \frac{1}{m} \sin\frac{m\pi x}{L} e^{-m^2\pi^2 Dt/L^2} \tag{10.15}$$

For small time, this converges, but very slowly. For large time, the convergence is very fast, often needing only one or two terms. As the time approaches infinity, the interior temperature approaches the surface temperature of zero. The graph shows the temperature profile at a sequence of times.

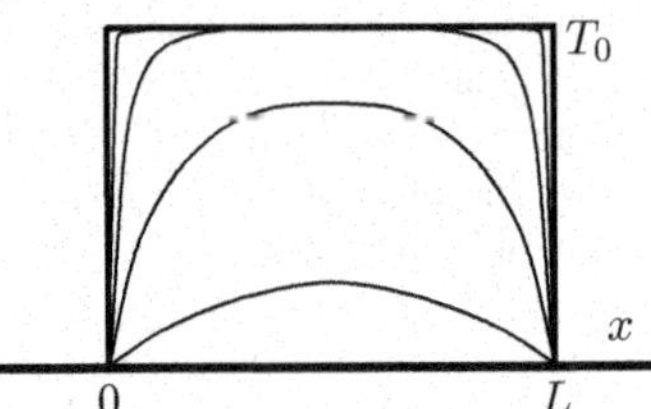

The curves show the temperature dropping very quickly for points near the surface ($x = 0$ or L). It drops more gradually near the center but eventually goes to zero.

You can see that the boundary conditions on the temperature led to these specific boundary conditions on the sines and cosines. This is exactly what happened in the general development of Fourier series when the fundamental relationship, Eq. (5.15), required certain boundary conditions in order to get the orthogonality of the solutions of the harmonic oscillator differential equation. That the function vanishes at the boundaries was one of the possible ways to insure orthogonality.

10.3 Oscillating Temperatures

Take a very thick slab of material and assume that the temperature on one side of it is oscillating. Let the material occupy the space $0 < x < \infty$ and at the coordinate $x = 0$ the temperature is varying in time as $T_1 \cos\omega t$. Is there any real situation in which this happens? Yes, the surface temperature of the Earth varies periodically from summer to winter (even in Florida). What happens to the temperature underground?

The differential equation for the temperature is still Eq. (10.3), and assume that the temperature inside the material approaches $T = 0$ far away from the surface. Separation of variables is the same as before, Eq. (10.9), but this time you know the time dependence at the surface. It's typical in cases involving oscillations that it is easier to work with complex exponentials than it is to work with sines and cosines. For that reason, specify that the surface temperature is $T_1 e^{-i\omega t}$ instead of a cosine, understanding that at the end of the problem you must take the real part of the result and throw away the imaginary part. The imaginary part corresponds to solving the problem for a surface temperature of $\sin\omega t$ instead of cosine. It's easier to solve the two problems together then either one separately. (The minus sign in the exponent of $e^{-i\omega t}$ is arbitrary; you could use a plus instead.)

The equation (10.9) says that the time dependence that I expect is

$$\frac{1}{f}\frac{df}{dt} = \alpha = \frac{1}{e^{-i\omega t}}\big(-i\omega e^{-i\omega t}\big) = -i\omega$$

The equation for the x-dependence is then

$$D\frac{d^2g}{dx^2} = \alpha g = -i\omega g$$

This is again a simple exponential solution, say $e^{\beta x}$. Substitute and you have

$$D\beta^2 e^{\beta x} = -i\omega e^{\beta x}, \qquad \text{implying} \qquad \beta = \pm\sqrt{-i\omega/D} \tag{10.16}$$

Evaluate this as

$$\sqrt{-i} = \left(e^{-i\pi/2}\right)^{1/2} = e^{-i\pi/4} = \frac{1-i}{\sqrt{2}}$$

Let $\beta_0 = \sqrt{\omega/2D}$, then the solution for the x-dependence is

$$g(x) = Ae^{(1-i)\beta_0 x} + Be^{-(1-i)\beta_0 x} \tag{10.17}$$

Look at the behavior of these two terms. The first has a factor that goes as e^{+x} and the second goes as e^{-x}. The temperature at large distances is supposed to approach zero, so that says that $A = 0$. The solutions for the temperature is now

$$Be^{-i\omega t}e^{-(1-i)\beta_0 x} \tag{10.18}$$

The further condition is that at $x = 0$ the temperature is $T_1 e^{-i\omega t}$, so that tells you that $B = T_1$.

$$T(x,t) = T_1 e^{-i\omega t}e^{-(1-i)\beta_0 x} = T_1 e^{-\beta_0 x}e^{i(-\omega t+\beta_0 x)} \tag{10.19}$$

When you remember that I'm solving for the real part of this solution, the final result is

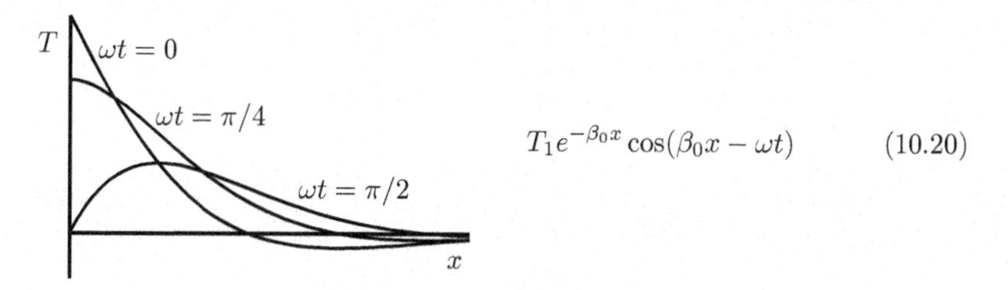

$$T_1 e^{-\beta_0 x}\cos(\beta_0 x - \omega t) \tag{10.20}$$

This has the appearance of a temperature wave moving into the material, albeit a very strongly damped one. In a half wavelength of this wave, $\beta_0 x = \pi$, and at that point the amplitude coming from the exponential factor out in front is down by a factor of $e^{-\pi} = 0.04$. That's barely noticeable. This is why wine cellars are cellars. Also, you can see that at a distance where $\beta_0 x > \pi/2$ the temperature change is reversed from the value at the surface. Some distance underground, summer and winter are reversed. This same sort of equation comes up with the study of eddy currents in electromagnetism, so the same sort of results obtain.

10.4 Spatial Temperature Distributions

The governing equation is Eq. (10.5). For an example of a problem that falls under this heading, take a cube that is heated on one side and cooled on the other five sides. What is the temperature distribution within the cube? How does it vary in time?

I'll take a simpler version of this problem to start with. First, I'll work in two dimensions instead of three; make it a very long rectangular shaped rod, extending in the z-direction. Second, I'll look for the equilibrium solution, for which the time derivative is zero. These restrictions reduce the equation (10.5) to

$$\nabla^2 T = \frac{\partial^2 T}{\partial x^2} + \frac{\partial^2 T}{\partial y^2} = 0 \tag{10.21}$$

I'll specify the temperature $T(x,y)$ on the surface of the rod to be zero on three faces and T_0 on the fourth. Place the coordinates so that the length of the rod is along the z-axis and the origin is in one corner of the rectangle.

$$\begin{aligned} T(0,y) &= 0 \quad (0 < y < b), \qquad & T(x,0) &= 0 \quad (0 < x < a) \\ T(a,y) &= 0 \quad (0 < y < b), \qquad & T(x,b) &= T_0 \quad (0 < x < a) \end{aligned} \tag{10.22}$$

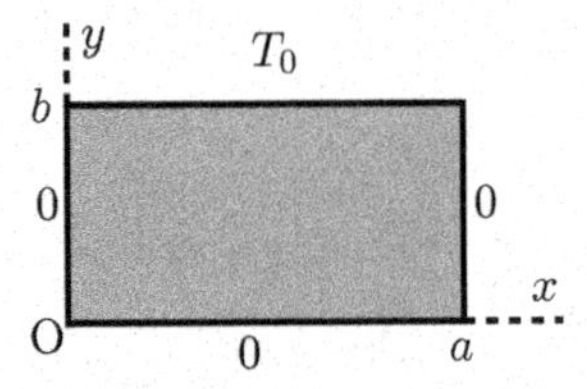

Look at this problem from several different angles, tear it apart, look at a lot of special cases, and see what can go wrong. In the process you'll see different techniques and especially a lot of applications of Fourier series. This single problem will illustrate many of the methods used to understand boundary value problems.

Use the same method used before for heat flow in one dimension: separation of variables. Assume a solution to be the product of a function of x and a function of y, then plug into the equation.

$$T(x,y) = f(x)g(y), \qquad \text{then} \qquad \nabla^2 T = \frac{d^2 f(x)}{dx^2} g(y) + f(x)\frac{d^2 g(y)}{dy^2} = 0$$

Just as in Eq. (10.8), when you divide by fg the resulting equation is separated into a term involving x only and one involving y only.

$$\frac{1}{f}\frac{d^2 f(x)}{dx^2} + \frac{1}{g}\frac{d^2 g(y)}{dy^2} = 0$$

Because x and y can be varied independently, these must be constants adding to zero.

$$\frac{1}{f}\frac{d^2 f(x)}{dx^2} = \alpha, \qquad \text{and} \qquad \frac{1}{g}\frac{d^2 g(y)}{dy^2} = -\alpha \tag{10.23}$$

As before, the separation constant can be any real or complex number until you start applying boundary conditions. You recognize that the solutions to these equations can be sines or cosines or exponentials or hyperbolic functions or linear functions, depending on what α is.

The boundary conditions state that the surface temperature is held at zero on the surfaces $x = 0$ and $x = a$. This suggests looking for solutions that vanish there, and that in turn says you should work with sines of x. In the other direction the surface temperature vanishes on only one side so you don't need sines in that case. The $\alpha = 0$ case gives linear functions is x and in y, and the fact that the temperature vanishes on $x = 0$ and $x = a$ kills these terms. (It does doesn't it?) Pick α to be a negative real number: call it $\alpha = -k^2$.

$$\frac{d^2 f(x)}{dx^2} = -k^2 f \quad \Longrightarrow \quad f(x) = A \sin kx + B \cos kx$$

The accompanying equation for g is now

$$\frac{d^2 g(y)}{dy^2} = +k^2 g \quad \Longrightarrow \quad g(y) = C \sinh ky + D \cosh ky$$

(Or exponentials if you prefer.) The combined, separated solution to $\nabla^2 T = 0$ is

$$(A \sin kx + B \cos kx)(C \sinh ky + D \cosh ky) \tag{10.24}$$

The general solution will be a sum of these, summed over various values of k. This is where you have to apply the boundary conditions to determine the allowed k's.

$$\text{left:} \quad T(0, y) = 0 = B(C \sinh ky + D \cosh ky), \qquad \text{so} \qquad B = 0$$

(This holds for all y in $0 < y < b$, so the second factor can't vanish unless both C and D vanish. If that is the case then *everything* vanishes.)

$$\text{right:} \quad T(a, y) = 0 = A \sin ka(C \sinh ky + D \cosh ky), \qquad \text{so} \qquad \sin ka = 0$$

(The factor with y can't vanish or everything vanishes. If $A = 0$ then everything vanishes. All that's left is $\sin ka = 0$.)

$$\text{bottom:} \quad T(x, 0) = 0 = A \sin kx\, D, \qquad \text{so} \qquad D = 0$$

(If $A = 0$ everything is zero, so it has to be D.)

You can now write a general solution that satisfies three of the four boundary conditions. Combine the coefficients A and C into one, and since it will be different for different values of k, call it γ_n.

$$T(x, y) = \sum_{n=1}^{\infty} \gamma_n \sin \frac{n\pi x}{a} \sinh \frac{n\pi y}{a} \tag{10.25}$$

The $n\pi/a$ appears because $\sin ka = 0$, and the limits on n omit the negative n because they are redundant.

Now to find all the unknown constants γ_n, and as before that's where Fourier techniques come in. The fourth side, at $y = b$, has temperature T_0 and that implies

$$\sum_{n=1}^{\infty} \gamma_n \sin\frac{n\pi x}{a}\sinh\frac{n\pi b}{a} = T_0$$

On this interval $0 < x < a$ these sine functions are orthogonal, so you take the scalar product of both side with the sine.

$$\int_0^a dx \sin\frac{m\pi x}{a}\sum_{n=1}^{\infty} \gamma_n \sin\frac{n\pi x}{a}\sinh\frac{n\pi b}{a} = \int_0^a dx \sin\frac{m\pi x}{a} T_0$$

$$\frac{a}{2}\gamma_m \sinh\frac{m\pi b}{a} = T_0\frac{a}{m\pi}\big[1-(-1)^m\big]$$

Just the odd m terms are present, $m = 2\ell + 1$, so the result is

$$T(x,y) = \frac{4}{\pi}T_0\sum_{\ell=0}^{\infty}\frac{1}{2\ell+1}\frac{\sinh\big((2\ell+1)\pi y/a\big)}{\sinh\big((2\ell+1)\pi b/a\big)}\sin\frac{(2\ell+1)\pi x}{a} \tag{10.26}$$

You're not done.

Does this make sense? The dimensions are clearly correct, but after that it takes some work. There's really just one parameter that you have to play around with, and that's the ratio b/a. If it's either very big or very small you may be able to check the result.

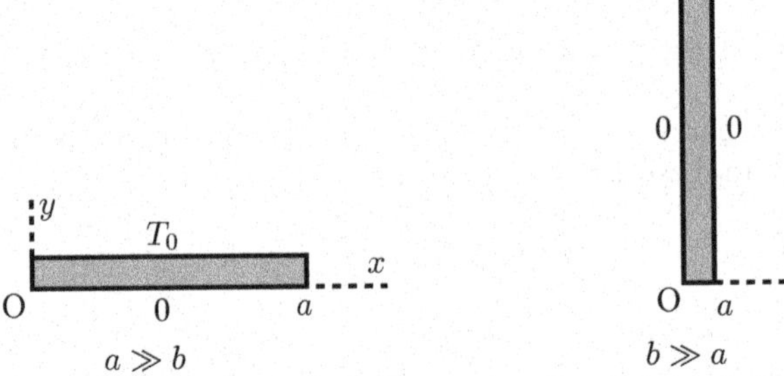

If $a \gg b$, it looks almost like a one-dimensional problem. It is a thin slab with temperature T_0 on one side and zero on the other. There's little variation along the x-direction, so the equilibrium equation is

$$\nabla^2 T = 0 = \frac{\partial^2 T}{\partial x^2} + \frac{\partial^2 T}{\partial y^2} \approx \frac{\partial^2 T}{\partial y^2}$$

This simply says that the second derivative with respect to y vanishes, so the answer is the straight line $T = A + By$, and with the condition that you know the temperature at $y = 0$ and at $y = b$ you easily find

$$T(x,y) \approx T_0 y/b$$

Does the exact answer look like this? It doesn't seem to, but look closer. If $b \ll a$ then because $0 < y < b$ you also have $y \ll a$. The hyperbolic function factors in Eq. (10.26) will have very small arguments, proportional to b/a. Recall the power series expansion of the hyperbolic sine: $\sinh x = x + \cdots$. These factors become approximately

$$\frac{\sinh\big((2\ell+1)\pi y/a\big)}{\sinh\big((2\ell+1)\pi b/a\big)} \approx \frac{(2\ell+1)\pi y/a}{(2\ell+1)\pi b/a} = \frac{y}{b}$$

The temperature solution is then

$$T(x,y) \approx \frac{4}{\pi}T_0 \sum_{\ell=0}^{\infty} \frac{1}{2\ell+1}\frac{y}{b}\sin\frac{(2\ell+1)\pi x}{a} = T_0\frac{y}{b}$$

Where did that last equation come from? The coefficient of y/b is just the Fourier series of the constant T_0 in terms of sines on $0 < x < a$.

What about the opposite extreme, for which $b \gg a$? This is the second picture just above. Instead of being short and wide it is tall and narrow. For this case, look again at the arguments of the hyperbolic sines. Now $\pi b/a$ is large and you can approximate the hyperbolic functions by going back to their definition.

$$\sinh x = \frac{e^x + e^{-x}}{2} \approx \frac{1}{2}e^x, \quad \text{for } x \gg 1$$

The denominators in all the terms of Eq. (10.26) are large, $\approx e^{\pi b/a}$ (or larger still because of the $(2\ell+1)$). This will make all the terms in the series extremely small *unless* the numerators are correspondingly large. This means that the temperature stays near zero unless y is large. That makes sense. It's only for y near the top end that you are near to the wall with temperature T_0.

You now have the case for which $b \gg a$ and $y \gg a$. This means that I can use the approximate form of the hyperbolic function for large arguments.

$$\frac{\sinh\big((2\ell+1)\pi y/a\big)}{\sinh\big((2\ell+1)\pi b/a\big)} \approx \frac{e^{(2\ell+1)\pi y/a}}{e^{(2\ell+1)\pi b/a}} = e^{(2\ell+1)\pi(y-b)/a}$$

The temperature distribution is now approximately

$$T(x,y) \approx \frac{4}{\pi}T_0 \sum_{\ell=0}^{\infty} \frac{1}{2\ell+1} e^{-(2\ell+1)\pi(b-y)/a} \sin\frac{(2\ell+1)\pi x}{a} \tag{10.27}$$

As compared to the previous approximation where $a \gg b$, you can't as easily tell whether this is plausible or not. You can however learn from it. See also problem 10.30.

At the very top, where $y = b$ this reduces to the constant T_0 that you're supposed to have at that position. Recall the Fourier series for a constant on $0 < x < a$.

Move down from $y = b$ by the distance a, so that $b - y = a$. That's a distance from the top equal to the width of the rectangle. It's still rather close to the end, but look at the series for that position.

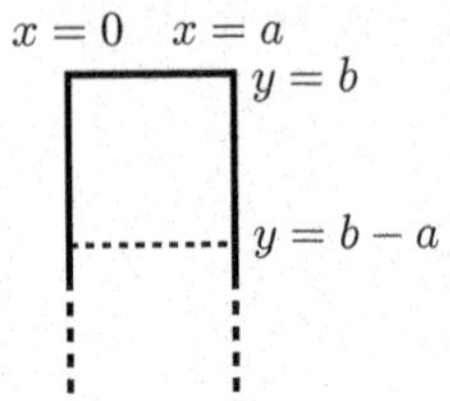

$x = 0 \quad x = a$, $y = b$, $y = b - a$

$$T(x, b-a) \approx \frac{4}{\pi} T_0 \sum_{\ell=0}^{\infty} \frac{1}{2\ell+1} e^{-(2\ell+1)\pi} \sin\frac{(2\ell+1)\pi x}{a}$$

For $\ell = 0$, the exponential factor is $e^{-\pi} = 0.043$, and for $\ell = 1$ this factor is $e^{-3\pi} = 0.00008$. This means that measured from the T_0 end, within the very short distance equal to the width, the temperature has dropped 95% of the way down to its limiting value of zero. The temperature in the rod is quite uniform until you are very close to the heated end.

The Heat Flow into the Box

All the preceding analysis and discussion was intended to make this problem and its solution sound oh-so-plausible. There's more, and it isn't pretty.

The temperature on one of the four sides was given as different from the temperatures on the other three sides. What will the heat flow into the region be? That is, what power must you supply to maintain the temperature T_0 on the single wall?

At the beginning of this chapter, Eq. (10.1), you have the equation for the power through an area A, but that equation assumed that the temperature gradient $\partial T/\partial x$ is the same all over the area A. If it isn't, you simply turn it into a density.

$$\Delta P = -\kappa \Delta A \frac{\partial T}{\partial x}, \qquad \text{and then} \qquad \frac{\Delta P}{\Delta A} \to \frac{dP}{dA} = -\kappa \frac{\partial T}{\partial x} \tag{10.28}$$

Equivalently, just use the vector form from Eq. (10.6), $\vec{H} = -\kappa \nabla T$. In Eq. (10.22) the temperature is T_0 along $y = b$, and the power density (energy / (time · area)) flowing in the $+y$ direction is $-\kappa \partial T/\partial y$, so the power density flowing *into* this area has the reversed sign,

$$+\kappa\, \partial T/\partial y \tag{10.29}$$

The total power flow is the integral of this over the area of the top face.

Let L be the length of this long rectangular rod, its extent in the z-direction. The element of area along the surface at $y = b$ is then $dA = L dx$, and the power flow into this face is

$$\int_0^a L\, dx \kappa \left.\frac{\partial T}{\partial y}\right|_{y=b}$$

The temperature function is the solution Eq. (10.26), so differentiate that equation with respect to y.

$$\int_0^a L\, dx \kappa \frac{4}{\pi} T_0 \sum_{\ell=0}^{\infty} \frac{[(2\ell+1)\pi/a]}{2\ell+1} \frac{\cosh\left((2\ell+1)\pi y/a\right)}{\sinh\left((2\ell+1)\pi b/a\right)} \sin\frac{(2\ell+1)\pi x}{a} \quad \text{at} \quad y = b$$

$$= \frac{4L\kappa T_0}{a} \int_0^a dx \sum_{\ell=0}^{\infty} \sin\frac{(2\ell+1)\pi x}{a}$$

and this sum does not converge. I'm going to push ahead anyway, temporarily pretending that I didn't notice this minor difficulty with the series. Just go ahead and integrate the series term by term and hope for the best.

$$= \frac{4L\kappa T_0}{a}\sum_{\ell=0}^{\infty}\frac{a}{\pi(2\ell+1)}\big[-\cos\big((2\ell+1)\pi\big)+1\big]$$

$$= \frac{4L\kappa T_0}{\pi}\sum_{\ell=0}^{\infty}\frac{2}{2\ell+1}=\infty$$

This infinite series for the total power entering the top face is infinite. The series doesn't converge (use the integral test).

This innocuous-seeming problem is suddenly pathological because it would take an infinite power source to maintain this temperature difference. Why should that be? Look at the corners. You're trying to maintain a non-zero temperature difference ($T_0 - 0$) between two walls that are touching. This can't happen, and the equations are telling you so! It means that the boundary conditions specified in Eq. (10.22) are impossible to maintain. The temperature on the boundary at $y = b$ can't be constant all the way to the edge. It must drop off to zero as it approaches $x = 0$ and $x = a$. This makes the problem more difficult, but then reality is typically more complicated than our simple, idealized models.

Does this make the solution Eq. (10.26) valueless? No, it simply means that you can't push it too hard. This solution will be good until you get near the corners, where you can't possibly maintain the constant-temperature boundary condition. In other regions it will be a good approximation to the physical problem.

10.5 Specified Heat Flow

In the previous examples, I specified the temperature on the boundaries and from that I determined the temperature inside. In the particular example, the solution was not physically plausible all the way to the edge, though the mathematics were (I hope) enlightening. Instead, I'll reverse the process and try to specify the size of the heat flow, computing the resulting temperature from that. This time perhaps the results will be a better reflection of reality.

Equation (10.29) tells you the power density at the surface, and I'll examine the case for which this is a constant. Call it F_0. (There's not a conventional symbol, so this will do.) The plus sign occurs because the flow is into the box.

$$+\kappa\frac{\partial T}{\partial y}(x,b)=F_0$$

The other three walls have the same zero temperature conditions as Eq. (10.22). Which forms of the separated solutions must I use now? The same ones as before or different ones?

Look again at the $\alpha = 0$ solutions to Eqs. (10.23). That solution is

$$(A+Bx)(C+Dy)$$

In order to handle the fact that the temperature is zero at $y = 0$ and that the derivative with respect to y is given at $y = b$,

$$(A + Bx)(C) = 0 \qquad \text{and} \qquad (A + Bx)(D) = F_0/\kappa,$$

$$\text{implying} \qquad C = 0 = B, \qquad \text{then} \qquad AD = F_0/\kappa \implies \frac{F_0}{\kappa}y \tag{10.30}$$

This matches the boundary conditions at both $y = 0$ and $y = b$. All that's left is to make everything work at the other two faces.

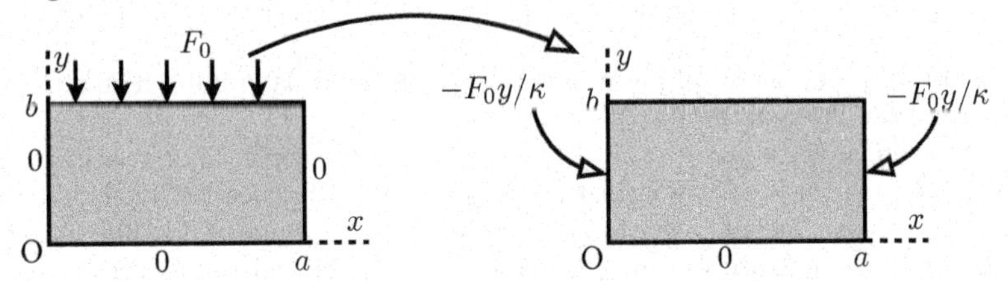

If I can find a solution that equals $-F_0y/\kappa$ on the left and right faces then it will cancel the $+F_0y/\kappa$ that Eq. (10.30) provides. *But* I can't disturb the top and bottom boundary conditions. The way to do that is to find functions that equal zero at $y = 0$ and whose derivative equals zero at $y = b$. This is a familiar sort of condition that showed up several times in chapter five on Fourier series. It is equivalent to saying that the top surface is insulated so that heat can't flow through it. You then use superposition to combine the solution with uniform heat flow and the solution with an insulated boundary.

Instead of Eq. (10.24), use the opposite sign for α, so the solutions are of the form

$$(A\sin ky + B\cos ky)(C\sinh kx + D\cosh kx)$$

I require that this equals zero at $y = 0$, so that says

$$(0 + B)(C\sinh kx + D\cosh kx) = 0$$

so $B = 0$. Now require that the derivative equals zero at $y = b$, so

$$Ak\cos kb = 0, \qquad \text{or} \qquad kb = (n + 1/2)\pi \qquad \text{for} \qquad n = 0,\, 1,\, 2\ldots$$

The value of the temperature is the same on the left that it is on the right, so

$$C\sinh k0 + D\cosh k0 = C\sinh ka + D\cosh ka \implies C = D(1 - \cosh ka)/\sinh ka \tag{10.31}$$

This is starting to get messy, so it's time to look around and see if I've missed anything that could simplify the calculation. There's no guarantee that there is any simpler way, but it is always worth looking. The fact that the system is the same on the left as on the right means that the temperature will be symmetric about the central axis of the box, about $x = a/2$. That it is even about this point implies that the hyperbolic functions of x should be even about $x = a/2$. You can do this simply by using a cosh about that point.

$$A\sin ky\big(D\cosh k(x - a/2)\big)$$

Put these together and you have a sum

$$\sum_{n=0}^{\infty} a_n \sin\left(\frac{(n+{}^1\!/\!_2)\pi y}{b}\right)\cosh\left(\frac{(n+{}^1\!/\!_2)\pi(x-{}^a\!/\!_2)}{b}\right) \tag{10.32}$$

Each of these terms satisfies Laplace's equation, satisfies the boundary conditions at $y=0$ and $y=b$, and is even about the centerline $x=a/2$. It is now a problem in Fourier series to match the conditions at $x=0$. They're then automatically satisfied at $x=a$.

$$\sum_{n=0}^{\infty} a_n \sin\left(\frac{(n+{}^1\!/\!_2)\pi y}{b}\right)\cosh\left(\frac{(n+{}^1\!/\!_2)\pi a}{2b}\right) = -F_0\frac{y}{\kappa} \tag{10.33}$$

The sines are orthogonal by the theorem Eq. (5.15), so you can pick out the component a_n by the orthogonality of these basis functions.

$$u_n = \sin\left(\frac{(n+{}^1\!/\!_2)\pi y}{b}\right), \qquad \text{then} \qquad \langle u_m, \text{left side}\rangle = \langle u_m, \text{right side}\rangle$$

$$\text{or,} \qquad a_m\langle u_m, u_m\rangle \cosh\left(\frac{(m+{}^1\!/\!_2)\pi a}{2b}\right) = -\frac{F_0}{\kappa}\langle u_m, y\rangle$$

Write this out; do the integrals, add the linear term, and you have

$$T(x,y) = F_0\frac{y}{\kappa} - \frac{8F_0 b}{\kappa\pi^2}\sum_{n=0}^{\infty}\frac{(-1)^n}{(2n+1)^2}\times \tag{10.34}$$
$$\sin\left(\frac{(n+{}^1\!/\!_2)\pi y}{b}\right)\cosh\left(\frac{(n+{}^1\!/\!_2)\pi(x-{}^a\!/\!_2)}{b}\right)\operatorname{sech}\left(\frac{(n+{}^1\!/\!_2)\pi a}{2b}\right)$$

Now analyze this to see if it makes sense. I'll look at the same cases as the last time: $b \ll a$ and $a \ll b$. The simpler case, where the box is short and wide, has $b \ll a$. This makes the arguments of the cosh and sech large, with an a/b in them. For large argument you can approximate the cosh by

$$\cosh x \approx e^x/2, \qquad x \gg 1$$

Now examine a typical term in the sum (10.34), and I have to be a little more specific and choose x on the left or right of $a/2$. The reason for that is the preceding equation requires x large and positive. I'll take x on the right, as it makes no difference. The hyperbolic functions in (10.34) are approximately

$$\frac{\exp\left((n+{}^1\!/\!_2)\pi(x-{}^a\!/\!_2)/b\right)}{\exp\left((n+{}^1\!/\!_2)\pi a/2b\right)} = e^{((2n+1)\pi(x-a)/2b)}$$

As long as x is not near the end, that is, not near $x=a$, the quantity in the exponential is large and negative for all n. The exponential in turn makes this extremely small so that the entire sum becomes negligible. The temperature distribution is then the single term

$$T(x,y) \approx F_0\frac{y}{\kappa}$$

It's essentially a one dimensional problem, with the heat flow only along the $-y$ direction.

In the reverse case for which the box is tall and thin, $a \ll b$, the arguments of the hyperbolic functions are small. This invites a power series expansion, but that approach doesn't work. The analysis of this case is quite tricky, and I finally concluded that it's not worth the trouble to write it up. It leads to a rather complicated integral.

10.6 Electrostatics

The equation for the electrostatic potential in a vacuum is exactly the same as Eq. (10.21) for the temperature in static equilibrium, $\nabla^2 V = 0$, with the electric field $\vec{E} = -\nabla V$. The same equation applies to the gravitational potential, Eq. (9.42).

Perhaps you've looked into a microwave oven. You can see inside it, but the microwaves aren't supposed to get out. How can this be? Light is just another form of electromagnetic radiation, so why does one EM wave get through while the other one doesn't? I won't solve the whole electromagnetic radiation problem here, but I'll look at the static analog to get some general idea of what's happening.

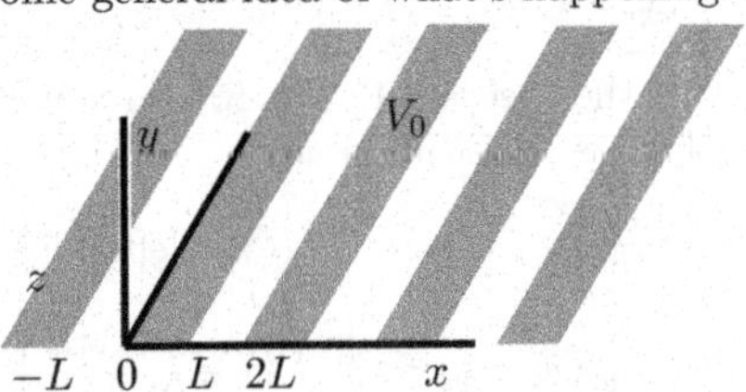

Arrange a set of conducting strips in the x-y plane and with insulation between them so that they don't quite touch each other. Now apply voltage V_0 on every other one so that the potentials are alternately zero and V_0. This sets the potential in the $z = 0$ plane to be independent of y and

$$z = 0: \quad V(x,y) = \begin{cases} V_0 & (0 < x < L) \\ 0 & (L < x < 2L) \end{cases} \qquad V(x+2L, y) = V(x,y), \text{ all } x,\ y \tag{10.35}$$

What is then the potential above the plane, $z > 0$? Above the plane V satisfies Laplace's equation,

$$\nabla^2 V = \frac{\partial^2 V}{\partial x^2} + \frac{\partial^2 V}{\partial y^2} + \frac{\partial^2 V}{\partial z^2} = 0 \tag{10.36}$$

The potential is independent of y in the plane, so it will be independent of y everywhere. Separate variables in the remaining coordinates.

$$V(x,z) = f(x)g(z) \Longrightarrow \frac{d^2 f}{dx^2} g + f \frac{d^2 g}{dz^2} = 0 \Longrightarrow \frac{1}{f}\frac{d^2 f}{dx^2} + \frac{1}{g}\frac{d^2 g}{dz^2} = 0$$

This is separated as a function of x plus a function of y, so the terms are constants.

$$\frac{1}{f}\frac{d^2 f}{dx^2} = -\alpha^2, \qquad \frac{1}{g}\frac{d^2 g}{dz^2} = +\alpha^2 \tag{10.37}$$

I've chosen the separation constant in this form because the boundary condition is periodic in x, and that implies that I'll want oscillating functions there, not exponentials.

$$\begin{aligned} f(x) &= e^{i\alpha x} \qquad \text{and} \qquad f(x+2L) = f(x) \\ \Longrightarrow e^{2Li\alpha} &= 1, \qquad \text{or} \qquad 2L\alpha = 2n\pi,\ n = 0, \pm 1, \pm 2, \ldots \end{aligned}$$

The separated solutions are then

$$f(x)g(z) = e^{n\pi i x/L}\left(Ae^{n\pi z/L} + Be^{-n\pi z/L}\right) \tag{10.38}$$

The solution for $z > 0$ is therefore the sum

$$V(x,z) = \sum_{n=-\infty}^{\infty} e^{n\pi ix/L}\big(A_n e^{n\pi z/L} + B_n e^{-n\pi z/L}\big) \tag{10.39}$$

The coefficients A_n and B_n are to be determined by Fourier techniques. First however, look at the z-behavior. As you move away from the plane toward positive z, the potential should not increase without bound. Terms such as $e^{\pi z/L}$ however increase with z. This means that the coefficients of the terms that increase exponentially in z cannot be there.

$$A_n = 0 \text{ for } n > 0, \qquad \text{and} \qquad B_n = 0 \text{ for } n < 0$$

$$V(x,z) = A_0 + B_0 + \sum_{n=1}^{\infty} e^{n\pi ix/L} B_n e^{-n\pi z/L} + \sum_{n=-\infty}^{-1} e^{n\pi ix/L} A_n e^{n\pi z/L} \tag{10.40}$$

The combined constant $A_0 + B_0$ is really one constant; you can call it C_0 if you like. Now use the usual Fourier techniques given that you know the potential at $z = 0$.

$$V(x,0) = C_0 + \sum_{n=1}^{\infty} B_n e^{n\pi ix/L} + \sum_{n=-\infty}^{-1} A_n e^{n\pi ix/L}$$

The scalar product of $e^{m\pi ix/L}$ with this equation is

$$\big\langle e^{m\pi ix/L}, V(x,0)\big\rangle = \begin{cases} 2LC_0 & (m=0) \\ 2LB_m & (m>0) \\ 2LA_m & (m<0) \end{cases} \tag{10.41}$$

Now evaluate the integral on the left side. First, $m \neq 0$:

$$\begin{aligned}
\big\langle e^{m\pi ix/L}, V(x,0)\big\rangle &= \int_{-L}^{L} dx\, e^{-m\pi ix/L} \begin{cases} 0 & (-L < x < 0) \\ V_0 & (0 < x < L) \end{cases} \\
&= V_0 \int_0^L dx\, e^{-m\pi ix/L} = V_0 \frac{L}{-m\pi i}\, e^{-m\pi ix/L}\Big|_0^L \\
&= V_0 \frac{L}{-m\pi i}\big[(-1)^m - 1\big]
\end{aligned}$$

Then evaluate it separately for $m = 0$, and you have $\langle 1, V(x,0)\rangle = V_0 L$.

Now assemble the result. Before plunging in, look at what will happen.
The $m = 0$ term sits by itself.
For the other terms, only odd m have non-zero values.

$$\begin{aligned}
V(x,z) &= \frac{1}{2}V_0 + V_0 \sum_{m=1}^{\infty} \frac{1}{-2m\pi i}\big[(-1)^m - 1\big] e^{m\pi ix/L} e^{-m\pi z/L} \\
&\quad + V_0 \sum_{m=-\infty}^{-1} \frac{1}{-2m\pi i}\big[(-1)^m - 1\big] e^{m\pi ix/L} e^{+m\pi z/L}
\end{aligned} \tag{10.42}$$

To put this into a real form that is easier to interpret, change variables, letting $m = -n$ in the second sum and $m = n$ in the first, finally changing the sum so that it is over just the odd terms.

$$\begin{aligned} V(x,z) &= \frac{1}{2}V_0 + V_0\sum_{n=1}^{\infty}\frac{1}{-2n\pi i}\big[(-1)^n - 1\big]e^{n\pi ix/L}e^{-n\pi z/L} \\ &\quad + V_0\sum_{1}^{\infty}\frac{1}{+2n\pi i}\big[(-1)^n - 1\big]e^{-n\pi ix/L}e^{-n\pi z/L} \\ &= \frac{1}{2}V_0 + V_0\sum_{n=1}^{\infty}\big[(-1)^n - 1\big]\frac{1}{-n\pi}\sin(n\pi x/L)e^{-n\pi z/L} \\ &= \frac{1}{2}V_0 + \frac{2}{\pi}V_0\sum_{\ell=0}^{\infty}\frac{1}{2\ell+1}\sin\big((2\ell+1)\pi x/L\big)e^{-(2\ell+1)\pi z/L} \end{aligned} \tag{10.43}$$

Having done all the work to get to the answer, what can I learn from it?
What does it look like?
Are there any properties of the solution that are unexpected?
Should I have anticipated the form of the result?
Is there an easier way to get to the result?

To see what it looks like, examine some values of z, the distance above the surface. If $z = L$, the coefficient for successive terms is

$$\ell = 0:\ \frac{2}{\pi}e^{-\pi} = 0.028 \qquad\qquad \ell = 1:\ \frac{2}{3\pi}e^{-3\pi} = 1.7\times10^{-5} \tag{10.44}$$

The constant term is the average potential, and the $\ell = 0$ term adds only a modest ripple, about 5% of the constant average value. If you move up to $z = 2L$ the first factor is 0.0012 and that's a little more than 0.2% ripple. The sharp jumps from $+V_0$ to zero and back disappear rapidly. That the oscillations vanish so quickly with distance is perhaps not what you would guess until you have analyzed such a problem.

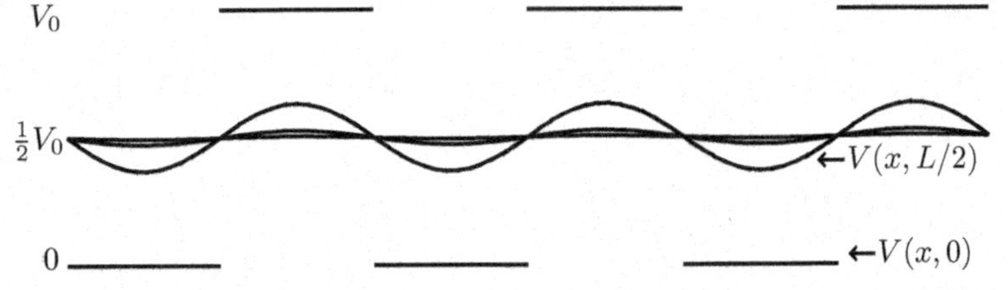

The graph shows the potential function at the surface, $z = 0$, as it oscillates between V_0 and zero. It then shows successive graphs of Eq. (10.43) at $z = L/2$, then at $z = L$, then at $z = 1.5L$. The ripple is barely visible at the that last distance. The radiation through the screen of a microwave oven is filtered in much the same way because the wavelength of the radiation is large compared to the size of the holes in the screen.

When you write the form of the series for the potential, Eq. (10.40), you can see this coming if you look for it. The oscillating terms in x are accompanied by exponential terms in z, and the rapid damping with distance is already apparent: $e^{-n\pi z/L}$. You don't have to solve for a single coefficient to see that the oscillations vanish very rapidly with distance.

The original potential on the surface was neither even nor odd, but except for the constant average value, it *is* an odd function.

$$z = 0: \quad V(x, y) = \frac{1}{2}V_0 + \begin{cases} +V_0/2 & (0 < x < L) \\ -V_0/2 & (L < x < 2L) \end{cases} \qquad V(x + 2L, y) = V(x, y) \quad (10.45)$$

Solve the potential problem for the constant $V_0/2$ and you have a constant. Solve it for the remaining odd function on the boundary and you should expect an odd function for $V(x, z)$. If you make these observations *before* solving the problem you can save yourself some algebra, as it will lead you to the form of the solution faster.

The potential is periodic on the x-y plane, so periodic boundary conditions are the appropriate ones. You can express these in more than one way, taking as a basis for the expansion either complex exponentials or sines and cosines.

$$e^{n\pi ix/L}, \; n = 0, \pm 1, \ldots$$

$$\text{or the combination} \qquad \cos(n\pi x/L), \; n = 0, 1, \ldots \qquad \sin(n\pi x/L), \; n = 1, 2, \ldots \qquad (10.46)$$

For a random problem with no special symmetry the exponential choice typically leads to easier integrals. In this case the boundary condition has some symmetry that you can take advantage of: it's almost odd. The constant term in Eq. (10.30) is the $n = 0$ element of the cosine set, and that's necessarily orthogonal to all the sines. For the rest, you do the expansion

$$\begin{cases} +V_0/2 & (0 < x < L) \\ -V_0/2 & (L < x < 2L) \end{cases} = \sum_1^\infty a_n \sin(n\pi x/L)$$

The odd term in the boundary condition (10.45) is necessarily a sum of sines, with no cosines. The cosines are orthogonal to an odd function. See problem 10.11.

More Electrostatic Examples

Specify the electric potential in the x-y plane to be an array, periodic in both the x and the y-directions. $V(x, y, z = 0)$ is V_0 on the rectangle $(0 < x < a,\ 0 < y < b)$ as well as in the darkened boxes in the picture; it is zero in the white boxes. What is the potential for $z > 0$?

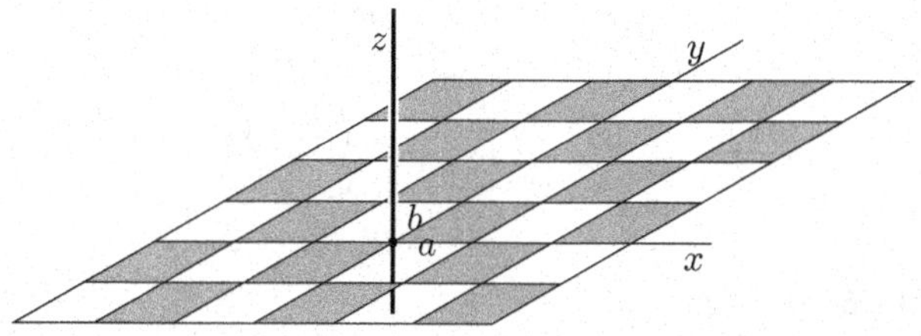

The equation is still Eq. (10.36), but now you have to do the separation of variables along all three coordinates, $V(x, y, z) = f(x)g(y)h(z)$. Substitute into the Laplace equation and divide by fgh.

$$\frac{1}{f}\frac{d^2f}{dx^2} + \frac{1}{g}\frac{d^2g}{dy^2} + \frac{1}{h}\frac{d^2h}{dz^2} = 0$$

These terms are functions of the single variables x, y, and z respectively, so the only way this can work is if they are separately constant.

$$\frac{1}{f}\frac{d^2 f}{dx^2} = -k_1^2, \qquad \frac{1}{g}\frac{d^2 g}{dy^2} = -k_2^2, \qquad \frac{1}{h}\frac{d^2 h}{dz^2} = k_1^2 + k_2^2 = k_3^2$$

I made the choice of the signs for these constants because the boundary function is periodic in x and in y, so I expect sines and cosines along those directions. The separated solution is

$$(A\sin k_1 x + B\cos k_1 x)(C\sin k_2 y + D\cos k_2 y)(Ee^{k_3 z} + Fe^{-k_3 z}) \tag{10.47}$$

What about the case for separation constants of zero? Yes, that's needed too; the average value of the potential on the surface is $V_0/2$, so just as with the example leading to Eq. (10.43) this will have a constant term of that value. The periodicity in x is $2a$ and in y it is $2b$, so this determines

$$k_1 = n\pi/a, \qquad k_2 = m\pi/b \qquad \text{then} \qquad k_3 = \sqrt{\frac{n^2\pi^2}{a^2} + \frac{m^2\pi^2}{b^2}}, \qquad n,\, m = 1,\, 2, \ldots$$

where n and m are *independent* integers. Use the experience that led to Eq. (10.45) to write V on the surface as a sum of the constant $V_0/2$ and a function that is odd in both x and in y. As there, the odd function in x will be represented by a sum of sines in x, and the same statement will hold for the y coordinate. This leads to the form of the sum

$$V(x,y,z) = \frac{1}{2}V_0 + \sum_{n=1}^{\infty}\sum_{m=1}^{\infty} \alpha_{nm} \sin\left(\frac{n\pi x}{a}\right)\sin\left(\frac{m\pi y}{b}\right)e^{-k_{nm}z}$$

where k_{nm} is the k_3 of the preceding equation. What happened to the other term in z, the one with the positive exponent? Did I say that I'm looking for solutions in the domain $z > 0$?

At $z = 0$ this must match the boundary conditions stated, and as before, the orthogonality of the sines on the two domains allows you to determine the coefficients. You simply have to do two integrals instead of one. See problem 10.19.

$$V(x,y,z>0) = \frac{1}{2}V_0 + \frac{8V_0}{\pi^2}\sum_{\text{odd } n}^{\infty}\sum_{\text{odd } m}^{\infty}\frac{1}{nm}\sin\left(\frac{n\pi x}{a}\right)\sin\left(\frac{m\pi y}{b}\right)e^{-k_{nm}z} \tag{10.48}$$

10.7 Cylindrical Coordinates

Rectangular coordinates are not always the right choice. Cylindrical, spherical, and other choices are often needed. For cylindrical coordinates, the gradient and divergence are, from Eqs. (9.24) and (9.15)

$$\nabla V = \hat{r}\frac{\partial V}{\partial r} + \hat{\phi}\frac{1}{r}\frac{\partial V}{\partial \phi} + \hat{z}\frac{\partial V}{\partial z} \qquad \text{and} \qquad \nabla\cdot\vec{v} = \frac{1}{r}\frac{\partial(rv_r)}{\partial r} + \frac{1}{r}\frac{\partial v_\phi}{\partial \phi} + \frac{\partial v_z}{\partial z}$$

Combine these to get the Laplacian in cylindrical coordinates.

$$\nabla\cdot\nabla V = \nabla^2 V = \frac{1}{r}\frac{\partial}{\partial r}\left(r\frac{\partial V}{\partial r}\right) + \frac{1}{r^2}\frac{\partial^2 V}{\partial\phi^2} + \frac{\partial^2 V}{\partial z^2} \tag{10.49}$$

For electrostatics, the equation remains $\nabla^2 V = 0$, and you can approach it the same way as before, using separation of variables. I'll start with the special case for which everything is independent of z. Assume a solution of the form $V = f(r)g(\phi)$, then

$$\nabla^2 V = g\frac{1}{r}\frac{\partial}{\partial r}\left(r\frac{\partial f}{\partial r}\right) + f\frac{1}{r^2}\frac{\partial^2 g}{\partial\phi^2} = 0$$

Multiply this by r^2 and divide by $f(r)g(\phi)$ to get

$$\frac{r}{f}\frac{\partial}{\partial r}\left(r\frac{\partial f}{\partial r}\right) + \frac{1}{g}\frac{\partial^2 g}{\partial\phi^2} = 0$$

This is separated. The first terms depends on r alone, and the second term on ϕ alone. For this to hold the terms must be constants.

$$\frac{r}{f}\frac{d}{dr}\left(r\frac{df}{dr}\right) = \alpha \qquad \text{and} \qquad \frac{1}{g}\frac{d^2 g}{d\phi^2} = -\alpha \tag{10.50}$$

The second equation, in ϕ, is familiar. If α is positive this is a harmonic oscillator, and that is the most common way this solution is applied. I'll then look at the case for $\alpha \geq 0$, for which the substitution $\alpha = n^2$ makes sense.

$$\alpha = 0: \qquad \frac{d^2 g}{d\phi^2} = 0 \implies g(\phi) = A + B\phi$$

$$r\frac{d}{dr}\left(r\frac{df}{dr}\right) = 0 \implies f(r) = C + D\ln r$$

$$\alpha = n^2 > 0 \qquad \frac{d^2 g}{d\phi^2} = -n^2 g \implies g(\phi) = A\cos n\phi + B\sin n\phi$$

$$r^2\frac{d^2 f}{dr^2} + r\frac{df}{dr} - n^2 f = 0 \implies f(r) = Cr^n + Dr^{-n}$$

There's not yet a restriction that n is an integer, though that's often the case. Verifying the last solution for f is easy.

A general solution that is the sum of all these terms is

$$V(r,\phi) = (C_0 + D_0\ln r)(A_0 + B_0\phi) + \sum_n (C_n r^n + D_n r^{-n})(A_n\cos n\phi + B_n\sin n\phi) \tag{10.51}$$

Some of these terms should be familiar:
C_0A_0 is just a constant potential.
$D_0A_0\ln r$ is the potential of a uniform line charge; $d\ln r/dr = 1/r$, and that is the way

that the electric field drops off with distance from the axis.
$C_1A_1r\cos\phi$ is the potential of a uniform field (as is the $r\sin\phi$ term). Write this in the form $C_1A_1r\cos\phi = C_1A_1x$, and the gradient of this is $C_1A_1\,\hat{x}$. The sine gives $\hat{y}$.
See problem 10.24.

Example
A conducting wire, radius R, is placed in a uniform electric field $\vec{E}_0$, and perpendicular to it. Put the wire along the z-axis and call the positive x-axis the direction that the field points. That's $\phi = 0$. In the absence of the wire, the potential for the uniform field is $V = -E_0x = -E_0r\cos\phi$, because $-\nabla V = E_0\,\hat{x}$. The total solution will be in the form of Eq. (10.51).

Now turn the general form of the solution into a particular one for this problem. The entire range of ϕ from 0 to 2π appears here; you can go all the way around the origin and come back to where you started. The potential is a function, meaning that it's single valued, and that eliminates $B_0\phi$. It also implies that all the n are integers. The applied field has a potential that is even in ϕ. That means that you should expect the solution to be even in ϕ too. Does it really? You also have to note that the physical system and its attendant boundary conditions are even in ϕ — it's a cylinder. Then too, the first few times that you do this sort of problem you should see what happens to the odd terms; what makes them vanish? I won't eliminate the $\sin\phi$ terms from the start, but I'll calculate them and show that they do vanish.

$$V(r,\phi) = (C_0 + D_0\ln r)B_0 + \sum_{n=1}^{\infty}(C_nr^n + D_nr^{-n})(A_n\cos n\phi + B_n\sin n\phi)$$

Carrying along all these products of (still unknown) factors such as D_nA_n is awkward. It makes it look neater and it is easier to follow if I combine and rename some of them.

$$V(r,\phi) = C_0 + D_0\ln r + \sum_{n=1}^{\infty}(C_nr^n + D_nr^{-n})\cos n\phi + \sum_{n=1}^{\infty}(C'_nr^n + D'_nr^{-n})\sin n\phi \quad (10.52)$$

As $r\to\infty$, the potential looks like $-E_0r\cos\phi$. That implies that $C_n = 0$ for $n > 1$, and that $C'_n = 0$ for all n, and that $C_1 = -E_0$.

Now use the boundary conditions on the cylinder. It is a conductor, so in this static case the potential is a constant all through it, in particular on the surface. I may as well take that constant to be zero, though it doesn't really matter.

$$V(R,\phi) = 0 = C_0 + D_0\ln R + \sum_{n=1}^{\infty}(C_nR^n + D_nR^{-n})\cos n\phi + \sum_{n=1}^{\infty}(C'_nR^n + D'_nR^{-n})\sin n\phi$$

Multiply by $\sin m\phi$ and integrate over ϕ. The trigonometric functions are orthogonal, so all that survives is

$$0 = (C'_mR^m + D'_mR^{-m})\pi \qquad \text{all } m \geq 1$$

That gets rid of all the rest of the sine terms as promised: $D'_m = 0$ for all m because $C'_m = 0$ for all m. Now repeat for $\cos m\phi$.

$$0 = C_0 + D_0\ln R \quad (m = 0) \qquad \text{and} \qquad 0 = (C_mR^m + D_mR^{-m})\pi \quad (m > 0)$$

All of the $C_m = 0$ for $m > 1$, so this says that the same applies to D_m. The $m = 1$ equation determines D_1 in terms of C_1 and then E_0.

$$D_1 = -C_1 R^2 = +E_0 R^2$$

Only the C_0 and D_0 terms are left, and that requires another boundary condition. When specifying the problem initially, I didn't say whether or not there is any charge on the wire. In such a case you could naturally assume that it is zero, but you have to say so explicitly because that affects the final result. Make it zero. That kills the $\ln r$ term. The reason for that goes back to the interpretation of this term. Its negative gradient is the electric field, and that would be $-1/r$, the field of a uniform line charge. If I assume there isn't one, then $D_0 = 0$ and so the same for C_0. Put this all together and

$$V(r,\phi) = -E_0 r \cos\phi + E_0 \frac{R^2}{r} \cos\phi \tag{10.53}$$

The electric field that this presents is, from Eq. (9.24)

$$\begin{aligned} \vec{E} = -\nabla V &= E_0\left(\hat{r}\cos\phi - \hat{\phi}\sin\phi\right) - E_0 R^2 \left(-\hat{r}\frac{1}{r^2}\cos\phi - \hat{\phi}\frac{1}{r^2}\sin\phi\right) \\ &= E_0\,\hat{x} + E_0 \frac{R^2}{r^2}\left(\hat{r}\cos\phi + \hat{\phi}\sin\phi\right) \end{aligned}$$

As a check to see what this looks like, what is the electric field at the surface of the cylinder?

$$\vec{E}(R,\phi) = E_0\left(\hat{r}\cos\phi - \hat{\phi}\sin\phi\right) - E_0 R^2 \left(-\hat{r}\frac{1}{R^2}\cos\phi - \hat{\phi}\frac{1}{R^2}\sin\phi\right) = 2E_0\,\hat{r}\cos\phi$$

It's perpendicular to the surface, as it should be. At the left and right, $\phi = 0, \pi$, it is twice as large as the uniform field alone would be at those points.

Problems

10.1 The specific heat of a particular type of stainless steel (CF8M) is 500 J/kg·K. Its thermal conductivity is 16.2 W/m·K and its density is 7750 kg/m^3. A slab of this steel 1.00 cm thick is at a temperature 100°C and it is placed into ice water. Assume the simplest boundary condition that its surface temperature stays at zero, and find the internal temperature at later times. When is the 2nd term in the series, Eq. (10.15), only 5% of the 1st? Sketch the temperature distribution then, indicating the scale correctly.

10.2 In Eq. (10.13) I eliminated the $n = 0$ solution by a fallacious argument. What is α in this case? This gives one more term in the sum, Eq. (10.14). Show that with the boundary conditions stated, this extra term is zero anyway (this time).

10.3 In Eq. (10.14) you have the sum of many terms. Does it still satisfy the original differential equation, Eq. (10.3)?

10.4 In the example Eq. (10.15) the final temperature was zero. What if the final temperature is T_1? Or what if I use the Kelvin scale, so that the final temperature is 273°? Add the appropriate extra term, making sure that you still have a solution to the original differential equation and that the boundary conditions are satisfied.

10.5 In the example Eq. (10.15) the final temperature was zero on both sides. What if it's zero on just the side at $x = 0$ while the side at $x = L$ stays at T_0? What is the solution now?
Ans: $T_0 x/L + (2T_0/\pi)\sum_1^\infty (1/n)\sin(n\pi x/L)e^{-n^2\pi^2 Dt/L^2}$

10.6 You have a slab of material of thickness L and at a uniform temperature T_0. The side at $x = L$ is insulated so that heat can't flow in or out of that surface. By Eq. (10.1), this tells you that $\partial T/\partial x = 0$ at that surface. Plunge the other side into ice water at temperature $T = 0$ and find the temperature inside at later time. The boundary condition on the $x = 0$ surface is the same as in the example in the text, $T(0, t) = 0$. (**a**) Separate variables and find the appropriate separated solutions for these boundary conditions. Are the separated solutions orthogonal? Use the techniques of Eq. (5.15). (**b**) When the lowest order term has dropped to where its contribution to the temperature at $x = L$ is $T_0/2$, how big is the next term in the series? Sketch the temperature distribution in the slab at that time.
Ans: $(4T_0/\pi)\sum_0^\infty (1/_{2n+1})\sin\left[(n+1/2)\pi x/L\right]e^{-(n+1/2)^2\pi^2 Dt/L^2}$, $\;-9.43\times 10^{-5}T_0$

10.7 In the analysis leading to Eq. (10.26) the temperature at $y = b$ was set to T_0. If instead, you have the temperature at $x = a$ set to T_0 with all the other sides at zero, write down the answer for the temperature within the rod. Now use the fact that Eq. (10.21) is linear to write down the solution if *both* the sides at $y = b$ and $x = a$ are set to T_0.

10.8 In leading up to Eq. (10.25) I didn't examine the third possibility for the separation constant, that it's zero. Do so.

10.9 Look at the boundary condition of Eq. (10.22) again. Another way to solve this problem is to use the solution for which the separation constant is zero, and to use it to

satisfy the conditions at $y = 0$ and $y = b$. You will then have one term in the separated solution that is $T_0 y/b$, and that means that in Eq. (10.23) you will have to choose the separation variable to be positive instead of negative. Why? Because now all the rest of the terms in the sum over separated solutions must vanish at $y = 0$ and $y = b$. You've already satisfied the boundary conditions on those surfaces by using the $T_0 y/b$ term. Now you have to satisfy the boundary conditions on $x = 0$ and $x = a$ because the total temperature there must be zero. That in turn means that the sum over all the rest of the separated terms must add to $-T_0 y/b$ at $x = 0$ and $x = a$. When you analyze this solution in the same spirit as the analysis of Eq. (10.26), compare the convergence properties of that solution to your new one. In particular, look at $a \ll b$ and $a \gg b$ to see which version converges better in each case.
Ans: $T_0 y/b + (2T_0/\pi)\sum_1^\infty \left[(-1)^n/n\right]\sin(n\pi y/b)\cosh\left(n\pi(x-a/2)/b\right)/\cosh(n\pi a/2b)$

10.10 Finish the re-analysis of the electrostatic boundary value problem Eq. (10.45) starting from Eq. (10.46). This will get the potential for $z \neq 0$ with perhaps less work than before.

10.11 Examine the solution Eq. (10.42) at $z = 0$ in the light of problem 5.11.

10.12 A thick slab of material is alternately heated and cooled at its surface so the its surface temperature oscillates as

$$T(0,t) = \begin{cases} T_1 & (0 < t < t_0) \\ -T_1 & (t_0 < t < 2t_0) \end{cases} \qquad T(0, t + 2t_0) = T(0,t)$$

That is, the period of the oscillation is $2t_0$. Find the temperature inside the material, for $x > 0$. How does this behavior differ from the solution in Eq. (10.20)?
Ans: $\frac{4T_1}{\pi}\sum_{k=0}^\infty \left(1/(2k+1)\right)e^{-\beta_k x}\sin\left((2k+1)\omega t - \beta_k x\right)$

10.13 Fill in the missing steps in finding the solution, Eq. (10.34).

10.14 A variation on the problem of the alternating potential strips in section 10.6. Place a grounded conducting sheet parallel to the x-y plane at a height $z = d$ above it. The potential there is then $V(x, y, z = d) = 0$. Solve for the potential in the gap between $z = 0$ and $z = d$. A suggestion: you may find it easier to turn the coordinate system over so that the grounded sheet is at $z = 0$ and the alternating strips are at $z = d$. This switch of coordinates is in no way essential, but it is a bit easier. Also, I want to point out that you will need to consider the case for which the separation constant in Eq. (10.37) is zero.

10.15 Starting from Eq. (10.52) and repeat the example there, but assume that the conducting wire is in an external electric field $E_0\,\hat{y}$ instead of $E_0\,\hat{x}$. Repeat the calculation for the potential and for the electric field, filling in the details of the missing steps.

10.16 A very long conducting cylindrical shell of radius R is split in two along lines parallel to its axis. The two halves are wired to a circuit that places one half at potential

V_0 and the other half at potential $-V_0$. (**a**) What is the potential everywhere inside the cylinder? Use the results of section 10.7 and assume a solution of the form

$$V(r,\phi) = \sum_0^\infty r^n \left(a_n \cos n\phi + b_n \sin n\phi\right)$$

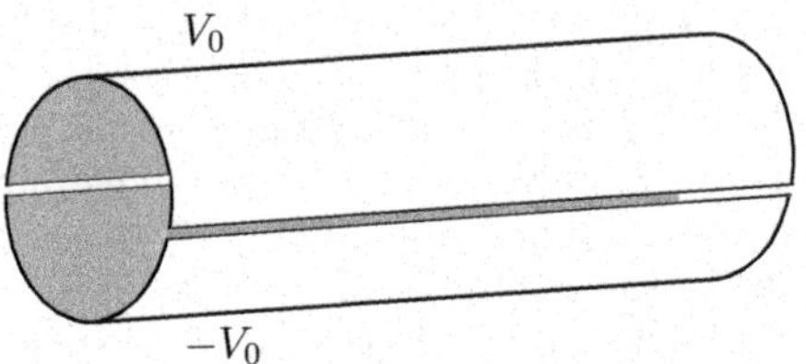

Match the boundary condition that

$$V(R,\phi) = \begin{cases} V_0 & (0 < \phi < \pi) \\ -V_0 & (\pi < \phi < 2\pi) \end{cases}$$

I picked the axis for $\phi = 0$ pointing toward the split between the cylinders. No particular reason, but you have to make a choice. I make the approximation that the cylinder is infinitely long so that z dependence doesn't enter. Also, the two halves of the cylinder almost touch so I'm neglecting the distance between them.
(**b**) What is the electric field, $-\nabla V$ on the central axis? Is this answer more or less what you would estimate before solving the problem? Ans: (**b**) $E = 4V_0/\pi R$.

10.17 Solve the preceding problem *outside* the cylinder. The integer n can be either positive or negative, and this time you'll need the negative values. (And *why* must n be an integer?)
Ans: $(4V_0/\pi)\sum_{n\text{ odd}}(1/n)(R/r)^n \sin n\phi$

10.18 In the split cylinder of problem 10.16, insert a coaxial wire of radius $R_1 < R$. It is at zero potential. Now what is the potential in the domain $R_1 < r < R$? You will need *both* the positive and negative n values, $\sum(A_n r^n + B_n r^{-n}) \sin n\phi$
Ans: $(4V_0/\pi)\sum_{m\text{ odd}} \sin m\phi[-R_1^{-m}r^m + R_1^m r^{-m}]/m[R^{-m}R_1^m - R_1^{-m}R^m]$

10.19 Fill in the missing steps in deriving Eq. (10.48).

10.20 Analyze how rapidly the solution Eq. (10.48) approaches a constant as z increases from zero. Compare Eq. (10.44).

10.21 A broad class of second order linear homogeneous differential equations can, with some manipulation, be put into the form (Sturm-Liouville)

$$(p(x)u')' + q(x)u = \lambda w(x)u$$

Assume that the functions p, q, and w are real, and use manipulations much like those that led to the identity Eq. (5.15). Derive the analogous identity for this new differential equation. When you use separation of variables on equations involving the Laplacian you will commonly come to an ordinary differential equation of exactly this form. The precise details will depend on the coordinate system you are using as well as other aspects of the PDE.

10.22 Carry on from Eq. (10.31) and deduce the separated solution that satisfies these boundary condition. Show that it is equivalent to Eq. (10.32).

10.23 The Laplacian in cylindrical coordinates is Eq. (10.49). Separate variables for the equation $\nabla^2 V = 0$ and you will see that the equations in z and ϕ are familiar. The equation in the r variable is less so, but you've seen it (almost) in Eqs. (4.18) and (4.22). Make a change of variables in the r-differential equation, $r = kr'$, and turn it into exactly the form described there.

10.24 In the preceding problem suppose that there's no z-dependence. Look at the case where the separation constant is zero for both the r and ϕ functions, finally assembling the product of the two for another solution of the whole equation.
These results provide four different solutions, a constant, a function of r alone, a function of ϕ alone, and a function of both. In each of these cases, assume that these functions are potentials V and that $\vec{E} = -\nabla V$ is the electric field from each potential. Sketch equipotentials for each case, then sketch the corresponding vector fields that they produce (a lot of arrows).

10.25 Do problem 8.23 and now solve it, finding all solutions to the wave equation.
Ans: $f(x - vt) + g(x + vt)$

10.26 Use the results of problem 10.24 to find the potential in the corner between two very large metal plates set at right angles. One at potential zero, the other at potential V_0. Compute the electric field, $-\nabla V$ and draw the results.
Ans: $-2V_0\,\hat{\phi}/\pi r$

10.27 A thin metal sheet has a straight edge for one of its boundaries. Another thin metal sheet is cut the same way. The two straight boundaries are placed in the same plane and almost, but not quite touching. Now apply a potential difference between them, putting one at a voltage V_0 and the other at $-V_0$. *In the region of space near to the almost touching boundary,* what is the electric potential? From that, compute and draw the electric field.

10.28 A slab of heat conducting material lies between coordinates $x = -L$ and $x = +L$, which are at temperatures T_1 and T_2 respectively. In the steady state ($\partial T/\partial t \equiv 0$), what is the temperature distribution inside? Now express the result in cylindrical coordinates around the z-axis and show how it matches the sum of cylindrical coordinate solutions of $\nabla^2 T = 0$ from problem 10.15. What if the surfaces of the slab had been specified at $y = -L$ and $y = +L$ instead?

10.29 The result of problem 10.16 has a series of terms that look like $(x^n/n)\sin n\phi$ (odd n). You can use complex exponentials, do a little rearranging and factoring, and sum this series. Along the way you will have to figure out what the sum $z + z^3/3 + z^5/5 + \cdots$ is. Refer to section 2.7. Finally of course, the answer is real, and if you look hard you may find a simple interpretation for the result. Be sure you've done problem 10.24 before trying this last step. Ans: $2V_0(\theta_1 + \theta_2)/\pi$. You still have to decipher what θ_1 and θ_2 are.

10.30 Sum the series Eq. (10.27) to get a closed-form analytic expression for the temperature distribution. You will find the techniques of section 5.7 useful, but there are still a lot of steps. Recall also $\ln(r\,e^{i\theta}) = \ln r + i\theta$. Ans: $(2T_0/\pi)\tan^{-1}\big[\sin(\pi x/a)/\sinh\big(\pi(b-y)/a\big)\big]$

10.31 A generalization of the problem specified in Eq. (10.22). Now the four sides have temperatures given respectively to be the constants T_1, T_2, T_3, T_4. Note: with a little bit of foresight, you won't have to work very hard at all to solve this.

10.32 Use the electrostatic equations from problem 9.21 and assume that the electric charge density is given by $\rho = \rho_0 a/r$, where this is in cylindrical coordinates. (**a**) What cylindrically symmetric electric field comes from this charge distribution? (**b**) From $\vec{E} = -\nabla V$ what potential function V do you get?

10.33 Repeat the preceding problem, but now interpret r as referring to spherical coordinates. What is $\nabla^2 V$?

10.34 The Laplacian in spherical coordinates is Eq. (9.43). The electrostatic potential equation is $\nabla^2 V = 0$ just as before, but now take the special case of azimuthal symmetry so the the potential function is independent of ϕ. Apply the method of separation of variables to find solutions of the form $f(r)g(\theta)$. You will get two ordinary differential equations for f and g. The second of these equations is much simpler if you make the change of independent variable $x = \cos\theta$. Use the chain rule a couple of times to do so, showing that the two differential equations are

$$(1-x^2)\frac{d^2 g}{dx^2} - 2x\frac{dg}{dx} + Cg = 0 \qquad \text{and} \qquad r^2\frac{d^2 f}{dr^2} + 2r\frac{df}{dr} - Cf = 0$$

10.35 For the preceding equations show that there are solutions of the form $f(r) = Ar^n$, and recall the analysis in section 4.11 for the g solutions. What values of the separation constant C will allow solutions that are finite as $x \to \pm 1$ $(\theta \to 0, \pi)$? What are the corresponding functions of r? Don't forget that there are two solutions to the second order differential equation for f — two roots to a quadratic equation.

10.36 Write out the separated solutions to the preceding problem (the ones that are are finite as θ approaches 0 or π) for the two smallest allowed values of the separation constant C: 0 and 2. In each of the four cases, interpret and sketch the potential and its corresponding electric field, $-\nabla V$. How do you sketch a potential? Draw equipotentials.

10.37 From the preceding problem you can have a potential, a solution of Laplace's equation, in the form $\big(Ar + B/r^2\big)\cos\theta$. Show that by an appropriate choice of A and B, this has an electric field that for large distances from the origin looks like $E_0\,\hat{z}$, and that on the sphere $r = R$ the total potential is zero — a grounded, conducting sphere. What does the total electric field look like for $r > R$; sketch some field lines. Start by asking what the electric field is as $r \to R$.

10.38 Similar to problem 10.16, but the potential on the cylinder is

$$V(R,\phi) = \begin{cases} V_0 & (0 < \phi < \pi/2 \text{ and } \pi < \phi < 3\pi/2) \\ -V_0 & (\pi/2 < \phi < \pi \text{ and } 3\pi/2 < \phi < 2\pi) \end{cases}$$

Draw the electric field in the region near $r = 0$.

10.39 What is the area charge density on the surface of the cylinder where the potential is given by Eq. (10.53)?

Numerical Analysis

You could say that some of the equations that you encounter in describing physical systems can't be solved in terms of familiar functions and that they require numerical calculations to solve. It would be misleading to say this however, because the reality is quite the opposite. *Most* of the equations that describe the real world are sufficiently complex that your only hope of solving them is to use numerical methods. The simple equations that you find in introductory texts are there because they *can* be solved in terms of elementary calculations. When you start to add reality, you quickly reach a point at which no amount of clever analytical ability will get you a solution. That becomes the subject of this chapter. In all of the examples that I present I'm just showing you a taste of the subject, but I hope that you will see the essential ideas of how to extend the concepts.

11.1 Interpolation

Given equally spaced tabulated data, the problem is to find a value between the tabulated points, and to estimate the error in doing so. As a first example, to find a value midway between given points use a linear interpolation:

$$f(x_0 + h/2) \approx \frac{1}{2}\big[f(x_0) + f(x_0 + h)\big]$$

This gives no hint of the error. To compute an error estimate, it is convenient to transform the variables so that this equation reads

$$f(0) \approx \frac{1}{2}\big[f(k) + f(-k)\big],$$

where the interval between data points is now $2k$. Use a power series expansion of f to find an estimate of the error.

$$f(k) = f(0) + kf'(0) + \frac{1}{2}k^2 f''(0) + \cdots$$
$$f(-k) = f(0) - kf'(0) + \frac{1}{2}k^2 f''(0) + \cdots$$

Then

$$\frac{1}{2}\big[f(k) + f(-k)\big] \approx f(0) + \big[\tfrac{1}{2}k^2 f''(0)\big], \tag{11.1}$$

where the last term is your error estimate:

$$\text{Error} = \text{Estimate} - \text{Exact} \quad = +k^2 f''(0)/2 = +h^2 f''(0)/8$$

And the relative error is (Estimate − Exact)/Exact.

As an example, interpolate the function $f(x) = 2^x$ between 0 and 1. Here $h = 1$.

$$2^{1/2} \approx \frac{1}{2}\big[2^0 + 2^1\big] = 1.5$$

The error term is

$$\text{error} \approx (\ln 2)^2 2^x/8 \qquad \text{for} \quad x = .5$$
$$= (.693)^2(1.5)/8 = .090,$$

and of course the true error is $1.5 - 1.414 = .086$

You can write a more general interpolation method for an arbitrary point between x_0 and $x_0 + h$. The solution is a simple extension of the above result.

The line passing through the two points of the graph is

$$y - f_0 = (x - x_0)(f_1 - f_0)/h,$$

$$f_0 = f(x_0), \qquad f_1 = f(x_0 + h).$$

At the point $x = x_0 + ph$ you have

$$y = f_0 + (ph)(f_1 - f_0)/h = f_0(1 - p) + f_1 p$$

As before, this approach doesn't suggest the error, but again, the Taylor series allows you to work it out to be $\left[h^2 p(1-p)f''(x_0 + ph)/2\right]$.

The use of only two points to do an interpolation ignores the data available in the rest of the table. By using more points, you can greatly improve the accuracy. The simplest example of this method is the 4-point interpolation to find the function halfway between the data points. Again, the independent variable has an increment $h = 2k$, so the problem can be stated as one of finding the value of $f(0)$ given $f(\pm k)$ and $f(\pm 3k)$.

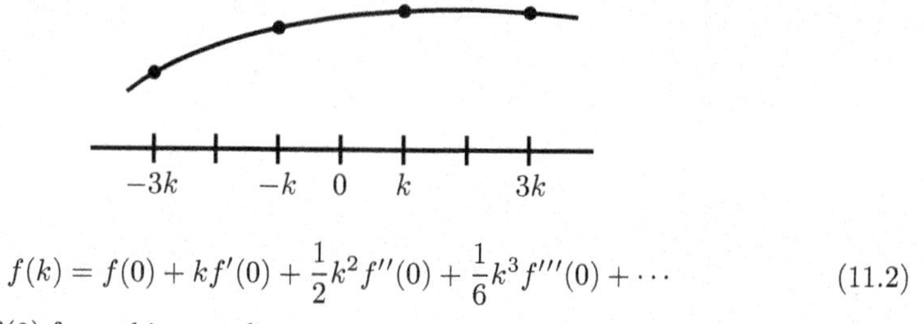

$$f(k) = f(0) + kf'(0) + \frac{1}{2}k^2 f''(0) + \frac{1}{6}k^3 f'''(0) + \cdots \tag{11.2}$$

I want to isolate $f(0)$ from this, so take

$$f(k) + f(-k) = 2f(0) + k^2 f''(0) + \frac{1}{12}k^4 f''''(0) + \cdots$$
$$f(3k) + f(-3k) = 2f(0) + 9k^2 f''(0) + \frac{81}{12}k^4 f''''(0) + \cdots$$

The biggest term after the $f(0)$ is in $k^2 f''(0)$, so I'll eliminate this.

$$\left[f(3k) + f(-3k)\right] - 9\left[f(k) - f(-k)\right] \approx -16f(0) + \left[\frac{81}{12} - \frac{9}{12}\right] k^4 f''''(0)$$

$$f(0) \approx \frac{1}{16}\big[-f(-3k) + 9f(-k) + 9f(k) - f(3k)\big] - \big[-\frac{3}{8}k^4 f''''(0)\big]. \tag{11.3}$$

The error estimate is then $-3h^4 f''''(0)/128$.

To apply this, take the same example as before, $f(x) = 2^x$ at $x = .5$

$$2^{1/2} \approx \frac{1}{16}\left[-2^{-1} + 9\cdot 2^0 + 9\cdot 2^1 - 2^2\right] = \frac{45}{32} = 1.40625,$$

and the error is $1.40625 - 1.41421 = -.008$, a tenfold improvement over the previous interpolation despite the fact that the function changes markedly in this interval and you shouldn't expect interpolation to work very well here.

11.2 Solving equations

Example: $\sin x - x/2 = 0$

From the first graph, the equation clearly has three real solutions, but finding them is the problem. The first method for solving $f(x) = 0$ is Newton's method. The basic idea is that over a small enough region, *everything* is more or less linear. This isn't true of course, so don't be surprised that this method doesn't always work. But then, *nothing* always works.

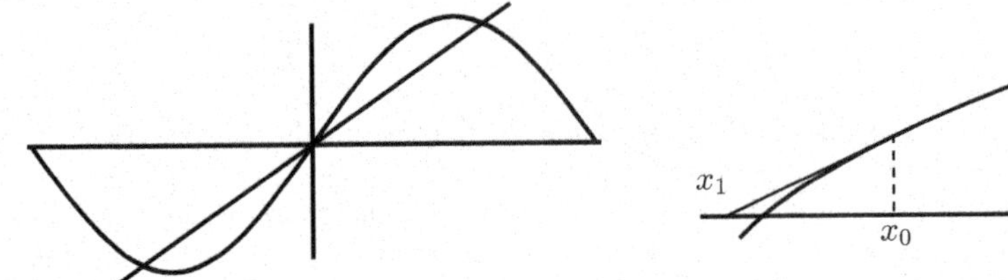

A general picture of a function with a root is the second graph. In that case, observe that if x_0 is a first approximation to the root of f, the straight line tangent to the curve can be used to calculate an improved approximation. The equation of this line is

$$y - f(x_0) = f'(x_0)(x - x_0)$$

The root of this line is defined by $y = 0$, with solution

$$x = x_0 - f(x_0)/f'(x_0)$$

Call this solution x_1. You can use this in an iterative procedure to find

$$x_2 = x_1 - f(x_1)/f'(x_1), \tag{11.4}$$

and in turn x_3 is defined in terms of x_2 etc.

Example: Solve $\sin x - x/2 = 0$. From the graph, a plausible guess for a root is $x_0 = 2$.

$$\begin{aligned}
x_1 &= x_0 - (\sin x_0 - x_0/2)/(\cos x_0 - 1/2) \\
&= 1.900995594 && f(x_1) = .00452 \\
x_2 &= x_1 - (\sin x_1 - x_1/2)/(\cos x_1 - 1/2) \\
&= 1.895511645 && f(x_2) = -.000014 \\
x_3 &= x_2 - (\sin x_2 - x_2/2)/(\cos x_2 - 1/2) \\
&= 1.895494267 && f(x_3) = 2 \times 10^{-10}
\end{aligned}$$

Such iterative procedures are ideal for use on a computer, but use them with caution, as a simple example shows:

$$f(x) = x^{1/3}.$$

Instead of the root $x = 0$, the iterations in this first graph carry the supposed solution infinitely far away. This happens here because the higher derivatives neglected in the straight line approximation are large near the root.

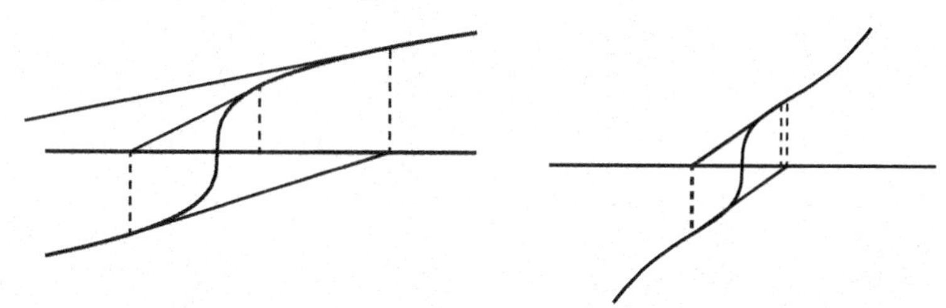

A milder form of non-convergence can occur if at the root the curvature changes sign and is large, as in the second graph. This can lead to a limit cycle where the iteration simply oscillates from one side of the root to the other without going anywhere. As I said earlier this doesn't *always* work.

A non-graphical derivation of this method starts from a Taylor series: If z_0 is an approximate root and $z_0 + \epsilon$ is a presumed exact root, then

$$f(z_0 + \epsilon) = 0 = f(z_0) + \epsilon f'(z_0) + \cdots$$

Neglecting higher terms then,

$$\epsilon = -f(z_0)/f'(z_0), \qquad \text{and} \qquad z_1 = z_0 + \epsilon = z_0 - f(z_0)/f'(z_0), \tag{11.5}$$

as before. Here z appears instead of x to remind you that this method is just as valid for complex functions as for real ones (and has as many pitfalls).

There is a simple variation on this method that can be used to speed convergence where it is poor or to bring about convergence where the technique would otherwise break down.

$$x_1 = x_0 - wf(x_0)/f'(x_0) \tag{11.6}$$

W is a factor that can be chosen greater than one to increase the correction or less than one to decrease it. Which one to do is more an art than a science (1.5 and 0.5 are common choices). You can easily verify that any choice of w between 0 and 2/3 will cause convergence for the solution of $x^{1/3} = 0$. You can also try this method on the solution of $f(x) = x^2 = 0$. A straight-forward iteration will certainly converge, but with painful slowness. The choice of $w > 1$ improves this considerably.

When Newton's method works well, it will typically double the number of significant figures at each iteration.

A drawback to this method is that it requires knowledge of $f'(x)$, and that may not be simple. An alternate approach that avoids this starts from the picture in which a secant through the curve is used in place of a tangent at a point.

Given $f(x_1)$ and $f(x_2)$, construct a straight line

$$y - f(x_2) = \left[\frac{f(x_2) - f(x_1)}{x_2 - x_1}\right](x - x_2)$$

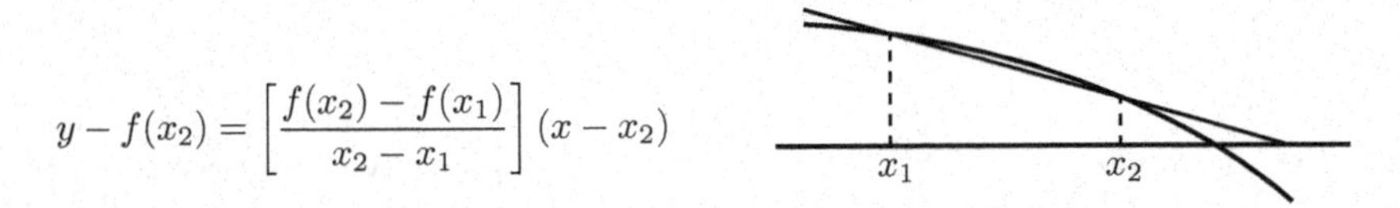

This has its root at $y = 0$, or

$$x = x_2 - f(x_2)\frac{x_2 - x_1}{f(x_2) - f(x_1)} \tag{11.7}$$

This root is taken as x_3 and the method is iterated, substituting x_2 and x_3 for x_1 and x_2. As with Newton's method, when it works, it works very well, but you must look out for the same type of non-convergence problems. This is called the secant method.

11.3 Differentiation

Given tabular or experimental data, how can you compute its derivative?

Approximating the tangent by a secant, a good estimate for the derivative of f at the midpoint of the (x_1, x_2) interval is

$$\big[f(x_2) - f(x_1)\big]/(x_2 - x_1)$$

As usual, the geometric approach doesn't indicate the size of the error, so it's back to Taylor's series.

Given data at points $x = 0,\ \pm h,\ \pm 2h,\ \ldots.$ I want the derivative $f'(0)$.

$$f(h) = f(0) + hf'(0) + \frac{1}{2}h^2 f''(0) + \frac{1}{6}h^3 f'''(0) + \cdots$$
$$f(-h) = f(0) - hf'(0) + \frac{1}{2}h^2 f''(0) - \frac{1}{6}h^3 f'''(0) + \cdots$$

In order to isolate the term in $f'(0)$, it's necessary to eliminate the larger term $f(0)$, so subtract:

$$f(h) - f(-h) = 2hf'(0) + \frac{1}{3}h^3 f'''(0) + \cdots,$$

giving

$$\frac{1}{2h}\big[f(h) - f(-h)\big] \approx f'(0) + \left[\frac{1}{6}h^2 f'''(0)\right] \tag{11.8}$$

and the last term, in brackets, estimates the error in the straight line approximation.

The most obvious point about this error term is that it varies as h^2, and so indicates by how much the error should decrease as you decrease the size of the interval. (How to estimate the factor $f'''(0)$ I'll come to presently.) This method evaluates the derivative at one of the data points; you can make it more accurate if you evaluate it between the points, so that the distance from where the derivative is being taken to where the data is available is smaller. As before, let $h = 2k$, then

$$\frac{1}{2k}\big[f(k) - f(-k)\big] = f'(0) + \frac{1}{6}k^2 f'''(0) + \cdots,$$

or, in terms of h with a shifted origin,

$$\frac{1}{h}\big[f(h) - f(0)\big] \approx f'(h/2) + \frac{1}{24}h^2 f'''\big(\frac{h}{2}\big), \tag{11.9}$$

and the error is only 1/4 as big.

As with interpolation methods, you can gain accuracy by going to higher order in the Taylor series,

$$\begin{aligned} f(h) - f(-h) &= 2hf'(0) + \frac{1}{3}h^3 f'''(0) + \frac{1}{60}h^5 f^v(0) + \cdots \\ f(2h) - f(-2h) &= 4hf'(0) + \frac{8}{3}h^3 f'''(0) + \frac{32}{60}h^5 f^v(0) + \cdots \end{aligned}$$

To eliminate the largest source of error, the h^3 term, multiply the first equation by 8 and subtract the second.

$$8\big[f(h) - f(-h)\big] - \big[f(2h) - f(-2h)\big] = 12hf'(0) - \frac{24}{60}h^5 f^v(0) + \cdots,$$

or

$$f'(0) \approx \frac{1}{12h}\big[f(-2h) - 8f(-h) + 8f(h) - f(2h)\big] - \Big[-\frac{1}{30}h^4 f^v(0)\Big] \tag{11.10}$$

with an error term of order h^4.

As an example of this method, let $f(x) = \sin x$ and evaluate the derivative at $x = 0.2$ by the 2-point formula and the 4-point formula with h=0.1:

$$\begin{aligned} \text{2-point:} &\quad \frac{1}{0.2}[0.2955202 - 0.0998334] = 0.9784340 \\ \text{4-point:} &\quad \frac{1}{1.2}[0.0 - 8 \times 0.0998334 + 8 \times 0.2955202 - 0.3894183] \\ &\qquad = 0.9800633 \\ &\qquad \cos 0.2 = 0.9800666 \end{aligned}$$

Again, you have a more accurate formula by evaluating the derivative between the data points: $h = 2k$

$$\begin{aligned} f(k) - f(-k) &= 2kf'(0) + \frac{1}{3}k^3 f'''(0) + \frac{1}{60}k^5 f^v(0) \\ f(3k) - f(-3k) &= 6kf'(0) + \frac{27}{3}k^3 f'''(0) + \frac{243}{60}k^5 f^v(0) \\ 27\big[f(k) - f(-k)\big] - \big[f(3k) - f(-3k)\big] &= 48kf'(0) - \frac{216}{60}k^5 f^v(0) \end{aligned}$$

Changing k to $h/2$ and translating the origin gives

$$\frac{1}{24h}\big[f(-h) - 27f(0) + 27f(h) - f(2h)\big] = f'(h/2) - \frac{3}{640}h^4 f^v(h/2), \tag{11.11}$$

and the coefficient of the error term is much smaller.

The previous example of the derivative of $\sin x$ at $x = 0.2$ with $h = 0.1$ gives, using this formula:

$$\frac{1}{2.4}[0.0499792 - 27 \times 0.1494381 + 27 \times 0.2474040 - 0.3428978] = 0.9800661,$$

and the error is less by a factor of about 7.

You can find higher derivatives the same way.

$$\begin{aligned} f(h) &= f(0) + hf'(0) + \frac{1}{2}h^2 f''(0) + \frac{1}{6}h^3 f'''(0) + \frac{1}{24}h^4 f''''(0) \\ f(h) + f(-h) &= 2f(0) + h^2 f''(0) + \frac{1}{12}h^4 f''''(0) + \cdots \\ f''(0) &= \frac{f(-h) - 2f(0) + f(h)}{h^2} - \frac{1}{12}h^2 f''''(0) + \cdots \end{aligned} \tag{11.12}$$

Notice that the numerical approximation for $f''(0)$ is even in h because the second derivative is unchanged if x is changed to $-x$.

You can get any of these expressions for higher derivatives recursively, though finding the error estimates requires the series method. The above expression for $f''(0)$ can be viewed as a combination of first derivative formulas:

$$\begin{aligned} f''(0) &\approx \big[f'(h/2) - f'(-h/2)\big]/h \\ &\approx \frac{1}{h}\left[\frac{f(h) - f(0)}{h} - \frac{f(0) - f(-h)}{h}\right] \\ &= \big[f(h) - 2f(0) + f(-h)\big]/h^2 \end{aligned} \tag{11.13}$$

Similarly, the third and higher derivatives can be computed. The numbers that appear in these numerical derivatives are simply the binomial coefficients, Eq. (2.18).

11.4 Integration

The basic definition of an integral is the limit of a sum as the intervals approach zero.

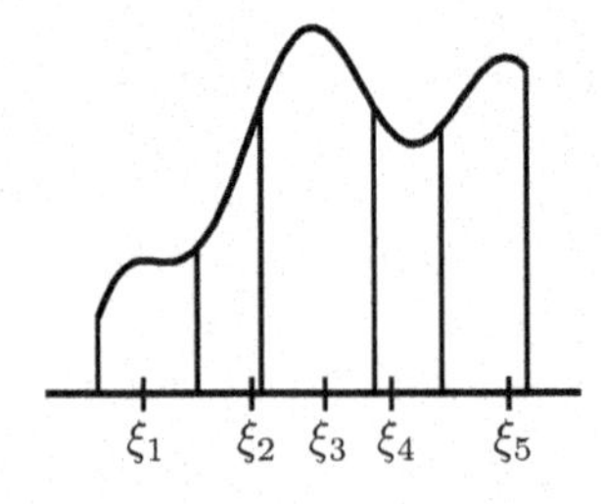

$$\sum f(\xi_i)(x_{i+1} - x_i) \qquad (x_i \le \xi_i \le x_{i+1}), \tag{11.14}$$

This is more fully explained in section 1.6, and it is the basis for the various methods of numerical evaluations of integrals.

The simplest choices to evaluate the integral of $f(x)$ over the domain x_0 to $x_0 + h$ would be to take the position of ξ at one of the endpoints or maybe in the middle (here h is small).

$$\begin{aligned} \int_{x_0}^{x_0+h} f(x)\,dx \approx & f(x_0)h && && \text{(a)} \\ \text{or} \quad & f(x_0+h)h && && \text{(b)} \qquad (11.15) \\ \text{or} \quad & f(x_0+h/2)h && \text{(midpoint rule)} && \text{(c)} \\ \text{or} \quad & \big[f(x_0)+f(x_0+h)\big]h/2 && \text{(trapezoidal rule)} && \text{(d)} \end{aligned}$$

The last expression is the average of the first two.

I can now compare the errors in all of these approximations. Set $x_0 = 0$.

$$\begin{aligned} \int_0^h dx\, f(x) &= \int_0^h dx\big[f(0) + xf'(0) + \frac{1}{2}x^2 f''(0) + \frac{1}{6}x^3 f'''(0) + \cdots\big] \\ &= hf(0) + \frac{1}{2}h^2 f'(0) + \frac{1}{6}h^3 f''(0) + \frac{1}{24}h^4 f'''(0) + \cdots \end{aligned}$$

This immediately gives the error in formula (a):

$$\text{error (a)} = hf(0) - \int_0^h dx\, f(x) \approx -\frac{1}{2}h^2 f'(0). \tag{11.16}$$

The error for expression (b) requires another expansion,

$$\begin{aligned} \text{error (b)} &= hf(h) - \int_0^h dx\, f(x) \\ &= h\big[f(0) + hf'(0) + \cdots\big] - \big[hf(0) + \frac{1}{2}h^2 f'(0) + \cdots\big] \\ &\approx \frac{1}{2}h^2 f'(0) \end{aligned} \tag{11.17}$$

Since this is the opposite sign from the previous error, it is immediately clear that the error in (d) will be less, because (d) is the average of (a) and (b).

$$\begin{aligned} \text{error (d)} = \big[f(0) + f(0) + hf'(0) + \frac{1}{2}h^2 f''(0) + \cdots\big]\frac{h}{2} \\ - \big[hf(0) + \frac{1}{2}h^2 f'(0) + \frac{1}{6}h^3 f''(0) + \cdots\big] \\ \approx \left(\frac{1}{4} - \frac{1}{6}\right) h^3 f''(0) = \frac{1}{12}h^3 f''(0) \end{aligned} \tag{11.18}$$

Similarly, the error in (c) is

$$\begin{aligned}\text{error (c)} &= h\big[f(0) + \frac{1}{2}hf'(0) + \frac{1}{8}h^2 f''(0) + \cdots\big] \\ &\qquad\qquad - \big[hf(0) + \frac{1}{2}h^2 f'(0) + \frac{1}{6}h^2 f''(0) + \cdots\big] \\ &\approx -\frac{1}{24}h^3 f''(0)\end{aligned} \tag{11.19}$$

The errors in the (c) and (d) formulas are both therefore the order of h^3.

Notice that just as the errors in formulas (a) and (b) canceled to highest order when you averaged them, the same thing can happen between formulas (c) and (d). Here however you need a weighted average, with twice as much of (c) as of (d). [1/12 2/24 = 0]

$$\frac{1}{3}(\text{d}) + \frac{2}{3}(\text{c}) = \big[f(x_0) + f(x_0 + h)\big]\frac{h}{6} + f(x_0 + h/2)\frac{4}{6}h \tag{11.20}$$

This is known as Simpson's rule.

Simpson's Rule

Before applying this last result, I'll go back and derive it in a more systematic way, putting it into the form you'll see most often.

Integrate Taylor's expansion over a symmetric domain to simplify the algebra:

$$\int_{-h}^{h} dx\, f(x) = 2hf(0) + \frac{2}{6}h^3 f''(0) + \frac{2}{120}h^5 f''''(0) + \cdots \tag{11.21}$$

I'll try to approximate this by a three point formula $\alpha f(-h) + \beta f(0) + \gamma f(h)$ where α, β, and γ, are unknown. Because of the symmetry of the problem, you can anticipate that $\alpha = \gamma$, but let that go for now and it will come out of the algebra.

$$\begin{aligned}\alpha f(-h) + \beta f(0) + \gamma f(h) = \\ &\alpha\big[f(0) - hf'(0) + \frac{1}{2}h^2 f''(0) - \frac{1}{6}h^3 f'''(0) + \frac{1}{24}h^4 f''''(0) + \cdots\big] \\ &+\beta f(0) \\ &+\gamma\big[f(0) + hf'(0) + \frac{1}{2}h^2 f''(0) + \frac{1}{6}h^3 f'''(0) + \frac{1}{24}h^4 f''''(0) + \cdots\big]\end{aligned}$$

You now determine the three constants by requiring that the two series for the same integral agree to as high an order as is possible for any f.

$$\begin{aligned} 2h &= \alpha + \beta + \gamma \\ 0 &= -\alpha h + \gamma h \\ \frac{1}{3}h^3 &= \frac{1}{2}(\alpha + \gamma)h^2 \end{aligned} \qquad \Longrightarrow \qquad \alpha = \gamma = h/3, \qquad \beta = 4h/3$$

$$\text{and so,} \qquad \int_{-h}^{h} dx\, f(x) \approx \frac{h}{3}\big[f(-h) + 4f(0) + f(h)\big]. \tag{11.22}$$

The error term (the "truncation error") is

$$\frac{h}{3}\big[f(-h)+4f(0)+f(-h)\big]-\int_{-h}^{h}dx\,f(x)\approx\frac{1}{12}\cdot\frac{1}{3}h^5f''''(0)-\frac{1}{60}h^5f''''(0)=\frac{1}{90}h^5f''''(0) \tag{11.23}$$

Simpson's rule is exact up through cubics, because the fourth and higher derivatives vanish in that case. It's worth noting that there is also an elementary derivation of Simpson's rule: Given three points, there is a unique quadratic in x that passes through all of them. Take the three points to be $\big(-h, f(-h)\big)$, $\big(0, f(0)\big)$, and $\big(h, f(h)\big)$, then integrate the resulting polynomial. Express your answer for the integral in terms of the values of f at the three points, and you get the above Simpson's rule. This has the drawback that it gives no estimate of the error.

To apply Simpson's rule, it is necessary to divide the region of integration into an even number of pieces and apply the above formula to each pair.

$$\begin{aligned}\int_a^b dx\,f(x)\approx{}&\frac{h}{3}\big[f(x_0)+4f(x_1)+f(x_2)\big]+\frac{h}{3}\big[f(x_2)+4f(x_3)+f(x_4)\big]+\cdots\\&+\frac{h}{3}\big[f(x_{N-2})+4f(x_{N-1})+f(x_N)\big]\\={}&\frac{h}{3}\big[f(x_0)+4f(x_1)+2f(x_2)+4f(x_3)+\cdots+4f(x_{N-1})+f(x_N)\big]\end{aligned} \tag{11.24}$$

Example:

$$\int_0^1\frac{4}{1+x^2}dx=4\tan^{-1}x\Big|_0^1=\pi$$

Divide the interval 0 to 1 into four pieces, then

$$\int_0^1\frac{4}{1+x^2}dx\approx\frac{4}{12}\left[1+4\frac{1}{1+(1/4)^2}+2\frac{1}{1+(1/2)^2}+4\frac{1}{1+(3/4)^2}+\frac{1}{1+1}\right]=3.1415686$$

as compared to $\pi=3.1415927\ldots$.

When the function to be integrated is smooth, this gives very accurate results.

Integration

If the integrand is known at all points of the interval and not just at discrete locations as for tabulated or experimental data, there is more freedom that you can use to gain higher accuracy even if you use only a two point formula:

$$\int_{-h}^{h}f(x)\,dx\approx\alpha\big[f(\beta)+f(-\beta)\big]$$

I could try picking two arbitrary points, not symmetrically placed in the interval, but the previous experience with Simpson's rule indicates that the result will come out as indicated. (Though it's easy to check what happens if you pick two general points in the interval.)

$$2hf(0)+\frac{1}{3}h^3f''(0)+\frac{1}{60}h^5f''''(0)+\cdots=\alpha\big[2f(0)+\beta^2f''(0)+\frac{1}{12}\beta^4f''''(0)+\cdots\big]$$

To make this an equality through the low orders implies

$$\begin{aligned} 2h &= 2\alpha & \frac{1}{3}h^3 &= \alpha\beta^2 \\ \alpha &= h & \beta &= h/\sqrt{3} \end{aligned} \tag{11.25}$$

with an error term

$$\frac{1}{12}\cdot\frac{1}{9}h^5 f''''(0) - \frac{1}{60}h^5 f''''(0) = -\frac{1}{135}h^5 f''''(0),$$

and

$$\int_{-h}^{h} f(x)\,dx \approx h\left[f\left(h/\sqrt{3}\right) + f\left(-h/\sqrt{3}\right)\right] - \left[-\frac{1}{135}h^5 f''''(0)\right] \tag{11.26}$$

With just two points, this expression yields an accuracy equal to the three point Simpson formula.

Notice that the two points found in this way are roots of a certain quadratic

$$\left(x - \frac{1}{\sqrt{3}}\right)\left(x + \frac{1}{\sqrt{3}}\right) = x^2 - 1/3,$$

which is proportional to

$$\frac{3}{2}x^2 - \frac{1}{2} = P_2(x), \tag{11.27}$$

the Legendre polynomial of second order. Recall section 4.11.

This approach to integration, called Gaussian integration, or Gaussian quadrature, can be extended to more points, as for example

$$\int_{-h}^{h} f(x)\,dx \approx \alpha f(-\beta) + \gamma f(0) + \alpha f(\beta)$$

The same expansion procedure leads to the result

$$\frac{h}{9}\left[5f\left(-h\sqrt{\frac{3}{5}}\right) + 8f(0) + f\left(h\sqrt{\frac{3}{5}}\right)\right], \tag{11.28}$$

with an error proportional to $h^7 f^{(6)}(0)$. The polynomial with roots $0, \pm\sqrt{3/5}$ is

$$\frac{5}{2}x^3 - \frac{3}{2}x = P_3(x), \tag{11.29}$$

the third order Legendre polynomial.

For an integral $\int_a^b f(x)\,dx$, let $x = [(a+b)/2] + z[(b-a)/2]$. Then for the domain $-1 < z < 1$, x covers the whole integration interval.

$$\int_a^b f(x)\,dx = \frac{b-a}{2}\int_{-1}^{1} dz\, f(x)$$

When you use an integration scheme such as Gauss's, it is in the form of a weighted sum over points. The weights and the points are defined by equations such as (11.26) or (11.28).

$$\int_{-1}^{1} dz\, f(z) \;\to\; \sum_k w_k f(z_k) \tag{11.30}$$

$$\text{or} \qquad \int_a^b f(x)\,dx = \frac{b-a}{2}\sum_k w_k f(x_k), \quad x_k = [(a+b)/2] + z_k[(b-a)/2]$$

Many other properties of Gaussian integration are discussed in the two books by C. Lanczos, "Linear Differential Operators," "Applied Analysis," both available in Dover reprints. The general expressions for the integration points as roots of Legendre polynomials and expressions for the coefficients are there. The important technical distinction he points out between the Gaussian method and generalizations of Simpson's rule involving more points is in the divergences for large numbers of points. Gauss's method does not suffer from this defect. In practice, there is rarely any problem with using the ordinary Simpson rule as indicated above, though it will require more points than the more elegant Gauss's method. When problems do arise with either of these methods, they often occur because the function is ill-behaved, and the high derivatives are very large. In this case it can be more accurate to use a method with a lower order derivative for the truncation error. In an extreme case such a integrating a non-differentiable function, the apparently worst method,Eq. (11.15)(a), can be the best.

11.5 Differential Equations

To solve the first order differential equation

$$y' = f(x,y) \qquad y(x_0) = y_0, \tag{11.31}$$

the simplest algorithm is Euler's method. The initial conditions are $y(x_0) = y_0$, and $y'(x_0) = f(x_0, y_0)$, and a straight line extrapolation is

$$y(x_0 + h) = y_0 + hf(x_0, y_0). \tag{11.32}$$

You can now iterate this procedure using this newly found value of y as a new starting condition to go from $x_0 + h$ to $x_0 + 2h$.

Runge-Kutta

Euler's method is not very accurate. For an improvement, change from a straight line extrapolation to a parabolic one. Take $x_0 = 0$ to keep the algebra down, and try a solution near 0 in the form $y(x) = \alpha + \beta x + \gamma x^2$; evaluate α, β, and γ so that the differential equation is satisfied near $x = 0$,

$$y' = \beta + 2\gamma x = f(x, \alpha + \beta x + \gamma x^2).$$

Recall the Taylor series expansion for a function of two variables, section 2.5:

$$\begin{aligned} f(x,y) = f(x_0,y_0) &+ (x-x_0)D_1 f(x_0,y_0) + (y-y_0)D_2 f(x_0,y_0) + \frac{1}{2}(x-x_0)^2 D_1 D_1 f(x_0,y_0) \\ &+ \frac{1}{2}(y-y_0)^2 D_2 D_2 f(x_0,y_0) + (x-x_0)(y-y_0)D_1 D_2 f(x_0,y_0) + \cdots \end{aligned} \tag{11.33}$$

$$\Longrightarrow \beta + 2\gamma x = f(0,\alpha) + xD_1 f(0,\alpha) + (\beta x + \gamma x^2)D_2 f(0,\alpha) + \cdots. \tag{11.34}$$

The initial condition is at $x = 0, y = y_0$, so $\alpha = y_0$. Equate coefficients of powers of x as high as is possible (here through x^1).

$$\beta = f(0,\alpha) \qquad 2\gamma = D_1 f(0,\alpha) + \beta D_2 f(0,\alpha).$$

(If you set $\gamma = 0$, this is Euler's method.)

$$y(h) = y_0 + hf(0,y_0) + \frac{h^2}{2}\big[D_1 f(0,y_0) + f(0,y_0)D_2 f(0,y_0)\big]. \tag{11.35}$$

The next problem is to evaluate these derivatives. If you can easily do them analytically, you can choose to do that. Otherwise, since they appear in a term that is multiplied by h^2, it is enough to use the simplest approximation for the numerical derivative,

$$D_1 f(0,y_0) = \big[f(h,y_0) - f(0,y_0)\big]/h \tag{11.36}$$

You cannot expect to use the same interval, h, for the y variable — it might not even have the same dimensions,

$$D_2 f(0,y_0) = \big[f(j,y_0+k) - f(j,y_0)\big]/k. \tag{11.37}$$

where j and k are the order of h. Note that because this term appears in an expression multiplied by h^2, it doesn't matter what j is. You can choose it for convenience. Possible values for these are

$$(1)\ j=0 \quad k = hf(0,y_0) \qquad (3)\ j=h \quad k = hf(0,y_0)$$
$$(2)\ j=0 \quad k = hf(h,y_0) \qquad (4)\ j=h \quad k = hf(h,y_0).$$

The third of these choices for example gives

$$\begin{aligned} y &= y_0 + hf(0,y_0) + \frac{h^2}{2}\left[\frac{1}{h}\big[f(h,y_0) - f(0,y_0)\big] + f(0,y_0)\frac{f(h,y_0+k) - f(h,y_0)}{hf(0,y_0)}\right] \\ &= y_0 + \frac{h}{2}f(0,y_0) + \frac{h}{2}f\big(h, y_0 + hf(0,y_0)\big) \end{aligned} \tag{11.38}$$

This procedure, a second order Runge-Kutta method, is a moderately accurate method for advancing from one point to the next in the solution of a differential equation. It requires evaluating the function twice for each step of the iteration.
Example: $y' = 1 + y^2 \qquad y(0) = 0.$ Let h=0.1

x	y(11.32)	y(11.38)	y(11.41)	tan x
0.	0.	0.	0.	0.
0.1	0.10	0.10050	0.10025	0.10053
0.2	0.20100	0.20304	0.20252	0.20271
0.3	0.30504	0.30981	0.30900	0.30934
0.4	0.41435	0.42341	0.42224	0.42279
0.5	0.53151	0.54702	0.54539	0.54630

(11.39)

The fractional error at $x = 0.5$ with the second order method of equation (11.38) is 0.13% and with Euler it is −2.7% (first column). The method in Eq. (11.41) has error −0.17%. The fourth order version of Eq. (11.42) has an error -3.3×10^{-7}, more places than are shown in the table.

The equation (11.38) is essentially a trapezoidal integration, like Eq. (11.15) (d). You would like to evaluate the function y at $x = h$ by

$$y(h) = y_0 + \int_0^h dx\, f\big(x, y(x)\big) \tag{11.40}$$

A trapezoidal integration would be

$$y(h) \approx y_0 + \frac{h}{2}\Big[f\big(0, y(0)\big) + f\big(h, y(h)\big)\Big]$$

BUT, you don't know the $y(h)$ that you need to evaluate the last term, so you estimate it by using the approximate result that you would have gotten from the simple Euler method, Eq. (11.32). The result is exactly equation (11.38).

If you can interpret this Runge-Kutta method as just an application of trapezoidal integration, why not use one of the other integration methods, such as the midpoint rule, Eq. (11.15) (c)? This says that you would estimate the integral in Eq. (11.40) by estimating the value of f at the midpoint $0 < x < h$.

$$y(h) \approx y_0 + hf\big(\tfrac{h}{2}, y(\tfrac{h}{2})\big) \approx y_0 + hf\Big(\frac{h}{2}, y_0 + \frac{h}{2}f\big(0, y_0\big)\Big) \tag{11.41}$$

This is the same order of accuracy as the preceding method, and it takes the same number of function evaluations as that one (two).

A commonly used version of this is the fourth order Runge-Kutta method:

$$y = y_0 + \frac{1}{6}\big[k_1 + 2k_2 + 2k_3 + k_4\big] \tag{11.42}$$

$$\begin{aligned} k_1 &= hf\big(0, y_0\big) & k_2 &= hf\big(h/2, y_0 + k_1/2\big) \\ k_3 &= hf\big(h/2, y_0 + k_2/2\big) & k_4 &= hf\big(h, y_0 + k_3\big). \end{aligned}$$

You can look up a fancier version of this called the Runge-Kutta-Fehlberg method. It's one of the better techniques around.

Higher Order Equations

How can you use either the Euler or the Runge-Kutta method to solve a second order differential equation? Answer: Turn it into a pair of first order equations.

$$y'' = f(x, y, y') \longrightarrow y' = v, \qquad \text{and} \qquad v' = f(x, y, v) \tag{11.43}$$

The Euler method, Eq. (11.32) becomes

$$y(x_0 + h) = y(x_0) + hv(x_0), \qquad \text{and} \qquad v(x_0 + h) = v(x_0) + hf\big(x_0, y(x_0), v(x_0)\big)$$

The construction for Runge-Kutta is essentially the same.

Adams Methods

The Runge-Kutta algorithm has the advantage that it is self-starting; it requires only the initial condition to go on to the next step. It has the disadvantage that it is inefficient. In going from one step to the next, it ignores all the information available from any previous steps. The opposite approach leads to the Adams methods, though these are not as commonly used any more. I'm going to develop a little of the subject mostly to show that the methods that I've used so far can lead to disaster if you're not careful.

Shift the origin to be the point at which you want the new value of y. Assume that you already know y at $-h$, $-2h$, $\ldots$, $-Nh$. Because of the differential equation $y' = f(x,y)$, you also know y' at these points.

Assume

$$y(0) = \sum_1^N \alpha_k y(-kh) + \sum_1^N \beta_k y'(-kh) \tag{11.44}$$

With $2N$ parameters, you can get this accurate to order h^{2N-1},

$$y(-kh) = \sum_0^\infty (-kh)^n \frac{y^{(n)}(0)}{n!}$$

Substitute this into the equation for $y(0)$:

$$y(0) = \sum_{k=1}^N \alpha_k \sum_{n=0}^\infty (-kh)^n \frac{y^{(n)}(0)}{n!} + h \sum_{k=1}^N \beta_k \sum_{n=0}^\infty (-kh)^n \frac{y^{(n+1)}(0)}{n!}$$

This should be an identity to as high an order as possible. The coefficient of h^0 gives

$$1 = \sum_{k=1}^N \alpha_k \tag{11.45}$$

The next orders are

$$\begin{aligned} 0 &= \sum_k \alpha_k(-kh) + h \sum_k \beta_k \\ 0 &= \sum_k \frac{1}{2}\alpha_k(-kh)^2 + h \sum_k \beta_k(-kh) \\ &\;\vdots \end{aligned} \tag{11.46}$$

$N = 1$ is Euler's method again.

$N = 2$ gives

$$\begin{aligned} \alpha_1 + \alpha_2 &= 1 & \alpha_1 + 2\alpha_2 &= \beta_1 + \beta_2 \\ \alpha_1 + 4\alpha_2 &= 2(\beta_1 + 2\beta_2) & \alpha_1 + 8\alpha_2 &= 3(\beta_1 + 4\beta_2) \end{aligned}$$

The solution of these equations is

$$\alpha_1 = -4 \qquad \alpha_2 = +5 \qquad \beta_1 = +4 \qquad \beta_2 = +2$$

$$y(0) = -4y(-h) + 5y(-2h) + h\big[4y'(-h) + 2y'(-2h)\big] \tag{11.47}$$

To start this algorithm off, you need two pieces of information: the values of y at $-h$ and at $-2h$. This is in contrast to Runge-Kutta, which needs only one point.

Example: Solve $y' = y \quad y(0) = 1 \qquad (h = 0.1)$

I could use Runge-Kutta to start and then switch to Adams as soon as possible. For the purpose of this example, I'll just take the exact value of y at $x = 0.1$.

$$\begin{aligned} e^{.1} &= 1.105170918 \\ y(.2) &= -4y(.1) + 5y(0) + .1\big[4f\big(.1, y(.1)\big) + 2f\big(0, y(0)\big)\big] \\ &= -4y(.1) + 5y(0) + .4y(.1) + .2y(0) \\ &= -3.6y(.1) + 5.2y(0) \\ &= 1.221\underline{3}84695 \end{aligned}$$

The exact value is $e^{.2} = 1.221402758$; the first error is in the underlined term. Continuing the calculation to higher values of x,

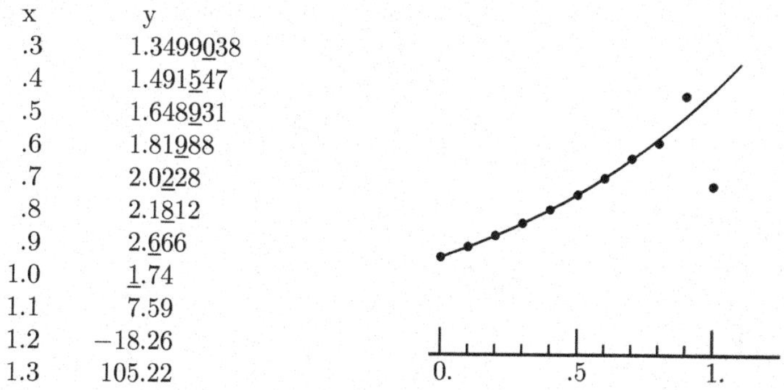

x	y
.3	$1.3499\underline{0}38$
.4	$1.491\underline{5}47$
.5	$1.648\underline{9}31$
.6	$1.81\underline{9}88$
.7	$2.0\underline{2}28$
.8	$2.1\underline{8}12$
.9	$2.\underline{6}66$
1.0	$\underline{1}.74$
1.1	7.59
1.2	−18.26
1.3	105.22

Everything is going very smoothly for a while, though the error is creeping up. At around $x = 1$, the numerical solution goes into wild oscillation and is completely unstable. The reason for this is in the coefficients -4 and $+5$ of $y(-h)$ and $y(-2h)$. Small errors are magnified by these large factors. (The coefficients of y' are not any trouble because of the factor h in front.)

Instability

You can compute the growth of this error explicitly in this simple example. The equation (11.47) together with $y' = y$ is

$$y(0) = -3.6y(-h) + 5.2y(-2h),$$

or in terms of an index notation

$$y_n = -3.6y_{n-1} + 5.2y_{n-2}$$

This is a linear, constant coefficient, difference equation, and the method for solving it is essentially the same as for a linear differential equation — assume an exponential form $y_n = k^n$.

$$k^n = -3.6k^{n-1} + 5.2k^{n-2}$$
$$k^2 + 3.6k - 5.2 = 0$$
$$k = 1.11 \qquad \text{and} \qquad -4.71$$

Just as with the differential equation, the general solution is a linear combination of these two functions of n:

$$y_n = A(1.11)^n + B(-4.71)^n,$$

where A and B are determined by two conditions, typically specifying y_1 and y_2. If $B = 0$, then y_n is proportional to 1.11^n and it is the well behaved exponential solution that you expect. If, however, there is even a little bit of B present (perhaps because of roundoff errors), that term will eventually dominate and cause the large oscillations. If B is as small as 10^{-6}, then when $n = 9$ the unwanted term is greater than 1.

When I worked out the coefficients in Eq. (11.47) the manipulations didn't look all that different from those leading to numerical derivatives or integrals, but the result was useless. This is a warning. You're in treacherous territory here; tread cautiously.

Are Adams-type methods useless? No, but you have to modify the development in order to get a stable algorithm. The difficulty in assuming the form

$$y(0) = \sum_1^N \alpha_k y(-kh) + \sum_1^N \beta_k y'(-kh)$$

is that the coefficients α_k are too large. To cure this, you can give up some of the $2N$ degrees of freedom that the method started with, and pick the α_k *a priori* to avoid instability. There are two common ways to do this, consistent with the constraint that must be kept on the α's,

$$\sum_{k=1}^N \alpha_k = 1$$

One way is to pick all the α_k to equal $1/N$. Another way is to pick $\alpha_1 = 1$ and all the others $= 0$, and both of these methods are numerically stable. The book by Lanczos in the bibliography goes into these techniques, and there are tabulations of these and other methods in Abramowitz and Stegun.

Backwards Iteration

Before leaving the subject, there is one more kind of instability that you can encounter. If you try to solve $y'' = +y$ with $y(0) = 1$ and $y'(0) = -1$, the solution is e^{-x}. If you use any stable numerical algorithm to solve this problem, it will soon deviate arbitrarily far from the desired one. The reason is that the general solution of this equation is $y = Ae^x + Be^{-x}$. Any numerical method will, through rounding errors, generate a little bit of the undesired solution, e^{+x}. Eventually, this must overwhelm the correct solution. No algorithm, no matter how stable, can get around this.

There is a clever trick that sometimes works in cases like this: backwards iteration. Instead of going from zero up, start at some large value of x and iterate downward. In this direction it is the desired solution, e^{-x}, that is unstable, and the e^{+x} is damped out. Pick an arbitrary value, say $x = 10$, and assign an arbitrary value to $y(10)$, say 0. Next, pick an arbitrary value for $y'(10)$, say 1. Use these as initial conditions (terminal conditions?) and solve the differential equation moving left; necessarily the dominant term will be the unstable one, e^{-x}, and independent of the choice of initial conditions, it will be *the* solution. At the end it is only necessary to multiply all the terms by a scale factor to reduce the value at $x = 0$ to the desired one; automatically, the value of $y'(0)$ will be correct. What you are really doing by this method is to replace the initial value problem by a two point boundary value problem. You require that the function approach zero for large x.

11.6 Fitting of Data

If you have a set of data in the form of independent and dependent variables $\{x_i, y_i\}$ $(i = 1, \ldots, N)$, and you have proposed a model that this data is to be represented by a linear combination of some set of functions, $f_\mu(x)$

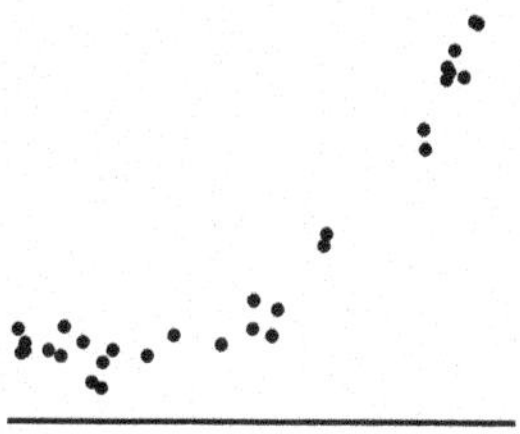

$$y = \sum_{\mu=1}^{M} \alpha_\mu f_\mu(x), \tag{11.48}$$

what values of α_μ will represent the observations in the best way? There are several answers to this question depending on the meaning of the word "best." The most commonly used one, largely because of its simplicity, is Gauss's method of least squares.

Here there are N data and there are M functions that I will use to fit the data. You have to pick the functions for yourself. You can choose them because they are the implications of a theoretical calculation; you can choose them because they are simple; you can choose them because your daily horoscope suggested them. The sum of functions, $\sum \alpha_\mu f_\mu$, now depends on only the M parameters α_μ. The fs are fixed. The difference between this sum and the data points y_i is what you want to be as small as possible. You can't use the differences themselves because they will as likely be negative as positive. The least squares method uses the sum of the squares of the differences between your sum of functions and the data. This criterion for best fit is that the sum

$$\sum_{i=1}^{N} \left[y_i - \sum_{\mu=1}^{M} \alpha_\mu f_\mu(x_i) \right]^2 = N\sigma^2 \tag{11.49}$$

be a minimum. The mean square deviation of the theory from the experiment is to be least. This quantity σ^2 is called the variance.

Some observations to make here: $N \geq M$, for otherwise there are more free parameters than data to fit them, and almost any theory with enough parameters can be forced to fit any data. Also, the functions f_μ must be linearly independent; if not, then you can throw away some and not alter the result — the solution is not unique. A further point: there is no requirement that all of the x_i are different; you may have repeated the measurements at some points.

Minimizing this is now a problem in ordinary calculus with M variables.

$$\frac{\partial}{\partial\alpha_\nu}\sum_{i=1}^{N}\left[y_i - \sum_{\mu=1}^{M}\alpha_\mu f_\mu(x_i)\right]^2 = -2\sum_i\left[y_i - \sum_\mu \alpha_\mu f_\mu(x_i)\right] f_\nu(x_i) = 0$$

$$\text{rearrange:}\qquad \sum_\mu\left[\sum_i f_\nu(x_i)f_\mu(x_i)\right]\alpha_\mu = \sum_i y_i f_\nu(x_i) \tag{11.50}$$

These linear equations are easily expressed in terms of matrices.

$$Ca = b,$$

where

$$C_{\nu\mu} = \sum_{i=1}^{N} f_\nu(x_i)f_\mu(x_i) \tag{11.51}$$

a is the column matrix with components α_μ and b has components $\sum_i y_i f_\nu(x_i)$.

The solution for a is

$$a = C^{-1}b. \tag{11.52}$$

If C turned out singular, so this inversion is impossible, the functions f_μ were not independent.

Example: Fit to a straight line

$$f_1(x) = 1 \qquad\qquad f_2(x) = x$$

Then $Ca = b$ is

$$\begin{pmatrix} N & \sum x_i \\ \sum x_i & \sum x_i^2 \end{pmatrix}\begin{pmatrix}\alpha_1\\ \alpha_2\end{pmatrix} = \begin{pmatrix}\sum y_i \\ \sum y_i x_i\end{pmatrix}$$

The inverse is

$$\begin{pmatrix}\alpha_1\\ \alpha_2\end{pmatrix} = \frac{1}{\left[N\sum x_i^2 - \left(\sum x_i\right)^2\right]}\begin{pmatrix}\sum x_i^2 & -\sum x_i \\ -\sum x_i & N\end{pmatrix}\begin{pmatrix}\sum y_i \\ \sum x_i y_i\end{pmatrix} \tag{11.53}$$

and the best fit line is

$$y = \alpha_1 + \alpha_2 x$$

11.7 Euclidean Fit

In fitting data to a combination of functions, the least squares method used Eq. (11.49) as a measure of how far the proposed function is from the data. If you're fitting data to a straight line (or plane if you have more variables) there's another way to picture the distance. Instead of measuring the distance from a point to the curve *vertically* using only y, measure it as the *perpendicular* distance to the line. Why should this be any better? It's not, but it does have different uses, and a primary one is data compression.

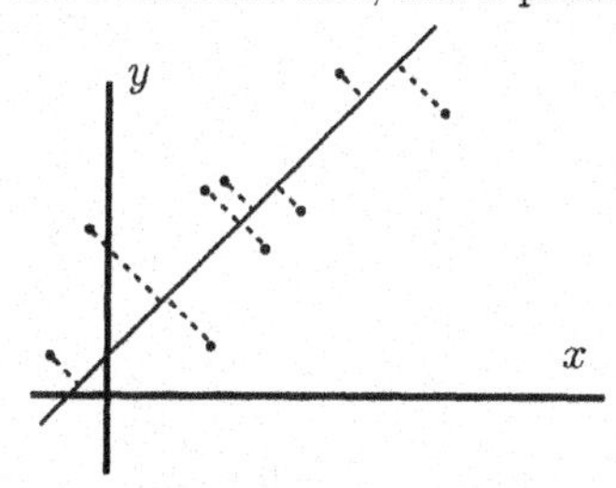

Do this in two dimensions, fitting the given data to a straight line, and to describe the line I'll use vector notation, where the line is $\vec{u} + \alpha\vec{v}$ and the parameter α varies over the reals. First I need to answer the simple question: what is the distance from a point to a line? The perpendicular distance from $\vec{w}$ to this line requires that

$$d^2 = \left(\vec{w} - \vec{u} - \alpha\vec{v}\right)^2$$

be a minimum. Differentiate this with respect to α and you have

$$\left(\vec{w} - \vec{u} - \alpha\vec{v}\right) \cdot \left(-\vec{v}\right) = 0 \qquad \text{implying} \qquad \alpha v^2 = \left(\vec{w} - \vec{u}\right) \cdot \vec{v}$$

For this value of α what is d^2?

$$\begin{aligned} d^2 &= \left(\vec{w} - \vec{u}\right)^2 + \alpha^2 v^2 - 2\alpha\vec{v} \cdot \left(\vec{w} - \vec{u}\right) \\ &= \left(\vec{w} - \vec{u}\right)^2 - \frac{1}{v^2}\left[\left(\vec{w} - \vec{u}\right) \cdot \vec{v}\right]^2 \end{aligned} \tag{11.54}$$

Is this plausible? (1) It's independent of the size of $\vec{v}$, depending on its direction only. (2) It depends on only the *difference* vector between $\vec{w}$ and $\vec{u}$, not on any other aspect of the vectors. (3) If I add any multiple of $\vec{v}$ to $\vec{u}$, the result is unchanged. See problem 11.37. Also, *can you find an easier way to get the result?* Perhaps one that simply requires some geometric insight?

The data that I'm trying to fit will be described by a set of vectors $\vec{w}_i$, and the sum of the distances squared to the line is

$$D^2 = \sum_1^N \left(\vec{w}_i - \vec{u}\right)^2 - \sum_1^N \frac{1}{v^2}\left[\left(\vec{w}_i - \vec{u}\right) \cdot \vec{v}\right]^2$$

Now to minimize this among all $\vec{u}$ and $\vec{v}$ I'll first take advantage of some of the observations from the preceding paragraph. Because the magnitude of $\vec{v}$ does not matter, I'll make it a unit vector.

$$D^2 = \sum \left(\vec{w}_i - \vec{u}\right)^2 - \sum \left[\left(\vec{w}_i - \vec{u}\right) \cdot \hat{v}\right]^2 \tag{11.55}$$

Now to figure out $\vec{u}$: Note that I expect the best fit line to go somewhere through the middle of the set of data points, so move the origin to the "center of mass" of the points.

$$\vec{w}_{\text{mean}} = \sum \vec{w}_i/N \qquad \text{and let} \qquad \vec{w}_i' = \vec{w}_i - \vec{w}_{\text{mean}} \qquad \text{and} \qquad \vec{u}\,' = \vec{u} - \vec{w}_{\text{mean}}$$

then the sum $\sum \vec{w}_i' = 0$ and

$$D^2 = \sum w_i'^2 + Nu'^2 - \sum (\vec{w}_i' \cdot \hat{v})^2 - N(\vec{u}\,' \cdot \hat{v})^2 \tag{11.56}$$

This depends on four variables, u_x', u_y', v_x and v_y. If I have to do derivatives with respect to all of them, so be it, but maybe some geometric insight will simplify the calculation. I can still add any multiple of $\hat{v}$ to $\vec{u}$ without changing this expression. That means that for a given $\vec{v}$ the derivative of D^2 as $\vec{u}\,'$ changes in *that* particular direction is zero. It's only as $\vec{u}\,'$ changes perpendicular to the direction of $\vec{v}$ that D^2 changes. The second and fourth term involve $u'^2 - (\vec{u}\,' \cdot \hat{v}\,)^2 = u'^2(1 - \cos^2\theta) = u'^2 \sin^2\theta$, where this angle θ is the angle between $\vec{u}\,'$ and $\vec{v}$. This *is* the perpendicular distance to the line (squared). Call it $u'_\perp = u' \sin\theta$.

$$D^2 = \sum w_i'^2 - \sum (\vec{w}_i' \cdot \hat{v})^2 + Nu'^2 - N(\vec{u}\,' \cdot \hat{v})^2 = \sum w_i'^2 - \sum (\vec{w}_i' \cdot \hat{v})^2 + Nu'^2_\perp$$

The minimum of this obviously occurs for $\vec{u}\,'_\perp = 0$. Also, because the component of $\vec{u}'$ along the direction of $\vec{v}$ is arbitrary, I may as well take it to be zero. That makes $\vec{u}' = 0$. Remember now that this is for the shifted $\vec{w}\,'$ data. For the original $\vec{w}_i$ data, $\vec{u}$ is shifted to $\vec{u} = \vec{w}_{\text{mean}}$.

$$D^2 = \sum w_i'^2 - \sum (\vec{w}_i' \cdot \hat{v})^2 \tag{11.57}$$

I'm not done. What is the *direction* of $\hat{v}$? That is, I have to find the minimum of D^2 subject to the constraint that $|\hat{v}| = 1$. Use Lagrange multipliers (section 8.12).

$$\text{Minimize} \quad D^2 = \sum w_i'^2 - \sum (\vec{w}_i' \cdot \vec{v})^2 \qquad \text{subject to} \qquad \phi = v_x^2 + v_y^2 - 1 = 0$$

The independent variables are v_x and v_y, and the problem becomes

$$\nabla\big(D^2 + \lambda\phi\big) = 0, \qquad \text{with} \qquad \phi = 0$$

Differentiate with respect to the independent variables and you have linear equations for v_x and v_y,

$$-\frac{\partial}{\partial v_x} \sum \big(w_{xi}' v_x + w_{yi}' v_y\big)^2 + \lambda 2 v_x = 0 \qquad \text{or} \qquad \begin{aligned} -\sum 2\big(w_{xi}' v_x + w_{yi}' v_y\big) w_{xi} + \lambda 2 v_x = 0 \\ -\sum 2\big(w_{xi}' v_x + w_{yi}' v_y\big) w_{yi} + \lambda 2 v_y = 0 \end{aligned} \tag{11.58}$$

Correlation, Principal Components

The correlation matrix of this data is

$$(C) = \frac{1}{N} \begin{pmatrix} \sum w_{xi}'^2 & \sum w_{xi} w_{yi}' \\ \sum w_{yi}' w_{xi}' & \sum w_{yi}'^2 \end{pmatrix}$$

The equations (11.58) are

$$\begin{pmatrix} C_{xx} & C_{xy} \\ C_{yx} & C_{yy} \end{pmatrix} \begin{pmatrix} v_x \\ v_y \end{pmatrix} = \lambda' \begin{pmatrix} v_x \\ v_y \end{pmatrix} \tag{11.59}$$

where $\lambda' = \lambda/N$. This is a traditional eigenvector equation, and there is a non-zero solution only if the determinant of the coefficients equals zero. Which eigenvalue to pick? There are two of them, and one will give the best fit while the other gives the *worst* fit. Just because the first derivative is zero doesn't mean you have a minimum of D^2; it could be a maximum or a saddle. Here the answer is that you pick the largest eigenvalue. You can see why this is plausible by looking at the special case for which all the data lie along the x-axis, then $C_{xx} > 0$ and all the other components of the matrix $= 0$. The eigenvalues are C_{xx} and zero, and the corresponding eigenvectors are $\hat{x}$ and $\hat{y}$ respectively. Clearly the best fit corresponds to the former, and the best fit line is the x-axis. The general form of the best fit line is (now using the original coordinate system for the data)

$$\alpha\hat{v} + \frac{1}{N}\sum \vec{w}_i = \alpha\hat{v} + \vec{w}_{\text{mean}}$$

and this $\hat{v}$ is the eigenvector having the largest eigenvalue. More generally, look at Eq. (11.57) and you see that that lone negative term is biggest if the $\vec{w}$s are in the same direction (or opposite) as $\hat{v}$.

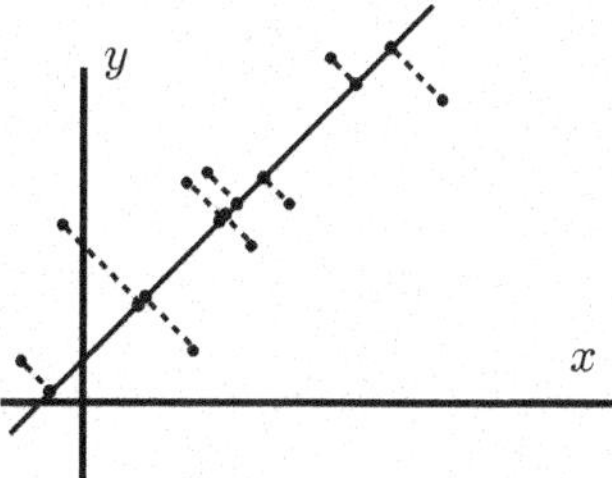

This establishes the best fit to the line in the Euclidean sense. What good is it? It leads into the subject of Principal Component Analysis and of Data Reduction. The basic idea of this scheme is that if this fit is a good one, and the original points lie fairly close to the line that I've found, I can replace the original data with the points on this line. The nine points in this figure require $9 \times 2 = 18$ coordinates to describe their positions. The nine points that approximate the data, but that lie on the line and are closest to the original points require $9 \times 1 = 9$ coordinates along this line. Of course you have some overhead in the data storage because you need to know the line. That takes three more data ($\vec{u}$ and the angle of $\hat{v}$), so the total data storage is 12 numbers. See problem 11.38

This doesn't look like much of a saving, but if you have 10^6 points you go from 2 000 000 numbers to 1 000 003 numbers, and that starts to be significant. Remember too that this is a two dimensional problem, with two numbers for each point. With more coordinates you will sometimes achieve far greater savings. You can easily establish the equation to solve for the values of α for each point, problem 11.38. The result is

$$\alpha_i = \big(\vec{w}_i - \vec{u}\big) \cdot \hat{v}$$

11.8 Differentiating noisy data

Differentiation involves dividing a small number by another small number. Any errors in the numerator will be magnified by this process, and if you have to differentiate experimental data this will always present a difficulty. If it is data from the output of a Monte Carlo calculation the same problem will arise.

Here is a method for differentiation that minimizes the sensitivity of the result to the errors in the input. Assume equally spaced data where each value of the dependent variable $f(x)$ is a random variable with mean $\langle f(x)\rangle$ and variance σ^2. Follow the procedure for differentiating smooth data and expand in a power series. Let $h = 2k$ and obtain the derivative between data points.

$$\begin{aligned}
f(k) &= f(0) + kf'(0) + \frac{1}{2}k^2 f''(0) + \frac{1}{6}k^3 f'''(0) + \cdots \\
f(k) - f(-k) &= 2kf'(0) + \frac{1}{3}k^3 f'''(0) + \cdots \\
f(3k) - f(-3k) &= 6kf'(0) + \frac{27}{3}k^3 f'''(0) + \cdots
\end{aligned}$$

I'll seek a formula of the form

$$f'(0) = \alpha\big[f(k) - f(-k)\big] + \beta\big[f(3k) - f(-3k)\big] \tag{11.60}$$

I am assuming that the variance of f at each point is the same, σ^2, and that the fluctuations in f at different points are uncorrelated. The last statement is, for random variables f_1 and f_2,

$$\big\langle\big(f_1 - \langle f_1\rangle\big)\big(f_2 - \langle f_2\rangle\big)\big\rangle = 0 \qquad \text{which expands to} \qquad \big\langle f_1 f_2\big\rangle = \big\langle f_1\big\rangle\big\langle f_2\big\rangle \tag{11.61}$$

Insert the preceding series expansions into Eq. (11.60) and match the coefficients of $f'(0)$. This gives an equation for α and β:

$$2k\alpha + 6k\beta = 1 \tag{11.62}$$

One way to obtain another equation for α and β is to require that the $k^3 f'''(0)$ term vanish; this leads back to the old formulas for differentiation, Eq. (11.11). Instead, require that the variance of $f'(0)$ be a minimum.

$$\begin{aligned}
\big\langle\big(f'(0) - \langle f'(0)\rangle\big)^2\big\rangle &= \big\langle\big[\alpha\big(f(k) - \langle f(k)\rangle\big) + \alpha\big(f(-k) - \langle f(-k)\rangle\big) + \cdots\big]^2\big\rangle \\
&= 2\sigma^2\alpha^2 + 2\sigma^2\beta^2
\end{aligned} \tag{11.63}$$

This comes from the fact that the correlation between say $f(k)$ and $f(-3k)$ vanishes, and that all the individual variances are σ^2. That is,

$$\Big\langle\big(f(k) - \langle f(k)\rangle\big)\big(f(-k) - \langle f(-k)\rangle\big)\Big\rangle = 0$$

along with all the other cross terms. Problem: minimize $2\sigma^2(\alpha^2 + \beta^2)$ subject to the constraint $2k\alpha + 6k\beta = 1$. It's hardly necessary to resort to Lagrange multipliers for this problem.

Eliminate α:

$$\frac{d}{d\beta}\left[\left(\frac{1}{2k}-3\beta\right)^2+\beta^2\right]=0 \quad\Longrightarrow\quad -6\left(\frac{1}{2k}-3\beta\right)+2\beta=0$$
$$\Longrightarrow\quad \beta=3/20k, \alpha \quad = 1/20k$$

$$f'(.5h)\approx\frac{-3f(-h)-f(0)+f(h)+3f(2h)}{10h}, \tag{11.64}$$

and the variance is $2\sigma^2(\alpha^2+\beta^2)=\sigma^2/5h^2$. In contrast, the formula for the variance in the standard four point differentiation formula Eq. (11.10), where the truncation error is least, is $65\sigma^2/72h^2$, which is 4.5 times larger.

When the data is noisy, and most data is, this expression will give much better results for this derivative. Can you do even better? Of course. You can for example go to higher order and both decrease the truncation error and minimize the statistical error. See problem 11.22.

11.9 Partial Differential Equations

I'll illustrate the ideas involved here and the difficulties that occur in even the simplest example of a PDE, a first order constant coefficient equation in one space dimension

$$\partial u/\partial t + c\,\partial u/\partial x = u_t + cu_x = 0, \tag{11.65}$$

where the subscript denotes differentiation with respect to the respective variables. This is a very simple sort of wave equation. Given the initial condition that at $t=0$, $u(0,x)=f(x)$, you can easily check that the solution is

$$u(t,x)=f(x-ct) \tag{11.66}$$

The simplest scheme to carry data forward in time from the initial values is a generalization of Euler's method for ordinary differential equations

$$\begin{aligned} u(t+\Delta t,x) &= u(t,x)+u_t(t,x)\Delta t \\ &= u(t,x)-u_x(t,x)c\Delta t \\ &= u(t,x)-\frac{c\Delta t}{2\Delta x}\big[u(t,x+\Delta x)-u(t,x-\Delta x)\big], \end{aligned} \tag{11.67}$$

Here, to evaluate the derivative, I used the two point differentiation formula.

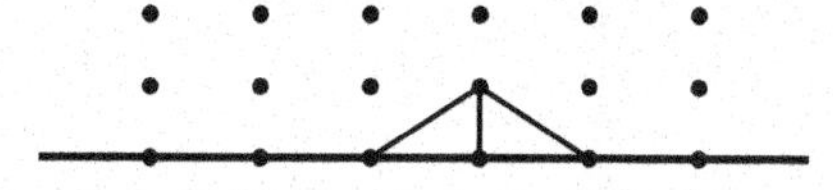

In this equation, the value of u at point $(\Delta t, 4\Delta x)$ depends on the values at $(0,3\Delta x)$, $(0,4\Delta x)$, and $(0,5\Delta x)$. This diagram shows the scheme as a picture, with the horizontal axis being x and the vertical axis t. You march the values of u at the grid points forward in time (or backward) by a set of simple equations.

The difficulties in this method are the usual errors, and more importantly, the instabilities that can occur. The errors due to the approximations involved can be classified in this case by how they manifest themselves on wavelike solutions. They can lead to dispersion or dissipation.

Analyze the dispersion first. Take as initial data $u(t,x) = A\cos kx$ (or if you prefer, e^{ikx}). The exact solution will be $A\cos(kx-\omega t)$ where $\omega = ck$. Now analyze the effect of the numerical scheme. If Δx is very small, using the discrete values of Δx in the iteration give an approximate equation

$$u_t = -\frac{c}{2\Delta x}\big[u(t,x+\Delta x) - u(t,x-\Delta x)\big]$$

A power series expansion in Δx gives, for the first two non-vanishing terms

$$u_t = -c\left[u_x + \frac{1}{6}(\Delta x)^2 u_{xxx}\right] \tag{11.68}$$

So, though I started off solving one equation, the numerical method more nearly represents quite a different equation. Try a solution of the form $A\cos(kx-\omega t)$ in this equation and you get

$$\omega = c\left[k - \frac{1}{6}(\Delta x)^2 k^3\right], \tag{11.69}$$

and you have dispersion of the wave. The velocity of the wave, ω/k, depends on k and so it depends on its wavelength or frequency.

The problem of instabilities is more conveniently analyzed by the use of an initial condition $u(0,x) = e^{ikx}$, then Eq. (11.67) is

$$\begin{aligned} u(\Delta t,x) &= e^{ikx} - \frac{c\Delta t}{2\Delta x}\left[e^{ik(x+\Delta x)} - e^{ik(x-\Delta x)}\right] \\ &= e^{ikx}\left[1 - \frac{ic\Delta t}{\Delta x}\sin k\Delta x\right]. \end{aligned} \tag{11.70}$$

The n-fold iteration of this, therefore involves just the nth power of the bracketed expression; that's why the exponential form is easier to use in this case. If $k\Delta x$ is small, the first term in the expansion of the sine says that this is approximately

$$e^{ikx}\big[1 - ikc\Delta t\big]^n,$$

and with small Δt and $n = t/\Delta t$ a large number, this is

$$e^{ikx}\left[1 - \frac{ikct}{n}\right]^n \approx e^{ik(x-ct)}$$

Looking more closely though, the object in brackets in Eq. (11.70) has magnitude

$$r = \left[1 + \frac{c^2(\Delta t)^2}{(\Delta x)^2}\sin^2 k\Delta x\right]^{1/2} > 1 \tag{11.71}$$

so the magnitude of the solution grows exponentially. This instability can be pictured as a kind of negative dissipation. This growth is reduced by requiring $kc\Delta t \ll 1$.

Given a finite fixed time interval, is it possible to get there with arbitrary accuracy by making Δt small enough? With n steps $= t/\Delta t$, r^n is

$$\begin{aligned} r &= \left[1 + \frac{c^2(\Delta t)^2}{(\Delta x)^2}\sin^2 k\Delta x\right]^{t/2\Delta t} = [1+\alpha]^\beta \\ &= \left[[1+\alpha]^{1/\alpha}\right]^{\alpha\beta} \approx e^{\alpha\beta} \\ &= \exp\left[\frac{c^2 t\Delta t}{2(\Delta x)^2}\sin^2 k\Delta x\right], \end{aligned}$$

so by shrinking Δt sufficiently, this is arbitrarily close to one.

There are several methods to avoid some of these difficulties. One is the Lax-Friedrichs method:

$$u(t+\Delta t, x) = \frac{1}{2}\big[u(t, x+\Delta x) + u(t, x-\Delta x)\big] - \frac{c\Delta t}{2\Delta x}\big[u(t, x+\Delta x) - u(t, x-\Delta x)\big] \tag{11.72}$$

By appropriate choice of Δt and Δx, this will have $r \le 1$, causing a dissipation of the wave. Another scheme is the Lax-Wendroff method.

$$\begin{aligned} u(t+\Delta t, x) = u(t, x) &- \frac{c\Delta t}{2\Delta x}\big[u(t, x+\Delta x) - u(t, x-\Delta x)\big] \\ &+ \frac{c^2(\Delta t)^2}{2(\Delta x)^2}\big[u(t, x+\Delta x) - 2u(t, x) + u(t, x-\Delta x)\big] \end{aligned} \tag{11.73}$$

This keeps one more term in the power series expansion.

Exercises

1 Use the four-point interpolation formula Eq. (11.3) to estimate $e^{3/4}$ from the values of e^x at 0, 1/2, 1, 3/2. From the known value of the number, compute the relative error.

2 Find a root of the equation $\cos x = x$. Start with a graph of course.

3 Find the values of α and of x for which $e^x = \alpha x$ has a single root.

4 Find the roots of $e^x = \alpha x$ when α is twice the value found in the preceding exercise. (and where is your graph?)

5 Use (a) midpoint formula and (b) Simpson's rule, with two intervals in each case, to evaluate $4\int_0^1 dx\, 1/(1+x^2)$.

6 Use Newton's method to solve the equation $\sin x = 0$, starting with the initial guess $x_0 = 3$.

Problems

11.1 Show that a two point extrapolation formula is

$$f(0) \approx 2f(-h) - f(-2h) + h^2 f''(0).$$

11.2 Show that a three point extrapolation formula is

$$f(0) \approx 3f(-h) - 3f(-2h) + f(-3h) + h^3 f'''(0).$$

11.3 Solve $x^2 - a = 0$ by Newton's method, showing graphically that in this case, no matter what the initial guess is (positive or negative), the sequence will always converge. Draw graphs. Find $\sqrt{2}$. (This is the basis for the library square root algorithm on some computers.)

11.4 Find all real roots of $e^{-x} = \sin x$ to $\pm 10^{-4}$. Ans: 0.588533, $\pi - 0.045166$, $2\pi + 0.00187\ldots$

11.5 The first root r_1 of $e^{-ax} = \sin x$ is a function of the variable $a > 0$. Find dr_1/da at $a = 1$ by two means. (**a**) First find r_1 for some values of a near 1 and use a four-point differentiation formula. (**b**) Second, use analytical techniques on the equation to solve for dr_1/da and evaluate the derivative in terms of the known value of the root from the previous problem.

11.6 Evaluate $\operatorname{erf}(1) = \frac{2}{\sqrt{\pi}} \int_0^1 dt\, e^{-t^2}$ Ans: 0.842736 (more exact: 0.842700792949715)

11.7 The principal value of an integral is ($a < x_0 < b$)

$$\mathsf{P}\int_a^b \frac{f(x)}{x - x_0}\,dx = \lim_{\epsilon \to 0}\left[\int_a^{x_0 - \epsilon} \frac{f(x)}{x - x_0}\,dx + \int_{x_0+\epsilon}^b \frac{f(x)}{x - x_0}\,dx\right].$$

(**a**) Show that an equal spaced integration scheme to evaluate such an integral is (using points $0, \pm h$)

$$\mathsf{P}\int_{-h}^{+h} \frac{f(x)}{x}\,dx = f(h) - f(-h) - \frac{2}{9}h^3 f'''(0).$$

(**b**) Also, an integration scheme of the Gaussian type is

$$\sqrt{3}\big[f(h/\sqrt{3}) - f(-h/\sqrt{3})\big] + \frac{h^5}{675} f^{v}(0).$$

11.8 Devise a two point Gaussian integration with errors for the class of integrals

$$\int_{-\infty}^{+\infty} dx\, e^{-x^2} f(x).$$

Find what standard polynomial has roots at the points where f is to be evaluated.
Ans: $\frac{1}{2}\sqrt{\pi}\big[f(-1/\sqrt{2})+f(1/\sqrt{2})\big]$

11.9 Same as the preceding problem, but make it a three point method.
Ans: $\sqrt{\pi}\big[\frac{1}{6}f\big(-\frac{1}{2}\sqrt{6}\big)+\frac{2}{3}f(0)+\frac{1}{6}f\big(+\frac{1}{2}\sqrt{6}\big)\big]$

11.10 Find one and two point Gauss methods for

$$\int_0^\infty dx\, e^{-x} f(x).$$

(**a**) Solve the one point method completely.
(**b**) For the two point case, the two points are roots of the equation $1-2x+\frac{1}{2}x^2=0$. Use that as given to find the weights. Look up Laguerre polynomials.

11.11 In numerical differentiation it is possible to choose the interval *too* small. Every computation is done to a finite precision. (**a**) Do the simplest numerical differentiation of some specific function and take smaller and smaller intervals. What happens when the interval gets very small? (**b**) To analyze the reason for this behavior, assume that every number in the two point differentiation formula is kept to a fixed number of significant figures (perhaps 7 or 8). How does the error vary with the interval? What interval gives the most accurate answer? Compare this theoretical answer with the experimental value found in the first part of the problem.

11.12 Just as in the preceding problem, the same phenomenon caused by roundoff errors occurs in integration. For any of the integration schemes discussed here, analyze the dependence on the number of significant figures kept and determine the most accurate interval. (Surprise?)

11.13 Compute the solution of $y'=1+y^2$ and check the numbers in the table where that example was given, Eq. (11.39).

11.14 If in the least square fit to a linear combination of functions, the result is constrained to pass through one point, so that $\sum \alpha_\mu f_\mu(x_0)=K$ is a requirement on the α's, show that the result becomes

$$a=C^{-1}\big[b+\lambda f_0\big],$$

where f_0 is the vector $f_\mu(x_0)$ and λ satisfies

$$\lambda\langle f_0, C^{-1}f_0\rangle = K-\langle f_0, C^{-1}b\rangle.$$

11.15 Find the variances in the formulas Eq. (11.8) and (11.10) for f', assuming noisy data.
Ans: $\sigma^2/2h^2,\ \ 65\sigma^2/72h^2$

11.16 Derive Eqs. (11.61), (11.62), and (11.63).

11.17 The Van der Pol equation arises in (among other places) nonlinear circuits and leads to self-exciting oscillations as in multi-vibrators

$$\frac{d^2x}{dt^2} - \epsilon(1 - x^2)\frac{dx}{dt} + x = 0.$$

Take $\epsilon = .3$ and solve subject to any non-zero initial conditions. Solve over many periods to demonstrate the development of the oscillations.

11.18 Find a formula for the numerical third derivative. Cf. Eq. (2.18)

11.19 The equation resulting from the secant method, Eq. (11.7), can be simplified by placing everything over a common denominator, $\big(f(x_2) - f(x_1)\big)$. Explain why this is a bad thing to do, how it can lead to inaccuracies.

11.20 Rederive the first Gauss integration formula Eq. (11.25) without assuming the symmetry of the result

$$\int_{-h}^{+h} f(x)\,dx \approx \alpha f(\beta) + \gamma f(\delta).$$

11.21 Derive the coefficients for the stable two-point Adams method.

11.22 By putting in one more parameter in the differentiation algorithm for noisy data, it is possible both to minimize the variance in f' *and* to eliminate the error terms in $h^2 f'''$. Find such a 6-point formula for the derivatives halfway between data points OR one for the derivatives at the data points (with errors and variance).
Ans: $f'(0) = \big[58\big(f(h) - f(-h)\big) + 67\big(f(2h) - f(-2h)\big) - 22\big(f(3h) - f(-3h)\big)\big]/(252h)$

11.23 In the same spirit as the method for differentiating noisy data, how do you *interpolate* noisy data? That is, use some extra points to stabilize the interpolation against random variations in the data. To be specific, do a midpoint interpolation for equally spaced points. Compare the variance here to that in Eq. (11.3). Ans: $f(0) \approx [f(-3k) + f(-k) + f(k) + f(3k)]/4$, σ^2 is 4.8 times smaller

11.24 Find the dispersion resulting from the use of a four point formula for u_x in the numerical solution of the PDE $u_t + cu_x = 0$.

11.25 Find the exact dispersion resulting from the equation

$$u_t = -c\big[u(t, x + \Delta x) - u(t, x - \Delta x)\big]/2\Delta x.$$

That is, don't do the series expansion on Δx.

11.26 Compute the dispersion and the dissipation in the Lax-Friedrichs and in the Lax-Wendroff methods.

11.27 In the simple iteration method of Eq. (11.71), if the grid points are denoted $x = m\Delta x$, $t = n\Delta t$, where n and m are integers ($-\infty < n, m < +\infty$), the result is a linear, constant-coefficient, partial difference equation. Solve subject to the initial condition

$$u(0, m) = e^{ikm\Delta x}.$$

11.28 Lobatto integration is like Gaussian integration, except that you require the end-points of the interval to be included in the sum. The interior points are left free. Three point Lobatto is the same as Simpson; find the four point Lobatto formula. The points found are roots of P'_{n-1}.

11.29 From the equation $y' = f(x, y)$, one derives $y'' = f_x + ff_y$. Derive a two point Adams type formula using the first and second derivatives, with error of order h^5 as for the standard four-point expression. This is useful when the analytic derivatives are easy. The form is

$$y(0) = y(-h) + \beta_1 y'(-h) + \beta_2 y'(-2h) + \gamma_1 y''(-h) + \gamma_2 y''(-2h)$$

Ans: $\beta_1 = -h/2$, $\beta_2 = 3h/2$, $\gamma_1 = 17h^2/12$, $\gamma_2 = 7h^2/12$

11.30 Using the same idea as in the previous problem, find a differential equation solver in the spirit of the original Euler method, (11.32), but doing a parabolic extrapolation instead of a linear one. That is, start from (x_0, y_0) and fit the initial data to $y = \alpha + \beta(x - x_0) + \gamma(x - x_0)^2$ in order to take a step. Ans: $y(h) = y_0 + hf(0, y_0) + (h^2/2)\left[f_x(0, y_0) + f_y(0, y_0)f(0, y_0)\right]$

11.31 Show that the root finding algorithm of Eq. (11.7) is valid for analytic functions of a complex variable with complex roots.

11.32 In the Runge-Kutta method, pick one of the other choices to estimate $D_2 f(0, y_0)$ in Eq. (11.37). How many function evaluations will it require at each step?

11.33 Sometimes you want an integral where the data is known outside the domain of integration. Find an integration scheme for $\int_0^h f(x)\,dx$ in terms of $f(h)$, $f(0)$, and $f(-h)$. Ans: $[-f(-h) + 8f(0) + 5f(h)]h/12$, error $\propto h^4$

11.34 When you must subtract two quantities that are almost the same size, you can find yourself trying to carry ridiculously many significant figures in intermediate steps. If a and b are very close and you want to evaluate $\sqrt{a} - \sqrt{b}$, devise an algorithm that does not necessitate carrying square roots out to many more places than you want in the final answer. Write $a = b + \epsilon$.
Ans: $\epsilon/2\sqrt{b}$, error: $\epsilon^2/8b^{3/2}$

11.35 Repeat the preceding problem but in a more symmetric fashion. Write $a = x + \epsilon$ and $b = x - \epsilon$. Compare the sizes of the truncation errors. Ans: $\epsilon/\sqrt{x}$, $-\epsilon^3/8x^{5/2}$

11.36 The value of π was found in the notes by integrating $4/(1 + x^2)$ from zero to one using Simpson's rule and five points. Do the same calculation using Gaussian integration and two points. Ans: 3.14754 (Three points give 3.14107)

11.37 (**a**) Derive Eq. (11.54).
(**b**) Explain why the plausibility arguments that follow it actually say something.

11.38 After you've done the Euclidean fit of data to a straight line and you want to do the data reduction described after Eq. (11.59), you have to find the coordinate along the line of the best fit to each point. This is essentially the problem: Given the line ($\vec{u}$ and $\hat{v}$) and a point ($\vec{w}$), the new reduced coordinate is the α in $\vec{u} + \alpha\hat{v}$ so that this point is closest to $\vec{w}$. What is it? You can do this the hard way, with a lot of calculus and algebra, or you can draw a picture and write the answer down.

11.39 Data is given as $(x_i, y_i) = \{(1,1), (2,2), (3,2)\}$. Compute the Euclidean best fit line. Also find the coordinates, α_i, along this line and representing the reduced data set.
Ans: $\vec{u} = (2, 5/3)$ $\hat{v} = (0.88167, 0.47186)$ $\alpha_1 = -1.1962$ $\alpha_2 = 0.1573$ $\alpha_3 = 1.0390$
The approximate points are $(0.945, 1.102)$, $(2.139, 1.741)$, $(2.916, 2.157)$
[It may not warrant this many significant figures, but it should make it easier to check your work.]

11.40 In the paragraph immediately following Eq. (11.23) there's mention of an alternate way to derive Simpson's rule. Carry it out, though you already know the answer.

11.41 (**a**) Derive the formula for the second derivative in terms of function values at three equally spaced points. (**b**) Use five points to get the second derivative, but using the extra data to minimize the sensitivity to noise in the data. Ans: (a) $\big[f(-h) - 2f(0) + f(h)\big]/h^2$
(b) $\big[-4f(0) - 5\big(f(h) + f(-h)\big) + 7\big(f(2h) + f(-2h)\big)\big]/23h^2$

Tensors

You can't walk across a room without using a tensor (the pressure tensor). You can't align the wheels on your car without using a tensor (the inertia tensor). You definitely can't understand Einstein's theory of gravity without using tensors (many of them).

This subject is often presented in the same language in which it was invented in the 1890's, expressing it in terms of transformations of coordinates and saturating it with such formidable-looking combinations as $\partial x^i/\partial \bar{x}^j$. This is really a sideshow to the subject, one that I will steer around, though a connection to this aspect appears in section 12.8.

Some of this material overlaps that of chapter 7, but I will extend it in a different direction. The first examples will then be familiar.

12.1 Examples

A tensor is a particular type of function. Before presenting the definition, some examples will clarify what I mean. Start with a rotating rigid body, and compute its angular momentum. Pick an origin and assume that the body is made up of N point masses m_i at positions described by the vectors $\vec{r}_i$ $(i = 1, 2, \ldots, N)$. The angular velocity vector is $\vec{\omega}$. For each mass the angular momentum is $\vec{r}_i \times \vec{p}_i = \vec{r}_i \times (m_i \vec{v}_i)$. The velocity $\vec{v}_i$ is given by $\vec{\omega} \times \vec{r}_i$ and so the angular momentum of the ith particle is $m_i \vec{r}_i \times (\vec{\omega} \times \vec{r}_i)$. The total angular momentum is therefore

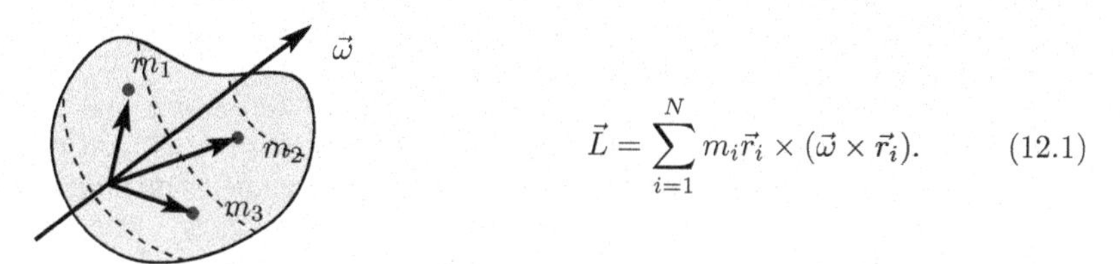

$$\vec{L} = \sum_{i=1}^{N} m_i \vec{r}_i \times (\vec{\omega} \times \vec{r}_i). \tag{12.1}$$

The angular momentum, $\vec{L}$, will depend on the distribution of mass within the body and upon the angular velocity. Write this as

$$\vec{L} = I(\vec{\omega}),$$

where the function I is called the tensor of inertia.

For a second example, take a system consisting of a mass suspended by six springs. At equilibrium the springs are perpendicular to each other. If now a (small) force $\vec{F}$ is applied to the mass it will undergo a displacement $\vec{d}$. Clearly, if $\vec{F}$ is along the direction of any of the springs then the displacement $\vec{d}$ will be in the same direction as $\vec{F}$. Suppose however that $\vec{F}$ is halfway between the k_1 and k_2 springs, and further that the spring k_2 was taken from a railroad locomotive while k_1 is a watch spring. Obviously in this case $\vec{d}$ will be mostly in the x direction (k_1) and is not aligned with $\vec{F}$. In any case there is a relation between $\vec{d}$ and $\vec{F}$,

$$\vec{d} = f(\vec{F}). \tag{12.2}$$

The function f is a tensor.

In both of these examples, the functions involved were *vector valued functions of vector variables.* They have the further property that they are linear functions, *i.e.* if α and β are real numbers,

$$I(\alpha\vec{\omega}_1 + \beta\vec{\omega}_2) = \alpha I(\vec{\omega}_1) + \beta I(\vec{\omega}_2), \qquad f(\alpha\vec{F}_1 + \beta\vec{F}_2) = \alpha f(\vec{F}_1) + \beta f(\vec{F}_2)$$

These two properties are the first *definition* of a tensor. (A generalization will come later.) There's a point here that will probably cause some confusion. Notice that in the equation $\vec{L} = I(\vec{\omega})$, the tensor is the function I. I didn't refer to "the function $I(\vec{\omega})$" as you commonly see. The reason is that $I(\vec{\omega})$, which equals $\vec{L}$, is a vector, not a tensor. It is the output of the function I after the independent variable $\vec{\omega}$ has been fed into it. For an analogy, retreat to the case of a real valued function of a real variable. In common language, you would look at the equation $y = f(x)$ and say that $f(x)$ is a function, but it's better to say that f is a function, and that $f(x)$ is the single number obtained by feeding the number x to f in order to obtain the number $f(x)$. In this language, f is regarded as containing a vast amount of information, all the relations between x and y. $f(x)$ however is just a single number. Think of f as the whole graph of the function and $f(x)$ as telling you one point on the graph. This apparently trivial distinction will often make no difference, but there are a number of cases (particularly here) where a misunderstanding of this point will cause confusion.

Definition of "Function"

This is a situation in which a very abstract definition of an idea will allow you to understand some fairly concrete applications far more easily.

> Let X and Y denote sets (possibly the same set) and x and y are elements of these sets ($x \in$ X, $y \in$ Y). Form a new set F consisting of some collection of ordered pairs of elements, one from X and one from Y. That is, a typical element of the set F is (x_1, y_1) where $x_1 \in$ X and $y_1 \in$ Y. Such a set is called a "relation" between X and Y.

If X is the set of real numbers and Y is the set of complex numbers, examples of relations are the sets

$$\begin{aligned} \mathsf{F}_1 &= \{(1.0, 7.3 - 2.1i),\ (-\pi, e + i\sqrt{2.}),\ (3.2\,\text{googol},\ 0. + 0.i),\ (1.0, e - i\pi)\} \\ \mathsf{F}_2 &= \{(x, z) \big|\ z^2 = 1 - x^2 \text{ and } -2 < x < 1\} \end{aligned}$$

There are four elements in the first of these relations and an infinite number in the second. A relation is not necessarily a function, as you need one more restriction. To define a function, you need it to be single-valued. That is the requirement that if $(x, y_1) \in$ F and $(x, y_2) \in$ F then $y_1 = y_2$. The ordinary notation for a function is $y = F(x)$, and in the language of sets we say $(x, y) \in$ F. The set F is the function. You can picture it as a graph, containing all the information about the function; it is by definition single-valued. You can check that F_1 above is a function and F_2 is not.

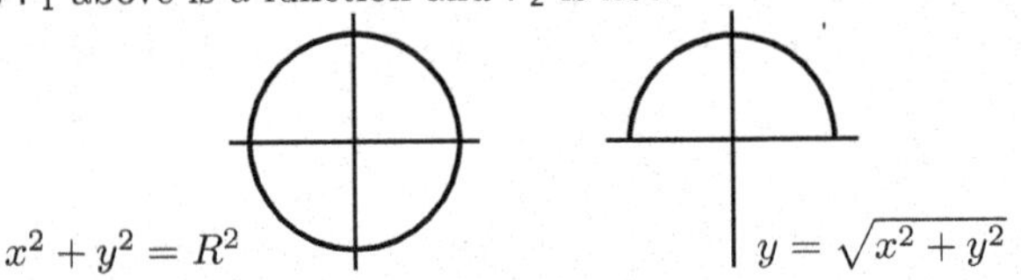

$$x^2 + y^2 = R^2 \qquad\qquad y = \sqrt{x^2 + y^2}$$

For the real numbers x and y, $x^2+y^2=R^2$ defines a *relation* between X and Y, but $y=\sqrt{R^2-x^2}$ is a *function*. In the former case for each x in the interval $-R<x<R$ you have two y's, $\pm\sqrt{R^2-x^2}$. In the latter case there is only one y for each x. The domain of a function is the set of elements x such that there is a y with $(x,y)\in \mathsf{F}$. The range is the set of y such that there is an x with $(x,y)\in \mathsf{F}$. For example, $-R\le x\le R$ is the domain of $y=\sqrt{R^2-x^2}$ and $0\le y\le R$ is its range.

Equation (12.1) defines a function I. The set X is the set of angular velocity vectors, and the set Y is the set of angular momentum vectors. For each of the former you have exactly one of the latter. The *function* is the set of all the pairs of input and output variables, so you can see why I don't want to call $I(\vec\omega\,)$ a function — it's a vector, $\vec L$.

Another physical example of a tensor is the polarizability tensor relating the electric dipole moment density vector $\vec P$ of matter to an applied electric field vector $\vec E$:

$$\vec P=\alpha(\vec E\,)$$

For the vacuum this is zero. More generally, for an isotropic linear medium, this function is nothing more than multiplication by a scalar,

$$\vec P=\alpha\vec E$$

In a crystal however the two fields $\vec P$ and $\vec E$ are not in the same direction, though the relation between them is still linear for small fields. This is analogous to the case above with a particle attached to a set of springs. The electric field polarizes the crystal more easily in some directions than in others.

The stress-strain relation in a crystal is a more complex situation that can also be described in terms of tensors. When a stress is applied, the crystal will distort slightly and this relation of strain to stress is, for small stress, a linear one. You will be able to use the notion of a tensor to describe what happens. In order to do this however it will be necessary to expand the notion of "tensor" to include a larger class of functions. This generalization will require some preliminary mathematics.

Functional

Terminology: A *functional* is a real (scalar) valued function of one or more vector variables. In particular, a linear functional is a function of one vector variable satisfying the linearity requirement.

$$f(\alpha\vec v_1+\beta\vec v_2)=\alpha f(\vec v_1)+\beta f(\vec v_2). \tag{12.3}$$

A simple example of such a functional is

$$f(\vec v\,)=\vec A\cdot\vec v, \tag{12.4}$$

where $\vec A$ is a fixed vector. In fact, because of the existence of a scalar product, all linear functionals are of this form, a result that is embodied in the following theorem, the representation theorem for linear functionals in finite dimensions.

> Let f be a linear functional: that is, f is a scalar valued function of one vector variable and is linear in that variable, $f(\vec v\,)$ is a real number and
>
> $$f(\alpha\vec v_1+\beta\vec v_2)=\alpha f(\vec v_1)+\beta f(\vec v_2)\qquad\text{then} \tag{12.5}$$
>
> there is a unique vector, $\vec A$, such that $\qquad f(\vec v\,)=\vec A\cdot\vec v\qquad$ for all $\vec v$.

Now obviously the function defined by $\vec{A}\cdot\vec{v}$, where $\vec{A}$ is a fixed vector, is a linear. The burden of this theorem is that all linear functionals are of precisely this form.

There are various approaches to proving this. The simplest is to write the vectors in components and to compute with those. There is also a more purely geometric method that avoids using components. The latter may be more satisfying, but it's harder. I'll pick the easy way.

To show that $f(\vec{v})$ can always be written as $\vec{A}\cdot\vec{v}$, I have to construct the vector $\vec{A}$. If this is to work it has to work for all vectors $\vec{v}$, and that means that it has to work for every particular vector such as $\hat{x}$, $\hat{y}$, and $\hat{z}$. It must be that

$$f(\hat{x}) = \vec{A}\cdot\hat{x} \qquad \text{and} \qquad f(\hat{y}) = \vec{A}\cdot\hat{y} \qquad \text{and} \qquad f(\hat{z}) = \vec{A}\cdot\hat{z}$$

The right side of these equations are just the three components of the vector A, so if this theorem is to be true the only way possible is that its components have the values

$$A_x = f(\hat{x}) \qquad \text{and} \qquad A_y = f(\hat{y}) \qquad \text{and} \qquad A_z = f(\hat{z}) \tag{12.6}$$

Now to find if the vectors with these components does the job.

$$f(\vec{v}) = f(v_x\hat{x} + v_y\hat{y} + v_z\hat{z}) = v_x f(\hat{x}) + v_y f(\hat{y}) + v_z f(\hat{z})$$

This is simply the property of linearity, Eq. (12.3). Now use the proposed values of the components from the preceding equation and this is exactly what's needed: $A_x v_x + A_y v_y + A_z v_z = \vec{A}\cdot\vec{v}$.

Multilinear Functionals

Functionals can be generalized to more than one variable. A bilinear functional is a scalar valued function of two vector variables, linear in each

$$\begin{aligned} T(\vec{v}_1,\, \vec{v}_2) &= \text{a scalar} \\ T(\alpha\vec{v}_1 + \beta\vec{v}_2,\, \vec{v}_3) &= \alpha T(\vec{v}_1,\, \vec{v}_3) + \beta T(\vec{v}_2,\, \vec{v}_3) \\ T(\vec{v}_1,\, \alpha\vec{v}_2 + \beta\vec{v}_3) &= \alpha T(\vec{v}_1,\, \vec{v}_2) + \beta T(\vec{v}_1,\, \vec{v}_3) \end{aligned} \tag{12.7}$$

Similarly for multilinear functionals, with as many arguments as you want.

Now apply the representation theorem for functionals to the subject of tensors. Start with a bilinear functional so that ${}^0_2T(\vec{v}_1, \vec{v}_2)$ is a scalar. This function of two variables can be looked on as a function of one variable by holding the other one temporarily fixed. Say $\vec{v}_2$ is held fixed, then ${}^0_2T(\vec{v}_1, \vec{v}_2)$ defines a linear functional on the variable $\vec{v}_1$. Apply the representation theorem now and the result is

$${}^0_2T(\vec{v}_1, \vec{v}_2) = \vec{v}_1\cdot\vec{A}$$

The vector $\vec{A}$ however will depend (linearly) on the choice of $\vec{v}_2$. It defines a new function that I'll call ${}^1_1T$

$$\vec{A} = {}^1_1T(\vec{v}_2) \tag{12.8}$$

This defines a tensor ${}^1_1T$, a linear, vector-valued function of a vector. That is, starting from a bilinear functional you can construct a linear vector-valued function. The reverse of this statement is easy to see because if you start with ${}^1_1T(\vec u\,)$ you can define a new function of two variables ${}^0_2T(\vec w,\vec u\,) = \vec w\cdot{}^1_1T(\vec u\,)$, and this is a bilinear functional, the same one you started with in fact.

With this close association between the two concepts it is natural to extend the definition of a tensor to include bilinear functionals. To be precise, I used a different name for the vector-valued function of one vector variable (${}^1_1T$) and for the scalar-valued function of two vector variables (${}^0_2T$). This is overly fussy, and it's common practice to use the same symbol (T) for both, with the hope that the context will make clear which one you actually mean. In fact it is so fussy that I will stop doing it. The *rank* of the tensor in either case is the sum of the number of vectors involved, two ($= 1 + 1 = 0 + 2$) in this case.

The next extension of the definition follows naturally from the previous reformulation. A tensor of nth rank is an n-linear functional, or any one of the several types of functions that can be constructed from it by the preceding argument. The meaning and significance of the last statement should become clear a little later. In order to clarify the meaning of this terminology, some physical examples are in order. The tensor of inertia was mentioned before:

$$\vec L = I(\vec\omega\,).$$

The dielectric tensor related $\vec D$ and $\vec E$:

$$\vec D = \varepsilon(\vec E\,)$$

The conductivity tensor relates current to the electric field:

$$\vec\jmath = \sigma(\vec E\,)$$

In general this is not just a scalar factor, and for the a.c. case σ is a function of frequency.

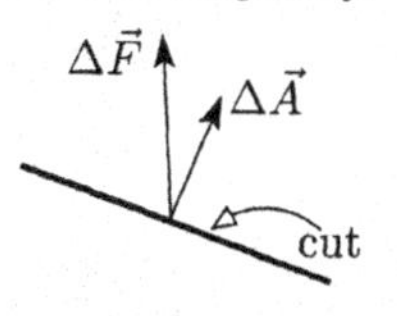

The stress tensor in matter is defined as follows: If a body has forces on it (compression or twisting or the like) or even internal defects arising from its formation, one part of the body will exert a force on another part. This can be made precise by the following device: Imagine making a cut in the material, then because of the internal forces, the two parts will tend to move with respect to each other. Apply enough force to prevent this motion. Call it $\Delta\vec F$. Typically for small cuts $\Delta\vec F$ will be proportional to the area of the cut. The area vector is perpendicular to the cut and of magnitude equal to the area. For small areas you have differential relation $d\vec F = S(d\vec A\,)$. This function S is called the stress tensor or pressure tensor. If you did problem 8.11 you saw a two dimensional special case of this, though in that case it was isotropic, leading to a scalar for the stress (also called the tension).

There is another second rank tensor called the strain tensor. I described it qualitatively in section 9.2 and I'll simply add here that it is a second rank tensor. When you apply *stress* to a solid body it will develop *strain*. This defines a function with a second rank tensor as input and a second rank tensor as output. It is the elasticity tensor and it has rank four.

So far, all the physically defined tensors except elasticity have been vector-valued functions of vector variables, and I haven't used the n-linear functional idea directly. However there is a very simple example of such a tensor:

$$\text{work} = \vec{F}\cdot\vec{d}$$

This is a scalar valued function of the two vectors $\vec{F}$ and $\vec{d}$. This is of course true for the scalar product of any two vectors $\vec{a}$ and $\vec{b}$

$$g(\vec{a},\vec{b}) = \vec{a}\cdot\vec{b} \tag{12.9}$$

g is a bilinear functional called the metric tensor. There are many other physically defined tensors that you will encounter later. In addition I would like to emphasize that although the examples given here will be in three dimensions, the formalism developed will be applicable to any number of dimensions.

12.2 Components

Up to this point, all that I've done is to make some rather general statements about tensors and I've given no techniques for computing with them. That's the next step. I'll eventually develop the complete apparatus for computation in an arbitrary basis, but for the moment it's a little simpler to start out with the more common orthonormal basis vectors, and even there I'll stay with rectangular coordinates for a while. (Recall that an orthonormal basis is an independent set of orthogonal unit vectors, such as $\hat{x}$, $\hat{y}$, $\hat{z}$.) Some of this material was developed in chapter seven, but I'll duplicate some of it. Start off by examining a second rank tensor, viewed as a vector valued function

$$\vec{u} = T(\vec{v}\,)$$

The vector $\vec{v}$ can be written in terms of the three basis vectors line $\hat{x}$, $\hat{y}$, $\hat{z}$. Or, as I shall denote them $\hat{e}_1$, $\hat{e}_2$, $\hat{e}_3$ where

$$|\hat{e}_1| = |\hat{e}_2| = |\hat{e}_3| = 1, \qquad \text{and} \qquad \hat{e}_1\cdot\hat{e}_2 = 0 \qquad \textit{etc.} \tag{12.10}$$

In terms of these independent vectors, $\vec{v}$ has components v_1, v_2, v_3:

$$\vec{v} = v_1\,\hat{e}_1 + v_2\,\hat{e}_2 + v_3\,\hat{e}_3 \tag{12.11}$$

The vector $\vec{u} = T(\vec{v}\,)$ can also be expanded in the same way:

$$\vec{u} = u_1\,\hat{e}_1 + u_2\,\hat{e}_2 + u_3\,\hat{e}_3 \tag{12.12}$$

Look at $T(\vec{v}\,)$ more closely in terms of the components

$$\begin{aligned} T(\vec{v}\,) &= T(v_1\,\hat{e}_1 + v_2\,\hat{e}_2 + v_3\,\hat{e}_3) \\ &= v_1\,T(\hat{e}_1) + v_2\,T(\hat{e}_2) + v_3\,T(\hat{e}_3) \end{aligned}$$

(by linearity). Each of the three objects $T(\hat{e}_1)$, $T(\hat{e}_2)$, $T(\hat{e}_3)$ is a vector, which means that you can expand each one in terms of the original unit vectors

$$\begin{aligned} T(\hat{e}_1) &= T_{11}\,\hat{e}_1 + T_{21}\,\hat{e}_2 + T_{31}\,\hat{e}_3 \\ T(\hat{e}_2) &= T_{12}\,\hat{e}_1 + T_{22}\,\hat{e}_2 + T_{32}\,\hat{e}_3 \\ T(\hat{e}_3) &= T_{13}\,\hat{e}_1 + T_{23}\,\hat{e}_2 + T_{33}\,\hat{e}_3 \end{aligned} \qquad \text{or more compactly,} \qquad T(\hat{e}_i) = \sum_j T_{ji}\,\hat{e}_j \tag{12.13}$$

The numbers T_{ij} ($i, j = 1, 2, 3$) are called the components of the tensor in the given basis. These numbers will depend on the basis chosen, just as do the numbers v_i, the components of the vector $\vec{v}$. The ordering of the indices has been chosen for later convenience, with the sum on the first index of the T_{ji}. This equation is *the fundamental equation* from which everything else is derived. (It will be modified when non-orthonormal bases are introduced later.)

Now, take these expressions for $T(\hat{e}_i)$ and plug them back into the equation $\vec{u} = T(\vec{v})$:

$$\begin{aligned} u_1\,\hat{e}_1 + u_2\,\hat{e}_2 + u_3\,\hat{e}_3 = T(\vec{v}\,) = \;& v_1\big[T_{11}\,\hat{e}_1 + T_{21}\,\hat{e}_2 + T_{31}\,\hat{e}_3\big] \\ &+v_2\big[T_{12}\,\hat{e}_1 + T_{22}\,\hat{e}_2 + T_{32}\,\hat{e}_3\big] \\ &+v_3\big[T_{13}\,\hat{e}_1 + T_{23}\,\hat{e}_2 + T_{33}\,\hat{e}_3\big] \\ = \;& \big[T_{11}v_1 + T_{12}v_2 + T_{13}v_3\big]\hat{e}_1 \\ &+\big[T_{21}v_1 + T_{22}v_2 + T_{23}v_3\big]\hat{e}_2 \\ &+\big[T_{31}v_1 + T_{32}v_2 + T_{33}v_3\big]\hat{e}_3 \end{aligned}$$

Comparing the coefficients of the unit vectors, you get the relations among the components

$$\begin{aligned} u_1 &= T_{11}v_1 + T_{12}v_2 + T_{13}v_3 \\ u_2 &= T_{21}v_1 + T_{22}v_2 + T_{23}v_3 \\ u_3 &= T_{31}v_1 + T_{32}v_2 + T_{33}v_3 \end{aligned} \tag{12.14}$$

More compactly:

$$u_i = \sum_{j=1}^{3} T_{ij}v_j \qquad \text{or} \qquad \begin{pmatrix} u_1 \\ u_2 \\ u_3 \end{pmatrix} = \begin{pmatrix} T_{11} & T_{12} & T_{13} \\ T_{21} & T_{22} & T_{23} \\ T_{31} & T_{32} & T_{33} \end{pmatrix} \begin{pmatrix} v_1 \\ v_2 \\ v_3 \end{pmatrix} \tag{12.15}$$

At this point it is convenient to use the summation convention (first* version). This convention says that if a given term contains a repeated index, then a summation over all the possible values of that index is understood. With this convention, the previous equation is

$$u_i = T_{ij}v_j. \tag{12.16}$$

Notice how the previous choice of indices has led to the conventional result, with the first index denoting the row and the second the column of a matrix.

---

* See section 12.5 for the later modification and generalization.

Now to take an example and tear it apart. Define a tensor by the equations

$$T(\hat{x}) = \hat{x} + \hat{y}, \qquad T(\hat{y}) = \hat{y}, \tag{12.17}$$

where $\hat{x}$ and $\hat{y}$ are given orthogonal unit vectors. These two expressions, combined with linearity, suffice to determine the effect of the tensor on all linear combinations of $\hat{x}$ and $\hat{y}$. (This is a two dimensional problem.)

To compute the components of the tensor pick a set of basis vectors. The obvious ones in this instance are

$$\hat{e}_1 = \hat{x}, \qquad \text{and} \qquad \hat{e}_2 = \hat{y}$$

By comparison with Eq. (12.13), you can read off the components of T.

$$\begin{array}{ll} T_{11} = 1 & T_{21} = 1 \\ T_{12} = 0 & T_{22} = 1. \end{array}$$

Write these in the form of a matrix as in Eq. (12.15)

$$\begin{pmatrix} T_{\text{row, column}} \end{pmatrix} = \begin{pmatrix} T_{11} & T_{12} \\ T_{21} & T_{22} \end{pmatrix} = \begin{pmatrix} 1 & 0 \\ 1 & 1 \end{pmatrix}$$

and writing the vector components in the same way, the components of the vectors $\hat{x}$ and $\hat{y}$ are respectively

$$\begin{pmatrix} 1 \\ 0 \end{pmatrix} \qquad \text{and} \qquad \begin{pmatrix} 0 \\ 1 \end{pmatrix}$$

The original equations (12.17), that defined the tensor become the components

$$\begin{pmatrix} 1 & 0 \\ 1 & 1 \end{pmatrix}\begin{pmatrix} 1 \\ 0 \end{pmatrix} = \begin{pmatrix} 1 \\ 1 \end{pmatrix} \qquad \text{and} \qquad \begin{pmatrix} 1 & 0 \\ 1 & 1 \end{pmatrix}\begin{pmatrix} 0 \\ 1 \end{pmatrix} = \begin{pmatrix} 0 \\ 1 \end{pmatrix}$$

12.3 Relations between Tensors

Go back to the fundamental representation theorem for linear functionals and see what it looks like in component form. Evaluate $f(\vec{v}\,)$, where $\vec{v} = v_i\,\hat{e}_i$. (The linear functional has one vector argument and a scalar output.)

$$f(\vec{v}\,) = f(v_i\,\hat{e}_i) = v_i\,f(\hat{e}_i) \tag{12.18}$$

Denote the set of numbers $f(\hat{e}_i)$ $(i = 1,\ 2,\ 3)$ by $A_i = f(\hat{e}_i)$, in which case,

$$f(\hat{v}\,) = A_i v_i = A_1 v_1 + A_2 v_2 + A_3 v_3$$

Now it is clear that the vector $\vec{A}$ of the theorem is just

$$\vec{A} = A_1\,\hat{e}_1 + A_2\,\hat{e}_2 + A_3\,\hat{e}_3 \tag{12.19}$$

Again, examine the problem of starting from a bilinear functional and splitting off one of the two arguments in order to obtain a vector valued function of a vector. I want to say

$$T(\vec{u},\, \vec{v}\,) = \vec{u}\cdot T(\vec{v}\,)$$

for all vectors $\vec{u}$ and $\vec{v}$. You should see that using the same symbol, T, for both functions doesn't cause any trouble. Given the bilinear functional, what is the explicit form for $T(\vec{v}\,)$? The answer is most readily found by a bit of trial and error until you reach the following result:

$$T(\vec{v}\,) = \hat{e}_i\, T(\hat{e}_i,\, \vec{v}\,) \tag{12.20}$$

(Remember the summation convention.) To verify this relation, multiply by an arbitrary vector, $\vec{u} = u_j\, \hat{e}_j$:

$$\vec{u}\cdot T(\vec{v}\,) = (u_j\, \hat{e}_j)\cdot \hat{e}_i\, T(\hat{e}_i,\, \vec{v}\,)$$

which is, by the orthonormality of the $\hat{e}$'s,

$$u_j\, \delta_{ji}\, T(\hat{e}_i,\, \vec{v}\,) = u_i\, T(\hat{e}_i,\, \vec{v}\,) = T(\vec{u},\, \vec{v}\,)$$

This says that the above expression is in fact the correct one. Notice also the similarity between this construction and the one in equation (12.19) for $\vec{A}$.

Now take $T(\vec{v}\,)$ from Eq. (12.20) and express $\vec{v}$ in terms of its components

$$\vec{v} = v_j\, \hat{e}_j, \qquad \text{then} \qquad T(\vec{v}\,) = \hat{e}_i\, T(\hat{e}_i, v_j\, \hat{e}_j) = \hat{e}_i\, T(\hat{e}_i, \hat{e}_j) v_j$$

The i component of this expression is

$$T(\hat{e}_i, \hat{e}_j) v_j = T_{ij} v_j$$

a result already obtained in Eq. (12.16).

There's a curiosity involved here; why should the left hand entry in $T(\ ,\ )$ be singled out to construct

$$\hat{e}_i\, T(\hat{e}_i, \vec{v}\,)?$$

Why not use the right hand one instead? Answer: No reason at all. It's easy enough to find out what happens when you do this. Examine

$$\hat{e}_i\, T(\vec{v},\, \hat{e}_i) \equiv \tilde{T}(\vec{v}\,) \tag{12.21}$$

Put $\vec{v} = v_j\, \hat{e}_j$, and you get

$$\hat{e}_i\, T(v_j\, \hat{e}_j,\, \hat{e}_i) = \hat{e}_i\, T(\hat{e}_j,\, \hat{e}_i) v_j$$

The ith component of which is

$$T_{ji} v_j$$

If you write this as a square matrix times a column matrix, the only difference between this result and that of Eq. (12.16) is that the matrix is transposed. This vector valued function $\tilde{T}$ is called the transpose of the tensor T. The nomenclature comes from the fact that in the matrix representation, the matrix of one equals the transpose of the other's matrix.

By an extension of the language, this applies to the other form of the tensor, T:

$$\tilde{T}(\vec{u},\, \vec{v}) = T(\vec{v},\, \vec{u}) \tag{12.22}$$

Symmetries

Two of the common and important classifications of matrices, symmetric and antisymmetric, have their reflections in tensors. A symmetric tensor is one that equals its transpose and an antisymmetric tensor is one that is the negative of its transpose. It is easiest to see the significance of this when the tensor is written in the bilinear functional form:

$$T_{ij} = T(\hat{e}_i,\, \hat{e}_j)$$

This matrix will equal its transpose if and only if

$$T(\vec{u},\, \vec{v}) = T(\vec{v},\, \vec{u})$$

for all $\vec{u}$ and $\vec{v}$. Similarly, if for all $\vec{u}$ and $\vec{v}$

$$T(\vec{u},\, \vec{v}) = -T(\vec{v},\, \vec{u})$$

then $T = -\tilde{T}$. Notice that it doesn't matter whether I speak of T as a scalar-valued function of two variables or as a vector-valued function of one; the symmetry properties are the same.

From these definitions, it is possible to take an arbitrary tensor and break it up into its symmetric part and its antisymmetric part:

$$\begin{aligned} T &= \frac{1}{2}(T+\tilde{T}) + \frac{1}{2}(T-\tilde{T}) = T_{\mathsf{S}} + T_{\mathsf{A}} \\ T_{\mathsf{S}}(\vec{u},\, \vec{v}) &= \frac{1}{2}\Big[T(\vec{u},\, \vec{v}) + T(\vec{v},\, \vec{u})\Big] \\ T_{\mathsf{A}}(\vec{u},\, \vec{v}) &= \frac{1}{2}\Big[T(\vec{u},\, \vec{v}) - T(\vec{v},\, \vec{u})\Big] \end{aligned} \tag{12.23}$$

Many of the common tensors such as the tensor of inertia and the dielectric tensor are symmetric. The magnetic field tensor in contrast, is antisymmetric. The basis of this symmetry in the case of the dielectric tensor is in the relation for the energy density in an electric field, $\int \vec{E}\cdot d\vec{D}$.* Apply an electric field in the x direction, then follow it by

---

* This can be proved by considering the energy in a plane parallel plate capacitor, which is, by definition of potential, $\int V\,dq$. The Potential difference V is the magnitude of the $\vec{E}$ field times the distance between the the capacitor plates. [$V = Ed$.] ($\vec{E}$ is perpendicular to the plates by $\nabla\times\vec{E} = 0$.) The normal component of $\vec{D}$ related to q by $\nabla\cdot\vec{D} = \rho$. [$A\vec{D}\cdot\hat{n} = q$.] Combining these, and dividing by the volume gives the energy density as $\int \vec{E}\cdot d\vec{D}$.

adding a field in the y direction; undo the field in the x direction and then undo the field in the y direction. The condition that the energy density returns to zero is the condition that the dielectric tensor is symmetric.

All of the above discussions concerning the symmetry properties of tensors were phrased in terms of second rank tensors. The extensions to tensors of higher rank are quite easy. For example in the case of a third rank tensor viewed as a 3-linear functional, it would be called completely symmetric if

$$T(\vec{u},\, \vec{v},\, \vec{w}\,) = T(\vec{v},\, \vec{u},\, \vec{w}\,) = T(\vec{u},\, \vec{w},\, \vec{v}\,) = etc.$$

for all permutations of $\vec{u}$, $\vec{v}$, $\vec{w}$, and for all values of these vectors. Similarly, if any interchange of two arguments changed the value by a sign,

$$T(\vec{u},\, \vec{v},\, \vec{w}\,) = -T(\vec{v},\, \vec{u},\, \vec{w}\,) = +T(\vec{v},\, \vec{w},\, \vec{u}\,) = etc.$$

then the T is completely antisymmetric. It is possible to have a mixed symmetry, where there is for example symmetry on interchange of the arguments in the first and second place and antisymmetry between the second and third.

Alternating Tensor

A curious (and very useful) result about antisymmetric tensors is that in three dimensions there is, up to a factor, exactly one totally antisymmetric third rank tensor; it is called the "alternating tensor." So, if you take any two such tensors, Λ and Λ', then one must be a multiple of the other. (The same holds true for the nth rank totally antisymmetric tensor in n dimensions.)

Proof: Consider the function $\Lambda - \alpha\Lambda'$ where α is a scalar. Pick any three independent vectors $\vec{v}_{10}$, $\vec{v}_{20}$, $\vec{v}_{30}$ as long as Λ' on this set is non-zero. Let

$$\alpha = \frac{\Lambda(\vec{v}_{10},\, \vec{v}_{20},\, \vec{v}_{30})}{\Lambda'(\vec{v}_{10},\, \vec{v}_{20},\, \vec{v}_{30})} \tag{12.24}$$

(If Λ' gives zero for every set of $\vec{v}$s then it's a trivial tensor, zero.) This choice of α guarantees that $\Lambda - \alpha\Lambda'$ will vanish for at least one set of values of the arguments. Now take a general set of three vectors $\vec{v}_1$, $\vec{v}_2$, and $\vec{v}_3$ and ask for the effect of $\Lambda - \alpha\Lambda'$ on them. $\vec{v}_1$, $\vec{v}_2$, and $\vec{v}_3$ can be expressed as linear combinations of the original $\vec{v}_{10}$, $\vec{v}_{20}$, and $\vec{v}_{30}$. Do so. Substitute into $\Lambda - \alpha\Lambda'$, use linearity and notice that all possible terms give zero.

The above argument is unchanged in a higher number of dimensions. It is also easy to see that you cannot have a totally antisymmetric tensor of rank $n+1$ in n dimensions. In this case, one of the $n+1$ variables would have to be a linear combination of the other n. Use linearity, and note that when any two variables equal each other, antisymmetry forces the result to be zero. These observations imply that the function must vanish identically. See also problem 12.17. If this sounds familiar, look back at section 7.7.

12.4 Birefringence

It's time to stop and present a serious calculation using tensors, one that leads to an interesting and not at all obvious result. This development assumes that you've studied Maxwell's electromagnetic field equations and are comfortable with vector calculus. If not then you will find this section obscure, maybe best left for another day. The particular problem that I will examine is how light passes through a transparent material, especially a crystal that has different electrical properties in different directions. The common and best known example of such a crystal is Iceland spar, a form of calcite ($CaCO_3$).

$$\nabla \times \vec{E} = -\frac{\partial \vec{B}}{\partial t} \qquad \nabla \times \vec{B} = \mu_0 \vec{j} + \mu_0 \epsilon_0 \frac{\partial \vec{E}}{\partial t} \qquad \vec{j} = \frac{\partial \vec{P}}{\partial t} \qquad \vec{P} = \alpha(\vec{E}) \tag{12.25}$$

$\vec{E}$ and $\vec{B}$ are the electric and magnetic fields. $\vec{P}$ is the polarization of the medium, the electric dipole moment density. α is the polarizability tensor, relating the amount of distortion of the charges in the matter to the applied electric field. The current density $\vec{j}$ appears because a time-varying electric field will cause a time-varying movement of the charges in the surrounding medium — that's a current.

Take the curl of the first equation and the time derivative of the second.

$$\nabla \times \nabla \times \vec{E} = -\frac{\partial}{\partial t}\nabla \times \vec{B} \qquad \nabla \times \frac{\partial \vec{B}}{\partial t} = \mu_0 \frac{\partial \vec{j}}{\partial t} + \mu_0 \epsilon_0 \frac{\partial^2 \vec{E}}{\partial t^2}$$

The two expressions involving $\vec{B}$ are the same, so eliminate $\vec{B}$.

$$\nabla \times \nabla \times \vec{E} = -\mu_0 \frac{\partial \vec{j}}{\partial t} - \mu_0 \epsilon_0 \frac{\partial^2 \vec{E}}{\partial t^2} = -\mu_0 \frac{\partial^2 \alpha(\vec{E})}{\partial t^2} - \mu_0 \epsilon_0 \frac{\partial^2 \vec{E}}{\partial t^2}$$

I make the assumption that α is time independent, so this is

$$\nabla \times \nabla \times \vec{E} = \nabla(\nabla \cdot \vec{E}) - \nabla^2 \vec{E} = -\mu_0 \alpha\left(\frac{\partial^2 \vec{E}}{\partial t^2}\right) - \mu_0 \epsilon_0 \frac{\partial^2 \vec{E}}{\partial t^2}$$

I am seeking a wave solution for the field, so assume a solution $\vec{E}(\vec{r}, t) = \vec{E}_0 e^{i\vec{k}\cdot\vec{r} - \omega t}$. Each ∇ brings down a factor of $i\vec{k}$ and each time derivative a factor $-i\omega$, so the equation is

$$-\vec{k}(\vec{k} \cdot \vec{E}_0) + k^2 \vec{E}_0 = \mu_0 \omega^2 \alpha(\vec{E}_0) + \mu_0 \epsilon_0 \omega^2 \vec{E}_0 \tag{12.26}$$

This is a linear equation for the vector $\vec{E}_0$.

A special case first: A vacuum, so there is no medium and $\alpha \equiv 0$. The equation is

$$(k^2 - \mu_0 \epsilon_0 \omega^2)\vec{E}_0 - \vec{k}(\vec{k} \cdot \vec{E}_0) = 0 \tag{12.27}$$

Pick a basis so that $\hat{z}$ is along $\vec{k}$, then

$$(k^2 - \mu_0 \epsilon_0 \omega^2)\vec{E}_0 - k^2\, \hat{z} \cdot \vec{E}_0 = 0 \qquad \text{or in matrix notation,}$$

$$\left[(k^2 - \mu_0 \epsilon_0 \omega^2)\begin{pmatrix} 1 & 0 & 0 \\ 0 & 1 & 0 \\ 0 & 0 & 1 \end{pmatrix} - k^2 \begin{pmatrix} 0 & 0 & 0 \\ 0 & 0 & 0 \\ 0 & 0 & 1 \end{pmatrix}\right]\begin{pmatrix} E_{0x} \\ E_{0y} \\ E_{0z} \end{pmatrix}$$

This is a set of linear homogeneous equations for the components of the electric field. One solution for $\vec{E}$ is identically zero, and this solution is unique *unless* the determinant of the coefficients vanishes. That is

$$\big(k^2 - \mu_0\epsilon_0\omega^2\big)^2\big(- \mu_0\epsilon_0\omega^2\big) = 0$$

This is a cubic equation for ω^2. One root is zero, the other two are $\omega^2 = k^2/\mu_0\epsilon_0$. The eigenvector corresponding to the zero root has components the column matrix $(\,0 \quad 0 \quad 1\,)$, or $E_{0x} = 0$ and $E_{0y} = 0$ with the z-component arbitrary. The field for this solution is

$$\vec{E} = E_{0z}e^{ikz}\,\hat{z}, \qquad \text{then} \qquad \nabla\cdot\vec{E} = \rho/\epsilon_0 = ikE_{0z}e^{ikz}$$

This is a static charge density, not a wave, so look to the other solutions. They are

$$\begin{gathered} E_{0z} = 0, \text{ with } E_{0x},\ E_{0y} \text{ arbitrary and } \vec{k}\cdot\vec{E}_0 = \nabla\cdot\vec{E} = 0 \\ \vec{E} = \big(E_{0x}\,\hat{x} + E_{0y}\,\hat{y}\big)e^{ikz-i\omega t} \qquad \text{and} \qquad \omega^2/k^2 = 1/\mu_0\epsilon_0 = c^2 \end{gathered}$$

This is a plane, transversely polarized, electromagnetic wave moving in the z-direction at velocity $\omega/k = 1/\sqrt{\mu_0\epsilon_0}$.

Now return to the case in which $\alpha \neq 0$. If the polarizability is a multiple of the identity, so that the dipole moment density is always along the direction of $\vec{E}$, all this does is to add a constant to ϵ_0 in Eq. (12.26). $\epsilon_0 \to \epsilon_0 + \alpha = \epsilon$, and the speed of the wave changes from $c = 1/\sqrt{\mu_0\epsilon_0}$ to $v = 1/\sqrt{\mu_0\epsilon}$

The more complicated case occurs when α is more than just multiplication by a scalar. If the medium is a crystal in which the charges can be polarized more easily in some directions than in others, α is a tensor. It is a symmetric tensor, though I won't prove that here. The proof involves looking at energy dissipation (rather, lack of it) as the electric field is varied. To compute I will pick a basis, and choose it so that the components of α form a diagonal matrix in this basis.

$$\text{Pick } \vec{e}_1 = \hat{x}, \quad \vec{e}_2 = \hat{y}, \quad \vec{e}_3 = \hat{z} \text{ so that } (\alpha) = \begin{pmatrix} \alpha_{11} & 0 & 0 \\ 0 & \alpha_{22} & 0 \\ 0 & 0 & \alpha_{33} \end{pmatrix}$$

The combination that really appears in Eq. (12.26) is $\epsilon_0 + \alpha$. The first term is a scalar, a multiple of the identity, so the matrix for this is

$$(\epsilon) = \epsilon_0 \begin{pmatrix} 1 & 0 & 0 \\ 0 & 1 & 0 \\ 0 & 0 & 1 \end{pmatrix} + (\alpha) = \begin{pmatrix} \epsilon_{11} & 0 & 0 \\ 0 & \epsilon_{22} & 0 \\ 0 & 0 & \epsilon_{33} \end{pmatrix} \tag{12.28}$$

I'm setting this up assuming that the crystal has three different directions with three different polarizabilities. When I get down to the details of the calculation I will take two of them equal — that's the case for calcite. The direction of propagation of the wave is $\vec{k}$, and Eq. (12.26) is

$$
\left[k^2\begin{pmatrix}1&0&0\\0&1&0\\0&0&1\end{pmatrix}-\mu_0\omega^2\begin{pmatrix}\epsilon_{11}&0&0\\0&\epsilon_{22}&0\\0&0&\epsilon_{33}\end{pmatrix}-\begin{pmatrix}k_1k_1&k_1k_2&k_1k_3\\k_2k_1&k_2k_2&k_2k_3\\k_3k_1&k_3k_2&k_3k_3\end{pmatrix}\right]\begin{pmatrix}E_{0x}\\E_{0y}\\E_{0z}\end{pmatrix}=0 \tag{12.29}
$$

In order to have a non-zero solution for the electric field, the determinant of this total 3×3 matrix must be zero. This is starting to get too messy, so I'll now make the simplifying assumption that two of the three directions in the crystal have identical properties: $\epsilon_{11}=\epsilon_{22}$. The makes the system cylindrically symmetric about the z-axis and I can then take the direction of the vector $\vec{k}$ to be in the x-z plane for convenience.

$$
\vec{k}=k\big(\vec{e}_1\sin\alpha+\vec{e}_3\cos\alpha\big)
$$

The matrix of coefficients in Eq. (12.29) is now

$$
\begin{aligned}
&k^2\begin{pmatrix}1&0&0\\0&1&0\\0&0&1\end{pmatrix}-\mu_0\omega^2\begin{pmatrix}\epsilon_{11}&0&0\\0&\epsilon_{11}&0\\0&0&\epsilon_{33}\end{pmatrix}-k^2\begin{pmatrix}\sin^2\alpha&0&\sin\alpha\cos\alpha\\0&0&0\\\sin\alpha\cos\alpha&0&\cos^2\alpha\end{pmatrix}\\
&=\begin{pmatrix}k^2\cos^2\alpha-\mu_0\omega^2\epsilon_{11}&0&-k^2\sin\alpha\cos\alpha\\0&k^2-\mu_0\omega^2\epsilon_{11}&0\\-k^2\sin\alpha\cos\alpha&0&k^2\sin^2\alpha-\mu_0\omega^2\epsilon_{33}\end{pmatrix}
\end{aligned} \tag{12.30}
$$

The determinant of this matrix is

$$
\begin{aligned}
&(k^2-\mu_0\omega^2\epsilon_{11})\Big[\big(k^2\cos^2\alpha-\mu_0\omega^2\epsilon_{11}\big)\big(k^2\sin^2\alpha-\mu_0\omega^2\epsilon_{33}\big)-\big(-k^2\sin\alpha\cos\alpha\big)^2\Big]\\
&\quad=(k^2-\mu_0\omega^2\epsilon_{11})\big[-\mu_0\omega^2k^2\big(\epsilon_{11}\sin^2\alpha+\epsilon_{33}\cos^2\alpha\big)+\mu_0^2\omega^4\epsilon_{11}\epsilon_{33}\big]=0
\end{aligned} \tag{12.31}
$$

This has a factor $\mu_0\omega^2=0$, and the corresponding eigenvector is $\vec{k}$ itself. In column matrix notation that is $(\sin\alpha\quad 0\quad\cos\alpha)$. It is another of those static solutions that aren't very interesting. For the rest, there are factors

$$
k^2-\mu_0\omega^2\epsilon_{11}=0 \qquad\text{and}\qquad k^2\big(\epsilon_{11}\sin^2\alpha+\epsilon_{33}\cos^2\alpha\big)-\mu_0\omega^2\epsilon_{11}\epsilon_{33}=0
$$

The first of these roots, $\omega^2/k^2=1/\mu_0\epsilon_{11}$, has an eigenvector $\vec{e}_2=\hat{y}$, or $(0\quad 1\quad 0)$. Then $\vec{E}=E_0\,\hat{y}e^{ikz-\omega t}$ and this is a normal sort of transverse wave that you commonly find in ordinary non-crystalline materials. The electric vector is perpendicular to the direction of propagation. The energy flow (the Poynting vector) is along the direction of $\vec{k}$, perpendicular to the wavefronts. It's called the "ordinary ray," and at a surface it obeys Snell's law. The reason it is ordinary is that the electric vector in this case is along one of the principal axes of the crystal. The polarization is along $\vec{E}$ just as it is in air or glass, so it behaves the same way as in those cases.

The second root has $E_{0y} = 0$, and an eigenvector computed as

$$\begin{aligned}
(k^2\cos^2\alpha - \mu_0\omega^2\epsilon_{11})E_{0x} - (k^2\sin\alpha\cos\alpha)E_{0z} &= 0 \\
(k^2\cos^2\alpha - k^2(\epsilon_{11}\sin^2\alpha + \epsilon_{33}\cos^2\alpha)/\epsilon_{33})E_{0x} - (k^2\sin\alpha\cos\alpha)E_{0z} &= 0 \\
\epsilon_{11}\sin\alpha E_{0x} + \epsilon_{33}\cos\alpha E_{0z} &= 0
\end{aligned}$$

If the two ϵ's are equal, this says that $\vec{E}$ is perpendicular to $\vec{k}$. If they aren't equal then $\vec{k}\cdot\vec{E} \neq 0$ and you can write this equation for the direction of $\vec{E}$ as

$$\begin{aligned}
&E_{0x} = E_0\cos\beta, \quad E_{0z} = -\sin\beta, \quad \text{then} \\
&\epsilon_{11}\sin\alpha\cos\beta - \epsilon_{33}\cos\alpha\sin\beta = 0 \\
&\text{and so} \quad \tan\beta = \frac{\epsilon_{11}}{\epsilon_{33}}\tan\alpha
\end{aligned}$$

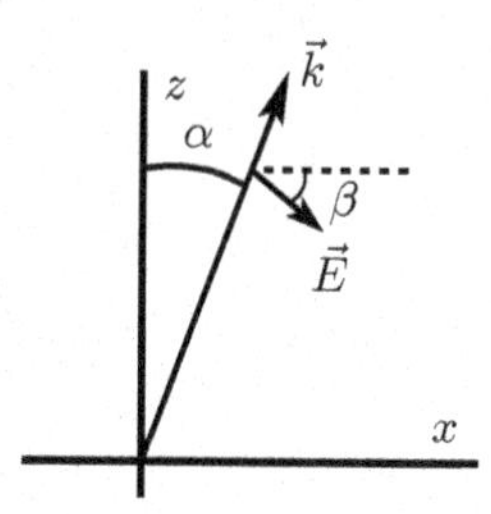

In calcite, the ratio $\epsilon_{11}/\epsilon_{33} = 1.056$, making β a little bigger than α.

The magnetic field is in the $\vec{k}\times\vec{E}$ direction, and the energy flow of the light is along the Poynting vector, in the direction $\vec{E}\times\vec{B}$. In this picture, that puts $\vec{B}$ along the $\hat{y}$-direction (into the page), and then the energy of the light moves in the direction perpendicular to $\vec{E}$ and $\vec{B}$. That is *not* along $\vec{k}$. This means that the propagation of the wave is not along the perpendicular to the wave fronts. Instead the wave skitters off at an angle to the front. The "extraordinary ray." Snell's law says that the wavefronts bend at a surface according to a simple trigonometric relation, *and they still do,* but the energy flow of the light does *not* follow the direction normal to the wavefronts. The light ray does not obey Snell's law

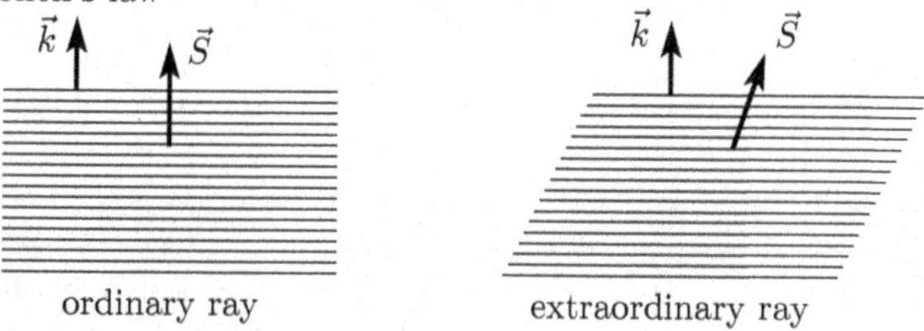

In calcite as it naturally grows, the face of the crystal is not parallel to any of the x-y or x-z or y-z planes. When a light ray enters the crystal perpendicular to the surface, the wave fronts are in the plane of the surface and what happens then depends on polarization. For the light polarized along one of the principle axes of the crystal (the y-axis in this sketch) the light behaves normally and the energy goes in an unbroken straight line. For the other polarization, the electric field has components along two different axes of the the crystal and the energy flows in an unexpected direction — disobeying Snell's law. The normal to the wave front is in the direction of the vector $\vec{k}$, and the direction of energy flow (the Poynting vector) is indicated by the vector $\vec{S}$.

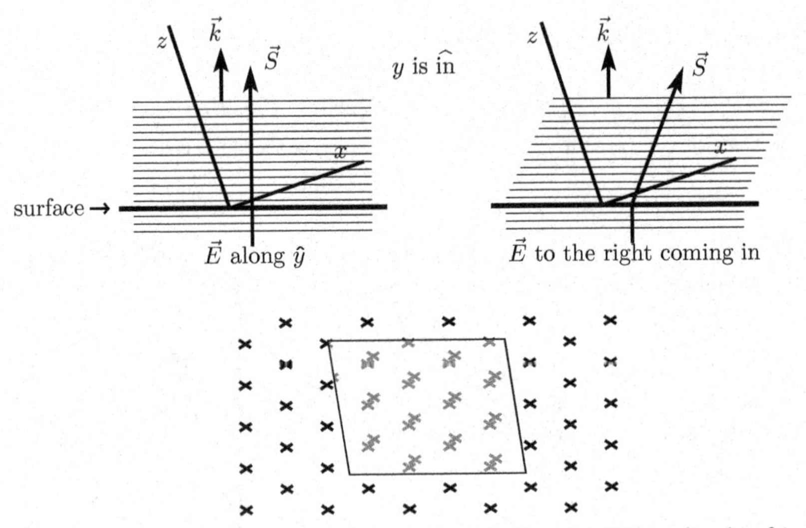

Simulation of a pattern of x's seen through Calcite. See also Wikipedia: birefringence.

12.5 Non-Orthogonal Bases

The next topic is the introduction of more general computational techniques. These will lift the restriction on the type of basis that can be used for computing components of various tensors. Until now, the basis vectors have formed an orthonormal set

$$|\hat{e}_i| = 1, \qquad \hat{e}_i \cdot \hat{e}_j = 0 \text{ if } i \neq j$$

Consider instead a more general set of vectors $\vec{e}_i$. These must be independent. That is, in three dimensions they are not coplanar. Other than this there is no restriction. Since by assumption the vectors $\vec{e}_i$ span the space you can write

$$\vec{v} = v^i \,\hat{e}_i.$$

with the numbers v^i being as before the components of the vector $\vec{v}$.

> NOTE: Here is a change in notation. Before, every index was a subscript. (It could as easily have been a superscript.) Now, be sure to make a careful distinction between sub- and superscripts. They will have different meanings.

Reciprocal Basis

Immediately, when you do the basic scalar product you find complications. If $\vec{u} = u^j\vec{e}_j$, then

$$\vec{u}\cdot\vec{v} = (u^j\vec{e}_j)\cdot(v^i\vec{e}_i) = u^j v^i \vec{e}_j \cdot \vec{e}_i.$$

But since the $\vec{e}_i$ aren't orthonormal, this is a much more complicated result than the usual scalar product such as

$$u_x v_y + u_y v_y + u_z v_z.$$

You can't assume that $\vec{e}_1 \cdot \vec{e}_2 = 0$ any more. In order to obtain a result that looks as simple as this familiar form, introduce an auxiliary basis: the *reciprocal basis.* (This trick will not really simplify the answer; it will be the same as ever. It will however be in a neater form and hence easier to manipulate.) The reciprocal basis is defined by the equation

$$\vec{e}_i \cdot \vec{e}^{\,j} = \delta_i^j = \begin{cases} 1 & \text{if } i = j \\ 0 & \text{if } i \neq j \end{cases} \tag{12.32}$$

The $\vec{e}^{\,j}$'s are vectors. The index is written as a superscript to distinguish it from the original basis, $\vec{e}_j$.

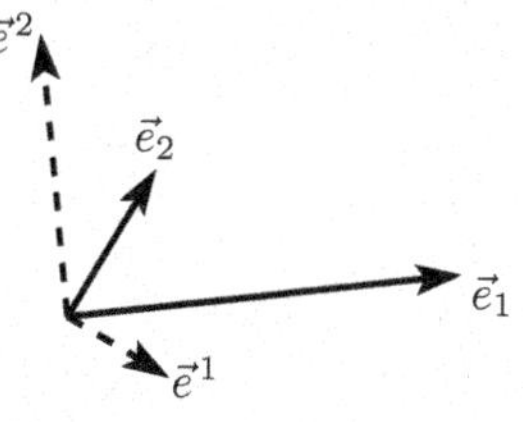

To elaborate on the meaning of this equation, $\vec{e}^{\,1}$ is perpendicular to the plane defined by $\vec{e}_2$ and $\vec{e}_3$ and is therefore more or less in the direction of $\vec{e}_1$. Its magnitude is adjusted so that the scalar product

$$\vec{e}^{\,1} \cdot \vec{e}_1 = 1.$$

The "direct basis" and "reciprocal basis" are used in solid state physics and especially in describing X-ray diffraction in crystallography. In that instance, the direct basis is the fundamental lattice of the crystal and the reciprocal basis would be defined from it. The reciprocal basis is used to describe the wave number vectors of scattered X-rays.

The basis reciprocal to the reciprocal basis is the direct basis.

Now to use these things: Expand the vector $\vec{u}$ in terms of the direct basis and $\vec{v}$ in terms of the reciprocal basis.

$$\vec{u} = u^i \vec{e}_i \qquad \text{and} \qquad \vec{v} = v_j \vec{e}^{\,j}. \qquad \text{Then} \qquad \begin{aligned} \vec{u} \cdot \vec{v} &= (u^i \vec{e}_i) \cdot (v_j \vec{e}^{\,j}) \\ &= u^i v_j \delta_i^j \\ &= u^i v_i = u^1 v_1 + u^2 v_2 + u^3 v_3. \end{aligned}$$

Notation: The superscript on the components (u^i) will refer to the components in the direct basis ($\vec{e}_i$); the subscripts (v_j) will come from the reciprocal basis ($\vec{e}^{\,j}$). You could also have expanded $\vec{u}$ in terms of the reciprocal basis and $\vec{v}$ in the direct basis, then

$$\vec{u} \cdot \vec{v} = u_i v^i = u^i v_i \tag{12.33}$$

Summation Convention

At this point, modify the previously established summation convention: Like indices in a given term are to be summed when one is a subscript and one is a superscript. Furthermore the notation is designed so that this is the only kind of sum that should occur. If you find a term such as $u_i v_i$ then this means that you made a mistake.

The scalar product now has a simple form in terms of components (at the cost of introducing an auxiliary basis set). Now for further applications to vectors and tensors.

Terminology: The components of a vector in the direct basis are called the contravariant components of the vector: v^i. The components in the reciprocal basis are called* the covariant components: v_i.

Examine the component form of the basic representation theorem for linear functionals, as in Eqs. (12.18) and (12.19).

$$f(\vec{v}) = \vec{A}\cdot\vec{v} \qquad \text{for all} \qquad \vec{v}.$$

$$\text{Claim:}\quad \vec{A} = \vec{e}^{\,i} f(\vec{e}_i) = \vec{e}_i f(\vec{e}^{\,i}) \tag{12.34}$$

The proof of this is as before: write $\vec{v}$ in terms of components and compute the scalar product $\vec{A}\cdot\vec{v}$.

$$\vec{v} = v^i\vec{e}_i. \qquad \text{Then} \qquad \begin{aligned}\vec{A}\cdot\vec{v} &= \big(\vec{e}^{\,j} f(\vec{e}_j)\big)\cdot\big(v^i\vec{e}_i\big)\\ &= v^i f(\vec{e}_j)\delta_i^j\\ &= v^i f(\vec{e}_i) = f(v^i\vec{e}_i) = f(\vec{v}).\end{aligned}$$

Analogous results hold for the expression of $\vec{A}$ in terms of the direct basis.

You can see how the notation forced you into considering this expression for $\vec{A}$. The summation convention requires one upper index and one lower index, so there is practically no other form that you could even consider in order to represent $\vec{A}$.

The same sort of computations will hold for tensors. Start off with one of second rank. Just as there were covariant and contravariant components of a vector, there will be covariant and contravariant components of a tensor. $T(\vec{u},\,\vec{v})$ is a scalar. Express $\vec{u}$ and $\vec{v}$ in contravariant component form:

$$\vec{u} = u^i\vec{e}_i \qquad \text{and} \qquad \vec{v} = v^j\vec{e}_j. \qquad \text{Then} \qquad \begin{aligned}T(\vec{u},\,\vec{v}) &= T(u^i\vec{e}_i,\, v^j\vec{e}_j)\\ &= u^i v^j\, T(\vec{e}_i,\,\vec{e}_j)\\ &= u^i v^j\, T_{ij}\end{aligned} \tag{12.35}$$

The numbers T_{ij} are called the covariant components of the tensor T.

Similarly, write $\vec{u}$ and $\vec{v}$ in terms of covariant components:

$$\vec{u} = u_i\vec{e}^{\,i} \qquad \text{and} \qquad \vec{v} = v_j\vec{e}^{\,j}. \qquad \text{Then} \qquad \begin{aligned}T(\vec{u},\,\vec{v}) &= T(u_i\vec{e}^{\,i},\, v_j\vec{e}^{\,j})\\ &= u_i v_j\, T(\vec{e}^{\,i},\,\vec{e}^{\,j})\\ &= u_i v_j\, T^{ij}\end{aligned} \tag{12.36}$$

And T^{ij} are the contravariant components of T. It is also possible to have mixed components:

$$\begin{aligned}T(\vec{u},\,\vec{v}) &= T(u_i\vec{e}^{\,i},\, v^j\vec{e}_j)\\ &= u_i v^j\, T(\vec{e}^{\,i},\,\vec{e}_j)\\ &= u_i v^j\, T^i{}_j\end{aligned}$$

---

* These terms are of more historical than mathematical interest.

As before, from the bilinear functional, a linear vector valued function can be formed such that

$$T(\vec{u},\vec{v}) = \vec{u}\cdot T(\vec{v}) \qquad \text{and} \qquad \begin{aligned} T(\vec{v}) &= \vec{e}^{\,i}T(\vec{e}_i,\vec{v}) \\ &= \vec{e}_i T(\vec{e}^{\,i},\vec{v}) \end{aligned}$$

For the proof of the last two lines, simply write $\vec{u}$ in terms of its contravariant or covariant components respectively.

All previous statements concerning the symmetry properties of tensors are unchanged because they were made in a way independent of basis, though it's easy to see that the symmetry properties of the tensor are reflected in the symmetry of the covariant or the contravariant components (but not usually in the mixed components).

Metric Tensor

Take as an example the metric tensor:

$$g(\vec{u},\vec{v}) = \vec{u}\cdot\vec{v}. \tag{12.37}$$

The linear function found by pulling off the $\vec{u}$ from this is the identity operator.

$$g(\vec{v}) = \vec{v}$$

This tensor is symmetric, so this must be reflected in its covariant and contravariant components. Take as a basis the vectors

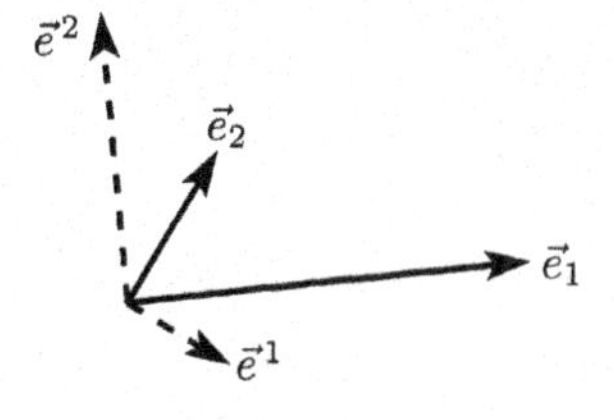

Let $|\vec{e}_2| = 1$ and $|e_1| = 2$; the angle between them being 45°. A little geometry shows that

$$|\vec{e}^{\,1}| = \frac{1}{\sqrt{2}} \qquad \text{and} \qquad |\vec{e}^{\,2}| = \sqrt{2}$$

Assume this problem is two dimensional in order to simplify things.

Compute the covariant components:

$$\begin{aligned} g_{11} &= g(\vec{e}_1,\vec{e}_1) = 4 \\ g_{12} &= g(\vec{e}_1,\vec{e}_2) = \sqrt{2} \\ g_{21} &= g(\vec{e}_2,\vec{e}_1) = \sqrt{2} \\ g_{22} &= g(\vec{e}_2,\vec{e}_2) = 1 \end{aligned} \qquad (g_{rc}) = \begin{pmatrix} 4 & \sqrt{2} \\ \sqrt{2} & 1 \end{pmatrix}$$

Similarly

$$\begin{aligned} g^{11} &= g(\vec{e}^{\,1},\vec{e}^{\,1}) = 1/2 \\ g^{12} &= g(\vec{e}^{\,1},\vec{e}^{\,2}) = -1/\sqrt{2} \\ g^{21} &= g(\vec{e}^{\,2},\vec{e}^{\,1}) = -1/\sqrt{2} \\ g^{22} &= g(\vec{e}^{\,2},\vec{e}^{\,2}) = 2 \end{aligned} \qquad (g^{rc}) = \begin{pmatrix} 1/2 & -1/\sqrt{2} \\ -1/\sqrt{2} & 2 \end{pmatrix}$$

The mixed components are

$$g^1{}_1 = g(\vec{e}^{\,1}, \vec{e}_1) = 1 \qquad g^1{}_2 = g(\vec{e}^{\,1}, \vec{e}_2) = 0 \qquad g^2{}_1 = g(\vec{e}^{\,2}, \vec{e}_1) = 0 \qquad g^2{}_2 = g(\vec{e}^{\,2}, \vec{e}_2) = 1 \qquad (g^r{}_c) = (\delta^r_c) = \begin{pmatrix} 1 & 0 \\ 0 & 1 \end{pmatrix} \tag{12.38}$$

I used r and c for the indices to remind you that these are the row and column variables. Multiply the first two matrices together and you obtain the third one — the unit matrix. The matrix (g_{ij}) is therefore the inverse of the matrix (g^{ij}). This last result is *not* general, but is due to the special nature of the tensor g.

12.6 Manifolds and Fields

Until now, all definitions and computations were done in one vector space. This is the same state of affairs as when you once learned vector algebra; the only things to do then were addition, scalar products, and cross products. Eventually however vector calculus came up and you learned about vector fields and gradients and the like. You have now set up enough apparatus to make the corresponding step here. First I would like to clarify just what is meant by a vector field, because there is sure to be confusion on this point no matter how clearly you think you understand the concept. Take a typical vector field such as the electrostatic field $\vec{E}$. $\vec{E}$ will be some function of position (presumably satisfying Maxwell's equations) as indicated at the six different points.

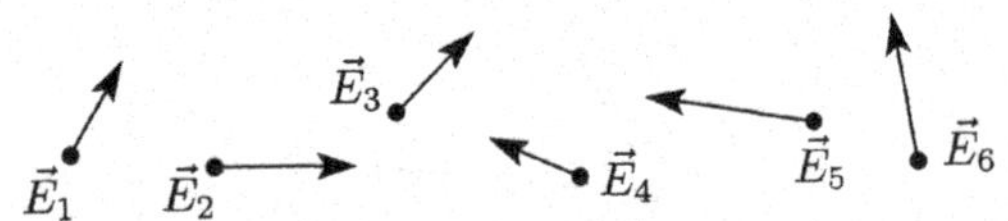

Does it make any sense to take the vector $\vec{E}_3$ and add it to the vector $\vec{E}_5$? These are after all, vectors; can't you always add one vector to another vector? Suppose there is also a magnetic field present, say with vectors $\vec{B}_1$, $\vec{B}_2$ *etc.* , at the same points. Take the magnetic vector at the point #3 and add it to the electric vector there. The reasoning would be exactly the same as the previous case; these are vectors, therefore they can be added. The second case is clearly nonsense, as should be the first. The electric vector is defined as the force per charge at a point. If you take two vectors at two different points, then the forces are on two different objects, so the sum of the forces is not a force on anything — it isn't even defined.

You can't add an electric vector at one point to an electric vector at another point. These two vectors occupy different vector spaces. At a single point in space there are many possible vectors; at this one point, the set of all possible electric vectors forms a vector space because they can be added to each other and multiplied by scalars while remaining at the same point. By the same reasoning the magnetic vectors at a point form a vector space. Also the velocity vectors. You could not add a velocity vector to an electric field vector even at the same point however. These too are in different vector spaces. You can picture all these vector spaces as attached to the points in the manifold and somehow sitting over them.

From the above discussion you can see that even to discuss one type of vector field, a vector space must be attached to each point of space. If you wish to make a drawing of such a system, It is at best difficult. In three dimensional space you could have a three dimensional vector space at each point. A crude way of picturing this is to restrict to two dimensions and draw a line attached to each point, representing the vector space attached to that point. This pictorial representation won't be used in anything to follow however, so you needn't worry about it.

The term "vector field" that I've been throwing around is just a prescription for selecting one vector out of each of the vector spaces. Or, in other words, it is a function that assigns to each point a vector in the vector space at that same point.

There is a minor confusion of terminology here in the use of the word "space." This could be space in the sense of the three dimensional Euclidean space in which we are sitting and doing computations. Each point of the latter will have a vector space associated with it. To reduce confusion (I hope) I shall use the word "manifold" for the space over which all the vector spaces are built. Thus: To each point of the manifold there is associated a vector space. A vector field is a choice of one vector from each of the vector spaces over the manifold. This is a vector field on the manifold. In short: The word "manifold" is substituted here for the phrase "three dimensional Euclidean space."

(A comment on generalizations. While using the word manifold as above, everything said about it will in fact be more general. For example it will still be acceptable in special relativity with four dimensions of space-time. It will also be correct in other contexts where the structure of the manifold is non-Euclidean.)

The point to emphasize here is that most of the work on tensors is already done and that the application to fields of vectors and fields of tensors is in a sense a special case. At each point of the manifold there is a vector space to which all previous results apply.

In the examples of vector fields mentioned above (electric field, magnetic field, velocity field) keep your eye on the velocity. It will play a key role in the considerations to come, even in considerations of other fields.

A word of warning about the distinction between a manifold and the vector spaces at each point of the manifold. You are accustomed to thinking of three dimensional Euclidean space (the manifold) as a vector space itself. That is, the displacement vector between two points is defined, and you can treat these as vectors just like the electric vectors at a point. *Don't!* Treating the manifold as a vector space will cause great confusion. Granted, it happens to be correct in this instance, but in attempting to understand these new concepts about vector fields (and tensor fields later), this additional knowledge will be a hindrance. For our purposes therefore the manifold will not be a vector space. The concept of a displacement vector is therefore *not defined.*

Just as vector fields were defined by picking a single vector from each vector space at various points of the manifold, a scalar field is similarly an assignment of a number (scalar) to each point. In short then, a scalar field is a function that gives a scalar (the dependent variable) for each point of the manifold (the independent variable).

For each vector space, you can discuss the tensors that act on that space and so, by picking one such tensor for each point of the manifold a tensor field is defined.

A physical example of a tensor field (of second rank) is stress in a solid. This will typically vary from point to point. But at each point a second rank tensor is given by the relation between infinitesimal area vectors and internal force vectors at that point. Similarly, the dielectric tensor in an inhomogeneous medium will vary with position and will therefore be expressed as a tensor field. Of course even in a homogeneous medium the dielectric tensor would be a tensor field relating $\vec{D}$ and $\vec{E}$ at the same point. It would however be a constant tensor field. Like a uniform electric field, the tensors at different points could be thought of as "parallel" to each other (whatever that means).

12.7 Coordinate Bases

In order to obtain a handle on this subject and in order to be able to do computations, it is necessary to put a coordinate system on the manifold. From this coordinate system there will come a natural way to define the basis vectors at each point (and so reciprocal basis vectors too). The orthonormal basis vectors that you are accustomed to in cylindrical and spherical coordinates are *not* "coordinate bases."

There is no need to restrict the discussion to rectangular or even to orthogonal coordinate systems. A coordinate system is a means of identifying different points of the manifold by different sets of numbers. This is done by specifying a set of functions: x^1, x^2, x^3, which are the coordinates. (There will be more in more dimensions of course.) These functions are real valued functions of points in the manifold. The coordinate axes are defined as in the drawing by

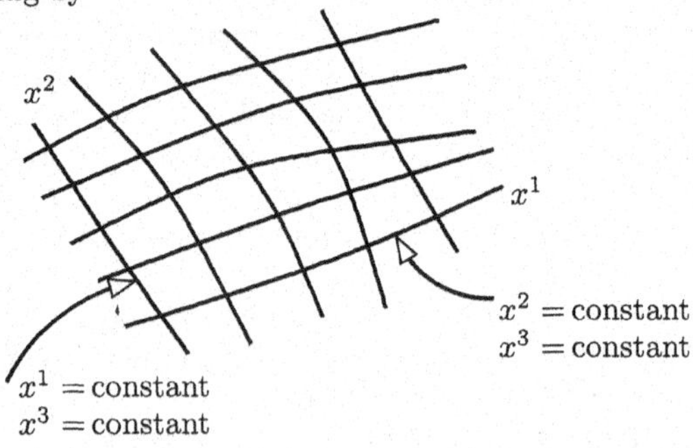

Specify the equations $x^2 = 0$ and $x^3 = 0$ for the x^1 coordinate axis. For example in rectangular coordinates $x^1 = x$, $x^2 = y$, $x^3 = z$, and the x-axis is the line $y = 0$, and $z = 0$. In plan polar coordinates $x^1 = r =$ a constant is a circle and $x^2 = \phi =$ a constant is a straight line starting from the origin.

Start with two dimensions and polar coordinates r and ϕ. As a basis in this system we routinely use the unit vectors $\hat{r}$ and $\hat{\phi}$, but is this the only choice? Is it the best choice? Not necessarily. Look at two vectors, the differential $d\vec{r}$ and the gradient ∇f.

$$d\vec{r} = \hat{r}\, dr + \hat{\phi}\, r d\phi \qquad \text{and} \qquad \nabla f = \hat{r}\frac{\partial f}{\partial r} + \hat{\phi}\frac{1}{r}\frac{\partial f}{\partial \phi} \tag{12.39}$$

And remember the chain rule too.

$$\frac{df}{dt} = \frac{\partial f}{\partial r}\frac{dr}{dt} + \frac{\partial f}{\partial \phi}\frac{d\phi}{dt} \tag{12.40}$$

In this basis the scalar product is simple, but you pay for it in that the components of the vectors have extra factors in them — the r and the $1/r$. An alternate approach is to place the complexity in the basis vectors and not in the components.

$$\vec{e}_1 = \hat{r}, \quad \vec{e}_2 = r\hat{\phi} \qquad \text{and} \qquad \vec{e}^{\,1} = \hat{r}, \quad \vec{e}^{\,2} = \hat{\phi}/r \tag{12.41}$$

These are reciprocal bases, as indicated by the notation. Of course the original $\hat{r}$-$\hat{\phi}$ is self-reciprocal. The preceding equations (12.39) are now

$$d\vec{r} = \vec{e}_1\, dr + \vec{e}_2 d\phi \qquad \text{and} \qquad \nabla f = \vec{e}^{\,1}\frac{\partial f}{\partial r} + \vec{e}^{\,2}\frac{\partial f}{\partial \phi} \tag{12.42}$$

Make another change in notation to make this appear more uniform. Instead of r-ϕ for the coordinates, call them x^1 and x^2 respectively, then

$$d\vec{r} = \vec{e}_1\, dx^1 + \vec{e}_2 dx^2 \qquad \text{and} \qquad \nabla f = \vec{e}^{\,1}\frac{\partial f}{\partial x^1} + \vec{e}^{\,2}\frac{\partial f}{\partial x^2} \tag{12.43}$$

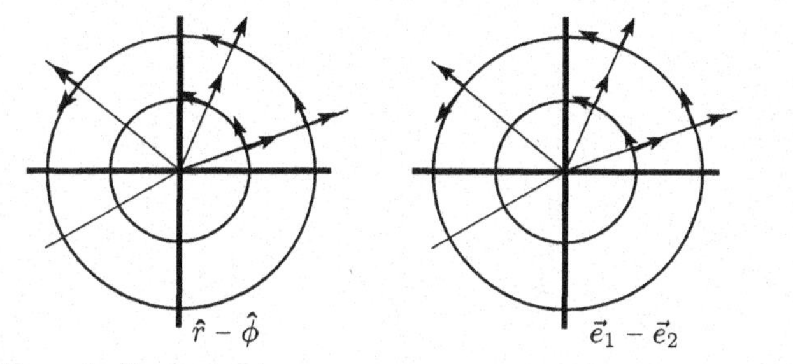

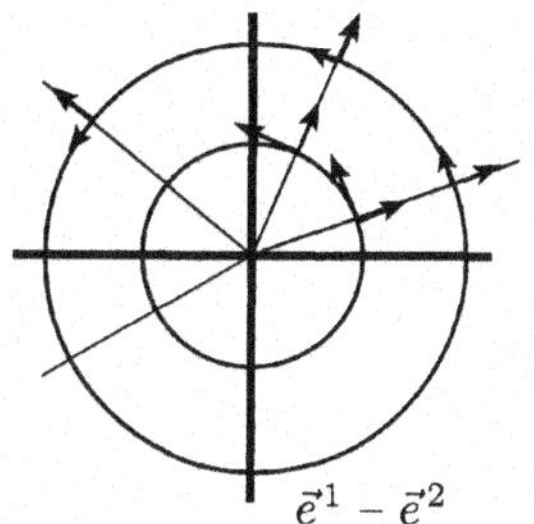

The velocity vector is

$$\frac{d\vec{r}}{dt} = \vec{e}_1\frac{dr}{dt} + \vec{e}_2\frac{d\phi}{dt} \qquad \text{and Eq. (12.40) is} \qquad \frac{df}{dt} = \frac{\partial f}{\partial x^i}\frac{dx^i}{dt} = (\nabla f)\cdot\vec{v}$$

This sort of basis has many technical advantages that aren't at all apparent here. At this point this "coordinate basis" is simply a way to sweep some of the complexity out of sight, but with further developments of the subject the sweeping becomes shoveling. When you go beyond the introduction found in this chapter, you find that using any basis other than a coordinate basis leads to equations that have complicated extra terms that you want nothing to do with.

In spherical coordinates $x^1 = r,\ x^2 = \theta,\ x^3 = \phi$

$$\vec{e}_1 = \hat{r}, \quad \vec{e}_2 = r\hat{\theta}, \quad \vec{e}_3 = r\sin\theta\,\hat{\phi} \qquad \text{and} \qquad \vec{e}^{\,1} = \hat{r}, \quad \vec{e}^{\,2} = \hat{\theta}/r, \quad \vec{e}^{\,3} = \hat{\phi}/r\sin\theta$$

The velocity components are now

$$dx^i/dt = \{dr/dt,\ d\theta/dt,\ d\phi/dt\}, \qquad \text{and} \qquad \vec{v} = \vec{e}_i\, dx^i/dt \tag{12.44}$$

This last equation is central to figuring out the basis vectors in an unfamiliar coordinate system.

The use of x^1, x^2, and x^3 (x^i) for the coordinate system makes the notation uniform. Despite the use of superscripts, these coordinates are *not* the components of any vectors, though their time derivatives are.

In ordinary rectangular coordinates. If a particle is moving along the x^1-axis (the x-axis) then $dx^2/dt = 0 = dx^3/dt$. Also, the velocity will be in the x direction and of size dx^1/dt.

$$\vec{e}_1 = \hat{x}$$

as you would expect. Similarly

$$\vec{e}_2 = \hat{y}, \qquad \vec{e}_3 = \hat{z}.$$

In a variation of rectangular coordinates in the plane the axes are not orthogonal to each other, but are rectilinear anyway.

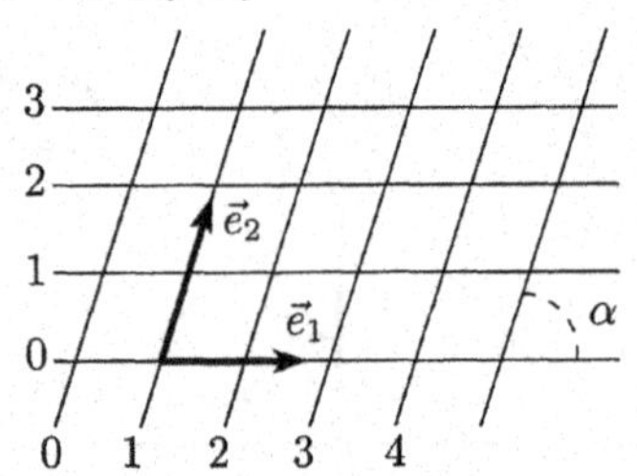

Still keep the requirement of Eq. (12.44)

$$\vec{v} = \vec{e}_i \frac{dx^i}{dt} = \vec{e}_1 \frac{dx^1}{dt} + \vec{e}_2 \frac{dx^2}{dt}. \tag{12.45}$$

If the particle moves along the x^1-axis (or parallel to it) then by the definition of the axes, x^2 is a constant and $dx^2/dt = 0$. Suppose that the coordinates measure centimeters, so that the *perpendicular* distance between the lines is one centimeter. The distance between the points $(0,0)$ and $(1,0)$ is then $1\,\text{cm}/\sin\alpha = \csc\alpha\,\text{cm}$. If in $\Delta t =$ one second, particle moves from the first to the second of these points, $\Delta x^1 =$ one cm, so $dx^1/dt = 1\,\text{cm/sec}$. The speed however, is $\csc\alpha\,\text{cm/sec}$ because the distance moved is greater by that factor. This means that

$$|\vec{e}_1| = \csc\alpha$$

and this is greater than one; it is not a unit vector. The magnitudes of $\vec{e}_2$ is the same. The dot product of these two vectors is $\vec{e}_1 \cdot \vec{e}_2 = \cos\alpha/\sin^2\alpha$.

Reciprocal Coordinate Basis

Construct the reciprocal basis vectors from the direct basis by the equation

$$\vec{e}^{\,i} \cdot \vec{e}_j = \delta^i_j$$

In rectangular coordinates the direct and reciprocal bases coincide because the basis is orthonormal. For the tilted basis of Eq. (12.45),

$$\vec{e}_2 \cdot \vec{e}^{\,2} = 1 = |\vec{e}_2|\,|\vec{e}^{\,2}|\cos\left(90^\circ - \alpha\right) = (\csc\alpha)|\vec{e}^{\,2}|\sin\alpha = |\vec{e}^{\,2}|$$

The reciprocal basis vectors in this case are unit vectors.

The direct basis is defined so that the components of velocity are as simple as possible. In contrast, the components of the gradient of a scalar field are equally simple provided that they are expressed in the reciprocal basis. If you try to use the same basis for both you can, but the resulting equations are a mess.

In order to compute the components of grad f (where f is a scalar field) start with its definition, and an appropriate definition should not depend on the coordinate system. It ought to be some sort of geometric statement that you can translate into any particular coordinate system that you want. One way to define grad f is that it is that vector pointing in the direction of maximum increase of f and equal in magnitude to $df/d\ell$ where ℓ is the distance measured in that direction. This is the first statement in section 8.5. While correct, this definition does not easily lend itself to computations.

Instead, think of the particle moving through the manifold. As a function of time it sees changing values of f. The time rate of change of f as felt by this particle is given by a scalar product of the particle's velocity and the gradient of f. This is essentially the same as Eq. (8.16), though phrased in a rather different way. Write this statement in terms of coordinates

$$\frac{d}{dt} f\big(x^1(t), x^2(t), x^3(t)\big) = \vec{v} \cdot \operatorname{grad} f$$

The left hand side is (by the chain rule)

$$\frac{\partial f}{\partial x^1}\bigg|_{x^2,x^3} \frac{dx^1}{dt} + \frac{\partial f}{\partial x^2}\bigg|_{x^1,x^3} \frac{dx^2}{dt} + \frac{\partial f}{\partial x^3}\bigg|_{x^1,x^2} \frac{dx^3}{dt} = \frac{\partial f}{\partial x^1}\frac{dx^1}{dt} + \frac{\partial f}{\partial x^2}\frac{dx^2}{dt} + \frac{\partial f}{\partial x^3}\frac{dx^3}{dt} \qquad (12.46)$$

$\vec{v}$ is expressed in terms of the direct basis by

$$\vec{v} = \vec{e}_i \frac{dx^i}{dt},$$

now express grad f in the reciprocal basis

$$\operatorname{grad} f = \vec{e}^{\,i} \big(\operatorname{grad} f\big)_i \qquad (12.47)$$

The way that the scalar product looks in terms of these bases, Eq. (12.33) is

$$\vec{v} \cdot \operatorname{grad} f = \vec{e}_i \frac{dx^i}{dt} \cdot \vec{e}^{\,j} \big(\operatorname{grad} f\big)_j = v^i \big(\operatorname{grad} f\big)_i \qquad (12.48)$$

Compare the two equations (12.46) and (12.48) and you see

$$\operatorname{grad} f = \vec{e}^{\,i} \frac{\partial f}{\partial x^i} \qquad (12.49)$$

For a formal proof of this statement consider three cases. When the particle is moving along the x^1 direction (x^2 & x^3 constant) only one term appears on each side of (12.46) and (12.48) and you can divide by $v^1 = dx^1/dt$. Similarly for x^2 and x^3. As usual with partial derivatives, the symbol $\partial f/\partial x^i$ assumes that the other coordinates x^2 and x^3 are constant.

For polar coordinates this equation for the gradient reads, using Eq. (12.41),

$$\operatorname{grad} f = \vec{e}^{\,1} \frac{\partial f}{\partial x^1} + \vec{e}^{\,2} \frac{\partial f}{\partial x^2} = (\hat{r}) \frac{\partial f}{\partial r} + \Big(\frac{1}{r}\hat{\phi}\Big) \frac{\partial f}{\partial \phi}$$

which is the standard result, Eq. (8.27). Notice again that the basis vectors are not dimensionless. They can't be because $\partial f/\partial r$ doesn't have the same dimensions as $\partial f/\partial \phi$.

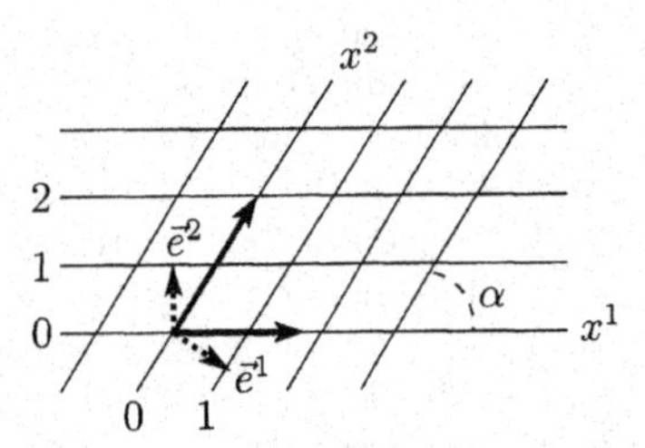

Example

I want an example to show that all this formalism actually gives the correct answer in a special case for which you can also compute all the results in the traditional way. Draw parallel lines a distance 1 cm apart and another set of parallel lines also a distance 1 cm apart intersecting at an angle α between them. These will be the constant values of the functions defining the coordinates, and will form a coordinate system labeled x^1 and x^2. The horizontal lines are the equations $x^2 = 0$, $x^2 = 1\,\text{cm}$, *etc.*

Take the case of the non-orthogonal rectilinear coordinates again. The components of grad f in the $\vec{e}^{\,1}$ direction is $\partial f/\partial x^1$, which is the derivative of f with respect to x^1 holding x_2 constant, and this derivative is *not* in the direction along $\vec{e}^{\,1}$, but in the direction where x^2 = a constant and that is along the x^1-axis, along $\vec{e}_1$. As a specific example to show that this makes sense, take a particular f defined by

$$f(x^1, x^2) = x^1$$

For this function $$\operatorname{grad} f = \vec{e}^{\,1}\frac{\partial f}{\partial x^1} + \vec{e}^{\,2}\frac{\partial f}{\partial x^2} = \vec{e}^{\,1}$$

$\vec{e}^{\,1}$ is perpendicular to the x^2-axis, the line x_1 =constant, (as it should be). Its magnitude is the magnitude of $\vec{e}^{\,1}$, which is one.

To verify that this magnitude is correct, calculate it directly from the definition. The magnitude of the gradient is the magnitude of $df/d\ell$ where ℓ is measured in the direction of the gradient, that is, in the direction $\vec{e}^{\,1}$.

$$\frac{df}{d\ell} = \frac{\partial f}{\partial x^1}\frac{dx^1}{d\ell} = 1\cdot 1 = 1$$

Why 1 for $dx^1/d\ell$? The coordinate lines in the picture are $x^1 = 0, 1, 2, \ldots$. When you move on the straight line perpendicular to x^2 = constant ($\vec{e}^{\,1}$), and go from $x^1 = 1$ to $x^2 = 2$, then both Δx^1 and Δs are one.

Metric Tensor

The simplest tensor field beyond the gradient vector above would be the metric tensor, which I've been implicitly using all along whenever a scalar product arose. It is defined at each point by

$$g(\vec{a}, \vec{b}) = \vec{a}\cdot\vec{b} \tag{12.50}$$

Compute the components of g in plane polar coordinates. The contravariant components of g are from Eq. (12.41)

$$g^{ij} = \vec{e}^{\,i}\cdot\vec{e}^{\,j} = \begin{pmatrix} 1 & 0 \\ 0 & 1/r^2 \end{pmatrix}$$

Covariant:

$$g_{ij} = \vec{e}_i\cdot\vec{e}_j = \begin{pmatrix} 1 & 0 \\ 0 & r^2 \end{pmatrix}$$

Mixed:

$$g^i{}_j = \vec{e}^{\,i} \cdot \vec{e}_j = \begin{pmatrix} 1 & 0 \\ 0 & 1 \end{pmatrix}$$

12.8 Basis Change

If you have two different sets of basis vectors you can compute the transformation on the components in going from one basis to the other, and in dealing with fields, a different set of basis vectors necessarily arises from a different set of coordinates on the manifold. It is convenient to compute the transformation matrices directly in terms of the different coordinate functions. Call the two sets of coordinates x^i and y^i. Each of them defines a set of basis vectors such that a given velocity is expressed as

$$\vec{v} = \vec{e}_i \frac{dx^i}{dt} = \vec{e}\,'_j \frac{dy^j}{dt} \tag{12.51}$$

What you need is an expression for $\vec{e}\,'_j$ in terms of $\vec{e}_i$ at each point. To do this, take a particular path for the particle — along the y^1-direction (y^2 & y^3 constant). The right hand side is then

$$\vec{e}\,'_1 \frac{dy^1}{dt}$$

Divide by dy^1/dt to obtain

$$\vec{e}\,'_1 = \vec{e}_i \frac{dx^i}{dt} \bigg/ \frac{dy^1}{dt}$$

But this quotient is just

$$\vec{e}\,'_1 = \vec{e}_i \left.\frac{\partial x^i}{\partial y^1}\right|_{y^2,y^3}$$

And in general

$$\vec{e}\,'_j = \vec{e}_i \frac{\partial x^i}{\partial y^j} \tag{12.52}$$

Do a similar calculation for the reciprocal vectors

$$\operatorname{grad} f = \vec{e}^{\,i} \frac{\partial f}{\partial x^i} = \vec{e}\,'^j \frac{\partial f}{\partial y^j}$$

Take the case for which $f = y^k$, then

$$\frac{\partial f}{\partial y^j} = \frac{\partial y^k}{\partial y^j} = \delta^k_j$$

which gives

$$\vec{e}\,'^k = \vec{e}^{\,i} \frac{\partial y^k}{\partial x^i} \tag{12.53}$$

The transformation matrices for the direct and the reciprocal basis are inverses of each other, In the present context, this becomes

$$
\begin{aligned}
e'^{k} \cdot \vec{e}\,'_j = \delta^k_j &= \vec{e}\,^{\ell} \frac{\partial y^k}{\partial x^\ell} \cdot \vec{e}_i \frac{\partial x^i}{\partial y^j} \\
&= \delta^\ell_i \frac{\partial y^k}{\partial x^\ell} \frac{\partial x^i}{\partial y^j} \\
&= \frac{\partial y^k}{\partial x^i} \frac{\partial x^i}{\partial y^j} = \frac{\partial y^k}{\partial y^j}
\end{aligned}
$$

The matrices $\partial x^i / \partial y^j$ and its inverse matrix, $\partial y^k / \partial x^i$ are called Jacobian matrices. When you do multiple integrals and have to change coordinates, the determinant of one or the other of these matrices will appear as a factor in the integral.

As an example, compute the change from rectangular to polar coordinates

$$
\begin{array}{cccc}
x^1 = x & y^1 = r & x^2 = y & y^2 = \phi \\
x = r\cos\phi & r = \sqrt{x^2 + y^2} & y = r\sin\phi & \phi = \tan^{-1} y/x
\end{array}
$$

$$
\begin{aligned}
\vec{e}\,'_j &= \vec{e}_i \frac{\partial x^i}{\partial y^j} \\
\vec{e}\,'_1 &= \vec{e}_1 \frac{\partial x^1}{\partial y^1} + \vec{e}_2 \frac{\partial x^2}{\partial y^1} = \hat{x} \frac{\partial x}{\partial r} + \hat{y} \frac{\partial y}{\partial r} \\
&= \hat{x}\cos\phi + \hat{y}\sin\phi = \hat{r} \\
\vec{e}\,'_2 &= \vec{e}_1 \frac{\partial x^1}{\partial y^2} + \vec{e}_2 \frac{\partial x^2}{\partial y^2} = \hat{x} \frac{\partial x}{\partial \phi} + \hat{y} \frac{\partial y}{\partial \phi} \\
&= \hat{x}(-r\sin\phi) + \hat{y}(r\cos\phi) = r\hat{\phi}
\end{aligned}
$$

Knowing the change in the basis vectors, the change of the components of any tensor by computing it in the new basis and using the linearity of the tensor itself. The new components will be linear combinations of those from the old basis.

A realistic example using non-orthonormal bases appears in special relativity. Here the manifold is four dimensional instead of three and the coordinate changes of interest represent Lorentz transformations. Points in space-time ("events") can be described by rectangular coordinates $(ct,\, x,\, y,\, z)$, which are concisely denoted by x^i

$$
i = (0, 1, 2, 3) \qquad \text{where} \qquad x^0 = ct,\; x^1 = x,\; x^2 = y,\; x^3 = z
$$

The introduction of the factor c into x^0 is merely a question of scaling. It also makes the units the same on all axes.

The basis vectors associated with this coordinate system point along the directions of the axes.

This manifold is *not* Euclidean however so that these vectors are not unit vectors in the usual sense. We have

$$
\vec{e}_0 \cdot \vec{e}_0 = -1 \qquad \vec{e}_1 \cdot \vec{e}_1 = 1 \qquad \vec{e}_2 \cdot \vec{e}_2 = 1 \qquad \vec{e}_3 \cdot \vec{e}_3 = 1
$$

and they are orthogonal pairwise. The reciprocal basis vectors are defined in the usual way,

$$\vec{e}^{\,i} \cdot \vec{e}_j = \delta^i_j$$

so that

$$\vec{e}^{\,0} = -\vec{e}_0 \qquad \vec{e}^{\,1} = \vec{e}_1 \qquad \vec{e}^{\,1} = \vec{e}_1 \qquad \vec{e}^{\,1} = \vec{e}_1$$

The contravariant (also covariant) components of the metric tensor are

$$g^{ij} = \begin{pmatrix} -1 & 0 & 0 & 0 \\ 0 & 1 & 0 & 0 \\ 0 & 0 & 1 & 0 \\ 0 & 0 & 0 & 1 \end{pmatrix} = g_{ij} \tag{12.54}$$

An observer moving in the $+x$ direction with speed v will have his own coordinate system with which to describe the events of space-time. The coordinate transformation is

$$\begin{aligned} x'^0 &= ct' = \frac{x^0 - \frac{v}{c}x^1}{\sqrt{1 - v^2/c^2}} = \frac{ct - \frac{v}{c}x}{\sqrt{1 - v^2/c^2}} \\ x'^1 &= x' = \frac{x^1 - \frac{v}{c}x^0}{\sqrt{1 - v^2/c^2}} = \frac{x - vt}{\sqrt{1 - v^2/c^2}} \\ x'^2 &= y' = x^2 = y \qquad x'^3 = z' = x^3 = z \end{aligned} \tag{12.55}$$

You can check that these equations represent the transformation to an observer moving in the $+x$ direction by asking where the moving observer's origin is as a function of time: It is at $x'^1 = 0$ or $x - vt = 0$, giving $x = vt$ as the locus of the moving observer's origin.

The graph of the coordinates is as usual defined by the equations (say for the x'^0-axis) that x'^1, x'^2, x'^3 are constants such as zero. Similarly for the other axes.

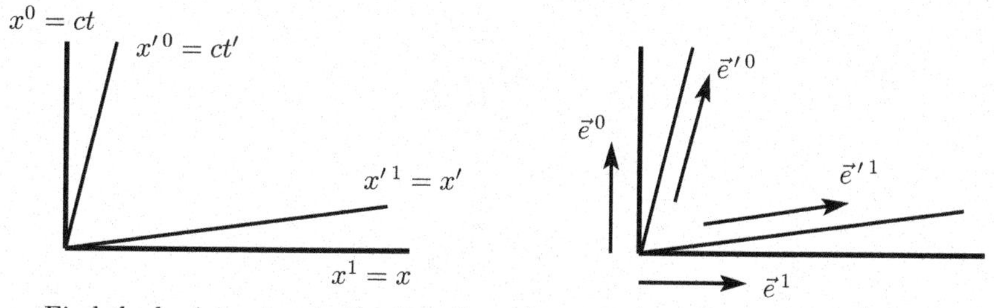

Find the basis vectors in the transformed system by using equation (12.52).

$$\vec{e}\,'_j = \vec{e}_i \frac{\partial x^i}{\partial y^j}$$

In the present case the y^j are x'^j and we need the inverse of the equations (12.55). They are found by changing v to $-v$ and interchanging primed and unprimed variables.

$$x^0 = \frac{x'^0 + \frac{v}{c}x'^1}{\sqrt{1 - v^2/c^2}} \qquad x^1 = \frac{x'^1 + \frac{v}{c}x'^0}{\sqrt{1 - v^2/c^2}}$$

$$\vec{e}\,'_0 = \vec{e}_i \frac{\partial x^i}{\partial x'^0} = \vec{e}_0 \frac{1}{\sqrt{1 - v^2/c^2}} + \vec{e}_1 \frac{v/c}{\sqrt{1 - v^2/c^2}}$$
$$\vec{e}\,'_1 = \vec{e}_i \frac{\partial x^i}{\partial x'^1} = \vec{e}_0 \frac{v/c}{\sqrt{1 - v^2/c^2}} + \vec{e}_1 \frac{1}{\sqrt{1 - v^2/c^2}} \tag{12.56}$$

It is easy to verify that these new vectors point along the primed axes as they should. They also have the property that they are normalized to plus or minus one respectively as are the original untransformed vectors. (How about the reciprocal vectors?)

As an example applying all this apparatus, do the transformation of the components of a second rank tensor, the electromagnetic field tensor. This tensor is the function that acts on the current density (four dimensional vector) and gives the force density (also a four vector). Its covariant components are

$$(F_{ij}) = F(\vec{e}_i, \vec{e}_j) = \begin{pmatrix} 0 & -E_x & -E_y & -E_z \\ E_x & 0 & B_z & -B_y \\ E_y & -B_z & 0 & B_x \\ E_z & B_y & -B_x & 0 \end{pmatrix} \tag{12.57}$$

where the E's and B's are the conventional electric and magnetic field components. Compute a sample component of this tensor in the primed coordinate system.

$$F'_{20} = F(\vec{e}\,'_2, \vec{e}\,'_0) = F\left(\vec{e}_2, \vec{e}_0 \frac{1}{\sqrt{1 - v^2/c^2}} + \vec{e}_1 \frac{v/c}{\sqrt{1 - v^2/c^2}}\right)$$
$$= \frac{1}{\sqrt{1 - v^2/c^2}} F_{20} + \frac{v/c}{\sqrt{1 - v^2/c^2}} F_{21}$$

or in terms of the E and B notation,

$$E'_y = \frac{1}{\sqrt{1 - v^2/c^2}} \left[E_y - \frac{v}{c} B_z\right]$$

Since v is a velocity in the $+x$ direction this in turn is

$$E'_y = \frac{1}{\sqrt{1 - v^2/c^2}} \left[E_y + \frac{1}{c}\left(\vec{v} \times \vec{B}\right)_y\right] \tag{12.58}$$

Except possibly for the factor in front of the brackets. this is a familiar, physically correct equation of elementary electromagnetic theory. A charge is always at rest *in its own reference system.* In its own system, the only force it feels is the electric force because its velocity with respect to itself is zero. The electric field that it experiences is $\vec{E}\,'$, not the $\vec{E}$ of the outside world. This calculation tells you that this force $q\vec{E}\,'$ is the same thing that I would expect if I knew the Lorentz force law, $\vec{F} = q\big[\vec{E} + \vec{v} \times \vec{B}\big]$. The factor of $\sqrt{1 - v^2/c^2}$ appears because force itself has some transformation laws that are not as simple as you would expect.

Problems

12.1 Does the function T defined by $T(v) = v + c$ with c a constant satisfy the definition of linearity?

12.2 Let the set X be the positive integers. Let the set Y be all real numbers. Consider the following sets and determine if they are relations between X and Y and if they are functions.

$$\{(0,0),\ (1,2.0),\ (3,-\pi),\ (0,1.0),\ (-1,e)\}$$
$$\{(0,0),\ (1,2.0),\ (3,-\pi),\ (0,1.0),\ (2,e)\}$$
$$\{(0,0),\ (1,2.0),\ (3,-\pi),\ (4,1.0),\ (2,e)\}$$
$$\{(0,0),\ (5,5.5),\ (5.,\pi)\ (3,-2.0)\ (7,8)\}$$

12.3 Starting from the definition of the tensor of inertia in Eq. (12.1) and using the defining equation for components of a tensor, compute the components of I.

12.4 Find the components of the tensor relating $\vec{d}$ and $\vec{F}$ in the example of Eq. (12.2)

12.5 The product of tensors is defined to be just the composition of functions for the second rank tensor viewed as a vector variable. If S and T are such tensors, then $(ST)(v) = S(T(v))$ (by definition) Compute the components of ST in terms of the components of S and of T. Express the result both in terms of index notation and matrices. Ans: This is matrix multiplication.

12.6 (**a**) The two tensors ${}^1_1T$ and ${}^1_1\tilde{T}$ are derived from the same bilinear functional ${}^0_2T$, in the Eqs. (12.20)–(12.22). Prove that for arbitrary $\vec{u}$ and $\vec{v}$,

$$\vec{u}\cdot{}^1_1T(\vec{v}) = {}^1_1\tilde{T}(\vec{u})\cdot\vec{v}$$

(If it's less confusing to remove all the sub- and superscripts, do so.)
(**b**) If you did this by writing everything in terms of components, do it again without components and just using the nature of these as functions. (If you did in without components, do it again using components.)

12.7 What is the significance of a tensor satisfying the relation $\tilde{T}[T(\vec{v})] = T[\tilde{T}(\vec{v})] = \vec{v}$ for all $\vec{v}$? Look at what effect it has on the scalar product of two vectors if you let T act on both. That is, what is the scalar product of $T\vec{u}$ and $T\vec{v}$?

12.8 Carry out the construction indicated in section 12.3 to show the dielectric tensor is symmetric.

12.9 Fill in the details of the proof that the alternating tensor is unique up to a factor.

12.10 Compute the components of such an alternating tensor in two dimensions.
Ans: The matrix is anti-symmetric.

12.11 Take an alternating tensor in three dimensions, pull off one of the arguments in order to obtain a vector valued function of two vector variables. See what the result looks like, and in particular, write it out in component form.
Ans: You should find a cross product here.

12.12 Take a basis $\vec{e}_1 = 2\hat{x}$, $\vec{e}_2 = \hat{x} + 2\hat{y}$. Compute the reciprocal basis. Find the components of the vectors $\vec{A} = \hat{x} - \hat{y}$ and $\vec{B} = \hat{y}$ in in each basis and compute $\vec{A} \cdot \vec{B}$ three different ways: in the $\vec{e}_1$-$\vec{e}_2$ basis, in its reciprocal basis, and mixed ($\vec{A}$ in one basis and $\vec{B}$ in the reciprocal basis). A fourth way uses the $\hat{x}$-$\hat{y}$ basis.

12.13 Show that if the direct basis vectors have unit length along the directions of $\overrightarrow{OA}$ and $\overrightarrow{OB}$ then the components of $\vec{v}$ in the direct basis are the lengths OA and OB. What are the components in the reciprocal basis?

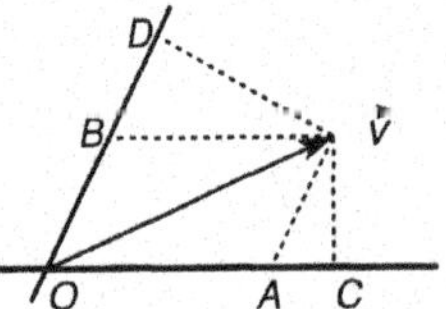

12.14 What happens to the components of the alternating tensor when a change of basis is made? Show that the only thing that can happen is that all the components are multiplied by the same number (which is the determinant of the transformation). Compute this explicitly in two dimensions.

12.15 A given tensor (viewed as a function of two vector variables) has the property that it equals zero whenever the two arguments are the same, $T(\vec{v}, \vec{v}) = 0$. Show that it is antisymmetric. This is also true if it is a function of more than two variables and the above equation holds on some pair of arguments. Consider $\vec{v} = \alpha\vec{u} + \beta\vec{w}$

12.16 If the components of the alternating tensor are (in three dimensions) e_{ijk} where $e_{123} = 1$, Compute e^{ijk}. Compute

$$e^{ijk}e_{\ell mk}, \qquad e^{ijk}e_{\ell jk}, \qquad e^{ijk}e_{ijk}$$

12.17 In three dimensions three non-collinear vectors from a point define a volume, that of the parallelepiped included between them. This defines a number as a function of three vectors. Show that if the volume is allowed to be negative when one of the vectors reverses that this defines a trilinear functional and that it is completely antisymmetric, an alternating tensor. (Note problem 12.15.) If the units of volume are chosen to correspond to the units of length of the basis vectors so that three one inch long perpendicular vectors enclose one cubic inch as opposed to $16.387\,\text{cm}^3$ then the functional is called "the" alternating tensor. Find all its components.

12.18 Find the direct coordinate basis in spherical coordinates, also the reciprocal basis.
Ans: One set is $\hat{r}$, $\hat{\theta}/r$, $\hat{\phi}/r\sin\theta$ (now which ones are these?)

12.19 Draw examples of the direct and reciprocal bases (to scale) for the example in Eq. (12.44). Do this for a wide range of angles between the axes.

12.20 Show that the area in two dimensions enclosed by the infinitesimal parallelogram between x^1 and $x^1 + dx^1$ and x^2 and $x^2 + dx^2$ is $\sqrt{g}dx^1\,dx^2$ where g is the determinant of (g_{ij}).

12.21 Compute the transformation laws for the other components of the electromagnetic field tensor.

12.22 The divergence of a tensor field is defined as above for a vector field. T is a tensor viewed as a vector valued function of a vector variable

$$\operatorname{div} T = \lim_{V \to 0} \frac{1}{V} \oint T(d\vec{A})$$

It is a vector. Note: As defined here it requires the addition of vectors at two different points of the manifold, so it must be assumed that you can move a vector over parallel to itself and add it to a vector at another point. Had I not made the point in the text about not doing this, perhaps you wouldn't have noticed the assumption involved, but now do it anyway. Compute $\operatorname{div} T$ in rectangular coordinates using the $\hat{x}$, $\hat{y}$, $\hat{z}$ basis.
Ans: $\partial T^{ij}/\partial x^j$ in coordinate basis.

12.23 Compute $\operatorname{div} T$ in cylindrical coordinates using both the coordinate basis and the usual unit vector $(\hat{r}, \hat{\phi}, \hat{z})$ basis. This is where you start to see why the coordinate basis has advantages.

12.24 Show that $g^{ij} g_{j\ell} = \delta^i_\ell$.

12.25 If you know what it means for vectors at two different points to be parallel, give a definition for what it means for two tensors to be parallel.

12.26 Fill in the missing steps in the derivation following Eq. (12.24), and show that the alternating tensor is unique up to a factor.

12.27 The divergence of a vector field computed in a coordinate basis is $\operatorname{div} \vec{F} = \partial F^i/\partial x^i$. The divergence computed in standard cylindrical coordinates is Eq. (9.15). Show that these results agree.

12.28 From Eq. (12.31) verify the stated properties of the $\omega = 0$ solution.

Vector Calculus 2

There's more to the subject of vector calculus than the material in chapter nine. There are a couple of types of line integrals and there are some basic theorems that relate the integrals to the derivatives, sort of like the fundamental theorem of calculus that relates the integral to the anti-derivative in one dimension.

13.1 Integrals

Recall the definition of the Riemann integral from section 1.6.

$$\int_a^b dx\, f(x) = \lim_{\Delta x_k \to 0} \sum_{k=1}^{N} f(\xi_k)\, \Delta x_k \tag{13.1}$$

This refers to a function of a single variable, integrated along that one dimension.

The basic idea is that you divide a complicated thing into little pieces to get an approximate answer. Then you refine the pieces into still smaller ones to improve the answer and finally take the limit as the approximation becomes perfect.

What is the length of a curve in the plane? Divide the curve into a lot of small pieces, then if the pieces are small enough you can use the Pythagorean Theorem to estimate the length of each piece.

$$\Delta \ell_k = \sqrt{(\Delta x_k)^2 + (\Delta y_k)^2}$$

The whole curve then has a length that you estimate to be the sum of all these intervals. Finally take the limit to get the exact answer.

$$\sum_k \Delta \ell_k = \sum \sqrt{(\Delta x_k)^2 + (\Delta y_k)^2} \longrightarrow \int d\ell = \int \sqrt{dx^2 + dy^2} \tag{13.2}$$

How do you actually *do* this? That will depend on the way that you use to describe the curve itself. Start with the simplest method and assume that you have a parametric representation of the curve:

$$x = f(t) \qquad \text{and} \qquad y = g(t)$$

Then $dx = \dot{f}(t)dt$ and $dy = \dot{g}(t)dt$, so

$$d\ell = \sqrt{\left(\dot{f}(t)dt\right)^2 + \left(\dot{g}(t)dt\right)^2} = \sqrt{\dot{f}(t)^2 + \dot{g}(t)^2}\, dt \tag{13.3}$$

and the integral for the length is

$$\int d\ell = \int_a^b dt \sqrt{\dot{f}(t)^2 + \dot{g}(t)^2}$$

where a and b are the limits on the parameter t. Think of this as $\int d\ell = \int v\,dt$, where v is the speed.

Do the simplest example first. What is the circumference of a circle? Use the parametrization

$$x = R\cos\phi, \quad y = R\sin\phi \qquad \text{then} \qquad d\ell = \sqrt{(-R\sin\phi)^2 + (R\cos\phi)^2}\,d\phi = R\,d\phi \tag{13.4}$$

The circumference is then $\int d\ell = \int_0^{2\pi} R\,d\phi = 2\pi R$. An ellipse is a bit more of a challenge; see problem 13.3.

If the curve is expressed in polar coordinates you may find another formulation preferable, though in essence it is the same. The Pythagorean Theorem is still applicable, but you have to see what it says in these coordinates.

$$\Delta\ell_k = \sqrt{(\Delta r_k)^2 + (r_k\Delta\phi_k)^2}$$

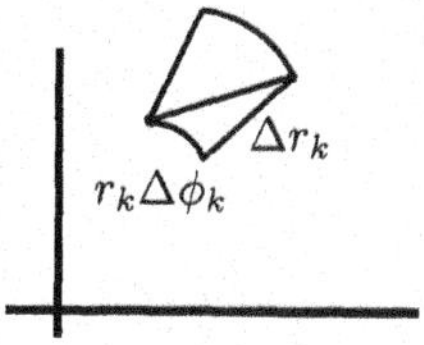

Δr_k

$r_k\Delta\phi_k$

If this picture doesn't seem to show much of a right triangle, remember there's a limit involved, as Δr_k and $\Delta\phi_k$ approach zero this becomes more of a triangle. The integral for the length of a curve is then

$$\int d\ell = \int \sqrt{dr^2 + r^2\,d\phi^2}$$

To actually do this integral you will pick a parameter to represent the curve, and that parameter may even be ϕ itself. For an example, examine one loop of a logarithmic spiral: $r = r_0\,e^{k\phi}$.

$$d\ell = \sqrt{dr^2 + r^2\,d\phi^2} = \sqrt{(dr/d\phi)^2 + r^2}\,d\phi$$

The length of the arc from $\phi = 0$ to $\phi = 2\pi$ is

$$\int \sqrt{\left(r_0 k\,e^{k\phi}\right)^2 + \left(r_0\,e^{k\phi}\right)^2}\,d\phi = \int_0^{2\pi} d\phi\, r_0\,e^{k\phi}\sqrt{k^2+1} = r_0\sqrt{k^2+1}\frac{1}{k}\left[e^{2k\pi} - 1\right]$$

If $k \to 0$ you can easily check that this give the correct answer. In the other extreme, for large k, you can also check that it is a plausible result, but it's a little harder to see.

Weighted Integrals

The time for a particle to travel along a short segment of a path is $dt = d\ell/v$ where v is the speed. The total time along a path is of course the integral of dt.

$$T = \int dt = \int \frac{d\ell}{v} \tag{13.5}$$

How much time does it take a particle to slide down a curve under the influence of gravity? If the speed is determined by gravity without friction, you can use conservation of energy to compute the speed. I'll use the coordinate y measured downward from the top point of the curve, then

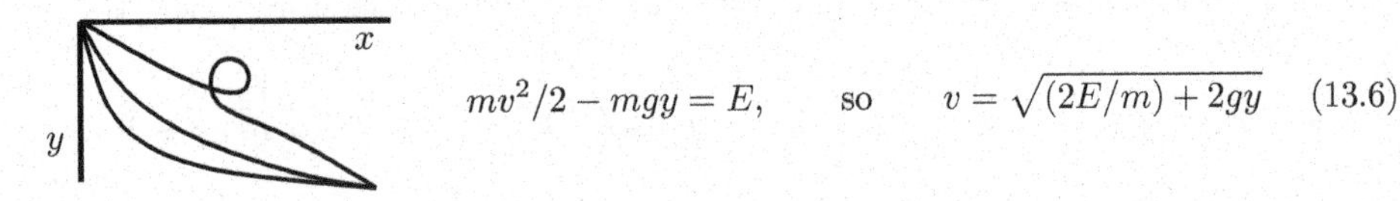

$$mv^2/2 - mgy = E, \qquad \text{so} \qquad v = \sqrt{(2E/m) + 2gy} \tag{13.6}$$

Suppose that this particle starts at rest from $y = 0$, then $E = 0$ and $v = \sqrt{2gy}$. Does the total time to reach a specific point depend on which path you take to get there? Very much so. This will lead to a classic problem called the "brachistochrone." See section 16.3 for that.

1 Take the straight-line path from $(0,0)$ to (x_0, y_0). The path is $x = y \cdot x_0/y_0$.

$$d\ell = \sqrt{dx^2 + dy^2} = dy\sqrt{1 + x_0^2/y_0^2}, \qquad \text{so}$$

$$T = \int \frac{d\ell}{v} = \int_0^{y_0} \frac{dy\sqrt{1 + x_0^2/y_0^2}}{\sqrt{2gy}} = \sqrt{1 + x_0^2/y_0^2}\frac{1}{\sqrt{2g}}\frac{1}{2}\sqrt{y_0} = \frac{1}{2}\frac{\sqrt{x_0^2 + y_0^2}}{\sqrt{2gy_0}} \tag{13.7}$$

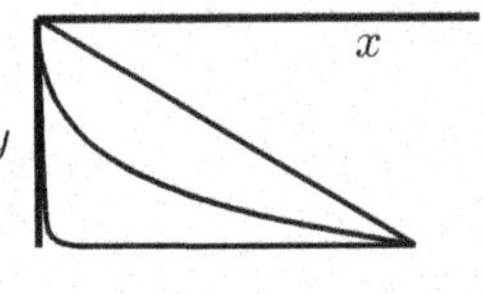

2 There are an infinite number of possible paths, and another choice of path can give a smaller or a larger time. Take another path for which it's easy to compute the total time. Drop straight down in order to pick up speed, then turn a sharp corner and coast horizontally. Compute the time along this path and it is the sum of two pieces.

$$\int_0^{y_0} \frac{dy}{\sqrt{2gy}} + \int_0^{x_0} \frac{dx}{\sqrt{2gy_0}} = \frac{1}{\sqrt{2g}}\left[\frac{1}{2}\sqrt{y_0} + \frac{x_0}{\sqrt{y_0}}\right] = \frac{1}{\sqrt{2gy_0}}[x_0 + y_0/2] \tag{13.8}$$

Which one takes a shorter time? See problem 13.9.

3 What if the path is a parabola, $x = y^2 \cdot x_0/y_0^2$? It drops rapidly at first, picking up speed, but then takes a more direct route to the end. Use y as the coordinate, then

$$dx = 2y \cdot x_0/y_0^2, \qquad \text{and} \qquad d\ell = \sqrt{(4y^2x_0^2/y_0^4) + 1}dy$$

$$T = \int \frac{dx}{v} = \int_0^{y_0} \frac{\sqrt{(4y^2x_0^2/y_0^4) + 1}}{\sqrt{2gy}}dy$$

This is not an integral that you're likely to have encountered yet. I'll refer you to a large table of integrals, where you can perhaps find it under the heading of elliptic integrals.

In more advanced treatments of optics, the time it takes light to travel along a path is of central importance because it is related to the phase of the light wave along that path. In that context however, you usually see it written with an extra factor of the speed of light.

$$cT = \int \frac{c\,d\ell}{v} = \int n\,d\ell \tag{13.9}$$

This last form, written in terms of the index of refraction, is called the optical path. Compare problems 2.35 and 2.39.

13.2 Line Integrals

Work done on a point mass in one dimension is an integral. If the mass moves in three dimensions and if the force happens to be a constant, then work is simply a dot product:

$$W = \int_{x_i}^{x_f} F_x(x)\,dx \qquad \text{respectively} \qquad W = \vec{F}\cdot\Delta\vec{r}$$

The general case for work on a particle moving along a trajectory in space is a line integral. It combines these two equations into a single expression for the work along an arbitrary path for an arbitrary force. There is not then any restriction to the elementary case of constant force.

The basic idea is a combination of Eqs. (13.1) and (13.2). Divide the specified curve into a number of pieces, at the points $\{\vec{r}_k\}$. Between points $k-1$ and k you had the estimate of the arc length as $\sqrt{(\Delta x_k)^2 + (\Delta y_k)^2}$, but here you need the whole vector from $\vec{r}_{k-1}$ to $\vec{r}_k$ in order to evaluate the work done as the mass moves from one point to the next. Let $\Delta\vec{r}_k = \vec{r}_k - \vec{r}_{k-1}$, then

$$\lim_{|\Delta\vec{r}_k|\to 0} \sum_{k=1}^{N} \vec{F}(\vec{r}_k)\cdot\Delta\vec{r}_k = \int \vec{F}(\vec{r})\cdot d\vec{r} \tag{13.10}$$

This is the definition of a line integral.

How do you evaluate these integrals? To repeat what happened with Eq. (13.2), that will depend on the way that you use to describe the curve itself. Start with the simplest method and assume that you have a parametric representation of the curve: $\vec{r}(t)$, then $d\vec{r} = \dot{\vec{r}}\,dt$ and the integral is

$$\int \vec{F}(\vec{r})\cdot d\vec{r} = \int \vec{F}\big(\vec{r}(t)\big)\cdot\dot{\vec{r}}\,dt$$

This is now an ordinary integral with respect to t. In many specific examples, you may find an easier way to represent the curve, but this is something that you can always fall back on.

In order to see exactly where this is used, start with $\vec{F} = m\vec{a}$, Take the dot product with $d\vec{r}$ and manipulate the expression.

$$\vec{F} = m\frac{d\vec{v}}{dt}, \qquad \text{so} \qquad \vec{F}\cdot d\vec{r} = m\frac{d\vec{v}}{dt}\cdot d\vec{r} = m\frac{d\vec{v}}{dt}\cdot\frac{d\vec{r}}{dt}dt = m\,d\vec{v}\cdot\frac{d\vec{r}}{dt} = m\vec{v}\cdot d\vec{v}$$
$$\text{or} \qquad \vec{F}\cdot d\vec{r} = \frac{m}{2}d(\vec{v}\cdot\vec{v}) \tag{13.11}$$

The integral of this from an initial point of the motion to a final point is

$$\int_{\vec{r}_i}^{\vec{r}_f} \vec{F}\cdot d\vec{r} = \int \frac{m}{2}d(\vec{v}\cdot\vec{v}) = \frac{m}{2}\left[v_f^2 - v_i^2\right] \tag{13.12}$$

This is a standard form of the work-energy theorem in mechanics. In most cases you have to specify the whole path, not just the endpoints, so this way of writing the theorem is somewhat misleading. Is it legitimate to manipulate $\vec{v}\cdot d\vec{v}$ as in Eq. (13.11)? Yes. Simply write it in rectangular components as $v_x dv_x + v_y dv_y + v_z dv_z$ and and you can integrate each term with no problem; then assemble the result as $v^2/2$.

Example
If $\vec{F} = Axy\hat{x} + B(x^2 + L^2)\hat{y}$, what is the work done going from point $(0,0)$ to (L,L) along the three different paths indicated.?

$$\int_{C_1} \vec{F}\cdot d\vec{r} = \int [F_x\hat{x} + F_y\hat{y}]\cdot[\hat{x}\,dx + \hat{y}\,dy]$$

$$= \int [F_x dx + F_y dy] = \int_0^L dx\,0 + \int_0^L dy\,B2L^2 = 2BL^3$$

$$\int_{C_2} \vec{F}\cdot d\vec{r} = \int_0^L dx\,Ax^2 + \int_0^L dy\,B(y^2 + L^2) = AL^3/3 + 4BL^3/3$$

$$\int_{C_3} \vec{F}\cdot d\vec{r} = \int_0^L dy\,B(0 + L^2) + \int_0^L dx\,AxL = BL^3 + AL^3/2$$

Gradient
What is the line integral of a gradient? Recall from section 8.5 and Eq. (8.16) that $df = \operatorname{grad} f\cdot d\vec{r}$. The integral of the gradient is then

$$\int_1^2 \operatorname{grad} f\cdot d\vec{r} = \int df = f_2 - f_1 \tag{13.13}$$

where the indices represent the initial and final points. When you integrate a gradient, you need the function only at its endpoints. The path doesn't matter. Well, almost. See problem 13.19 for a caution.

13.3 Gauss's Theorem

The original definition of the divergence of a vector field is Eq. (9.9),

$$\operatorname{div}\vec{v} = \lim_{V\to 0}\frac{1}{V}\frac{dV}{dt} = \lim_{V\to 0}\frac{1}{V}\oint \vec{v}\cdot d\vec{A}$$

Fix a closed surface and evaluate the surface integral of $\vec{v}$ over that surface.

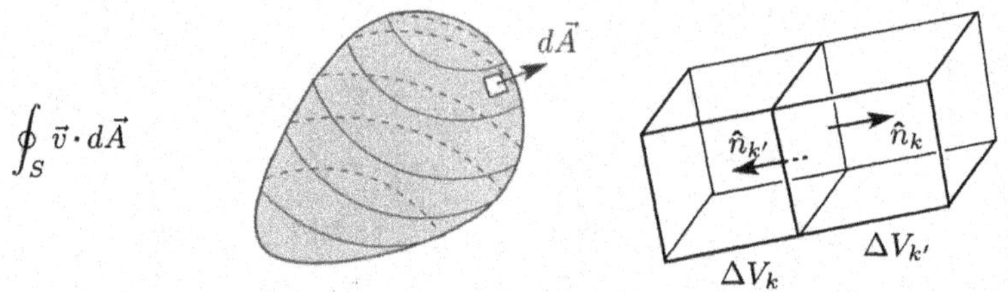

Now divide this volume into a lot of little volumes ΔV_k with individual bounding surfaces S_k. The picture on the right shows just two adjoining pieces of whole volume, but there are many more. If you do the surface integrals of $\vec{v}\cdot d\vec{A}$ over each of these pieces and add all of them, the result is the original surface integral.

$$\sum_k \oint_{S_k} \vec{v}\cdot d\vec{A} = \oint_S \vec{v}\cdot d\vec{A} \tag{13.14}$$

The reason for this is that each interior face of volume V_k is matched with the face of an adjoining volume $V_{k'}$. The latter face will have $d\vec{A}$ pointing in the opposite direction, $\hat{n}_{k'} = -\hat{n}_k$, so when you add all the interior surface integrals they cancel. All that's left is the surface on the outside and the sum over all *those* faces is the original surface integral.

In the equation (13.14) multiply and divide every term in the sum by the volume ΔV_k.

$$\sum_k \left[\frac{1}{\Delta V_k}\oint_{S_k} \vec{v}\cdot d\vec{A}\right]\Delta V_k = \oint_S \vec{v}\cdot d\vec{A}$$

Now increase the number of subdivisions, finally taking the limit as all the ΔV_k approach zero. The quantity inside the brackets becomes the definition of the divergence of $\vec{v}$ and you then get

$$\text{Gauss's Theorem:}\qquad \int_V \operatorname{div}\vec{v}\,dV = \oint_S \vec{v}\cdot d\vec{A} \tag{13.15}$$

This* is Gauss's theorem, the divergence theorem.

Example

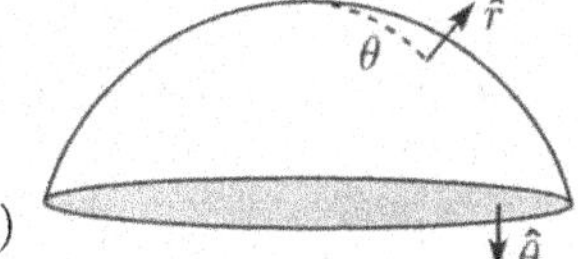

Verify Gauss's Theorem for the solid hemisphere, $r \le R$, $0 \le \theta \le \pi/2$, $0 \le \phi \le 2\pi$. Use the vector field

$$\vec{F} = \hat{r}\alpha r^2\sin\theta + \hat{\theta}\beta r\theta^2\cos^2\phi + \hat{\phi}\gamma r\sin\theta\cos^2\phi \tag{13.16}$$

Doing the surface integral on the hemisphere $\hat{n} = \hat{r}$, and on the bottom flat disk you have $\hat{n} = \hat{\theta}$. The surface integral is then assembled from two pieces,

$$\begin{aligned}\oint \vec{F}\cdot d\vec{A} &= \int_{\text{hemisph}} \hat{r}\alpha r^2\sin\theta\cdot\hat{r}\,dA + \int_{\text{disk}} \hat{\theta}\beta r\theta^2\cos^2\phi\cdot\hat{\theta}\,dA \\ &= \int_0^{\pi/2} R^2\sin\theta\,d\theta\int_0^{2\pi} d\phi\,\alpha R^2\sin\theta + \int_0^R r\,dr\int_0^{2\pi} d\phi\,\beta r(\pi/2)^2\cos^2\phi \\ &= \alpha\pi^2R^4/2 + \beta\pi^3R^3/12 \end{aligned}\tag{13.17}$$

Now do the volume integral of the divergence, using Eq. (9.16).

$$\begin{aligned}\operatorname{div}\vec{F} &= \frac{1}{r^2}\frac{\partial}{\partial r}\alpha r^4\sin\theta + \frac{1}{r\sin\theta}\frac{\partial}{\partial\theta}\beta r\theta^2\sin\theta\cos^2\phi + \frac{1}{r\sin\theta}\frac{\partial}{\partial\phi}\gamma r\sin\theta\cos^2\phi \\ &= 4\alpha r\sin\theta + \beta\cos^2\phi[2\theta + \theta^2\cot\theta] + 2\gamma\sin\phi\cos\phi\end{aligned}$$

The γ term in the volume integral is zero because the $2\sin\phi\cos\phi = \sin 2\phi$ factor averages to zero over all ϕ.

$$\begin{aligned}&\int_0^R dr\,r^2\int_0^{\pi/2}\sin\theta\,d\theta\int_0^{2\pi} d\phi\left[4\alpha r\sin\theta + \beta\cos^2\phi[2\theta + \theta^2\cot\theta]\right] \\ &\qquad = 4\alpha\cdot\frac{R^4}{4}\cdot 2\pi\cdot\frac{\pi}{4} + \beta\cdot\frac{R^3}{3}\cdot\pi\cdot\int_0^{\pi/2} d\theta\,\sin\theta[2\theta + \theta^2\cot\theta] \\ &\qquad = \alpha\pi^2R^2/2 + \beta\pi^3R^3/12\end{aligned}$$

---

* You will sometimes see the notation ∂V instead of S for the boundary surface surrounding the volume V. Also ∂A instead of C for the boundary curve surrounding the area A. It is probably a better and more consistent notation, but it isn't yet as common in physics books.

The last integration used parametric differentiation starting from $\int_0^{\pi/2} d\theta \cos k\theta$, with differentiation with respect to k.

13.4 Stokes' Theorem

The expression for the curl in terms of integrals is Eq. (9.17),

$$\operatorname{curl} \vec{v} = \lim_{V\to 0} \frac{1}{V} \oint d\vec{A} \times \vec{v} \tag{13.18}$$

Use the same reasoning that leads from the definition of the divergence to Eqs. (13.14) and (13.15) (see problem 13.6), and this leads to the analog of Gauss's theorem, but with cross products.

$$\oint_S d\vec{A} \times \vec{v} = \int_V \operatorname{curl} \vec{v}\, dV \tag{13.19}$$

This isn't yet in a form that is all that convenient, and a special case is both easier to interpret and more useful in applications. First apply it to a particular volume, one that is very thin and small. Take a tiny disk of height Δh, with top and bottom area ΔA_1. Let $\hat{n}_1$ be the unit normal vector out of the top area. For small enough values of these dimensions, the right side of Eq. (13.18) is simply the value of the vector $\operatorname{curl} \vec{v}$ inside the volume times the volume $\Delta A_1 \Delta h$ itself.

$$\oint_S d\vec{A} \times \vec{v} = \int_V \operatorname{curl} \vec{v}\, dV = \operatorname{curl} \vec{v}\; \Delta A_1 \Delta h$$

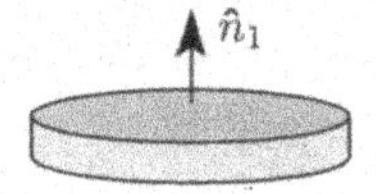

Take the dot product of both sides with $\hat{n}_1$, and the parts of the surface integral from the top and the bottom faces disappear. That's just the statement that on the top and the bottom, $d\vec{A}$ is in the direction of $\pm\hat{n}_1$, so the cross product makes $d\vec{A} \times \vec{v}$ perpendicular to $\hat{n}_1$.

I'm using the subscript $_1$ for the top surface and I'll use $_2$ for the surface around the edge. Otherwise it's too easy to get the notation mixed up.

Now look at $d\vec{A} \times \vec{v}$ around the thin edge. The element of area has height Δh and length $\Delta \ell$ along the arc. Call $\hat{n}_2$ the unit normal out of the edge.

$$\Delta \vec{A}_2 = \Delta h \Delta \ell \hat{n}_2$$

The product $\hat{n}_1 \cdot \Delta\vec{A}_2 \times \vec{v} = \hat{n}_1 \cdot \hat{n}_2 \times \vec{v}\, \Delta h \Delta \ell = \hat{n}_1 \times \hat{n}_2 \cdot \vec{v}\, \Delta h \Delta \ell$, using the property of the triple scalar product. The product $\hat{n}_1 \times \hat{n}_2$ is in the direction along the arc of the edge, so

$$\hat{n}_1 \times \hat{n}_2\, \Delta \ell = \Delta \vec{\ell} \tag{13.20}$$

Put all these pieces together and you have

$$\hat{n}_1 \cdot \oint_S d\vec{A} \times \vec{v} = \oint_C \vec{v} \cdot d\vec{\ell}\, \Delta h = \hat{n}_1 \cdot \operatorname{curl} \vec{v}\; \Delta A_1 \Delta h$$

Divide by $\Delta A_1 \Delta h$ and take the limit as $\Delta A_1 \to 0$. Recall that all the manipulations above work only under the assumption that you take this limit.

$$\hat{n}_1 \cdot \operatorname{curl} \vec{v} = \lim_{\Delta A \to 0} \frac{1}{\Delta A} \oint_C \vec{v} \cdot d\vec{\ell} \tag{13.21}$$

You will sometimes see this equation (13.21) taken as the definition of the curl, and it does have an intuitive appeal. The one drawback to doing this is that it isn't at all obvious that the thing on the right-hand side is the dot product of $\hat{n}_1$ with anything. It is, because I deduced that fact from the vectors in Eq. (13.19), but if you use Eq. (13.21) as your starting point you have some proving to do.

This form is easier to interpret than was the starting point with a volume integral. The line integral of $\vec{v} \cdot d\vec{\ell}$ is called the circulation of $\vec{v}$ around the loop. Divide this by the area of the loop and take the limit as the area goes to zero and you then have the "circulation density" of the vector field. The component of the curl along some direction is then the circulation density around that direction. Notice that the equation (13.20) dictates the right-hand rule that the direction of integration around the loop is related to the direction of the normal $\hat{n}_1$.

Stokes' theorem follows in a few lines from Eq. (13.21). Pick a surface A with a boundary C (or ∂A in the other notation). The surface doesn't have to be flat, but you have to be able to tell one side from the other.* From here on I'll imitate the procedure of Eq. (13.14). Divide the surface into a lot of little pieces ΔA_k, and do the line integral of $\vec{v} \cdot d\vec{\ell}$ around each piece. Add all these pieces and the result is the whole line integral around the outside curve.

$$\sum_k \oint_{C_k} \vec{v} \cdot d\vec{\ell} = \oint_C \vec{v} \cdot d\vec{\ell} \tag{13.22}$$

As before, on each interior boundary between area ΔA_k and the adjoining $\Delta A_{k'}$, the parts of the line integrals on the common boundary cancel because the directions of integration are opposite to each other. All that's left is the curve on the outside of the whole loop, and the sum over *those* intervals is the original line integral.

Multiply and divide each term in the sum (13.22) by ΔA_k and you have

$$\sum_k \left[\frac{1}{\Delta A_k} \oint_{C_k} \vec{v} \cdot d\vec{\ell} \right] \Delta A_k = \oint_C \vec{v} \cdot d\vec{\ell} \tag{13.23}$$

Now increase the number of subdivisions of the surface, finally taking the limit as all the $\Delta A_k \to 0$, and the quantity inside the brackets becomes the normal component of the curl of $\vec{v}$ by Eq. (13.21). The limit of the sum is the definition of an integral, so

$$\text{Stokes' Theorem:} \qquad \int_A \operatorname{curl} \vec{v} \cdot d\vec{A} = \oint_C \vec{v} \cdot d\vec{\ell} \tag{13.24}$$

---

* That means no Klein bottles or Möbius strips.

What happens if the vector field $\vec{v}$ is the gradient of a function, $\vec{v} = \nabla f$? By Eq. (13.13) the line integral in (13.24) depends on just the endpoints of the path, but in this integral the initial and final points are the same. That makes the integral zero: $f_1 - f_1$. That implies that the surface integral on the left is zero no matter what the surface spanning the contour is, and that can happen only if the thing being integrated is itself zero. $\operatorname{curl}\operatorname{grad} f = 0$. That's one of the common vector identities in problem 9.36. Of course this statement requires the usual assumption that there are no singularities of $\vec{v}$ within the area.

Example

Verify Stokes' theorem for that part of a spherical surface $r = R$, $0 \le \theta \le \theta_0$, $0 \le \phi < 2\pi$. Use for this example the vector field

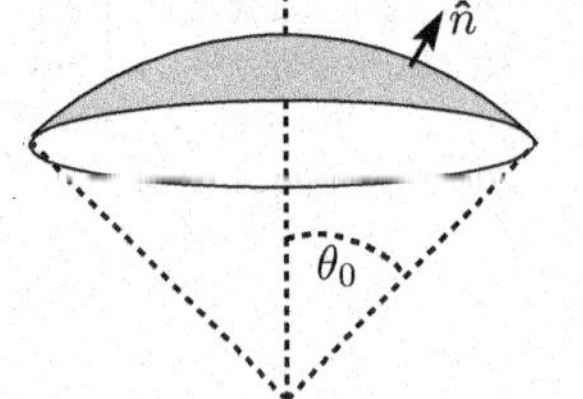

$$\vec{F} = \hat{r}\,Ar^2\sin\theta + \hat{\theta}\,Br\theta^2\cos\phi + \hat{\phi}Cr\sin\theta\cos^2\phi \tag{13.25}$$

To compute the curl of $\vec{F}$, use Eq. (9.33), getting

$$\begin{aligned}\nabla\times\vec{F} &= \hat{r}\frac{1}{r\sin\theta}\left(\frac{\partial}{\partial\theta}\big(\sin\theta\,Cr\sin\theta\cos^2\phi\big) - \frac{\partial}{\partial\phi}\big(Br\theta^2\cos\phi\big)\right) + \cdots \\ &= \hat{r}\frac{1}{r\sin\theta}\big(Cr\cos^2\phi\,2\sin\theta\cos\theta + Br\theta^2\sin\phi\big) + \cdots \end{aligned}\tag{13.26}$$

I need only the $\hat{r}$ component of the curl because the surface integral uses only the normal ($\hat{r}$) component. The surface integral of this has the area element $dA = r^2\sin\theta\,d\theta d\phi$.

$$\begin{aligned}\int \operatorname{curl}\vec{F}\cdot d\vec{A} &= \int_0^{\theta_0} R^2\sin\theta\,d\theta\int_0^{2\pi} d\phi\frac{1}{R\sin\theta}\big(CR\cos^2\phi\,2\sin\theta\cos\theta + BR\theta^2\sin\phi\big) \\ &= R^2\int_0^{\theta_0} d\theta\int_0^{2\pi} d\phi\,2C\cos^2\phi\sin\theta\cos\theta \\ &= R^2 2C\pi\sin^2\theta_0/2 = CR^2\pi\sin^2\theta_0\end{aligned}$$

The other side of Stokes' theorem is the line integral around the circle at angle θ_0.

$$\begin{aligned}\oint \vec{F}\cdot d\vec{\ell} &= \int_0^{2\pi} r\sin\theta_0\,d\phi\,Cr\sin\theta\cos^2\phi \\ &= \int_0^{2\pi} d\phi\,CR^2\sin^2\theta_0\cos^2\phi \\ &= CR^2\sin^2\theta_0\,\pi\end{aligned}\tag{13.27}$$

and the two sides of the theorem agree. Check! Did I get the overall signs right? The direction of integration around the loop matters. A further check: If $\theta_0 = \pi$, the length of the loop is zero and both integrals give zero as they should.

Conservative Fields

An immediate corollary of Stokes' theorem is that if the curl of a vector field is zero throughout a region then line integrals are independent of path in that region. To state it a bit more precisely, in a volume for which any closed path can be shrunk to a point without leaving the region, if the curl of $\vec{v}$ equals zero, then $\int_a^b \vec{F}\cdot d\vec{r}$ depends on the endpoints of the path, and not on how you get there.

To see why this follows, take two integrals from point a to point b.

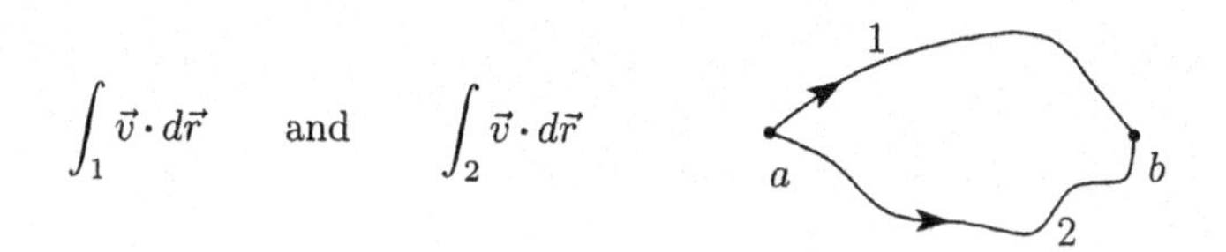

The difference of these two integrals is

$$\int_1 \vec{v}\cdot d\vec{r} - \int_2 \vec{v}\cdot d\vec{r} = \oint \vec{v}\cdot d\vec{r}$$

This equations happens because the minus sign is the same thing that you get by integrating in the reverse direction. For a field with $\nabla\times\vec{v} = 0$, Stokes' theorem says that this closed path integral is zero, and the statement is proved.

What was that fussy-sounding statement "for which any closed path can be shrunk to a point without leaving the region?" Consider the vector field in three dimensions, written in rectangular and cylindrical coordinates,

$$\vec{v} = A(x\hat{y} - y\hat{x})/(x^2+y^2) = A\hat{\phi}/r \tag{13.28}$$

You can verify (in either coordinate system) that its curl is zero — except for the z-axis, where it is singular. A closed loop line integral that doesn't encircle the z-axis will be zero, but if it does go around the axis then it is not. The vector's direction $\hat{\theta}$ always points counterclockwise around the axis. See problem 13.17. If you have a loop that encloses the singular line, then you can't shrink the loop without its getting hung up on the axis.

The converse of this theorem is also true. If every closed-path line integral of $\vec{v}$ is zero, and if the derivatives of $\vec{v}$ are continuous, then its curl is zero. Stokes' theorem tells you that every surface integral of $\nabla\times\vec{v}$ is zero, so you can pick a point and a small $\Delta\vec{A}$ at this point. For small enough area whatever the curl is, it won't change much. The integral over this small area is then $\nabla\times\vec{v}\cdot\Delta\vec{A}$, and by assumption this is zero. It's zero for all values of the area vector. The only vector whose dot product with all vectors is zero is itself the zero vector.

Potentials

The relation between the vanishing curl and the fact that the line integral is independent of path leads to the existence of potential functions.

If $\operatorname{curl}\vec{F} = 0$ in a simply-connected domain (that's one for which any closed loop can be shrunk to a point), then I can write $\vec{F}$ as a gradient, $-\operatorname{grad} f$. The minus sign is conventional. I've already constructed the answer (almost), and to complete the calculation note that line integrals are independent of path in such a domain, and that means that the integral

$$\int_{\vec{r}_0}^{\vec{r}} \vec{F} \cdot d\vec{r} \tag{13.29}$$

is a function of the two endpoints alone. Fix $\vec{r}_0$ and treat this as a function of the upper limit $\vec{r}$. Call it $-f(\vec{r})$. The defining equation for the gradient is Eq. (8.16),

$$df = \operatorname{grad} f \cdot d\vec{r}$$

How does the integral (13.29) change when you change $\vec{r}$ a bit?

$$\int_{\vec{r}_0}^{\vec{r}+d\vec{r}} \vec{F} \cdot d\vec{r} - \int_{\vec{r}_0}^{\vec{r}} \vec{F} \cdot d\vec{r} = \int_{\vec{r}}^{\vec{r}+d\vec{r}} \vec{F} \cdot d\vec{r} = F \cdot d\vec{r}$$

This is $-df$ because I called this integral $-f(\vec{r})$. Compare the last two equations and because $d\vec{r}$ is arbitrary you immediately get

$$\vec{F} = -\operatorname{grad} f \tag{13.30}$$

I used this equation in section 9.9, stating that the existence of the gravitational potential energy followed from the fact that $\nabla \times \vec{g} = 0$.

Vector Potentials

This is not strictly under the subject of conservative fields, but it's a convenient place to discuss it anyway. When a vector field has zero curl then it's a gradient. When a vector field has zero divergence then it's a curl. In both cases the converse is simple, and it's what you see first: $\nabla \times \nabla f = 0$ and $\nabla \cdot \nabla \times \vec{A} = 0$ (problem 9.36). In Eqs. (13.29) and (13.30) I was able to construct the function f because $\nabla \times \vec{F} = 0$. It is also possible, if $\nabla \cdot \vec{F} = 0$, to construct the function $\vec{A}$ such that $\vec{F} = \nabla \times \vec{A}$.

In both cases, there are extra conditions needed for the statements to be completely true. To conclude that a conservative field ($\nabla \times \vec{F} = 0$) is a gradient requires that the domain be simply-connected, allowing the line integral to be completely independent of path. To conclude that a field satisfying $\nabla \cdot \vec{F} = 0$ can be written as $\vec{F} = \nabla \times \vec{A}$ requires something similar: that all closed *surfaces* can be shrunk to a point. This statement is not so easy to prove, and the explicit construction of $\vec{A}$ from $\vec{F}$ is not very enlightening.

You can easily verify that $\vec{A} = \vec{B} \times \vec{r}/2$ is a vector potential for the uniform field $\vec{B}$. Neither the scalar potential nor the vector potential are unique. You can always add a constant to a scalar potential because the gradient of a scalar is zero and it doesn't change the result. For the vector potential you can add the gradient of an arbitrary function because that doesn't change the curl.

$$\vec{F} = -\nabla(f + C) = -\nabla f, \qquad \text{and} \qquad \vec{B} = \nabla \times (\vec{A} + \nabla f) = \nabla \times \vec{A} \tag{13.31}$$

13.5 Reynolds Transport Theorem

When an integral has limits that are functions of time, how do you differentiate it? That's pretty easy for one-dimensional integrals, as in Eqs. (1.19) and (1.21).

$$\frac{d}{dt}\int_{f_1(t)}^{f_2(t)} dx\, g(x,t) = \int_{f_1(t)}^{f_2(t)} dx\, \frac{\partial g(x,t)}{\partial t} + g(f_2(t),t)\frac{df_2(t)}{dt} - g(f_1(t),t)\frac{df_1(t)}{dt} \tag{13.32}$$

One of Maxwell's equations for electromagnetism is

$$\nabla\times\vec{E} = -\frac{\partial\vec{B}}{\partial t} \tag{13.33}$$

Integrate this equation over the surface S.

$$\int_S \nabla\times\vec{E}\cdot d\vec{A} = \int_C \vec{E}\cdot d\vec{\ell} = \int_S -\frac{\partial\vec{B}}{\partial t}\cdot d\vec{A} \tag{13.34}$$

This used Stokes' theorem, and I would like to be able to pull the time derivative out of the integral, but can I? If the surface is itself time independent then the answer is yes, but what if it isn't? What if the surface integral has a surface that is moving? Can this happen? That's how generators works, and you wouldn't be reading this now without the power they provide. The copper wire loops are rotating at high speed, and it is this motion that provides the EMF.

I'll work backwards and compute the time derivative of a surface integral, allowing the surface itself to move. To do this, I'll return to the definition of a derivative. The time variable appears in two places, so use the standard trick of adding and subtracting a term, just as in section 1.5. Call Φ the flux integral, $\int \vec{B}\cdot d\vec{A}$.

$$\begin{aligned}\Delta\Phi &= \int_{S(t+\Delta t)} \vec{B}(t+\Delta t)\cdot d\vec{A} - \int_{S(t)} \vec{B}(t)\cdot d\vec{A}\\ &= \int_{S(t+\Delta t)} \vec{B}(t+\Delta t)\cdot d\vec{A} - \int_{S(t+\Delta t)} \vec{B}(t)\cdot d\vec{A}\\ &\qquad + \int_{S(t+\Delta t)} \vec{B}(t)\cdot d\vec{A} - \int_{S(t)} \vec{B}(t)\cdot d\vec{A}\end{aligned} \tag{13.35}$$

$\vec{B}$ is a function of $\vec{r}$ too, but I won't write it. The first two terms have the same surface, so they combine to give

$$\int_{S(t+\Delta t)} \Delta\vec{B}\cdot d\vec{A}$$

and when you divide by Δt and let it approach zero, you get

$$\int_{S(t)} \frac{\partial\vec{B}}{\partial t}\cdot d\vec{A}$$

Now for the next two terms, which require some manipulation. Add and subtract the surface that forms the edge between the boundaries $C(t)$ and $C(t+\Delta t)$.

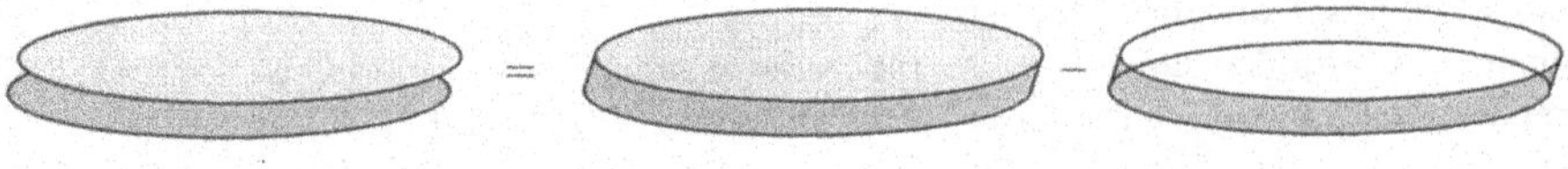

$$\int_{S(t+\Delta t)} \vec{B}(t)\cdot d\vec{A} - \int_{S(t)} \vec{B}(t)\cdot d\vec{A} = \oint \vec{B}(t)\cdot d\vec{A} - \int_{\text{edge}} \vec{B}\cdot d\vec{A} \tag{13.36}$$

The strip around the edge between the two surfaces make the surface integral closed, but I then have to subtract it as a separate term.

You can convert the surface integral to a volume integral with Gauss's theorem, but it's still necessary to figure out how to write the volume element. [Yes, $\nabla\cdot\vec{B} = 0$, but this result can be applied in other cases too, so don't use that fact here.] The surface is moving at velocity $\vec{v}$, so an area element $\Delta\vec{A}$ will in time Δt sweep out a volume $\Delta\vec{A}\cdot\vec{v}\Delta t$. Note: $\vec{v}$ isn't necessarily a constant in space and these surfaces aren't necessarily flat.

$$\Delta V = \Delta\vec{A}\cdot\vec{v}\Delta t \implies \oint \vec{B}(t)\cdot d\vec{A} = \int d^3r\,\nabla\cdot\vec{B} = \int_{S(t)} \nabla\cdot\vec{B}\,d\vec{A}\cdot\vec{v}\Delta t \tag{13.37}$$

To do the surface integral around the edge, use the same method as in deriving Stokes' theorem, Eq. (13.20).

$$\Delta\vec{A} = \Delta\vec{\ell}\times\vec{v}\Delta t$$

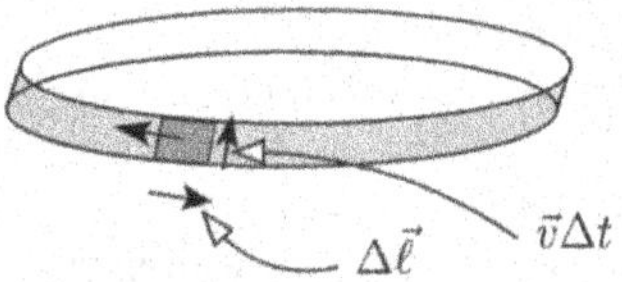

$$\int_{\text{edge}} \vec{B}\cdot d\vec{A} = \int_C \vec{B}\cdot d\vec{\ell}\times\vec{v}\Delta t = \int_C \vec{v}\times\vec{B}\cdot d\vec{\ell}\,\Delta t \tag{13.38}$$

Put Eqs. (13.37) and (13.38) into Eq. (13.36) and then into Eq. (13.35), divide by Δt and let $\Delta t\to 0$.

$$\frac{d}{dt}\int_{S(t)} \vec{B}\cdot d\vec{A} = \int_{S(t)} \frac{\partial\vec{B}}{\partial t}\cdot d\vec{A} + \int_{S(t)} \nabla\cdot\vec{B}\,\vec{v}\cdot d\vec{A} - \int_{C(t)} \vec{v}\times\vec{B}\cdot d\vec{\ell} \tag{13.39}$$

This transport theorem is the analog of Eq. (13.32) for a surface integral.

In order to check this equation, and to see what the terms do, try some example vector functions that isolate the terms, so that only one of the terms on the right side of (13.39) is non-zero at a time.

$$1: \qquad \vec{B} = B_0\,\hat{z}\,t, \qquad \text{with a surface} \qquad z = 0, \quad x^2+y^2 < R^2$$

For a constant B_0, and $\vec{v} = 0$, only the first term is present. The equation is $B_0\pi R^2 = B_0\pi R^2$.

Now take a static field

$$2: \qquad \vec{B} = Cz\,\hat{z}, \qquad \text{with a moving surface} \qquad z = vt, \quad x^2+y^2 < R^2$$

The first and third terms on the right vanish, and $\nabla \cdot \vec{B} = C$. The other terms are

$$\frac{d}{dt} Cz\,\hat{z} \cdot \pi R^2\,\hat{z}\Big|_{z=vt} = Cv\pi R^2 = \int (C)v\,\hat{z} \cdot d\vec{A} = Cv\pi R^2$$

Now take a uniform static field

$$3:\ \vec{B} = B_0\,\hat{z} \qquad \text{with a radially expanding surface} \qquad z = 0, \quad x^2 + y^2 < R^2, \quad R = vt$$

The first and second terms on the right are now zero, and

$$\frac{d}{dt} B_0\pi(vt)^2 = 2B_0\pi v^2 t = -\oint (v\hat{r} \times B_0\,\hat{z}) \cdot \hat{\theta} d\ell$$
$$= -\oint (-vB_0\,\hat{\theta}) \cdot \hat{\theta} d\ell = +vB_0 2\pi R\Big|_{R=vt} = 2B_0\pi v^2 t$$

Draw some pictures of these three cases to see if the pictures agree with the algebra.

Faraday's Law

If you now apply the transport theorem (13.39) to Maxwell's equation (13.34), and use the fact that $\nabla \cdot \vec{B} = 0$ you get

$$\int_{C(t)} (\vec{E} + \vec{v} \times \vec{B}) \cdot d\vec{\ell} = -\frac{d}{dt} \int_{S(t)} \vec{B} \cdot d\vec{A} \tag{13.40}$$

This is Faraday's law, saying that the force per charge integrated around a closed loop (called the EMF) is the negative time derivative of the magnetic flux through the loop.

Occasionally you will find an introductory physics text that writes Faraday's law without the $\vec{v} \times \vec{B}$ term. That's o.k. as long as the integrals involve only stationary curves and surfaces, but some will try to apply it to generators, with moving conductors. This results in amazing contortions to try to explain the results. For another of Maxwell's equations, see problem 13.30.

The electromagnetic force on a charge is $\vec{F} = q(\vec{E} + \vec{v} \times \vec{B})$. This means that if a charge inside a conductor is free to move, the force on it comes from both the electric and the magnetic fields in this equation. (The Lorentz force law.) The integral of this force $\cdot\, d\vec{\ell}$ is the work done on a charge along some specified path. If this integral is independent of path: $\nabla \times \vec{E} = 0$ and $\vec{v} = 0$, then this work divided by the charge is the potential difference, the voltage, between the initial and final points. In the more general case, where one or the other of these requirements is false, then it's given the somewhat antiquated name EMF, for "electromotive force." (It is often called "voltage" anyway, though if you're being fussy that's not really correct.)

13.6 Fields as Vector Spaces

It's sometimes useful to look back at the general idea of a vector space and to rephrase some common ideas in that language. Vector fields, such as $\vec{E}(x, y, z)$ can be added and multiplied by scalars. They form vector spaces, infinite dimensional of course. They even have a natural scalar product

$$\langle \vec{E}_1, \vec{E}_2 \rangle = \int d^3r\, \vec{E}_1(\vec{r}) \cdot \vec{E}_2(\vec{r}) \tag{13.41}$$

Here I'm assuming that the scalars are real numbers, though you can change that if you like. For this to make sense, you have to assume that the fields are square integrable, but for the case of electric or magnetic fields that just means that the total energy in the field is finite. Because these are supposed to satisfy some differential equations (Maxwell's), the derivative must also be square integrable, and I'll require that they go to zero at infinity faster than $1/r^3$ or so.

The curl is an operator on this space, taking a vector field into another vector field. Recall the definitions of symmetric and hermitian operators from section 7.14. The curl satisfies the identity

$$\langle \vec{E}_1, \nabla\times\vec{E}_2\rangle = \langle \nabla\times\vec{E}_1, \vec{E}_2\rangle \tag{13.42}$$

For a proof, just write it out and then find the vector identity that will allow you to integrate by parts.

$$\nabla\cdot\big(\vec{A}\times\vec{B}\big) = \vec{B}\cdot\nabla\times\vec{A} - \vec{A}\cdot\nabla\times\vec{B} \tag{13.43}$$

Equation (13.42) is

$$\int d^3r\, \vec{E}_1(\vec{r})\cdot\nabla\times\vec{E}_2(\vec{r}) = \int d^3r\, \big(\nabla\times\vec{E}_1(\vec{r})\big)\cdot\vec{E}_2(\vec{r}) - \int d^3r\, \nabla\cdot\big(\vec{E}_1\times\vec{E}_2\big)$$

The last integral becomes a surface integral by Gauss's theorem, $\oint d\vec{A}\cdot\big(\vec{E}_1\times\vec{E}_2\big)$, and you can now let the volume (and so the surface) go to infinity. The fields go to zero sufficiently fast, so this is zero and the result is proved: Curl is a symmetric operator. Its eigenvalues are real and its eigenvectors are orthogonal. This is not a result you will use often, but the next one is important.

Helmholtz Decomposition

There are subspaces in this vector space of fields: (1) The set of all fields that are gradients. (2) The set of all fields that are curls. These subspaces are orthogonal to each other; every vector in the first is orthogonal to every vector in the second. To prove this, just use the same vector identity (13.43) and let $\vec{A} = \nabla f$. I will first present a restricted version of this theorem because it's simpler. Assume that the domain is all space and that the fields and their derivatives all go to zero infinitely far away. A generalization to finite boundaries will be mentioned at the end.

$$\nabla f\cdot\nabla\times\vec{B} = \vec{B}\cdot\nabla\times\nabla f - \nabla\cdot\big(\nabla f\times\vec{B}\big)$$

Calculate the scalar product of one vector field with the other.

$$\begin{aligned}\langle\nabla f, \nabla\times B\rangle &= \int d^3r\, \nabla f\cdot\nabla\times B = \int d^3r\, \big[\vec{B}\cdot\nabla\times\nabla f - \nabla\cdot\big(\nabla f\times\vec{B}\big)\big] \\ &= 0 - \oint \big(\nabla f\times\vec{B}\big)\cdot d\vec{A} = 0 \end{aligned} \tag{13.44}$$

As usual, the boundary condition that the fields and their derivatives go to zero rapidly at infinity kills the surface term. This proves the result, that the two subspaces are mutually orthogonal.

Do these two cases exhaust all possible vector fields? In this restricted case with no boundaries short of infinity, the answer is yes. The general case later will add other possibilities. Here you have two orthogonal subspaces, and to show that these two fill out the whole vector space, I will ask the question: what are all the vector fields orthogonal to *both* of them? I will show first that whatever they are will satisfy Laplace's equation, and then the fact that the fields go to zero at infinity will be enough to show that this third case is identically zero. This statement is the Helmholtz theorem: Such vector fields can be written as the sum of two orthogonal fields: a gradient, and a curl.

To prove it, my plan of attack is to show that if a field $\vec F$ is orthogonal to all gradients and to all curls, then $\nabla^2\vec F$ is orthogonal to *all* square-integrable vector fields. The only vector that is orthogonal to everything is the zero vector, so $\vec F$ satisfies Laplace's equation. The assumption now is that for general f and $\vec v$,

$$\int d^3r\,\vec F\cdot\nabla f = 0 \qquad \text{and} \qquad \int d^3r\,\vec F\cdot\nabla\times\vec v = 0$$

I want to show that for a general vector field $\vec u$,

$$\int d^3r\,\vec u\cdot\nabla^2\vec F = 0$$

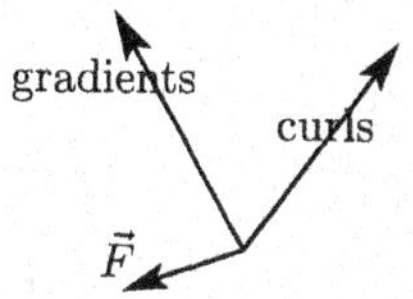

The method is essentially two partial integrals, moving two derivatives from $\vec F$ over to $\vec u$. Start with the $\partial^2/\partial z^2$ term in the Laplacian and hold off on the dx and dy integrals. Remember that all these functions go to zero at infinity. Pick the i-component of $\vec u$ and the j-component of $\vec F$.

$$\begin{aligned}\int_{-\infty}^{\infty} dz\,u_i\frac{\partial^2}{\partial z^2}F_j &= u_i\partial_z F_j\Big|_{-\infty}^{\infty} - \int dz\,(\partial_z u_i)(\partial_z F_j)\\ &= 0 - (\partial_z u_i)F_j\Big|_{-\infty}^{\infty} + \int dz\,(\partial_z^2 u_i)F_j = \int_{-\infty}^{\infty} dz\,(\partial_z^2 u_i)F_j\end{aligned}$$

Now reinsert the dx and dy integrals. Repeat this for the other two terms in the Laplacian, $\partial_x^2 F_j$ and $\partial_y^2 F_j$. The result is

$$\int d^3r\,\vec u\cdot\nabla^2\vec F = \int d^3r\,(\nabla^2\vec u)\cdot\vec F \tag{13.45}$$

If this looks familiar it is just the three dimensional version of the manipulations that led to Eq. (5.15).

Now use the identity

$$\nabla\times(\nabla\times\vec u) = \nabla(\nabla\cdot\vec u) - \nabla^2\vec u$$

in the right side of (13.45) to get

$$\int d^3r\,\vec u\cdot\nabla^2\vec F = \int d^3r\left[\big(\nabla(\nabla\cdot\vec u)\big)\cdot\vec F - \big(\nabla\times(\nabla\times\vec u)\big)\cdot\vec F\right] \tag{13.46}$$

The first term on the right is the scalar product of the vector field $\vec{F}$ with a gradient. The second term is the scalar product with a curl. Both are zero by the hypotheses of the theorem, thereby demonstrating that the Laplacian of $\vec{F}$ is orthogonal to everything, and so $\nabla^2\vec{F} = 0$.

When you do this in all of space, with the boundary conditions that the fields all go to zero at infinity, the only solutions to Laplace's equation are identically zero. In other words, the two vector spaces (the gradients and the curls) exhaust all the possibilities. How to prove this? Just pick a component, say F_x, treat it as simply a scalar function — call it f — and apply a vector identity, problem 9.36.

$$\nabla\cdot(\phi\vec{A}) = (\nabla\phi)\cdot\vec{A} + \phi(\nabla\cdot\vec{A})$$

$$\text{Let } \phi = f \quad \text{and} \quad \vec{A} = \nabla f, \qquad \text{then} \qquad \nabla\cdot(f\nabla f) = \nabla f\cdot\nabla f + f\nabla^2 f$$

Integrate this over all space and apply Gauss's theorem. (*I.e.* integrate by parts.)

$$\int d^3r\,\nabla\cdot(f\nabla f) = \oint f\nabla f\cdot d\vec{A} = \int d^3r\,\left[\nabla f\cdot\nabla f + f\nabla^2 f\right] \tag{13.47}$$

If f and its derivative go to zero fast enough at infinity (a modest requirement), the surface term, $\oint d\vec{A}$, goes to zero. The Laplacian term, $\nabla^2 f = 0$, and all that's left is

$$\int d^3r\,\nabla f\cdot\nabla f = 0$$

This is the integral of a quantity that can never be negative. The only way that the integral can be zero is that the integrand is zero. If $\nabla f = 0$, then f is a constant, and if it must also go to zero far away then that constant is zero.

This combination of results, the Helmholtz theorem, describes a field as the sum of a gradient and a curl, but is there a way to find these two components explicitly? Yes.

$$\vec{F} = \nabla f + \nabla\times B, \quad \text{so} \quad \nabla\cdot\vec{F} = \nabla^2 f, \quad \text{and} \quad \nabla\times\vec{F} = \nabla\times\nabla\times B = \nabla(\nabla\cdot\vec{B}) - \nabla^2\vec{B}$$

Solutions of these equations are

$$f(\vec{r}) = \frac{-1}{4\pi}\int d^3r'\,\frac{\nabla\cdot\vec{F}(\vec{r}\,')}{|\vec{r}-\vec{r}\,'|} \qquad \text{and} \qquad \vec{B}(\vec{r}) = \frac{1}{4\pi}\int d^3r'\,\frac{\nabla\times\vec{F}(\vec{r}\,')}{|\vec{r}-\vec{r}\,'|} \tag{13.48}$$

Generalization

In all this derivation, I assumed that the domain is all of three-dimensional space, and this made the calculations easier. A more general result lets you specify boundary conditions on some finite boundary and then a general vector field is the sum of as many as five classes of vector functions. This is the Helmholtz-Hodge decomposition theorem, and it has applications in the more complicated aspects of fluid flow (as if there are any simple ones), even in setting up techniques of numerical analysis for such problems. The details are involved, and I will simply refer you to a good review article* on the subject.

---

* Cantarella, DeTurck, and Gluck: The American Mathematical Monthly, May 2002. The paper is an unusual mix of abstract topological methods and very concrete examples. It thereby gives you a fighting chance at the subject.

Exercises

1 For a circle, from the definition of the integral, what is $\oint d\vec{\ell}$? What is $\oint d\ell$? What is $\oint d\vec{\ell} \times \vec{C}$ where $\vec{C}$ is a constant vector?

2 What is the work you must do in lifting a mass m in the Earth's gravitational field from a radius R_1 to a radius R_2. These are measured from the center of the Earth and the motion is purely radial.

3 Same as the preceding exercise but the motion is 1. due north a distance $R_1\theta_0$ then 2. radially out to R_2 then 3. due south a distance $R_2\theta_0$.

4 Verify Stokes' Theorem by separately calculating the left and the right sides of the theorem for the case of the vector field

$$\vec{F}(x,y) = \hat{x}\,Ay + \hat{y}\,Bx$$

around the rectangle $(a < x < b)$, $(c < y < d)$.

5 Verify Stokes' Theorem by separately calculating the left and the right sides of the theorem for the case of the vector field

$$\vec{F}(x,y) = \hat{x}\,Ay - \hat{y}\,Bx$$

around the rectangle $(a < x < b)$, $(c < y < d)$.

6 Verify Stokes' Theorem for the semi-cylinder $0 < z < h$, $0 < \phi < \pi$, $r = R$. The vector field is $\vec{F}(r,\phi,z) = \hat{r}Ar^2\sin\phi + \hat{\phi}Br\phi^2 z + \hat{z}Crz^2$

7 Verify Gauss's Theorem using the whole cylinder $0 < z < h$, $r = R$ and the vector field $\vec{F}(r,\phi,z) = \hat{r}Ar^2\sin\phi + \hat{\phi}Brz\sin^2\phi + \hat{z}Crz^2$.

8 What would happen if you used the volume of the preceding exercise and the field of the exercise before that one to check Gauss's law?

Problems

13.1 In the equation (13.4) what happens if you start with a different parametrization for x and y, perhaps $x = R\cos(\phi'/2)$ and $y = R\sin(\phi'/2)$ for $0 < \phi' < 4\pi$. Do you get the same answer?

13.2 What is the length of the arc of the parabola $y = (a^2 - x^2)/b$, $(-a < x < a)$?
But First draw a sketch and make a rough estimate of what the result ought to be. *Then* do the calculation and compare the answers. What limiting cases allow you to check your result? Ans: $(b/2)\big[\sinh^{-1} c + c\sqrt{1+c^2}\big]$ where $c = 2a/b$

13.3 (a) You can describe an ellipse as $x = a\cos\phi$, $y = b\sin\phi$. (Prove this.)
(b) Warm up by computing the area of the ellipse.
(c) What is the circumference of this ellipse? You will find a (probably) unfamiliar integral here, so to put this integral into a standard form, note that it is $4\int_0^{\pi/2}$. Then use $\cos^2\phi = 1 - \sin^2\phi$. Finally, look up chapter 17, Elliptic Integrals, of Abramowitz and Stegun. You will find the reference to this at the end of section 1.4. Notice in this integral that when you integrate, it will not matter whether you have a $\sin^2$ or a $\cos^2$.
Ans: $4aE(m)$

13.4 For another derivation of the work-energy theorem, one that doesn't use the manipulations of calculus as in Eq. (13.11), go back to basics.
(a) For a constant force, start from $\vec F = m\vec a$ and derive by elementary manipulations that

$$\vec F\cdot\Delta\vec r = \frac{m}{2}\left[v_{\mathrm f}^2 - v_{\mathrm i}^2\right]$$

All that you need to do is to note that the acceleration is a constant so you can get $\vec v$ and $\vec r$ as functions of time. Then eliminate t
(b) Along a specified curve Divide the curve at points

$$\vec r_{\mathrm i} = \vec r_0,\ \vec r_1,\ \vec r_2,\ \ldots\ \vec r_N = \vec r_{\mathrm f}$$

In each of these intervals apply the preceding equation. This makes sense in that if the interval is small the force won't change much in the interval.
(c) Add all these N equations and watch the kinetic energy terms telescope and (mostly) cancel. This limit as all the $\Delta\vec r_k \to 0$ is Eq. (13.12).

13.5 The manipulation in the final step of Eq. (13.12) seems almost *too* obvious. Is it? Well yes, but write out the definition of this integral as the limit of a sum to verify that it really is easy.

13.6 Mimic the derivation of Gauss's theorem, Eq. (13.15), and derive the identities

$$\oint_S d\vec A\times\vec v = \int_V \operatorname{curl}\vec v\, dV, \qquad \text{and} \qquad \oint_S f\, d\vec A = \int_V \operatorname{grad} f\, dV$$

13.7 The force by a magnetic field on a small piece of wire, length $d\ell$, and carrying a current I is $d\vec{F} = I\,d\vec{\ell} \times \vec{B}$. The total force on a wire carrying this current in a complete circuit is the integral of this. Let $\vec{B} = \hat{x}\,Ay - \hat{y}\,Ax$. The wire consists of the line segments around the rectangle $0 < x < a$, $0 < y < b$. The direction of the current is in the $+\hat{y}$ direction on the $x = 0$ line. What is the total force on the loop? Ans: 0

13.8 Verify Stokes' theorem for the field $\vec{F} = Axy\,\hat{x} + B(1+x^2y^2)\,\hat{y}$ and for the rectangular loop $a < x < b$, $c < y < d$.

13.9 Which of the two times in Eqs. (13.7) and (13.8) is shorter. (Compare their squares; it's easier.)

13.10 Write the equations (9.36) in an integral form.

13.11 Start with Stokes' theorem and shrink the boundary curve to a point. That doesn't mean there's no surface left; it's not flat, remember. The surface is pinched off like a balloon. It is now a closed surface, and what is the value of this integral? Now apply Gauss's theorem to it and what do you get? Ans: See Eq. (9.34)

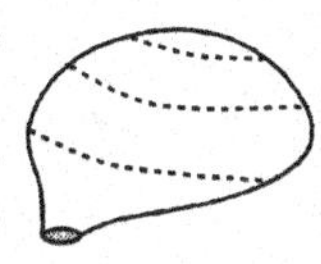

13.12 Use the same surface as in the example, Eq. (13.25), and verify Stokes' theorem for the vector field

$$\vec{F} = \hat{r}\,Ar^{-1}\cos^2\theta\sin\phi + \hat{\theta}\,Br^2\sin\theta\cos^2\phi + \hat{\phi}Cr^{-2}\cos^2\theta\sin^2\phi$$

13.13 Use the same surface as in the example, Eq. (13.25), and examine Stokes' theorem for the vector field

$$\vec{F} = \hat{r}\,f(r,\theta,\phi) + \hat{\theta}\,g(r,\theta,\phi) + \hat{\phi}\,h(r,\theta,\phi)$$

(**a**) Show from the line integral part that the answer can depend only on the function h, not f or g. (**b**) Now examine the surface integral over this cap and show the same thing.

13.14 For the vector field in the x-y plane: $\vec{F} = (x\hat{y} - y\hat{x})/2$, use Stokes' theorem to compute the line integral of $\vec{F}\cdot d\vec{r}$ around an arbitrary closed curve. What is the significance of the sign of the result? When you considered an "arbitrary" loop, did you consider the possibilities presented by these curves?

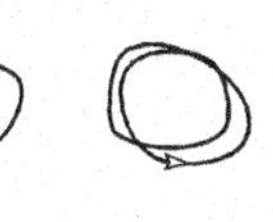

13.15 What is the (closed) surface integral of $\vec{F} = \vec{r}/3$ over an arbitrary closed surface? Ans: V.

13.16 What is the (closed) surface integral of $\vec{F} = \vec{r}/3$ over an arbitrary closed surface? This time however, the surface integral uses the cross product: $\oint d\vec{A} \times \vec{F}$. If in doubt, try drawing the picture for a special case first.

13.17 For the vector field Eq. (13.28) explicitly show that $\oint \vec{v}\cdot d\vec{r}$ is zero for a curve such as that in the picture and that it is not zero for a circle going around the singularity.

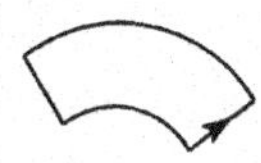

13.18 Refer to Eq. (13.27) and check it for small θ_0. Notice the combination $\pi(R\theta_0)^2$.

13.19 For the vector field, Eq. (13.28), use Eq. (13.29) to try to construct a potential function. Because within a certain domain the integral *is* independent of path, you can pick the most convenient possible path, the one that makes the integration easiest. What goes wrong?

13.20 Refer to problem 9.33 and construct the solutions by integration, using the methods of this chapter.

13.21 (**a**) Evaluate $\oint \vec{F}\cdot d\vec{r}$ for $\vec{F} = \hat{x}\,Axy + \hat{y}\,Bx$ around the circle of radius R centered at the origin.
(**b**) Now do it again, using Stokes' theorem this time.

13.22 Same as the preceding problem, but $\oint d\vec{r}\times\vec{F}$ instead.

13.23 Use the same field as the preceding two problems and evaluate the surface integral of $\vec{F}\cdot d\vec{A}$ over the hemispherical surface $x^2+y^2+z^2=R^2,\ z>0$.

13.24 The same field and surface as the preceding problem, but now the surface integral $d\vec{A}\times\vec{F}$. Ans: $\hat{z}\,2\pi Br^3/3$

13.25 (**a**) Prove the identity $\nabla\cdot\left(\vec{A}\times\vec{B}\right) = \vec{B}\cdot\nabla\times\vec{A} - \vec{A}\cdot\nabla\times\vec{B}$. (index mechanics?)
(**b**) Next apply Gauss's theorem to $\nabla\cdot\left(\vec{A}\times\vec{B}\right)$ and take the special case that $\vec{B}$ is an arbitrary constant to derive Eq. (13.19).

13.26 (**a**) Prove the identity $\nabla\cdot(f\vec{F}) = f\nabla\cdot\vec{F} + \vec{F}\cdot\nabla f$.
(**b**) Apply Gauss's theorem to $\nabla\cdot(f\vec{F})$ for an arbitrary constant $\vec{F}$ to derive a result found in another problem.
(**c**) Explain why the word "arbitrary" is necessary here.

13.27 The vector potential is not unique, as you can add an arbitrary gradient to it without affecting its curl. Suppose that $\vec{B} = \nabla\times\vec{A}$ with

$$\vec{A} = \hat{x}\,\alpha xyz + \hat{y}\,\beta x^2 z + \hat{z}\,\gamma xyz^2$$

Find a function $f(x,y,z)$ such that $\vec{A}' = \vec{A} + \nabla f$ has the z-component identically zero. Do you get the same $\vec{B}$ by taking the curl of $\vec{A}$ and of $\vec{A}'$?

13.28 Take the vector field

$$\vec{B} = \alpha xy\,\hat{x} + \beta xy\,\hat{y} + \gamma(xz+yz)\hat{z}$$

Write out the equation $\vec{B} = \nabla\times\vec{A}$ in rectangular components and figure out what functions $A_x(x,y,z)$, $A_y(x,y,z)$, and $A_z(x,y,z)$ will work. Note: From the preceding problem you see that you may if you wish pick any one of the components of $\vec{A}$ to be zero — that will cut down on the labor. Also, you should expect that this problem is impossible unless $\vec{B}$ has zero divergence. That fact should come *out* of your calculations, so don't put it in yet. Determine the conditions on α, β, and γ that make this problem solvable, and show that this is equivalent to $\nabla\cdot\vec{B} = 0$.

13.29 A magnetic monopole, if it exists, will have a magnetic field $\mu_0 q_m \hat{r}/4\pi r^2$. The divergence of this magnetic field is zero except at the origin, but that means that not every closed surface can be shrunk to a point without running into the singularity. The necessary condition for having a vector potential is not satisfied. Try to construct such a potential anyway. Assume a solution in spherical coordinates of the form $\vec{A} = \hat{\phi} f(r) g(\theta)$ and figure out what f and g will have this $\vec{B}$ for a curl. Sketch the resulting $\vec{A}$. You will run into a singularity (or two, depending). Ans: $\vec{A} = \hat{\phi}\mu_0 q_m (1 - \cos\theta)/(4\pi r^2 \sin\theta)$ (not unique)

13.30 Apply the Reynolds transport theorem to the other of Maxwell's equations.

$$\nabla \times \vec{B} = \mu_0 \vec{j} + \mu_0 \epsilon_0 \frac{\partial \vec{E}}{\partial t}$$

Don't simply leave the result in the first form that you find. Manipulate it into what seems to be the best form. Use $\mu_0 \epsilon_0 = 1/c^2$.
Ans: $\oint \left(\vec{B} - \vec{v} \times \vec{E}/c^2\right) \cdot d\vec{\ell} = \mu_0 \int \left(\vec{j} - \rho\vec{v}\right) \cdot d\vec{A} + \mu_0 \epsilon_0 (d/dt) \int \vec{E} \cdot d\vec{A}$

13.31 Derive the analog of the Reynolds transport theorem, Eq. (13.39), for a line integral around a closed loop.

(a) $$\frac{d}{dt}\oint_{C(t)} \vec{F}(\vec{r}, t) \cdot d\vec{\ell} = \oint_{C(t)} \frac{\partial \vec{F}}{\partial t} \cdot d\vec{\ell} - \oint_{C(t)} \vec{v} \times (\nabla \times \vec{F}) \cdot d\vec{\ell}$$

and for the surface integral of a scalar. You will need problem 13.6.

(b) $$\frac{d}{dt}\int_{S(t)} \phi(\vec{r}, t) d\vec{A} = \int_{S(t)} \frac{\partial \phi}{\partial t} d\vec{A} + \int_{S(t)} (\nabla\phi)\, d\vec{A} \cdot \vec{v} - \oint_{C(t)} \phi\, d\vec{\ell} \times \vec{v}$$

Make up examples that test the validity of individual terms in the equations. I recommend cylindrical coordinates for your examples.

13.32 Another transport theorem is more difficult to derive.

$$\frac{d}{dt}\oint_{C(t)} d\vec{\ell} \times \vec{F}(\vec{r}, t) = \oint_{C(t)} d\vec{\ell} \times \frac{\partial \vec{F}}{\partial t} + \oint_{C(t)} (\nabla \cdot \vec{F}) d\vec{\ell} \times \vec{v} - \int_{C(t)} (\nabla \vec{F}) \cdot d\vec{\ell} \times \vec{v}$$

I had to look up some vector identities, including one for $\nabla \times (\vec{A} \times \vec{B})$. A trick that I found helpful: At a certain point take the dot product of the whole equation with a fixed vector $\vec{B}$ and manipulate the resulting product, finally factoring out the arbitrary vector $\vec{B}\cdot$ at the end. Make up examples that test the validity of individual terms in the equations. Again, I recommend cylindrical coordinates for your examples.

13.33 Apply Eq. (13.39) to the velocity field itself. That is, let $\vec{B} = \vec{v}$. Suppose further the the fluid is incompressible with $\nabla \cdot \vec{v} = 0$ and that the flow is stationary (no time dependence). Explain the results.

13.34 Assume that the Earth's atmosphere obeys the density equation $\rho = \rho_0 e^{-z/h}$ for a height z above the surface. (**a**) Through what amount of air does sunlight have to travel when coming from straight overhead? Take the measure of this to be $\int \rho\, d\ell$ (called the "air mass"). (**b**) Through what amount of air does sunlight have to travel when coming from just on the horizon at sunset? Neglect the fact that light will refract in the atmosphere and that the path in the second case won't really be a straight line. Take $h = 10\,\text{km}$ and the radius of the Earth to be $6400\,\text{km}$. The integral you get for the second case is probably not familiar. You may evaluate it numerically for the numbers that I stated, or you may look it up in a big table of integrals such as Gradshteyn and Ryzhik, or you may use an approximation, $h \ll R$. (I recommend the last.) What is the numerical value of the ratio of these two air mass integrals? This goes far in explaining why you can look at the setting sun.
(**c**) If refraction in the atmosphere is included, the light will bend and pass through a still larger air mass. The overall refraction comes to about 0.5°, and calculating the path that light takes is hard, but you can find a bound on the answer by assuming a path that follows the surface of the Earth through this angle and then takes off on a straight line. What is the air mass ratio in this case? The real answer is somewhere between the two calculations. (The *really* real answer is a little bigger than either because the atmosphere is not isothermal and so the approximation $\rho = \rho_0 e^{-z/h}$ is not exact.)
Ans: $\approx \sqrt{R\pi/2h} = 32, \quad +R\theta/h \to 37$.

13.35 Work in a thermodynamic system is calculated from $dW = P\,dV$. Assume an ideal gas, so that $PV = nRT$. (**a**) What is the total work, $\oint dW$, done around this cycle as the pressure increases at constant volume, then decreases at constant temperature, finally the volume decreases at constant pressure.
(**b**) In the special case for which the changes in volume and pressure are very small, estimate from the graph approximately what to expect for the answer. Now do an expansion of the result of part (**a**) to see if it agrees with what you expect. Ans: $\approx \Delta P\, \Delta V/2$

13.36 Verify the divergence theorem for the vector field

$$\vec F = \alpha xyz\,\hat x + \beta x^2 z(1+y)\,\hat y + \gamma xyz^2\,\hat z$$

and for the volume $(0 < x < a)$, $(0 < y < b)$, $(0 < z < c)$.

13.37 Evaluate $\int \vec F \cdot d\vec A$ over the curved surface of the hemisphere $x^2 + y^2 + z^2 = R^2$ and $z > 0$. The vector field is given by $\vec F = \nabla \times \big(\alpha y\,\hat x + \beta x\,\hat y + \gamma xy\,\hat z\big)$. Ans: $(\beta - \alpha)\pi R^2$

13.38 A vector field is given in cylindrical coordinates to be $\vec F = \hat r \alpha r^2 z \sin^2\phi + \hat\phi \beta r z + \hat z \gamma z r \cos^2\phi$. Verify the divergence theorem for this field for the region $(0 < r < R)$, $(0 < \phi < 2\pi)$, $(0 < z < h)$.

13.39 For the function $F(r,\theta) = r^n(A + B\cos\theta + C\cos^2\theta)$, compute the gradient and then the divergence of this gradient. For what values of the constants A, B, C, and (positive, negative, or zero) integer n is this result, $\nabla \cdot \nabla F$, zero? These coordinates are

spherical, and this combination div grad is the Laplacian.
Ans: In part, $n = 2$, $C = -3A$, $B = 0$.

13.40 Repeat the preceding problem, but now interpret the coordinates as cylindrical (change θ to ϕ). And don't necessarily leave your answers in the first form that you find them.

13.41 Evaluate the integral $\int \vec{F} \cdot d\vec{A}$ over the surface of the hemisphere $x^2 + y^2 + z^2 = 1$ with $z > 0$. The vector field is $\vec{F} = A(1 + x + y)\hat{x} + B(1 + y + z)\hat{y} + C(1 + z + x)\hat{z}$. You may choose to do this problem the hard way or the easy way. Or both.
Ans: $\pi(2A + 2B + 5C)/3$

13.42 An electric field is known in cylindrical coordinates to be $\vec{E} = f(r)\hat{r}$, and the electric charge density is a function of r alone, $\rho(r)$. They satisfy the Maxwell equation $\nabla \cdot \vec{E} = \rho/\epsilon_0$. If the charge density is given as $\rho(r) = \rho_0\, e^{-r/r_0}$. Compute $\vec{E}$. Demonstrate the behavior of $\vec{E}$ is for large r and for small r.

13.43 Repeat the preceding problem, but now r is a spherical coordinate.

13.44 Find a vector field $\vec{F}$ such that $\nabla \cdot \vec{F} = \alpha x + \beta y + \gamma$ and $\nabla \times \vec{F} = \hat{z}$. Next, find an infinite number of such fields.

13.45 Gauss's law says that the total charge contained inside a surface is $\epsilon_0 \oint \vec{E} \cdot d\vec{A}$. For the electric field of problem 10.37, evaluate this integral over a sphere of radius $R_1 > R$ and centered at the origin.

13.46 (**a**) In cylindrical coordinates, for what n does the vector field $\vec{v} = r^n \hat{\phi}$ have curl equal to zero? Draw it.
(**b**) Also, for the same closed path as in problem 13.17 and for all n, compute $\oint \vec{v} \cdot d\vec{r}$

13.47 Prove the identity Eq. (13.43). Write it out in index notation first.

13.48 There an analog of Stokes' theorem for $\oint d\vec{\ell} \times \vec{B}$. This sort of integral comes up in computing the total force on the current in a circuit. Try multiplying (dot) the integral by a constant vector $\vec{C}$. Then manipulate the result by standard methods and hope that in the end you have the same constant $\vec{C}\cdot$ something.
Ans: $= \int \left[(\nabla\vec{B}) \cdot d\vec{A} - (\nabla \cdot \vec{B}) \cdot d\vec{A}\right]$ and the second term vanishes for magnetic fields.

13.49 In the example (13.16) using Gauss's theorem, the term in γ contributed zero to the surface integral (13.17). In the subsequent volume integral the same term vanished because of the properties of $\sin\phi\cos\phi$. *But* this term will vanish in the surface integral no matter what the function of ϕ is in the $\hat{\phi}$ component of the vector $\vec{F}$. How then is it always guaranteed to vanish in the volume integral?

13.50 Interpret the vector field $\vec{F}$ from problem 13.37 as an electric field $\vec{E}$, then use Gauss's law that $q_{\text{enclosed}} = \epsilon_0 \oint \vec{E} \cdot d\vec{A}$ to evaluate the charge enclosed within a sphere or radius R centered at the origin.

13.51 Derive the identity Eq. (13.32) starting from the definition of a derivative and doing the same sort of manipulation that you use in deriving the ordinary product rule for differentiation.

13.52 A right tetrahedron has three right triangular sides that meet in one vertex. Think of a corner chopped off of a cube. The sum of the squares of the areas of the three right triangles equals the square of the area of the fourth face. The results of problem 13.6 will be useful.

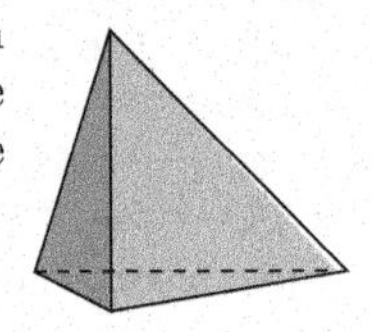

Complex Variables

In the calculus of functions of a complex variable there are three fundamental tools, the same fundamental tools as for real variables. Differentiation, Integration, and Power Series. I'll first introduce all three in the context of complex variables, then show the relations between them. The applications of the subject will form the major part of the chapter.

14.1 Differentiation

When you try to differentiate a continuous function is it always differentiable? If it's differentiable once is it differentiable again? The answer to both is no. Take the simple absolute value function of the real variable x.

$$f(x) = |x| = \begin{cases} x & (x \geq 0) \\ -x & (x < 0) \end{cases}$$

This has a derivative for all x except zero. The limit

$$\frac{f(x+\Delta x) - f(x)}{\Delta x} \longrightarrow \begin{cases} 1 & (x > 0) \\ -1 & (x < 0) \\ ? & (x = 0) \end{cases} \tag{14.1}$$

works for both $x > 0$ and $x < 0$. If $x = 0$ however, you get a different result depending on whether $\Delta x \to 0$ through positive or through negative values.

If you integrate this function,

$$\int_0^x |x'|\, dx' = \begin{cases} x^2/2 & (x \geq 0) \\ -x^2/2 & (x < 0) \end{cases}$$

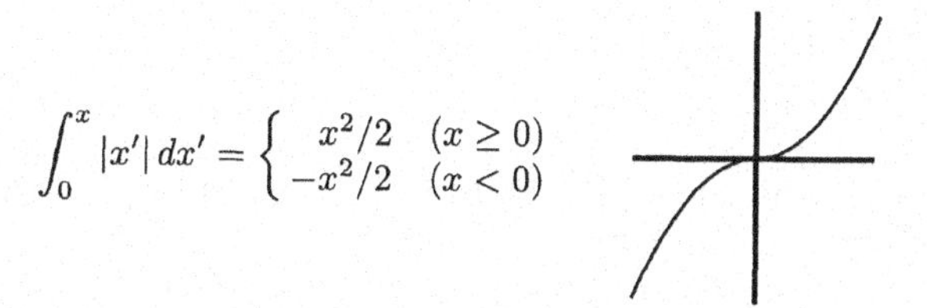

the result has a derivative everywhere, including the origin, but you can't differentiate it twice. A few more integrations and you can produce a function that you can differentiate 42 times but not 43.

There are functions that are continuous but with no derivative anywhere. They're harder* to construct, but if you grant their existence then you can repeat the preceding manipulation and create a function with any number of derivatives everywhere, but no more than that number anywhere.

For a derivative to exist at a point, the limit Eq. (14.1) must have the same value whether you take the limit from the right or from the left.

* Weierstrass surprised the world of mathematics with $\sum_0^\infty a^k \cos(b^k x)$. If $a < 1$ while $ab > 1$ this is continuous but has no derivative anywhere. This statement is much more difficult to prove than it looks.

Extend the idea of differentiation to complex-valued functions of complex variables. Just change the letter x to the letter $z = x + iy$. Examine a function such as $f(z) = z^2 = x^2 - y^2 + 2ixy$ or $\cos z = \cos x \cosh y + i \sin x \sinh y$. Can you differentiate these (yes) and what does that mean?

$$f'(z) = \lim_{\Delta z \to 0} \frac{f(z + \Delta z) - f(z)}{\Delta z} = \frac{df}{dz} \tag{14.2}$$

is the appropriate definition, but for it to exist there are even more restrictions than in the real case. For real functions you have to get the same limit as $\Delta x \to 0$ whether you take the limit from the right or from the left. In the complex case there are an infinite number of directions through which Δz can approach zero and you must get the same answer from all directions. This is such a strong restriction that it isn't obvious that *any* function has a derivative. To reassure you that I'm not talking about an empty set, differentiate z^2.

$$\frac{(z + \Delta z)^2 - z^2}{\Delta z} = \frac{2z\Delta z + (\Delta z)^2}{\Delta z} = 2z + \Delta z \longrightarrow 2z$$

It doesn't matter whether $\Delta z = \Delta x$ or $= i\Delta y$ or $= (1 + i)\Delta t$. As long as it goes to zero you get the same answer.

For a contrast take the complex conjugation function, $f(z) = z^* = x - iy$. Try to differentiate that.

$$\frac{(z + \Delta z)^* - z^*}{\Delta z} = \frac{(\Delta z)^*}{\Delta z} = \frac{\Delta r\, e^{-i\theta}}{\Delta r\, e^{i\theta}} = e^{-2i\theta}$$

The polar form of the complex number is more convenient here, and you see that as the distance Δr goes to zero, this difference quotient depends on the direction through which you take the limit. From the right and the left you get $+1$. From above and below $(\theta = \pm\pi/2)$ you get -1. The limits aren't the same, so this function has no derivative anywhere. Roughly speaking, the functions that you're familiar with or that are important enough to have names (sin, cos, tanh, Bessel, elliptic, ...) will be differentiable as long as you don't have an explicit complex conjugation in them. Something such as $|z| = \sqrt{z^* z}$ does not have a derivative for any z.

For functions of a real variable, having one or fifty-one derivatives doesn't guarantee you that it has two or fifty-two. The amazing property of functions of a complex variable is that if a function has a single derivative everywhere in the neighborhood of a point then you are guaranteed that it has a infinite number of derivatives. You will also be assured that you can do a power series expansions about that point and that the series will always converge to the function. There are important and useful integration methods that will apply to all these functions, and for a relatively small effort they will open impressively large vistas of mathematics.

For an example of the insights that you gain using complex variables, consider the function $f(x) = 1/\big(1 + x^2\big)$. This is a perfectly smooth function of x, starting at $f(0) = 1$ and slowing dropping to zero as $x \to \pm\infty$. Look at the power series expansion about $x = 0$ however. This is just a geometric series in $(-x^2)$, so

$$\big(1 + x^2\big)^{-1} = 1 - x^2 + x^4 - x^6 + \cdots$$

This converges only if $-1 < x < +1$. Why such a limitation? The function is infinitely differentiable for all x and is completely smooth throughout its domain. This remains mysterious as long as you think of x as a real number. If you expand your view and consider the function of the complex variable $z = x + iy$, then the mystery disappears. $1/(1+z^2)$ blows up when $z \to \pm i$. The reason that the series fails to converge for values of $|x| > 1$ lies in the complex plane, in the fact that at the distance $= 1$ in the i-direction there is a singularity, and in the fact that the domain of convergence is a disk extending out to the nearest singularity.

Definition: A function is said to be analytic at the point z_0 if it is differentiable for every point z in the disk $\lvert z - z_0\rvert < \varepsilon$. Here the positive number ε may be small, but it is not zero.

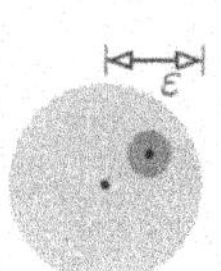

Necessarily if f is analytic at z_0 it will also be analytic at every point within the disk $|z - z_0| < \varepsilon$. This follows because at any point z_1 within the original disk you have a disk centered at z_1 and of radius $(\varepsilon - |z_1 - z_0|)/2$ on which the function is differentiable.

The common formulas for differentiation are exactly the same for complex variables as they are for real variables, and their proofs are exactly the same. For example, the product formula:

$$\begin{aligned}
&\frac{f(z+\Delta z)g(z+\Delta z) - f(z)g(z)}{\Delta z} \\
&= \frac{f(z+\Delta z)g(z+\Delta z) - f(z)g(z+\Delta z) + f(z)g(z+\Delta z) - f(z)g(z)}{\Delta z} \\
&= \frac{f(z+\Delta z) - f(z)}{\Delta z} g(z+\Delta z) + f(z)\frac{g(z+\Delta z) - g(z)}{\Delta z}
\end{aligned}$$

As $\Delta z \to 0$, this becomes the familiar $f'g + fg'$. That the numbers are complex made no difference.

For integer powers you can use induction, just as in the real case: $dz/dz = 1$ and

$$\text{If } \frac{dz^n}{dz} = nz^{n-1}, \qquad \text{then use the product rule}$$

$$\frac{dz^{n+1}}{dz} = \frac{d(z^n \cdot z)}{dz} = nz^{n-1} \cdot z + z^n \cdot 1 = (n+1)z^n$$

The other differentiation techniques are in the same spirit. They follow very closely from the definition. For example, how do you handle negative powers? Simply note that $z^n z^{-n} = 1$ and use the product formula. The chain rule, the derivative of the inverse of a function, all the rest, are close to the surface.

14.2 Integration

The standard Riemann integral of section 1.6 is

$$\int_a^b f(x)\,dx = \lim_{\Delta x_k \to 0} \sum_{k=1}^{N} f(\xi_k)\Delta x_k$$

The extension of this to complex functions is direct. Instead of partitioning the interval $a < x < b$ into N pieces, you have to specify a curve in the complex plane and partition *it* into N pieces. The interval is the complex number $\Delta z_k = z_k - z_{k-1}$.

$$\int_C f(z)\,dz = \lim_{\Delta z_k \to 0} \sum_{k=1}^{N} f(\zeta_k)\Delta z_k$$

Just as ξ_k is a point in the kth interval, so is ζ_k a point in the kth interval along the curve C.

How do you evaluate these integrals? Pretty much the same way that you evaluate line integrals in vector calculus. You can write this as

$$\int_C f(z)\,dz = \int \big(u(x,y) + iv(x,y)\big)\big(dx + idy\big) = \int \big[(u\,dx - v\,dy) + i(u\,dy + v\,dx)\big]$$

If you have a parametric representation for the values of $x(t)$ and $y(t)$ along the curve this is

$$\int_{t_1}^{t_2} \big[(u\,\dot{x} - v\,\dot{y}) + i(u\,\dot{y} + v\,\dot{x})\big]\,dt$$

For example take the function $f(z) = z$ and integrate it around a circle centered at the origin. $x = R\cos\theta$, $y = R\sin\theta$.

$$\begin{aligned}\int z\,dz &= \int \big[(x\,dx - y\,dy) + i(x\,dy + y\,dx)\big] \\ &= \int_0^{2\pi} d\theta R^2 \big[(-\cos\theta\sin\theta - \sin\theta\cos\theta) + i(\cos^2\theta - \sin^2\theta)\big] = 0\end{aligned}$$

Wouldn't it be easier to do this in polar coordinates? $z = re^{i\theta}$.

$$\int z\,dz = \int re^{i\theta}\big[e^{i\theta}dr + ire^{i\theta}d\theta\big] = \int_0^{2\pi} Re^{i\theta} iRe^{i\theta}\,d\theta = iR^2\int_0^{2\pi} e^{2i\theta}\,d\theta = 0 \tag{14.3}$$

Do the same thing for the function $1/z$. Use polar coordinates.

$$\oint \frac{1}{z}dz = \int_0^{2\pi} \frac{1}{Re^{i\theta}} iRe^{i\theta}\,d\theta = \int_0^{2\pi} i\,d\theta = 2\pi i \tag{14.4}$$

This is an important result! Do the same thing for z^n where n is any positive or negative integer, problem 14.1.

Rather than spending time on more examples of integrals, I'll jump to a different subject. The main results about integrals will follow after that (the residue theorem).

14.3 Power (Laurent) Series

The series that concern us here are an extension of the common Taylor or power series, and they are of the form

$$\sum_{-\infty}^{+\infty} a_k(z-z_0)^k \tag{14.5}$$

The powers can extend through all positive and negative integer values. This is sort of like the Frobenius series that appear in the solution of differential equations, except that here the powers are all integers and they can either have a finite number of negative powers or the powers can go all the way to minus infinity.

The common examples of Taylor series simply represent the case for which no negative powers appear.

$$\sin z = \sum_0^\infty (-1)^k \frac{z^{2k+1}}{(2k+1)!} \qquad \text{or} \qquad J_0(z) = \sum_0^\infty (-1)^k \frac{z^{2k}}{2^{2k}(k!)^2} \qquad \text{or} \qquad \frac{1}{1-z} = \sum_0^\infty z^k$$

If a function has a Laurent series expansion that has a finite number of negative powers, it is said to have a *pole*.

$$\frac{\cos z}{z} = \sum_0^\infty (-1)^k \frac{z^{2k-1}}{(2k)!} \qquad \text{or} \qquad \frac{\sin z}{z^3} = \sum_0^\infty (-1)^k \frac{z^{2k-2}}{(2k+1)!}$$

The *order* of the pole is the size of the largest negative power. These have respectively first order and second order poles.

If the function has an infinite number of negative powers, and the series converges all the way down to (but of course not at) the singularity, it is said to have an *essential singularity*.

$$e^{1/z} = \sum_0^\infty \frac{1}{k!\, z^k} \qquad \text{or} \qquad \sin\left[t\left(z+\frac{1}{z}\right)\right] = \cdots \qquad \text{or} \qquad \frac{1}{1-z} = \frac{1}{z}\frac{-1}{1-\frac{1}{z}} = -\sum_1^\infty z^{-k}$$

The first two have essential singularities; the third does not.

It's worth examining some examples of these series and especially in seeing what kinds of singularities they have. In analyzing these I'll use the fact that the familiar power series derived for real variables apply here too. The binomial series, the trigonometric functions, the exponential, many more.

$1/z(z-1)$ has a zero in the denominator for both $z=0$ and $z=1$. What is the full behavior near these two points?

$$\frac{1}{z(z-1)} = \frac{-1}{z(1-z)} = \frac{-1}{z}(1-z)^{-1} = \frac{-1}{z}\left[1+z+z^2+z^3+\cdots\right] = \frac{-1}{z} - 1 - z - z^2 - \cdots$$

$$\begin{aligned}\frac{1}{z(z-1)} &= \frac{1}{(z-1)(1+z-1)} = \frac{1}{z-1}\left[1+(z-1)\right]^{-1} \\ &= \frac{1}{z-1}\left[1+(z-1)+(z-1)^2+(z-1)^3+\cdots\right] = \frac{1}{z-1} + 1 + (z-1) + \cdots\end{aligned}$$

This shows the full Laurent series expansions near these points. Keep your eye on the coefficient of the inverse first power. That term alone plays a crucial role in what will follow.

$\csc^3 z$ near $z = 0$:

$$\begin{aligned}\frac{1}{\sin^3 z} &= \frac{1}{\left[z - \frac{z^3}{6} + \frac{z^5}{120} - \cdots\right]^3} = \frac{1}{z^3\left[1 - \frac{z^2}{6} + \frac{z^4}{120} - \cdots\right]^3} \\ &= \frac{1}{z^3}\left[1 + x\right]^{-3} = \frac{1}{z^3}\left[1 - 3x + 6x^2 - 10x^3 + \cdots\right] \\ &= \frac{1}{z^3}\left[1 - 3\left(-\frac{z^2}{6} + \frac{z^4}{120} - \cdots\right) + 6\left(-\frac{z^2}{6} + \frac{z^4}{120} - \cdots\right)^2 - \cdots\right] \\ &= \frac{1}{z^3}\left[1 + \frac{z^2}{2} + z^4\left(\frac{1}{6} - \frac{3}{120}\right) + \cdots\right] \\ &= \frac{1}{z^3}\left[1 + \frac{1}{2}z^2 + \frac{17}{120}z^4 + \cdots\right] \end{aligned} \tag{14.6}$$

This has a third order pole, and the coefficient of $1/z$ is $1/2$. Are there any other singularities for this function? Yes, every place that the sine vanishes you have a pole, at $n\pi$. (What is the order of these other poles?) As I commented above, you'll soon see that the coefficient of the $1/z$ term plays a special role, and if that's all that you're looking for you don't have to work this hard. Now that you've seen what various terms do in this expansion, you can stop carrying along so many terms and still get the $1/2z$ term. See problem 14.17

The structure of a Laurent series is such that it will converge in an annulus. Examine the absolute convergence of such a series.

$$\sum_{-\infty}^{\infty} \left|a_k z^k\right| = \sum_{-\infty}^{-1} \left|a_k z^k\right| + \sum_{0}^{\infty} \left|a_k z^k\right|$$

The ratio test on the second sum is

$$\text{if for large enough positive } k, \qquad \frac{|a_{k+1}||z|^{k+1}}{|a_k||z|^k} = \frac{|a_{k+1}|}{|a_k|}|z| \le x < 1 \tag{14.7}$$

then the series converges. The smallest such x defines the upper bound of the $|z|$ for which the sum of positive powers converges. If $|a_{k+1}|/|a_k|$ has a limit then $|z|_{\text{max}} = \lim |a_k|/|a_{k+1}|$.

Do the same analysis for the series of negative powers, applying the ratio test.

$$\text{if for large enough } \textit{negative } \text{k, you have} \qquad \frac{|a_{k-1}||z|^{k-1}}{|a_k||z|^k} = \frac{|a_{k-1}|}{|a_k|}\frac{1}{|z|} \le x < 1 \tag{14.8}$$

then the series converges. The largest such x defines the *lower* bound of those $|z|$ for which the sum of negative powers converges. If $|a_{k-1}|/|a_k|$ has a limit as $k \to -\infty$ then $|z|_{\text{min}} = \lim |a_{k-1}|/|a_k|$.

If $|z|_{\text{min}} < |z|_{\text{max}}$ then there is a range of z for which the series converges absolutely (and so of course it converges).

$$|z|_{\text{min}} < |z| < |z|_{\text{max}} \qquad \text{an annulus}$$

$|z|_{\text{min}}$

$|z|_{\text{max}}$

If either of these series of positive or negative powers is finite, terminating in a polynomial, then respectively $|z|_{\text{max}} = \infty$ or $|z|_{\text{min}} = 0$.

A major result is that when a function is analytic at a point (and so automatically in a neighborhood of that point) then it will have a Taylor series expansion there. The series will converge, and the series will converge to the given function. Is it possible for the Taylor series for a function to converge but not to converge to the expected function? Yes, for functions of a real variable it is. See problem 14.3. The important result is that for analytic functions of a complex variable this cannot happen, and all the manipulations that you would like to do will work. (Well, almost all.)

14.4 Core Properties

There are four closely intertwined facts about analytic functions. Each one implies the other three. For the term "neighborhood" of z_0, take it to mean all points satisfying $|z - z_0| < r$ for some positive r.

1. The function has a single derivative in a neighborhood of z_0.
2. The function has an infinite number of derivatives in a neighborhood of z_0.
3. The function has a power series (positive exponents) expansion about z_0 and the series converges to the specified function in a disk centered at z_0 and extending to the nearest singularity. You can compute the derivative of the function by differentiating the series term-by-term.
4. All contour integrals of the function around closed paths in a neighborhood of z_0 are zero.

Item 3 is a special case of the result about Laurent series. There are no negative powers when the function is analytic at the expansion point.

The second part of the statement, that it's the presence of a singularity that stops the series from converging, requires some computation to prove. The key step in the proof is to show that when the series converges in the neighborhood of a point then you *can* differentiate term-by-term and get the right answer. Since you won't have a derivative at a singularity, the series can't converge there. That important part of the proof is the one that I'll leave to every book on complex variables ever written. *E.g.* Schaum's outline on Complex Variables by Spiegel, mentioned in the bibliography. It's not hard, but it requires attention to detail.

Instead of a direct approach to all these ideas, I'll spend some time showing how they're related to each other. The proofs that these are valid are not all that difficult, but I'm going to spend time on their applications instead.

14.5 Branch Points

The function $f(z) = \sqrt{z}$ has a peculiar behavior. You are so accustomed to it that you may not think of it as peculiar, but simply an annoyance that you have to watch out for. It's double valued. The very definition of a function however says that a function is single valued, so what is this? I'll leave the answer to this until later, section 14.7, but for now I'll say that when you encounter this problem you have to be careful of the path along which you move, in order to avoid going all the way around such a point.

14.6 Cauchy's Residue Theorem

This is *the* fundamental result for applications in physics. If a function has a Laurent series expansion about the point z_0, the coefficient of the term $1/(z - z_0)$ is called the residue of f at z_0. The residue theorem tells you the value of a contour integral around a closed loop in terms of the residues of the function inside the loop.

$$\oint f(z)\,dz = 2\pi i \sum_k \operatorname{Res}(f)\big|_{z_k} \tag{14.9}$$

To make sense of this result I have to specify the hypotheses. The direction of integration is counter-clockwise. Inside and on the simple closed curve defining the path of integration, f is analytic except at isolated points of singularity, where there is a Laurent series expansion. There are no branch points inside the curve. It says that at each singularity z_k inside the contour, find the residue; add them; the result (times $2\pi i$) is the value of the integral on the left. The term "simple" closed curve means that it doesn't cross itself.

Why is this theorem true? The result depends on some of the core properties of analytic functions, especially that fact that you can distort the contour of integration as long as you don't pass over a singularity. If there are several isolated singularities inside a contour (poles or essential singularities), you can contract the contour C_1 to C_2 and then further to loops around the separate singularities. The parts of C_2 other than the loops are pairs of line segments that go in opposite directions so that the integrals along these pairs cancel each other.

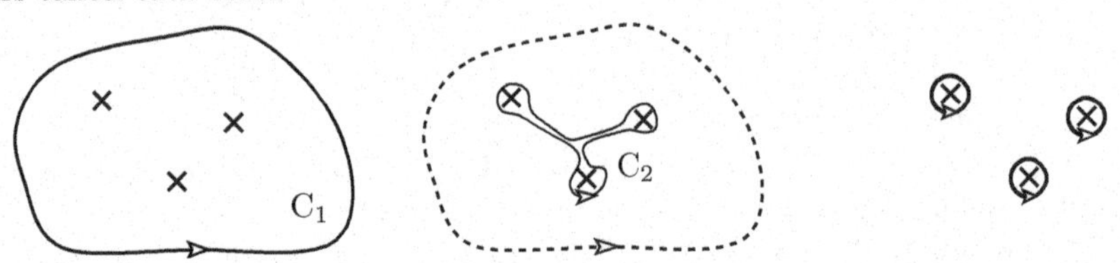

The problem is now to evaluate the integrals around the separate singularities of the function, then add them. The function being integrated is analytic in the neighborhoods of the singularities (by assumption they are isolated). That means there is a Laurent series expansion around each, and that it converges all the way down to the singularity itself (though not at it). Now at the singularity z_k you have

$$\oint \sum_{n=-\infty}^{\infty} a_n (z - z_k)^n$$

The lower limit may be finite, but that just makes it easier. In problem 14.1 you found that the integral of z^n around counterclockwise about the origin is zero unless $n = -1$ in which case it is $2\pi i$. Integrating the individual terms of the series then gives zero from all terms but one, and then it is $2\pi i a_{-1}$, which is $2\pi i$ times the residue of the function at z_k. Add the results from all the singularities and you have the Residue theorem.

Example 1

The integral of $1/z$ around a circle of radius R centered at the origin is $2\pi i$. The Laurent series expansion of this function is trivial — it has only one term. This reproduces Eq. (14.4). It also says that the integral around the same path of $e^{1/z}$ is $2\pi i$. Write out the series expansion of $e^{1/z}$ to determine the coefficient of $1/z$.

Example 2

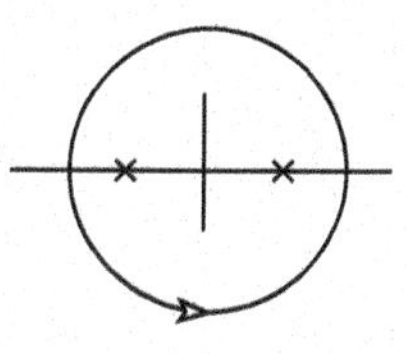

Another example. The integral of $1/(z^2 - a^2)$ around a circle centered at the origin and of radius $2a$. You can do this integral two ways. First increase the radius of the circle, pushing it out toward infinity. As there are no singularities along the way, the value of the integral is unchanged. The magnitude of the function goes as $1/R^2$ on a large ($R \gg a$) circle, and the circumference is $2\pi R$. the product of these goes to zero as $1/R$, so the value of the original integral (unchanged, remember) is zero.

Another way to do the integral is to use the residue theorem. There are two poles inside the contour, at $\pm a$. Look at the behavior of the integrand near these two points.

$$\frac{1}{z^2 - a^2} = \frac{1}{(z-a)(z+a)} = \frac{1}{(z-a)(2a+z-a)} \approx [\text{near } +a]\ \frac{1}{2a(z-a)}$$
$$= \frac{1}{(z+a)(z+a-2a)} \approx [\text{near } -a]\ \frac{1}{-2a(z+a)}$$

The integral is $2\pi i$ times the sum of the two residues.

$$2\pi i \left[\frac{1}{2a} + \frac{1}{-2a}\right] = 0$$

For another example, with a more interesting integral, what is

$$\int_{-\infty}^{+\infty} \frac{e^{ikx}\,dx}{a^4 + x^4} \tag{14.10}$$

If these were squares instead of fourth powers, and it didn't have the exponential in it, you could easily find a trigonometric substitution to evaluate it. *This* integral would be formidable though. To illustrate the method, I'll start with that easier example, $\int dx/(a^2 + x^2)$.

Example 3

The function $1/(a^2 + z^2)$ is singular when the denominator vanishes — when $z = \pm ia$. The integral is the contour integral along the x-axis.

$$\int_{C_1} \frac{dz}{a^2 + z^2} \tag{14.11}$$

The figure shows the two places at which the function has poles, $\pm ia$. The method is to move the contour around and to take advantage of the theorems about contour integrals. First remember that as long as it doesn't move across a singularity, you can distort a contour at will. I will push the contour C_1 up, but I have to leave the endpoints where they are in order not let the contour cross the pole at ia. Those are my sole constraints.

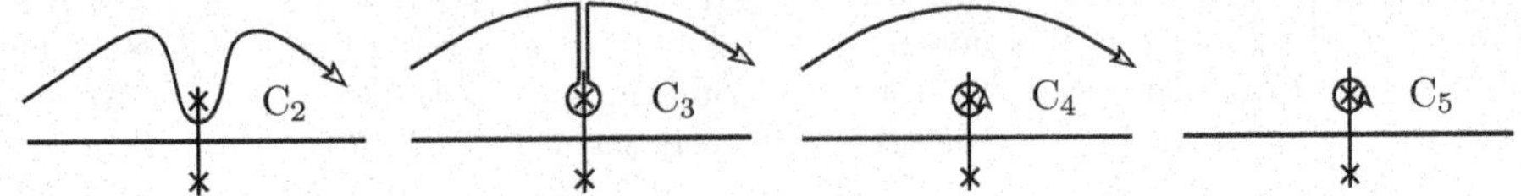

As I push the contour from C_1 up to C_2, nothing has changed, and the same applies to C_3. The next two steps however, requires some comment. In C_3 the two straight-line segments that parallel the y-axis are going in opposite directions, and as they are squeezed together, they cancel each other; they are integrals of the same function in reverse directions. In the final step, to C_5, I pushed the contour all the way to $+i\infty$ and eliminated it. How does that happen? On a big circle of radius R, the function $1/(a^2+z^2)$ has a magnitude approximately $1/R^2$. As you push the top curve in C_4 out, forming a big circle, its length is πR. The product of these is π/R, and that approaches zero as $R\to\infty$. All that is left is the single closed loop in C_5, and I evaluate that with the residue theorem.

$$\int_{C_1} = \int_{C_5} = 2\pi i \operatorname*{Res}_{z=ia} \frac{1}{a^2+z^2}$$

Compute this residue by examining the behavior near the pole at ia.

$$\frac{1}{a^2+z^2} = \frac{1}{(z-ia)(z+ia)} \approx \frac{1}{(z-ia)(2ia)}$$

Near the point $z=ia$ the value of $z+ia$ is nearly $2ia$, so the coefficient of $1/(z-ia)$ is $1/(2ia)$, and that is the residue. The integral is $2\pi i$ times this residue, so

$$\int_{-\infty}^{\infty} dx \frac{1}{a^2+x^2} = 2\pi i \cdot \frac{1}{2ia} = \frac{\pi}{a} \tag{14.12}$$

The most obvious check on this result is that it has the correct dimensions. $[dz/z^2] = \mathrm{L}/\mathrm{L}^2 = 1/\mathrm{L}$, a reciprocal length (assuming a is a length). What happens if you push the contour *down* instead of up? See problem 14.10

Example 4

How about the more complicated integral, Eq. (14.10)? There are more poles, so that's where to start. The denominator vanishes where $z^4=-a^4$, or at

$$z = a\big(e^{i\pi+2in\pi}\big)^{1/4} = ae^{i\pi/4}e^{in\pi/2}$$

$$\int_{C_1} \frac{e^{ikz}\,dz}{a^4+z^4}$$

C1

I'm going to use the same method as before, pushing the contour past some poles, but I have to be a bit more careful this time. The exponential, not the $1/z^4$, will play the dominant role in the behavior at infinity. If k is positive then if $z = iy$, the exponential $e^{i^2 ky} = e^{-ky} \to 0$ as $y \to +\infty$. It will blow up in the $-i\infty$ direction. Of course if k is negative the reverse holds.

Assume $k > 0$, then in order to push the contour into a region where I can determine that the integral along it is zero, I have to push it toward $+i\infty$. That's where the exponential drops rapidly to zero. It goes to zero faster than any inverse power of y, so even with the length of the contour going as πR, the combination vanishes.

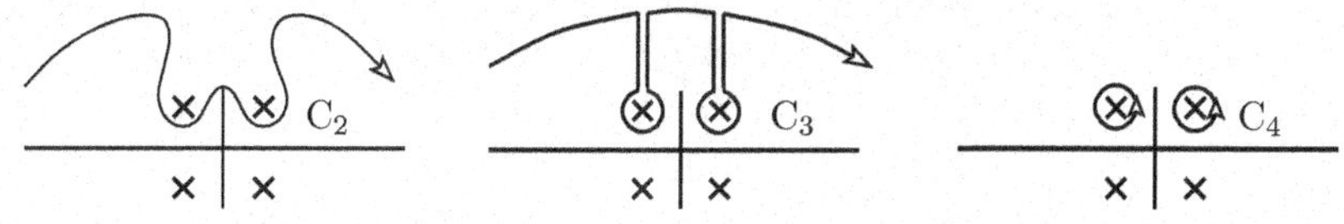

As before, when you push C_1 up to C_2 and to C_3, nothing has changed, because the contour has crossed no singularities. The transition to C_4 happens because the pairs of straight line segments cancel when they are pushed together and made to coincide. The large contour is pushed to $+i\infty$ where the negative exponential kills it. All that's left is the sum over the two residues at $ae^{i\pi/4}$ and $ae^{3i\pi/4}$.

$$\int_{C_1} = \int_{C_4} = 2\pi i \sum \text{Res}\, \frac{e^{ikz}}{a^4 + z^4}$$

The denominator factors as

$$a^4 + z^4 = (z - ae^{i\pi/4})(z - ae^{3i\pi/4})(z - ae^{5i\pi/4})(z - ae^{7i\pi/4})$$

The residue at $ae^{i\pi/4} = a(1+i)/\sqrt{2}$ is the coefficient of $1/(z - ae^{i\pi/4})$, so it is

$$\frac{e^{ika(1+i)/\sqrt{2}}}{(ae^{i\pi/4} - ae^{3i\pi/4})(ae^{i\pi/4} - ae^{5i\pi/4})(ae^{i\pi/4} - ae^{7i\pi/4})}$$

Do you have to do a lot of algebra to evaluate this denominator? Maybe you will prefer that to the alternative: *draw a picture.* The distance from the center to a corner of the square is a, so each side has length $a\sqrt{2}$. The first factor in the denominator of the residue is the line labelled "1" in the figure, so it is $a\sqrt{2}$. Similarly the second and third factors (labeled in the diagram) are $2a(1+i)/\sqrt{2}$ and $ia\sqrt{2}$. This residue is then

$$\underset{e^{i\pi/4}}{\text{Res}} = \frac{e^{ika(1+i)/\sqrt{2}}}{(a\sqrt{2})(2a(1+i)/\sqrt{2})(ia\sqrt{2})} = \frac{e^{ika(1+i)/\sqrt{2}}}{a^3 2\sqrt{2}(-1+i)} \tag{14.13}$$

For the other pole, at $e^{3i\pi/4}$, the result is

$$\underset{e^{3i\pi/4}}{\text{Res}} = \frac{e^{ika(-1+i)/\sqrt{2}}}{(-a\sqrt{2})(2a(-1+i)/\sqrt{2})(ia\sqrt{2})} = \frac{e^{ika(-1+i)/\sqrt{2}}}{a^3 2\sqrt{2}(1+i)} \tag{14.14}$$

The final result for the integral Eq. (14.10) is then the sum of these ($\times 2\pi i$)

$$\int_{-\infty}^{+\infty} \frac{e^{ikx}\,dx}{a^4+x^4} = 2\pi i\big[(14.13)+(14.14)\big] = \frac{\pi e^{-ka/\sqrt{2}}}{a^3}\cos[(ka/\sqrt{2})-\pi/4] \qquad (14.15)$$

This would be a challenge to do by other means, without using contour integration. It is probably possible, but would be much harder. Does the result make any sense? The dimensions work, because the $[dz/z^4]$ is the same as $1/a^3$. What happens in the original integral if k changes to $-k$? It's even in k of course. (Really? Why?) This result doesn't look even in k but then it doesn't have to because it applies only for the case that $k > 0$. If you have a negative k you can do the integral again (problem 14.42) and verify that it is even.

Example 5

Another example for which it's not immediately obvious how to use the residue theorem:

$$\int_{-\infty}^{\infty} dx\,\frac{\sin ax}{x}$$

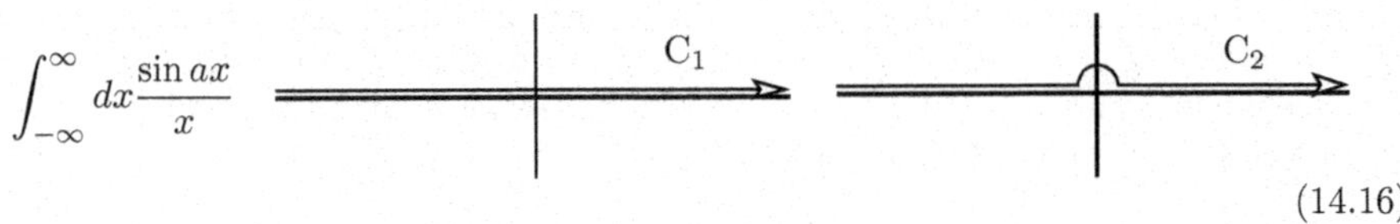

(14.16)

This function has no singularities. The sine doesn't, and the only place the integrand could have one is at zero. Near that point, the sine itself is linear in x, so $(\sin ax)/x$ is finite at the origin. The trick in using the residue theorem here is to create a singularity where there is none. Write the sine as a combination of exponentials, then the contour integral along C_1 is the same as along C_2, and

$$\int_{C_1} \frac{e^{iaz}-e^{-iaz}}{2iz} = \int_{C_2} \frac{e^{iaz}-e^{-iaz}}{2iz} = \int_{C_2} \frac{e^{iaz}}{2iz} - \int_{C_2} \frac{e^{-iaz}}{2iz}$$

I had to move the contour away from the origin in anticipation of this splitting of the integrand because I don't want to try integrating *through* this singularity that appears in the last two integrals. In the first form it doesn't matter because there is no singularity at the origin and the contour can move anywhere I want as long as the two points at $\pm\infty$ stay put. In the final two separated integrals it matters very much.

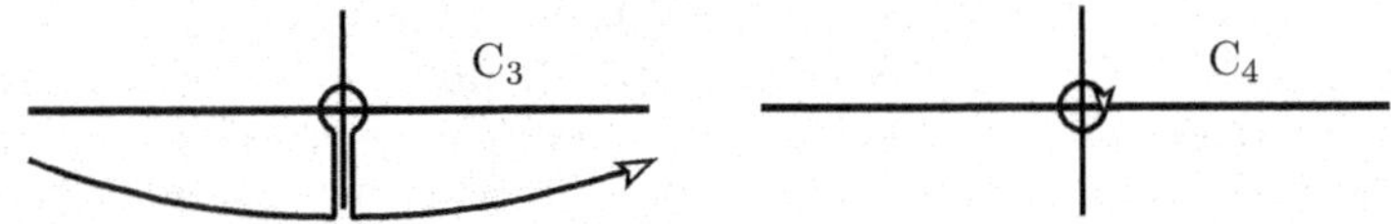

Assume that $a > 0$. In this case, $e^{iaz} \to 0$ as $z \to +i\infty$. For the other exponential, it vanishes toward $-i\infty$. This implies that I can push the contour in the first integral toward $+i\infty$ and the integral over the contour at infinity will vanish. As there are no singularities in the way, that means that the first integral is zero. For the second integral you have to push the contour toward $-i\infty$, and that hangs up on the pole at the origin. That integral is then

$$-\int_{C_2} \frac{e^{-iaz}}{2iz} = -\int_{C_4} \frac{e^{-iaz}}{2iz} = -(-2\pi i)\,\text{Res}\,\frac{e^{-iaz}}{2iz} = \pi$$

The factor $-2\pi i$ in front of the residue occurs because the integral is over a clockwise contour, thereby changing its sign. Compare the result of problem 5.29(b).

Notice that the result is independent of $a > 0$. (And what if $a < 0$?) You can check this fact by going to the original integral, Eq. (14.16), and making a change of variables. See problem 14.16.

Example 6

What is $\int_0^\infty dx/(a^2+x^2)^2$? The first observation I'll make is that by dimensional analysis alone, I expect the result to vary as $1/a^3$. Next: the integrand is even, so using the same methods as in the previous examples, extend the integration limits to the whole axis (times ½).

$$\frac{1}{2}\int_{C_1} \frac{dz}{(a^2+z^2)^2}$$

C1

As with Eq. (14.11), push the contour up and it is caught on the pole at $z = ia$. That's curve C_5 following that equation. This time however, the pole is second order, so it take a (little) more work to evaluate the residue.

$$\begin{aligned}
\frac{1}{2}\frac{1}{(a^2+z^2)^2} &= \frac{1}{2}\frac{1}{(z-ia)^2(z+ia)^2} = \frac{1}{2}\frac{1}{(z-ia)^2(z-ia+2ia)^2} \\
&= \frac{1}{2}\frac{1}{(z-ia)^2(2ia)^2\big[1+(z-ia)/2ia\big]^2} \\
&= \frac{1}{2}\frac{1}{(z-ia)^2(2ia)^2}\left[1-2\frac{(z-ia)}{2ia}+\cdots\right] \\
&= \frac{1}{2}\frac{1}{(z-ia)^2(2ia)^2}+\frac{1}{2}(-2)\frac{1}{(z-ia)(2ia)^3}+\cdots
\end{aligned}$$

The residue is the coefficient of the $1/(z-ia)$ term, so the integral is

$$\int_0^\infty dx/(a^2+x^2)^2 = 2\pi i\cdot(-1)\cdot\frac{1}{(2ia)^3} = \frac{\pi}{4a^3}$$

Is this plausible? The dimensions came out as expected, and to estimate the size of the coefficient, $\pi/4$, look back at the result Eq. (14.12). Set $a = 1$ and compare the π there to the $\pi/4$ here. The range of integration is half as big, so that accounts for a factor of two. The integrands are always less than one, so in the second case, where the denominator is squared, the integrand is always less than that of Eq. (14.12). The integral must be less, and it is. Why less by a factor of two? Dunno, but plot a few points and sketch a graph to see if you believe it. (Or use parametric differentiation to relate the two.)

Example 7

A trigonometric integral: $\int_0^{2\pi} d\theta/(a+b\cos\theta)$. The first observation is that unless $|a| > |b|$ then this denominator will go to zero somewhere in the range of integration (assuming that a and b are real). Next, the result can't depend on the relative sign of a and b, because the change of variables $\theta' = \theta+\pi$ changes the coefficient of b while the periodicity of the cosine means that you can leave the limits alone. I may as well assume that a and b are positive. The trick now is to use Euler's formula and express the cosine in terms of exponentials.

$$\text{Let } z = e^{i\theta}, \qquad \text{then} \qquad \cos\theta = \frac{1}{2}\left[z + \frac{1}{z}\right] \qquad \text{and} \qquad dz = i\,e^{i\theta}d\theta = iz\,d\theta$$

As θ goes from 0 to 2π, the complex variable z goes around the unit circle. The integral is then

$$\int_0^{2\pi} d\theta \frac{1}{(a + b\cos\theta)} = \int_C \frac{dz}{iz}\frac{1}{a + b\left(z + \frac{1}{z}\right)/2}$$

The integrand obviously has some poles, so the first task is to locate them.

$$2az + bz^2 + b = 0 \qquad \text{has roots} \qquad z = \frac{-2a \pm \sqrt{(2a)^2 - 4b^2}}{2b} = z_\pm$$

Because $a > b$, the roots are real. The important question is: Are they inside or outside the unit circle? The roots depend on the ratio $a/b = \lambda$.

$$z_\pm = \left[-\lambda \pm \sqrt{\lambda^2 - 1}\right] \tag{14.17}$$

As λ varies from 1 to ∞, the two roots travel from $-1 \to -\infty$ and from $-1 \to 0$, so z_+ stays inside the unit circle (problem 14.19). The integral is then

$$\begin{aligned} -\frac{2i}{b}\int_C \frac{dz}{z^2 + 2\lambda z + 1} &= -\frac{2i}{b}\int_C \frac{dz}{(z - z_+)(z - z_-)} = -\frac{2i}{b}2\pi i \operatorname*{Res}_{z=z_+} \\ &= -\frac{2i}{b}2\pi i\frac{1}{z_+ - z_-} = \frac{2\pi}{b\sqrt{\lambda^2 - 1}} = \frac{2\pi}{\sqrt{a^2 - b^2}} \end{aligned}$$

14.7 Branch Points

Before looking at any more uses of the residue theorem, I have to return to the subject of branch points. They are another type of singularity that an analytic function can have after poles and essential singularities. $\sqrt{z}$ provides the prototype.

The *definition* of the word function, as in section 12.1, requires that it be single-valued. The function $\sqrt{z}$ stubbornly refuses to conform to this. You can get around this in several ways: First, ignore it. Second, change the definition of "function" to allow it to be multiple-valued. Third, change the domain of the function.

You know I'm not going to ignore it. Changing the definition is not very fruitful. The third method was pioneered by Riemann and is the right way to go.

The complex plane provides a geometric picture of complex numbers, but when you try to handle square roots it becomes a hindrance. It isn't adequate for the task. There are several ways to develop the proper extension, and I'll show a couple of them.

The first is a sort of algebraic way, and the second is a geometric interpretation of the first way. There are other, even more general methods, leading into the theory of Riemann Surfaces and their topological structure, but I won't go into those.

Pick a base point, say $z_0 = 1$, from which to start. This will be a kind of fiduciary point near which I *know* the values of the function. Every other point needs to be referred to this base point. If I state that the square root of z_0 is one, then I haven't run into trouble yet. Take another point $z = re^{i\theta}$ and try to figure out the square root there.

$$\sqrt{z} = \sqrt{re^{i\theta}} = \sqrt{r}\, e^{i\theta/2} \qquad \text{or} \qquad \sqrt{z} = \sqrt{re^{i(\theta+2\pi)}} = \sqrt{r}\, e^{i\theta/2}e^{i\pi}$$

The key question: *How do you get from z_0 to z?* What path did you select?

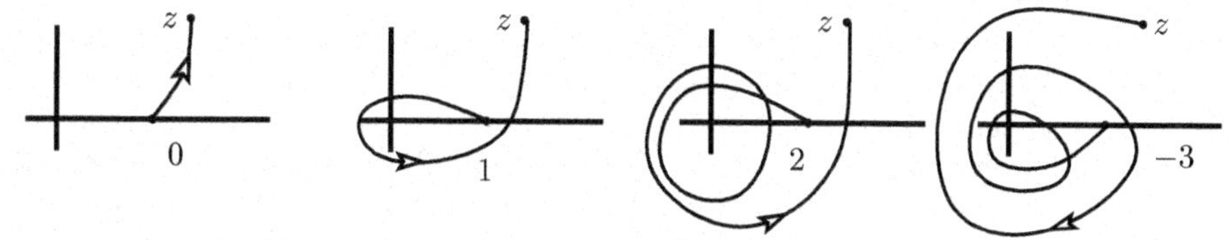

- In the picture, z appears to be at about $1.5e^{0.6i}$ or so.
- On the path labelled 0, the angle θ starts at zero at z_0 and increases to 0.6 radians, so $\sqrt{r}\, e^{i\theta/2}$ varies continuously from 1 to about $1.25e^{0.3i}$.
- On path labeled 1, angle θ again starts at zero and increases to $0.6 + 2\pi$, so $\sqrt{r}\, e^{i\theta/2}$ varies continuously from 1 to about $1.25e^{(\pi+0.3)i}$, which is minus the result along path #0.
- On the path labelled 2, angle θ goes from zero to $0.6 + 4\pi$, and $\sqrt{r}\, e^{i\theta/2}$ varies from 1 to $1.25e^{(2\pi+0.3)i}$ and that is back to the same value as path #0.
- For the path labeled −3, the angle is $0.6 - 6\pi$, resulting in the same value as path #1.

There are two classes of paths from z_0 to z, those that go around the origin an even number of times and those that go around an odd number of times. The "winding number" w is the name given to the number of times that a closed loop goes counterclockwise around a point (positive or negative), and if I take the path #1 and move it slightly so that it passes through z_0, you can more easily see that the only difference between paths 0 and 1 is the single loop around the origin. The value for the square root depends on two variables, z and the winding number of the path. Actually less than this, because it depends only on whether the winding number is even or odd: $\sqrt{z} \to \sqrt{(z, w)}$.

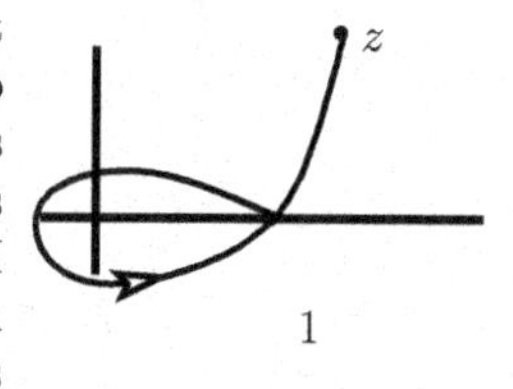

In this notation then $z_0 \to (z_0, 0)$ is the base point, and the square root of that is one. The square root of $(z_0, 1)$ is then minus one. Because the sole relevant question about the winding number is whether it is even or odd, it's convenient simply to say that the second argument can take on the values either 0 or 1 and be done with it.

Geometry of Branch Points

How do you picture such a structure? There's a convenient artifice that lets you picture and manipulate functions with branch points. In this square root example, picture two sheets and slice both along some curve starting at the origin and going to infinity. As it's a matter of convenience how you draw the cut I may as well make it a straight line along the x-axis, but any other line (or simple curve) from the origin will do. As these are mathematical planes I'll use mathematical scissors, which have the elegant property that as I cut starting from infinity on the right and proceeding down to the origin, the points that are actually *on* the x-axis are placed on the right side of the cut and the left side of the cut is left open. Indicate this with solid and dashed lines in the figure. (This is not an important point; don't worry about it.)

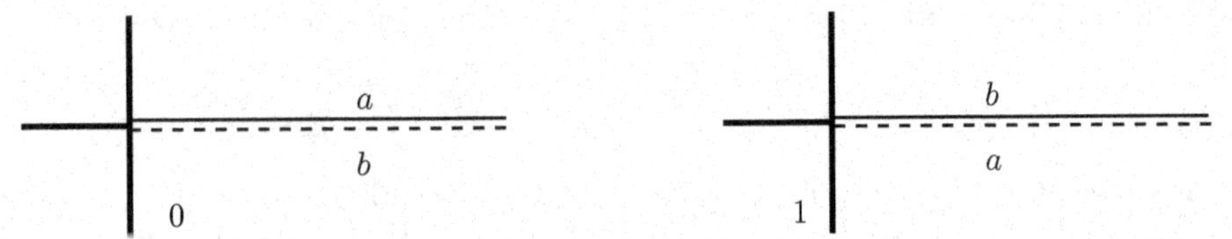

Now sew the sheets together along these cuts. Specifically, sew the top edge from sheet #0 to the bottom edge from sheet #1. I then sew the bottom edge of sheet #0 to the top edge of sheet #1. This sort of structure is called a Riemann surface. How to do this? Do it the same way that you read a map in an atlas of maps. If page 38 of the atlas shows a map with the outline of Brazil and page 27 shows a map with the outline of Bolivia, you can flip back and forth between the two pages and understand that the two maps* represent countries that are touching each other along their common border.

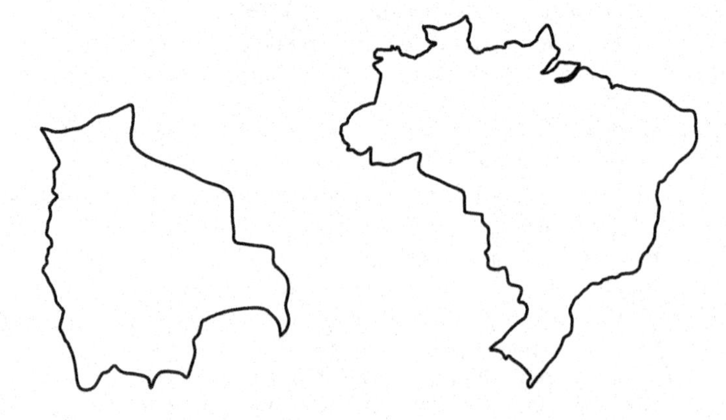

You can see where they fit even though the two countries are not drawn to the same scale. Brazil is a whole lot larger than Bolivia, but where the images fit along the Western border of Brazil and the Eastern border of Bolivia is clear. You are accustomed to doing this with maps, understanding that the right edge of the map on page 27 is the same as the left edge of the map on page 38 — you probably take it for granted. Now you get to do it with Riemann surfaces.

You have two cut planes (two maps), and certain edges are understood to be identified as identical, just as two borders of a geographic map are understood to represent the same line on the surface of the Earth. Unlike the maps above, you will usually draw both to the same scale, but you won't make the cut ragged (no pinking shears) so you

---

* www.worldatlas.com

need to use some notation to indicate what is attached to what. That's what the letters a and b are. Side a is the same as side a. The same for b. When you have more complicated surfaces, arising from more complicated functions of the complex variable with many branch points, you will have a fine time sorting out the shape of the surface.

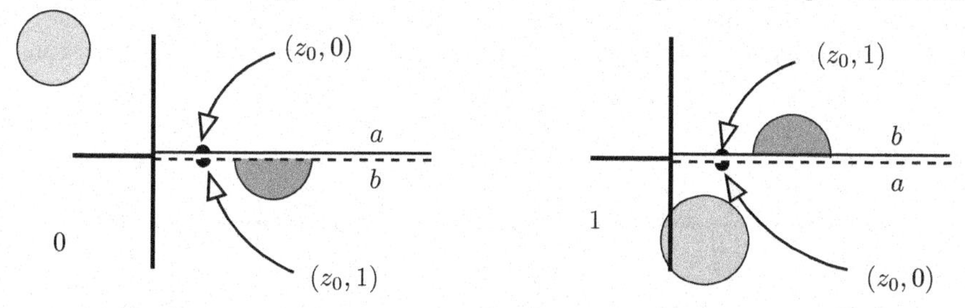

I drew three large disks on this Riemann surface. One is entirely within the first sheet (the first map); a second is entirely within the second sheet. The third disk straddles the two, but is is nonetheless a disk. On a political map this might be disputed territory. Going back to the original square root example, I also indicated the initial point at which to define the value of the square root, $(z_0, 0)$, and because a single dot would really be invisible I made it a little disk, which necessarily extends across both sheets.

Here is a picture of a closed loop on this surface. I'll probably not ask you to do contour integrals along such curves though.

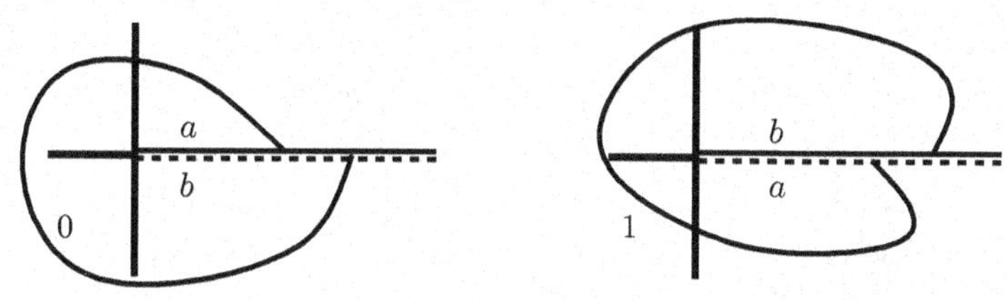

Other Functions

Cube Root Take the next simple step. What about the cube root? Answer: Do exactly the same thing, except that you need three sheets to describe the whole . Again, I'll draw a closed loop. As long as you have a single branch point it's no more complicated than this.

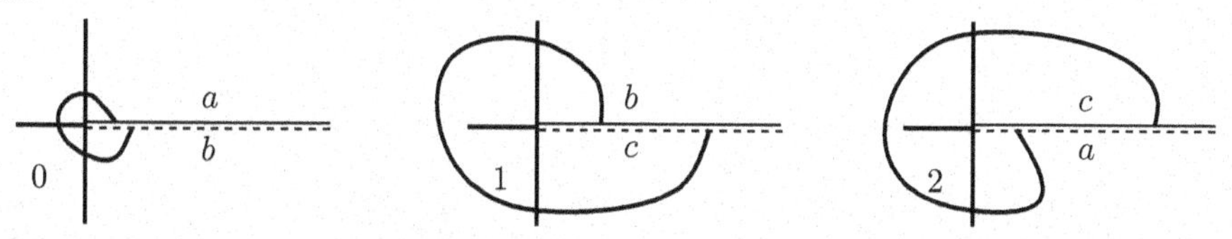

Logarithm How about a logarithm? $\ln z = \ln\left(re^{i\theta}\right) = \ln r + i\theta$. There's a branch point at the origin, but this time, as the angle keeps increasing you never come back to a previous value. This requires an infinite number of sheets. That number isn't any more difficult to handle — it's just like two, only bigger. In this case the whole winding number around the origin comes into play because every loop around the origin, taking you to the next sheet of the surface, adds another $2\pi i w$, and w is any integer from $-\infty$ to $+\infty$. The picture of the surface is like that for the cube root, but with infinitely many sheets instead of three. The complications start to come when you have several branch points.

Two Square Roots Take $\sqrt{z^2 - 1}$ for an example. Many other functions will do just as well. Pick a base point z_0; I'll take 2. (Not two base points, the number 2.) $f(z_0, 0) = \sqrt{3}$. Now follow the function around some loops. This repeats the development as for the single branch, but the number of possible paths will be larger. Draw a closed loop starting at z_0.

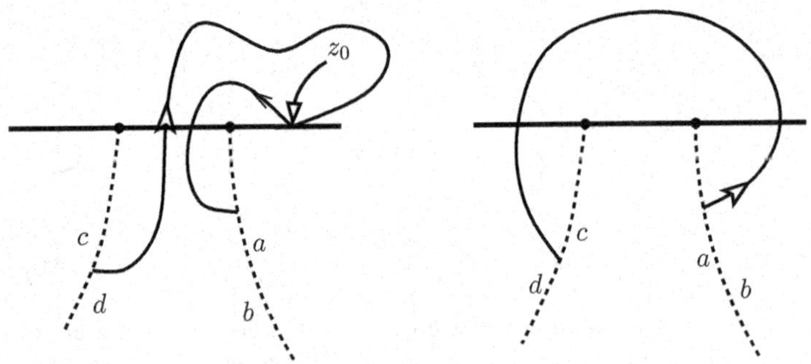

Despite the two square roots, you still need only two sheets to map out this surface. I drew the ab and cd cuts below to keep them out of the way, but they're very flexible. Start the base point and follow the path around the point $+1$; that takes you to the second sheet. You already know that if you go around $+1$ again it takes you back to where you started, so explore a different path: go around -1. Now observe that this function is the *product* of two square roots. Going around the first one introduced a factor of -1 into the function and going around the second branch point will introduce a second identical factor. As $(-1)^2 = +1$, then when you you return to z_0 the function is back at $\sqrt{3}$, you have returned to the base point and this whole loop is closed. If this were the sum of two square roots instead of their product, this wouldn't work. You'll need four sheets to map that surface. See problem 14.22.

These cuts are rather awkward, and now that I know the general shape of the surface it's possible to arrange the maps into a more orderly atlas. Here are two better ways to draw the maps. They're much easier to work with.

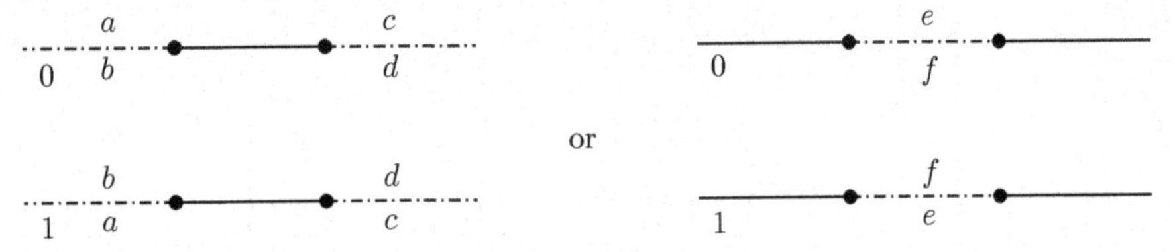

I used the dash-dot line to indicate the cuts. In the right pair, the base point is on the right-hand solid line of sheet #0. In the left pair, the base point is on the c part of sheet #0. See problem 14.20.

14.8 Other Integrals

There are many more integrals that you can do using the residue theorem, and some of these involve branch points. In some cases, the integrand you're trying to integrate has a branch point already built into it. In other cases you can pull some tricks and artificially introduce a branch point to facilitate the integration. That doesn't sound likely, but it can happen.

Example 8

The integral $\int_0^\infty dx\, x/(a+x)^3$. You can do this by elementary methods (very easily in fact), but I'll use it to demonstrate a contour method. This integral is from zero to infinity and it isn't even, so the previous tricks don't seem to apply. Instead, consider the integral $(a>0)$

$$\int_0^\infty dx\, \ln x\, \frac{x}{(a+x)^3}$$

and you see that right away, I'm creating a branch point where there wasn't one before.

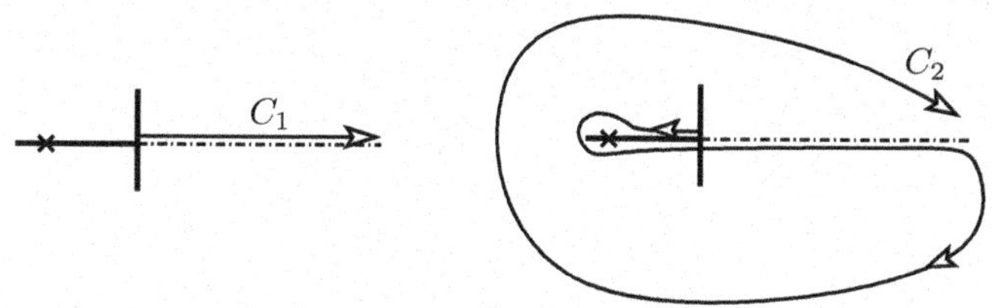

The fact that the logarithm goes to infinity at the origin doesn't matter because it is such a weak singularity that any positive power, even $x^{0.0001}$ times the logarithm, gives a finite limit as $x \to 0$. Take advantage of the branch point that this integrand provides.

$$\int_{C_1} dz\, \ln z \frac{z}{(a+z)^3} = \int_{C_2} dz\, \ln z \frac{z}{(a+z)^3}$$

On C_1 the logarithm is real. After the contour is pushed into position C_2, there are several distinct pieces. A part of C_2 is a large arc that I can take to be a circle of radius R. The size of the integrand is only as big as $(\ln R)/R^2$, and when I multiply this by $2\pi R$, the circumference of the arc, it will go to zero as $R \to \infty$.

The next pieces of C_2 to examine are the two straight lines between the origin and $-a$. The integrals along here are in opposite directions, and there's no branch point intervening, so these two segments simply cancel each other.

What's left is C_3.

$$\int_0^\infty dx\, \ln x\, \frac{x}{(a+x)^3} = \int_{C_1} = \int_{C_3}$$
$$= 2\pi i \operatorname*{Res}_{z=-a} + \int_0^\infty dx\, \big(\ln x + 2\pi i\big) \frac{x}{(a+x)^3}$$

Below the positive real axis, that is, below the cut, the logarithm differs from its original value by the constant $2\pi i$. Among all these integrals, the integral with the logarithm on the left side of the equation appears on the right side too. These terms cancel and you're left with

$$0 = 2\pi i \operatorname*{Res}_{z=-a} + \int_0^\infty dx\, 2\pi i \frac{x}{(a+x)^3} \qquad \text{or} \qquad \int_0^\infty dx\, \frac{x}{(a+x)^3} = -\operatorname*{Res}_{z=-a} \ln z \frac{z}{(a+z)^3}$$

This is a third-order pole, so it takes a bit of work. First expand the log around $-a$. Here it's probably easiest to plug into Taylor's formula for the power series and compute the derivatives of $\ln z$ at $-a$.

$$\ln z = \ln(-a) + (z+a)\frac{1}{-a} + \frac{(z+a)^2}{2!}\frac{-1}{(-a)^2} + \cdots$$

Which value of $\ln(-a)$ to take? That answer is dictated by how I arrived at the point $-a$ when I pushed the contour from C_1 to C_2. That is, $\ln a + i\pi$.

$$-\ln z\,\frac{z}{(a+z)^3} = -\left[\ln a + i\pi - \frac{1}{a}(z+a) - \frac{1}{a^2}\frac{(z+a)^2}{2} + \cdots\right]\big[(z+a) - a\big]\frac{1}{(z+a)^3}$$

I'm interested solely in the residue, so look only for the coefficient of the power $1/(z+a)$. That is

$$-\left[-\frac{1}{a} - \frac{1}{2a^2}(-a)\right] = \frac{1}{2a}$$

Did you have to do all this work to get this answer? Absolutely not. This falls under the classic heading of using a sledgehammer as a fly swatter. It does show the technique though, and in the process I had an excuse to show that third-order poles needn't be all that intimidating.

14.9 Other Results

Polynomials: There are some other consequences of looking in the complex plane that are very different from any of the preceding. If you did problem 3.11, you realize that the function $e^z = 0$ has no solutions, even in the complex plane. You are used to finding roots of equations such as quadratics and maybe you've encountered the cubic formula too. How do you know that every polynomial even has a root? Maybe there's an order-137 polynomial that has none. No, it doesn't happen. That every polynomial has a root (n of them in fact) is the Fundamental Theorem of Algebra. Gauss proved it, but after the advent of complex variable theory it becomes an elementary exercise.

A polynomial is $f(z) = a_n z^n + a_{n-1}z^{n-1} + \cdots + a_0$. Consider the integral

$$\int_C dz\,\frac{f'(z)}{f(z)}$$

around a large circle. $f'(z) = na_n z^{n-1} + \cdots$, so this is

$$\int_C dz\,\frac{na_n z^{n-1} + (n-1)a_{n-1}z^{n-2} + \cdots}{a_n z^n + a_{n-1}z^{n-1} + \cdots + a_0} = \int_C dz\,\frac{n}{z}\,\frac{1 + \frac{(n-1)a_{n-1}}{na_n z} + \cdots}{1 + \frac{a_{n-1}}{a_n z} + \cdots}$$

Take the radius of the circle large enough that only the first term in the numerator and the first term in the denominator are important. That makes the integral

$$\int_C dz\,\frac{n}{z} = 2\pi i n$$

It is certainly not zero, so that means that there is a pole inside the loop, and so a root of the denominator.

Function Determined by its Boundary Values: If a function is analytic throughout a simply connected domain and C is a simple closed curve in this domain, then the values of f inside C are determined by the values of f *on* C. Let z be a point inside the contour then I will show

$$\frac{1}{2\pi i}\int_{\mathrm{C}} dz\,\frac{f(z')}{z'-z} = f(z) \tag{14.18}$$

Because f is analytic in this domain I can shrink the contour to be an arbitrarily small curve C_1 around z, and because f is continuous, I can make the curve close enough to z that $f(z') = f(z)$ to any desired accuracy. That implies that the above integral is the same as

$$\frac{1}{2\pi i} f(z)\int_{\mathrm{C}_1} dz'\,\frac{1}{z'-z} = f(z)$$

Eq. (14.18) is Cauchy's integral formula, giving the analytic function in terms of its boundary values.

Derivatives: You can differentiate Cauchy's formula any number of times.

$$\frac{d^n f(z)}{dz^n} = \frac{n!}{2\pi i}\int_{\mathrm{C}} dz\,\frac{f(z')}{(z'-z)^{n+1}} \tag{14.19}$$

Entire Functions: An entire function is one that has no singularities anywhere. e^z, polynomials, sines, cosines are such. There's a curious and sometimes useful result about such functions. A bounded entire function is necessarily a constant. For a proof, take two points, z_1 and z_2 and apply Cauchy's integral theorem, Eq. (14.18).

$$f(z_1) - f(z_2) = \frac{1}{2\pi i}\int_{\mathrm{C}} dz\, f(z')\left[\frac{1}{z'-z_1} - \frac{1}{z'-z_2}\right] = \frac{1}{2\pi i}\int_{\mathrm{C}} dz\, f(z')\frac{z_1-z_2}{(z'-z_1)(z'-z_2)}$$

By assumption, f is bounded, $|f(z)| \le M$. A basic property of complex numbers is that $|u+v| \le |u| + |v|$ for any complex numbers u and v. This means that in the defining sum for an integral,

$$\left|\sum_k f(\zeta_k)\Delta z_k\right| \le \sum_k |f(\zeta_k)||\Delta z_k|, \qquad \text{so} \qquad \left|\int f(z)dz\right| \le \int |f(z)||dz| \tag{14.20}$$

Apply this.

$$|f(z_1) - f(z_2)| \le \int |dz||f(z')|\left|\frac{z_1-z_2}{(z'-z_1)(z'-z_2)}\right| \le M|z_1-z_2|\int |dz|\left|\frac{1}{(z'-z_1)(z'-z_2)}\right|$$

On a big enough circle of radius R, this becomes

$$|f(z_1) - f(z_2)| \le M|z_1-z_2| 2\pi R\frac{1}{R^2} \longrightarrow 0 \quad \text{as } R \to \infty$$

The left side doesn't depend on R, so $f(z_1) = f(z_2)$.

Exercises

1 Describe the shape of the function e^z of the complex variable z. That is, where in the complex plane is this function big? small? oscillating? what is its phase? Make crude sketches to help explain how it behaves as you move toward infinity in many and varied directions. Indicate not only the magnitude, but something about the phase. Perhaps words can be better than pictures?

2 Same as the preceding but for e^{iz}.

3 Describe the shape of the function z^2. Not just magnitude, other pertinent properties too such as phase, so you know how it behaves.

4 Describe the shape of i/z.

5 Describe the shape of $1/(a^2+z^2)$. Here you need to show what it does for large distances from the origin and for small. Also near the singularities.

6 Describe the shape of e^{iz^2}.

7 Describe the shape of $\cos z$.

8 Describe the shape of e^{ikz} where k is a real parameter, $-\infty < k < \infty$.

9 Describe the shape of the Bessel function $J_0(z)$. Look up for example Abramowitz and Stegun chapter 9, sections 1 and 6. $(I_0(z) = J_0(iz))$

Problems

14.1 Explicitly integrate $z^n\,dz$ around the circle of radius R centered at the origin, just as in Eq. (14.4). The number n is any positive, negative, or zero integer.

14.2 Repeat the analysis of Eq. (14.3) but change it to the integral of z^*dz.

14.3 For the real-valued function of a real variable,

$$f(x) = \begin{cases} e^{-1/x^2} & (x \neq 0) \\ 0 & (x = 0) \end{cases}$$

Work out all the derivatives at $x = 0$ and so find the Taylor series expansion about zero. Does it converge? Does it converge to f? You did draw a careful graph didn't you? Perhaps even put in some numbers for moderately small x.

14.4 (**a**) The function $1/(z - a)$ has a singularity (pole) at $z = a$. Assume that $|z| < |a|$, and write its series expansion in powers of z/a. Next assume that $|z| > |a|$ and write the series expansion in powers of a/z.
(**b**) In both cases, determine the set of z for which the series is absolutely convergent, replacing each term by its absolute value. Also sketch these sets.
(**c**) Does your series expansion in a/z imply that this function has an essential singularity at $z = 0$? Since you know that it doesn't, what happened?

14.5 The function $1/(1 + z^2)$ has a singularity at $z = i$. Write a Laurent series expansion about that point. To do so, note that $1 + z^2 = (z - i)(z + i) = (z - i)(2i + z - i)$ and use the binomial expansion to produce the desired series. (Or you can find another, more difficult method.) Use the ratio test to determine the domain of convergence of this series. Specifically, look for (and sketch) the set of z for which the absolute values of the terms form a convergent series.
Ans: $|z - i| < 2$ OR $|z - i| > 2$ depending on which way you did the expansion. If you did one, find the other. If you expanded in powers of $(z - i)$, try expanding in powers of $1/(z - i)$.

14.6 What is $\int_0^i dz/(1 - z^2)$? Ans: $i\pi/4$

14.7 (**a**) What is a Laurent series expansion about $z = 0$ with $|z| < 1$ to at least four terms for

$$\sin z/z^4 \qquad\qquad e^z/z^2(1 - z)$$

(**b**) What is the residue at $z = 0$ for each function?
(**c**) Then assume $|z| > 1$ and find the Laurent series.
Ans: $|z| > 1$: $\sum_{-\infty}^{+\infty} z^n f(n)$, where $f(n) = -e$ if $n < -3$ and $f(n) = -\sum_{n+3}^{\infty} 1/k!$ if $n \geq -3$.

14.8 By explicit integration, evaluate the integrals around the counterclockwise loops:

$$\int_{C_1} z^2\,dz \qquad \int_{C_2} z^3\,dz$$

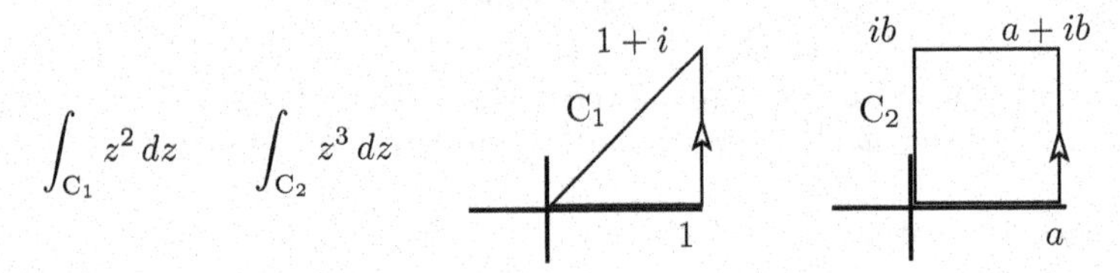

14.9 Evaluate the integral along the straight line from a to $a + i\infty$: $\int e^{iz}dz$. Take a to be real. Ans: ie^{ia}

14.10 (**a**) Repeat the contour integral Eq. (14.11), but this time push the contour *down*, not up.
(**b**) What happens to the same integral if a is negative? And be sure to explain your answer in terms of the contour integrals, even if you see an easier way to do it.

14.11 Carry out all the missing steps starting with Eq. (14.10) and leading to Eq. (14.15).

14.12 Sketch a graph of Eq. (14.15) and for $k < 0$ too. What is the behavior of this function in the neighborhood of $k = 0$? (Careful!)

14.13 In the integration of Eq. (14.16) the contour C_2 had a bump into the upper half-plane. What happens if the bump is into the lower half-plane?

14.14 For the function in problem 14.7, $e^z/z^2(1 - z)$, do the Laurent series expansion about $z = 0$, but this time assume $|z| > 1$. What is the coefficient of $1/z$ now? You should have no trouble summing the series that you get for this. Now explain why this result is as it is. Perhaps review problem 14.1.

14.15 In the integration of Eq. (14.16) the contour C_2 had a bump into the upper half-plane, but the original function had no singularity at the origin, so you can instead start with *this* curve and carry out the analysis. What answer do you get?

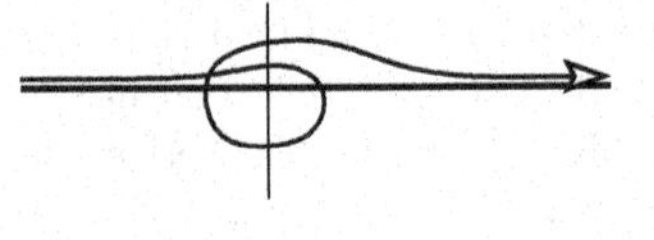

14.16 Use contour integration to evaluate Eq. (14.16) for the case that $a < 0$.
Next, independently of this, make a change of variables in the original integral Eq. (14.16) in order to see if the answer is independent of a. In this part, consider two cases, $a > 0$ and $a < 0$.

14.17 Recalculate the residue done in Eq. (14.6), but economize your labor. If all that all you really want is the coefficient of $1/z$, keep only the terms that you need in order to get it.

14.18 What is the order of all the other poles of the function $\csc^3 z$, and what is the residue at each pole?

14.19 Verify the location of the roots of Eq. (14.17).

14.20 Verify that the Riemann surfaces work as defined for the function $\sqrt{z^2-1}$ using the alternative maps in section 14.7.

14.21 Map out the Riemann surface for $\sqrt{z(z-1)(z-2)}$. You will need four sheets.

14.22 Map out the Riemann surface for $\sqrt{z}+\sqrt{z-1}$. You will need four sheets.

14.23 Evaluate

$$\int_C dz\, e^{-z} z^{-n}$$

where C is a circle of radius R about the origin.

14.24 Evaluate

$$\int_C dz\, \tan z$$

where C is a circle of radius πn about the origin. Ans: $-4\pi i n$

14.25 Evaluate the residues of these functions at their singularities. a, b, and c are distinct. Six answers: you should be able to do five of them in your head.

(**a**) $\dfrac{1}{(z-a)(z-b)(z-c)}$ (**b**) $\dfrac{1}{(z-a)(z-b)^2}$ (**c**) $\dfrac{1}{(z-a)^3}$

14.26 Evaluate the residue at the origin for the function

$$\frac{1}{z}e^{z+\frac{1}{z}}$$

The result will be an infinite series, though if you want to express the answer in terms of a standard function you will have to hunt. Ans: $I_0(2) = 2.2796$, a modified Bessel function.

14.27 Evaluate $\int_0^\infty dz/(a^4+x^4)$, and to check, compare it to the result of Eq. (14.15).

14.28 Show that

$$\int_0^\infty dx\, \frac{\cos bx}{a^2+x^2} = \frac{\pi}{2a}e^{-ab} \qquad (a,\, b>0)$$

14.29 Evaluate (a real)

$$\int_{-\infty}^{\infty} dx\, \frac{\sin^2 ax}{x^2}$$

Ans: $|a|\pi$

14.30 Evaluate

$$\int_{-\infty}^{\infty} dx\, \frac{\sin^2 bx}{x(a^2+x^2)}$$

14.31 Evaluate the integral $\int_0^\infty dx\,\sqrt{x}\,/(a+x)^2$. Use the ideas of example 8, but without the logarithm. $(a>0)$ Ans: $\pi/2\sqrt{a}$

14.32 Evaluate

$$\int_0^\infty dx \, \frac{\ln x}{a^2 + x^2}$$

(What happens if you consider $(\ln x)^2$?) Ans: $(\pi \ln a)/2a$

14.33 Evaluate ($\lambda > 1$) by contour integration

$$\int_0^{2\pi} \frac{d\theta}{\left(\lambda + \sin\theta\right)^2}$$

Ans: $2\pi\lambda/(\lambda^2 - 1)^{3/2}$

14.34 Evaluate

$$\int_0^\pi d\theta \, \sin^{2n}\theta$$

Recall Eq. (2.19). Ans: $\pi \, {}_{2n}C_n \, / 2^{2n-1} = \pi(2n-1)!!/(2n)!!$

14.35 Evaluate the integral of problem 14.33 another way. Assume λ is large and expand the integrand in a power series in $1/\lambda$. Use the result of the preceding problem to evaluate the individual terms and then sum the resulting infinite series. Will section 1.2 save you any work? Ans: Still $2\pi\lambda/(\lambda^2 - 1)^{3/2}$

14.36 Evaluate

$$\int_0^\infty dx \, \cos\alpha x^2 \qquad \text{and} \qquad \int_0^\infty dx \, \sin\alpha x^2 \qquad \text{by considering} \qquad \int_0^\infty dx \, e^{i\alpha x^2}$$

Push the contour of integration toward the 45° line. Ans: $\frac{1}{2}\sqrt{\pi/2\alpha}$

14.37

$$f(z) = \frac{1}{z(z-1)(z-2)} - \frac{1}{z^2(z-1)^2(z-2)^2}$$

What is $\int_C dz \, f(z)$ about the circle $x^2 + y^2 = 9$?

14.38 Derive

$$\int_0^\infty dx \frac{1}{a^3 + x^3} = 2\pi\sqrt{3}/9a^2$$

14.39 Go back to problem 3.45 and find the branch points of the inverse sine function.

14.40 What is the Laurent series expansion of $1/(1 + z^2)$ for small $|z|$? Again, for large $|z|$? What is the domain of convergence in each case?

14.41 Examine the power series $\sum_0^\infty z^{n!}$. What is its radius of convergence? What is its behavior as you move out from the origin along a radius at a rational angle? That is, $z = re^{i\pi p/q}$ for p and q integers. This result is called a natural boundary.

14.42 Evaluate the integral Eq. (14.10) for the case $k < 0$. Combine this with the result in Eq. (14.15) and determine if the overall function is even or odd in k (or neither).

14.43 At the end of section 14.1 several differentiation formulas are mentioned. Derive them.

14.44 Look at the criteria for the annulus of convergence of a Laurent series, Eqs. (14.7) and (14.8), and write down an example of a Laurent series that converges nowhere.

14.45 Verify the integral of example 8 using elementary methods. It will probably take at least three lines to do.

14.46 What is the power series representation for $f(z) = \sqrt{z}$ about the point $1+i$? What is the radius of convergence of this series? In the Riemann surface for this function as described in section 14.7, show the disk of convergence.

Fourier Analysis

Fourier series allow you to expand a function on a finite interval as an infinite series of trigonometric functions. What if the interval is infinite? That's the subject of this chapter. Instead of a sum over frequencies, you will have an integral.

15.1 Fourier Transform

For the finite interval you have to specify the boundary conditions in order to determine the particular basis that you're going to use. On the infinite interval you don't have this large set of choices. After all, if the boundary is infinitely far away, how can it affect what you're doing over a finite distance? But see section 15.6.

In section 5.3 you have several boundary condition listed that you can use on the differential equation $u'' = \lambda u$ and that will lead to orthogonal functions on your interval. For the purposes here the easiest approach is to assume periodic boundary conditions on the finite interval and then to take the limit as the length of the interval approaches infinity. On $-L < x < +L$, the conditions on the solutions of $u'' = \lambda u$ are then $u(-L) = u(+L)$ and $u'(-L) = u'(+L)$. The solution to this is most conveniently expressed as a complex exponential, Eq. (5.19)

$$u(x) = e^{ikx}, \qquad \text{where} \qquad u(-L) = e^{-ikL} = u(L) = e^{ikL}$$

This implies $e^{2ikL} = 1$, or $2kL = 2n\pi$, for integer $n = 0, \pm1, \pm2, \ldots$. With these solutions, the other condition, $u'(-L) = u'(+L)$ is already satisfied. The basis functions are then

$$u_n(x) = e^{ik_n x} = e^{n\pi i x/L}, \qquad \text{for} \qquad n = 0,\ \pm1,\ \pm2,\ \textit{etc.} \tag{15.1}$$

On this interval you have the Fourier series expansion

$$f(x) = \sum_{-\infty}^{\infty} a_n\, u_n(x), \qquad \text{and} \qquad \langle u_m, f\rangle = \langle u_m, \sum_{-\infty}^{\infty} a_n\, u_n\rangle = a_m \langle u_m, u_m\rangle \tag{15.2}$$

In the basis of Eq. (15.1) this normalization is $\langle u_m, u_m\rangle = 2L$.

Insert this into the series for f.

$$f(x) = \sum_{n=-\infty}^{\infty} \frac{\langle u_n, f\rangle}{\langle u_n, u_n\rangle} u_n(x) = \frac{1}{2L} \sum_{n=-\infty}^{\infty} \langle u_n, f\rangle u_n(x)$$

Now I have to express this in terms of the explicit basis functions in order to manipulate it. When you use the explicit form you have to be careful not to use the same symbol (x) for two different things in the same expression. Inside the $\langle u_n, f\rangle$ there is no "x" left over — it's the dummy variable of integration and it is not the same x that is in the $u_n(x)$ at the end. Denote $k_n = \pi n/L$.

$$f(x) = \frac{1}{2L} \sum_{n=-\infty}^{\infty} \int_{-L}^{L} dx'\, u_n(x')^* f(x')\, u_n(x) = \frac{1}{2L} \sum_{n=-\infty}^{\infty} \int_{-L}^{L} dx'\, e^{-ik_n x'} f(x')\, e^{ik_n x}$$

Now for some manipulation: As n changes by 1, k_n changes by $\Delta k_n = \pi/L$.

$$f(x) = \frac{1}{2\pi} \sum_{n=-\infty}^{\infty} \frac{\pi}{L} \int_{-L}^{L} dx'\, e^{-ik_n x'} f(x')\, e^{ik_n x}$$

$$= \frac{1}{2\pi} \sum_{n=-\infty}^{\infty} e^{ik_n x}\, \Delta k_n \int_{-L}^{L} dx'\, e^{-ik_n x'} f(x') \tag{15.3}$$

For a given value of k, define the integral

$$g_L(k) = \int_{-L}^{L} dx' e^{-ikx'} f(x')$$

If the function f vanishes sufficiently fast as $x' \to \infty$, this integral will have a limit as $L \to \infty$. Call that limit $g(k)$. Look back at Eq. (15.3) and you see that for large L the last factor will be approximately $g(k_n)$, where the approximation becomes exact as $L \to \infty$. Rewrite that expression as

$$f(x) \approx \frac{1}{2\pi} \sum_{n=-\infty}^{\infty} e^{ik_n x}\, \Delta k_n\, g(k_n) \tag{15.4}$$

As $L \to \infty$, you have $\Delta k_n \to 0$, and that turns Eq. (15.4) into an integral.

$$f(x) = \int_{-\infty}^{\infty} \frac{dk}{2\pi} e^{ikx} g(k), \qquad \text{where} \qquad g(k) = \int_{-\infty}^{\infty} dx\, e^{-ikx} f(x) \tag{15.5}$$

The function g is called* the Fourier transform of f, and f is the inverse Fourier transform of g.

Examples

For an example, take the step function

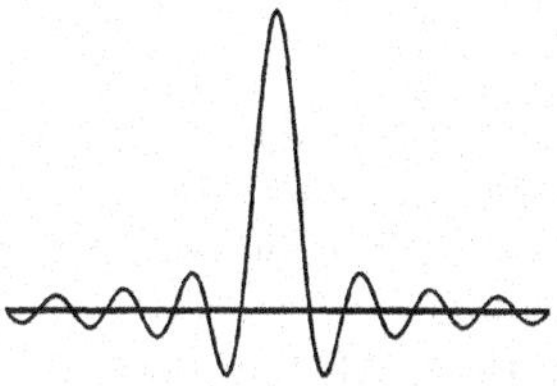

$$f(x) = \begin{cases} 1 & (-a < x < a) \\ 0 & (\text{elsewhere}) \end{cases} \qquad \text{then}$$

$$g(k) = \int_{-a}^{a} dx\, e^{-ikx}\, 1 \tag{15.6}$$

$$= \frac{1}{-ik}\left[e^{-ika} - e^{+ika}\right] = \frac{2\sin ka}{k}$$

The first observation is of course that the dimensions check: If dx is a length then so is $1/k$. After that, there is only one parameter that you can vary, and that's a. As a increases, obviously the width of the function f increases, but now look at g. The first

* Another common notation is to define g with an integral $dx/\sqrt{2\pi}$. That will require a corresponding $dk/\sqrt{2\pi}$ in the inverse relation. It's more symmetric that way, but I prefer the other convention.

place where $g(k) = 0$ is at $ka = \pi$. This value, π/a *decreases* as a increases. As f gets broader, g gets narrower (and taller). This is a general property of these Fourier transform pairs.

Can you invert this Fourier transform, evaluating the integral of g to get back to f? Yes, using the method of contour integration this is very easy. Without contour integration it would be extremely difficult, and that is typically the case with these transforms; complex variable methods are essential to get anywhere with them. The same statement holds with many other transforms (Laplace, Radon, Mellin, Hilbert, *etc.*)

The inverse transform is

$$\int_{-\infty}^{\infty} \frac{dk}{2\pi} e^{ikx} \frac{2\sin ka}{k} = \int_{C_1} \frac{dk}{2\pi} e^{ikx} \frac{e^{ika} - e^{-ika}}{ik}$$
$$= -i \int_{C_2} \frac{dk}{2\pi} \frac{1}{k} \left[e^{ik(x+a)} - e^{ik(x-a)} \right]$$

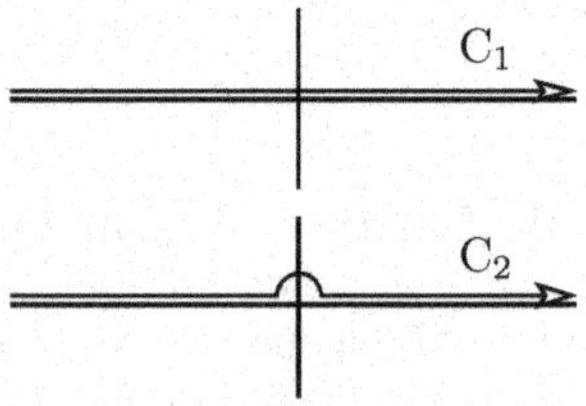

C₁

C₂

1. If $x > +a$ then both $x + a$ and $x - a$ are positive, which implies that both exponentials vanish rapidly as $k \to +i\infty$. Push the contour C_2 toward this direction and the integrand vanishes exponentially, making the integral zero.
2. If $-a < x < +a$, then only $x + a$ is positive. The integral of the first term is then zero by exactly the preceding reasoning, but the other term has an exponential that vanishes as $k \to -i\infty$ instead, implying that you must push the contour down toward $-i\infty$.

$$= i \int_{C_3} \frac{dk}{2\pi} \frac{1}{k} e^{ik(x-a)} = \int_{C_4}$$
$$= +i \frac{1}{2\pi} (-1) 2\pi i \operatorname*{Res}_{k=0} \frac{e^{ik(x-a)}}{k} = -i \frac{1}{2\pi} \cdot 2\pi i = 1$$

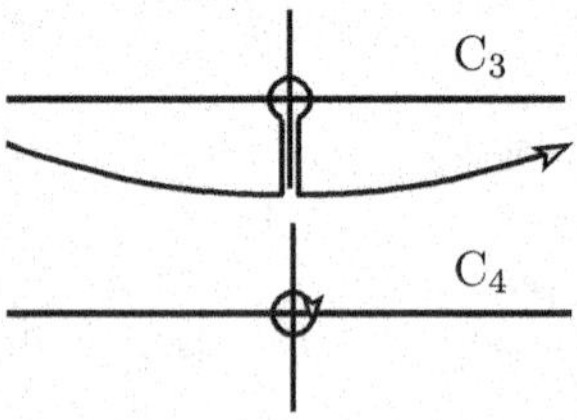

C₃

C₄

The extra (-1) factor comes because the contour is clockwise.
3. In the third domain, $x < -a$, both exponentials have the form e^{-ik}, requiring you to push the contour toward $-i\infty$. The integrand now has both exponentials, so it is analytic at zero and there is zero residue. The integral vanishes and the whole analysis takes you back to the original function, Eq. (15.6).

Another example of a Fourier transform, one that shows up often in quantum mechanics

$$f(x) = e^{-x^2/\sigma^2}, \qquad \text{so} \qquad g(k) = \int_{-\infty}^{\infty} dx\, e^{-ikx} e^{-x^2/\sigma^2} = \int_{-\infty}^{\infty} dx\, e^{-ikx - x^2/\sigma^2}$$

The trick to doing this integral is to complete the square inside the exponent.

$$-ikx - x^2/\sigma^2 = \frac{-1}{\sigma^2}\left[x^2 + \sigma^2 ikx - \sigma^4 k^2/4 + \sigma^4 k^2/4\right] = \frac{-1}{\sigma^2}\left[(x + ik\sigma^2/2)^2 + \sigma^4 k^2/4\right]$$

The integral of f is now

$$g(k) = e^{-\sigma^2k^2/4}\int_{-\infty}^{\infty} dx' e^{-x'^2/\sigma^2} \qquad \text{where} \qquad x' = x + ik\sigma/2$$

The change of variables makes this a standard integral, Eq. (1.10), and the other factor, with the exponential of k^2, comes outside the integral. The result is

$$g(k) = \sigma\sqrt{\pi}\, e^{-\sigma^2k^2/4} \tag{15.7}$$

This has the curious result that the Fourier transform of a Gaussian is* a Gaussian.

15.2 Convolution Theorem

What is the Fourier transform of the product of two functions? It is a convolution of the individual transforms. What that means will come out of the computation. Take two functions f_1 and f_2 with Fourier transforms g_1 and g_2.

$$\begin{aligned}\int_{-\infty}^{\infty} dx\, f_1(x)f_2(x)e^{-ikx} &= \int dx \int \frac{dk'}{2\pi} g_1(k')e^{ik'x} f_2(x)e^{-ikx}\\ &= \int \frac{dk'}{2\pi} g_1(k') \int dx\, e^{ik'x} f_2(x)e^{-ikx}\\ &= \int \frac{dk'}{2\pi} g_1(k') \int dx\, f_2(x)e^{-i(k-k')x}\\ &= \int_{-\infty}^{\infty} \frac{dk'}{2\pi} g_1(k')g_2(k-k') \end{aligned} \tag{15.8}$$

The last expression (except for the 2π) is called the convolution of g_1 and g_2.

$$\int_{-\infty}^{\infty} dx\, f_1(x)f_2(x)e^{-ikx} = \frac{1}{2\pi}(g_1 * g_2)(k) \tag{15.9}$$

The last line shows a common notation for the convolution of g_1 and g_2.

What is the integral of $|f|^2$ over the whole line?

$$\begin{aligned}\int_{-\infty}^{\infty} dx\, f^*(x)f(x) &= \int dx\, f^*(x) \int \frac{dk}{2\pi} g(k)e^{ikx}\\ &= \int \frac{dk}{2\pi} g(k) \int dx\, f^*(x)e^{ikx}\\ &= \int \frac{dk}{2\pi} g(k) \left[\int dx\, f(x)e^{-ikx}\right]^*\\ &= \int_{-\infty}^{\infty} \frac{dk}{2\pi} g(k)g^*(k) \end{aligned} \tag{15.10}$$

This is Parseval's identity for Fourier transforms. There is an extension to it in problem 15.10.

---

* Another function has this property: the hyperbolic secant. Look up the quantum mechanical harmonic oscillator solution for an infinite number of others.

15.3 Time-Series Analysis

Fourier analysis isn't restricted to functions of x, sort of implying position. They're probably more often used in analyzing functions of time. If you're presented with a complicated function of time, how do you analyze it? What information is present in it? If that function of time is a sound wave you may choose to analyze it with your ears, and if it is music, the frequency content is just what you will be listening for. That's Fourier analysis. The Fourier transform of the signal tells you its frequency content, and sometimes subtle periodicities will show up in the transformed function even though they aren't apparent in the original signal.

A function of time is $f(t)$ and its Fourier transform is

$$g(\omega) = \int_{-\infty}^{\infty} dt\, f(t)\, e^{i\omega t} \qquad \text{with} \qquad f(t) = \int_{-\infty}^{\infty} \frac{d\omega}{2\pi}\, g(\omega)\, e^{-i\omega t}$$

The sign convention in these equations appear backwards from the one in Eq. (15.5), and it is. One convention is as good as the other, but in the physics literature you'll find this pairing more common because of the importance of waves. A function $e^{i(kx-\omega t)}$ represents a wave with (phase) velocity ω/k, and so moving to the right. You form a general wave by taking linear combinations of these waves, usually an integral.

Example

When you hear a musical note you will perceive it as having a particular frequency. It doesn't, and if the note has a very short duration it becomes hard to tell its* pitch. Only if its duration is long enough do you have a real chance to discern what note you're hearing. This is a reflection of the facts of Fourier transforms.

If you hear what you think of as a single note, it will not last forever. It starts and it ends. Say it lasts from $t = -T$ to $t = +T$, and in that interval it maintains the frequency ω_0.

$$f(t) = Ae^{-i\omega_0 t} \qquad (-T < t < T) \tag{15.11}$$

The frequency analysis comes from the Fourier transform.

$$g(\omega) = \int_{-T}^{T} dt\, e^{i\omega t}\, Ae^{-i\omega_0 t} = A\frac{e^{i(\omega-\omega_0)T} - e^{-i(\omega-\omega_0)T}}{i(\omega-\omega_0)} = 2A\frac{\sin(\omega-\omega_0)T}{(\omega-\omega_0)}$$

This is like the function of Eq. (15.6) except that its center is shifted. It has a peak at $\omega = \omega_0$ instead of at the origin as in that case. The width of the function g is determined by the time interval T. As T is large, g is narrow and high, with a sharp localization near ω_0. In the reverse case of a short pulse, the range of frequencies that constitute the note is spread over a wide range of frequencies, and you will find it difficult to tell by listening to it just what the main pitch is supposed to be. This figure shows the frequency spectrum for two notes having the same nominal pitch, but one of them lasts three times as long as

---

* Think of a hemisemidemiquaver played at tempo prestissimo.

the other before being cut off. It therefore has a narrower spread of frequencies.

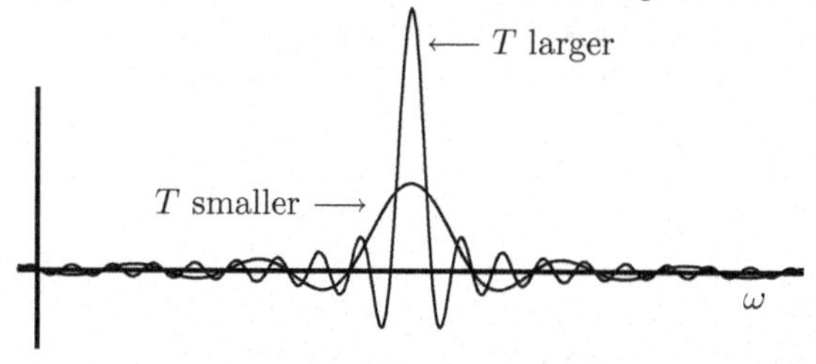

Example

Though you can do these integrals numerically, and when you are dealing with real data you will have to, it's nice to have some analytic examples to play with. I've already shown, Eq. (15.7), how the Fourier transform of a Gaussian is simple, so start from there.

$$\text{If} \qquad g(\omega) = e^{-(\omega-\omega_0)^2/\sigma^2} \qquad \text{then} \qquad f(t) = \frac{\sigma}{2\sqrt{\pi}} e^{-i\omega_0 t} e^{-\sigma^2 t^2/4}$$

If there are several frequencies, the result is a sum.

$$g(\omega) = \sum_n A_n \, e^{-(\omega-\omega_n)^2/\sigma_n^2} \qquad \Longleftrightarrow \qquad f(t) = \sum_n A_n \frac{\sigma_n}{2\sqrt{\pi}} e^{-i\omega_n t} e^{-\sigma_n^2 t^2/4}$$

In a more common circumstance you will have the time series, $f(t)$, and will want to obtain the frequency decomposition, $g(\omega)$, though for this example I worked backwards. The function of time is real, but the transformed function g is complex. Because f is real, it follows that g satisfies $g(-\omega) = g^*(\omega)$. See problem 15.13.

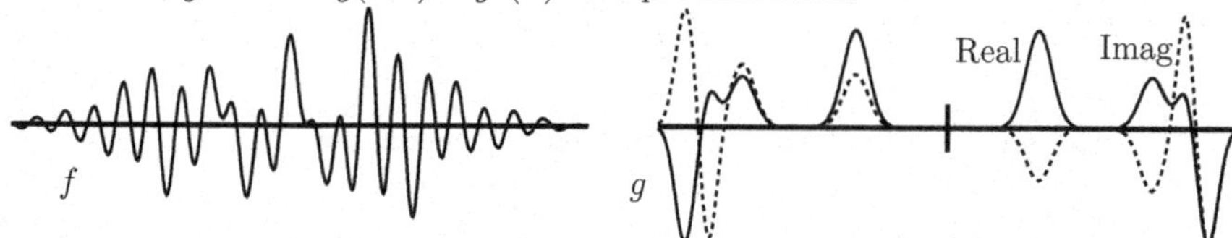

This example has four main peaks in the frequency spectrum. The real part of g is an even function and the imaginary part is odd.

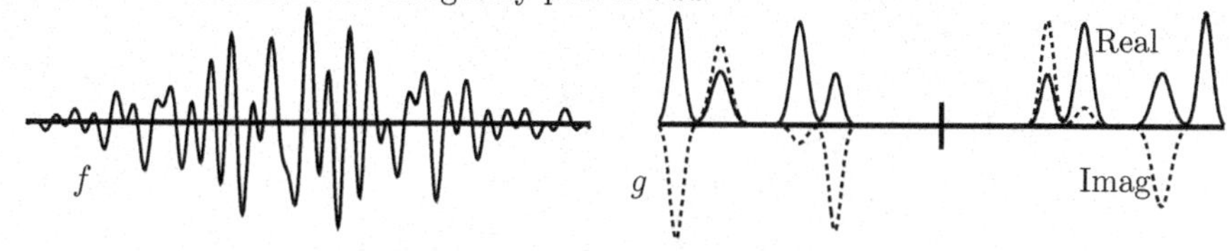

This is another example with four main peaks.

In either case, if you simply look at the function of time on the left it isn't obvious what sort of frequencies are present. That's why there are standard, well-developed computer programs to do the Fourier analysis.

15.4 Derivatives

There are a few simple, but important relations involving differentiation. What is the Fourier transform of the derivative of a function? Do some partial integration.

$$\mathcal{F}(\dot{f}) = \int dt\, e^{i\omega t}\frac{df}{dt} = e^{i\omega t} f(t)\Big|_{-\infty}^{\infty} - i\omega \int dt\, e^{i\omega t} f(t) = -i\omega\mathcal{F}(f) \tag{15.12}$$

Here I've introduced the occasionally useful notation that $\mathcal{F}(f)$ is the Fourier transform of f. The boundary terms in the partial integration will go to zero if you assume that the function f approaches zero at infinity.

The nth time derivative simply give you more factors: $(-i\omega)^n$ on the transformed function.

15.5 Green's Functions

This technique showed up in the chapter on ordinary differential equations, section 4.6, as a method to solve the forced harmonic oscillator. In that instance I said that you can look at a force as a succession of impulses, as if you're looking at the atomic level and visualizing a force as many tiny collisions by atoms. Here I'll get to the same sort of result as an application of transform methods. The basic technique is to Fourier transform everything in sight.

The damped, forced harmonic oscillator differential equation is

$$m\frac{d^2x}{dt^2} + b\frac{dx}{dt} + kx = F_0(t) \tag{15.13}$$

Multiply by $e^{i\omega t}$ and integrate over all time. You do the transforms of the derivatives by partial integration as in Eq. (15.12).

$$\int_{-\infty}^{\infty} dt\, e^{i\omega t}\,[\text{Eq. (15.13)}] = -m\omega^2\tilde{x} - ib\omega\tilde{x} + k\tilde{x} = \tilde{F}_0, \quad \text{where} \quad \tilde{x}(\omega) = \int_{-\infty}^{\infty} dt\, e^{i\omega t} x(t)$$

This is an algebraic equation that is easy to solve for the function $\tilde{x}(\omega)$.

$$\tilde{x}(\omega) = \frac{\tilde{F}_0(\omega)}{-m\omega^2 - ib\omega + k}$$

Now use the inverse transform to recover the function $x(t)$.

$$\begin{aligned}
x(t) &= \int_{-\infty}^{\infty} \frac{d\omega}{2\pi} e^{-i\omega t}\tilde{x}(\omega) = \int \frac{d\omega}{2\pi} e^{-i\omega t}\frac{\tilde{F}_0(\omega)}{-m\omega^2 - ib\omega + k} \\
&= \int \frac{d\omega}{2\pi}\frac{e^{-i\omega t}}{-m\omega^2 - ib\omega + k}\int dt'\, F_0(t')e^{i\omega t'} \\
&= \int dt'\, F_0(t') \int \frac{d\omega}{2\pi}\frac{e^{-i\omega t}}{-m\omega^2 - ib\omega + k} e^{i\omega t'}
\end{aligned} \tag{15.14}$$

In the last line I interchanged the order of integration, and in the preceding line I had to be sure to use another symbol t' in the second integral, not t. Now do the ω integral.

$$\int_{-\infty}^{\infty} \frac{d\omega}{2\pi} \frac{e^{-i\omega t}}{-m\omega^2 - ib\omega + k} e^{i\omega t'} = \int_{-\infty}^{\infty} \frac{d\omega}{2\pi} \frac{e^{-i\omega(t-t')}}{-m\omega^2 - ib\omega + k} \tag{15.15}$$

To do this, use contour integration. The singularities of the integrand are at the roots of the denominator, $-m\omega^2 - ib\omega + k = 0$. They are

$$\omega = \frac{-ib \pm \sqrt{-b^2 + 4km}}{2m} = \omega_\pm$$

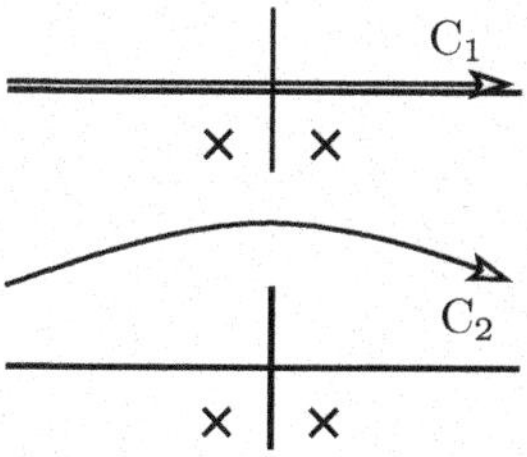

C1

C2

Both of these poles are in the lower half complex plane. The contour integral C_1 is along the real axis, and now I have to decide where to push the contour in order to use the residue theorem. This will be governed by the exponential, $e^{-i\omega(t-t')}$.

First take the case $t < t'$, then $e^{-i\omega(t-t')}$ is of the form $e^{+i\omega}$, so in the complex ω-plane its behavior in the $\pm i$ directions is as a decaying exponential toward $+i$ ($\propto e^{-|\omega|}$). It is a rising exponential toward $-i$ ($\propto e^{+|\omega|}$). This means that pushing the contour C_1 up toward C_2 and beyond will make this integral go to zero. I've crossed no singularities, so that means that Eq. (15.15) *is* zero for $t < t'$.

Next, the case that $t > t'$. Now $e^{-i\omega(t-t')}$ is of the form $e^{-i\omega}$, so its behavior is reversed from that of the preceding paragraph. It dies off rapidly toward $-i\infty$ and rises in the opposite direction. That means that I must push the contour in the opposite direction, down to C_3 and to C_4. Because of the decaying exponential, the large arc of the contour that is pushed down to $-i\infty$ gives zero for its integral; the two lines that parallel the i-axis cancel each other; only the two residues remain.

$$\int_{-\infty}^{\infty} \frac{d\omega}{2\pi} \frac{e^{-i\omega(t-t')}}{-m\omega^2 - ib\omega + k} = -2\pi i \sum_{\omega_\pm} \text{Res} \tag{15.16}$$

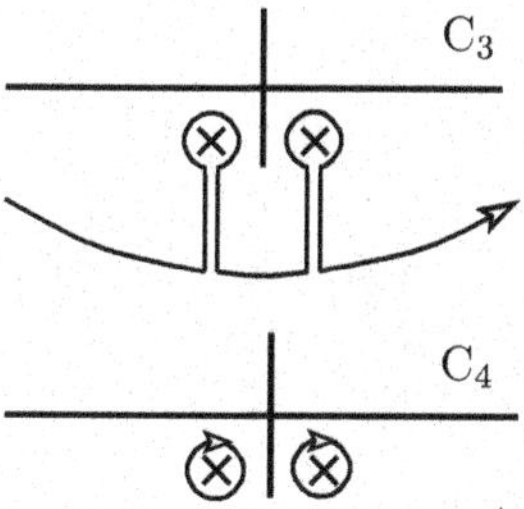

C3

C4

The denominator in Eq. (15.15) is $-m(\omega - \omega_+)(\omega - \omega_-)$. Use this form to compute the residues. Leave the $1/2\pi$ aside for the moment and you have

$$\frac{e^{-i\omega(t-t')}}{-m\omega^2 - ib\omega + k} = \frac{e^{-i\omega(t-t')}}{-m(\omega - \omega_+)(\omega - \omega_-)}$$

The residues of this at $\omega_\pm$ are the coefficients of these first order poles.

$$\text{at } \omega_+: \quad \frac{e^{-i\omega_+(t-t')}}{-m(\omega_+ - \omega_-)} \qquad \text{and at } \omega_-: \quad \frac{e^{-i\omega_-(t-t')}}{-m(\omega_- - \omega_+)}$$

The explicit values of $\omega_\pm$ are

$$\omega_+ = \frac{-ib + \sqrt{-b^2 + 4km}}{2m} \qquad \text{and} \qquad \omega_- = \frac{-ib - \sqrt{-b^2 + 4km}}{2m}$$

$$\text{Let} \qquad \omega' = \frac{\sqrt{-b^2 + 4km}}{2m} \qquad \text{and} \qquad \gamma = \frac{b}{2m}$$

The difference that appears in the preceding equation is then

$$\omega_+ - \omega_- = (\omega' - i\gamma) - (-\omega' - i\gamma) = 2\omega'$$

Eq. (15.16) is then

$$\begin{aligned}
\int_{-\infty}^{\infty} \frac{d\omega}{2\pi} \cdot \frac{e^{-i\omega(t-t')}}{-m\omega^2 - ib\omega + k} &= -i\left[\frac{e^{-i(\omega'-i\gamma)(t-t')}}{-2m\omega'} + \frac{e^{-i(-\omega'-i\gamma)(t-t')}}{+2m\omega'}\right] \\
&= \frac{-i}{2m\omega'} e^{-\gamma(t-t')}\left[-e^{-i\omega'(t-t')} + e^{+i\omega'(t-t')}\right] \\
&= \frac{1}{m\omega'} e^{-\gamma(t-t')} \sin\left(\omega'(t-t')\right)
\end{aligned}$$

Put this back into Eq. (15.14) and you have

$$x(t) = \int_{-\infty}^{t} dt'\, F_0(t') G(t-t'), \qquad \text{where} \qquad G(t-t') = \frac{1}{m\omega'} e^{-\gamma(t-t')} \sin\left(\omega'(t-t')\right) \tag{15.17}$$

If you eliminate the damping term, setting $b = 0$, this is exactly the same as Eq. (4.34). The integral stops at $t' = t$ because the Green's function vanishes beyond there. The motion at time t is determined by the force that was applied in the past, not the future.

Example

Apply a constant external force to a damped harmonic oscillator, starting it at time $t = 0$ and keeping it on. What is the resulting motion?

$$F_0(t) = \begin{cases} 0 & (t < 0) \\ F_1 & (t > 0) \end{cases}$$

where F_1 is a constant. The equation (15.17) says that the solution is $(t > 0)$

$$\begin{aligned}
x(t) &= \int_{-\infty}^{t} dt'\, F_0(t')G(t-t') = \int_0^t dt'\, F_1 G(t-t') \\
&= F_1 \int_0^t dt' \frac{1}{m\omega'} e^{-\gamma(t-t')} \sin\left(\omega'(t-t')\right) \\
&= \frac{F_1}{2im\omega'} \int_0^t dt' e^{-\gamma(t-t')} \left[e^{i\omega'(t-t')} - e^{-i\omega'(t-t')}\right] \\
&= \frac{F_1}{2im\omega'} \left[\frac{1}{\gamma - i\omega'} e^{(-\gamma+i\omega')(t-t')} - \frac{1}{\gamma + i\omega'} e^{(-\gamma-i\omega')(t-t')}\right]_{t'=0}^{t'=t} \\
&= \frac{F_1}{2im\omega'} \left[\frac{2i\omega'}{\gamma^2+\omega'^2} - \frac{e^{-\gamma t}}{\gamma^2+\omega'^2}\left[2i\gamma \sin\omega' t + 2i\omega' \cos\omega' t\right]\right] \\
&= \frac{F_1}{m(\gamma^2+\omega'^2)} \left[1 - e^{-\gamma t}\left[\cos\omega' t + \frac{\gamma}{\omega'}\sin\omega' t\right]\right]
\end{aligned}$$

Check the answer. If $t = 0$ it is correct; $x(0) = 0$ as it should.
If $t \to \infty$, $x(t)$ goes to $F_1/\big(m(\gamma^2 + \omega'^2)\big)$; is *this* correct? Check it out! And maybe simplify the result in the process. Is the small time behavior correct?

15.6 Sine and Cosine Transforms

Return to the first section of this chapter and look again at the derivation of the Fourier transform. It started with the Fourier series on the interval $-L < x < L$ and used periodic boundary conditions to define which series to use. Then the limit as $L \to \infty$ led to the transform.

What if you know the function only for positive values of its argument? If you want to write $f(x)$ as a series when you know it only for $0 < x < L$, it doesn't make much sense to start the way I did in section 15.1. Instead, pick the boundary condition at $x = 0$ carefully because this time the boundary won't go away in the limit that $L \to \infty$. The two common choices to define the basis are

$$u(0) = 0 = u(L), \qquad \text{and} \qquad u'(0) = 0 = u'(L) \tag{15.18}$$

Start with the first, then $u_n(x) = \sin(n\pi x/L)$ for positive n. The equation (15.2) is unchanged, save for the limits.

$$f(x) = \sum_1^\infty a_n\, u_n(x), \qquad \text{and} \qquad \big\langle u_m, f\big\rangle = \big\langle u_m, \sum_{n=1}^\infty a_n\, u_n\big\rangle = a_m\big\langle u_m, u_m\big\rangle$$

In this basis, $\big\langle u_m, u_m\big\rangle = L/2$, so

$$f(x) = \sum_{n=1}^\infty \frac{\big\langle u_n, f\big\rangle}{\big\langle u_n, u_n\big\rangle} u_n(x) = \frac{2}{L}\sum_{n=1}^\infty \langle u_n, f\rangle u_n(x)$$

Now explicitly use the sine functions to finish the manipulation, and as in the work leading up to Eq. (15.3), denote $k_n = \pi n/L$, and the difference $\Delta k_n = \pi/L$.

$$\begin{aligned} f(x) &= \frac{2}{L}\sum_1^\infty \int_0^L dx'\, f(x') \sin\frac{n\pi x'}{L} \sin\frac{n\pi x}{L} \\ &= \frac{2}{\pi}\sum_1^\infty \sin\frac{n\pi x}{L}\, \Delta k_n \int_0^L dx'\, f(x') \sin n\pi x'/L \end{aligned} \tag{15.19}$$

For a given value of k, define the integral

$$g_L(k) = \int_0^L dx' \,\sin(kx') f(x')$$

If the function f vanishes sufficiently fast as $x' \to \infty$, this integral will have a limit as $L \to \infty$. Call that limit $g(k)$. Look back at Eq. (15.19) and you see that for large L the last factor will be approximately $g(k_n)$, where the approximation becomes exact as $L \to \infty$. Rewrite that expression as

$$f(x) \approx \frac{2}{\pi}\sum_1^\infty \sin(k_n x)\Delta k_n\, g(k_n) \tag{15.20}$$

As $L \to \infty$, you have $\Delta k_n \to 0$, and that turns Eq. (15.20) into an integral.

$$f(x) = \frac{2}{\pi}\int_0^\infty dk\, \sin kx\, g(k), \qquad \text{where} \qquad g(k) = \int_0^\infty dx\, \sin kx\, f(x) \tag{15.21}$$

This is the Fourier Sine transform. For a parallel calculation leading to the Cosine transform, see problem 15.22, where you will find that the equations are the same except for changing sine to cosine.

$$f(x) = \frac{2}{\pi}\int_0^\infty dk\, \cos kx\, g(k), \qquad \text{where} \qquad g(k) = \int_0^\infty dx\, \cos kx\, f(x) \tag{15.22}$$

What is the sine transform of a derivative? Integrate by parts, remembering that f has to approach zero at infinity for any of this to make sense.

$$\int_0^\infty dx\, \sin kx\, f'(x) = \sin kx f(x)\Big|_0^\infty - k\int_0^\infty dx\, \cos kx\, f(x) = -k\int_0^\infty dx\, \cos kx\, f(x)$$

For the second derivative, repeat the process.

$$\int_0^\infty dx\, \sin kx\, f''(x) = kf(0) - k^2\int_0^\infty dx\, \sin kx\, f(x) \tag{15.23}$$

15.7 Wiener-Khinchine Theorem

If a function of time represents the pressure amplitude of a sound wave or the electric field of an electromagnetic wave the power received is proportional to the amplitude squared. By Parseval's identity, the absolute square of the Fourier transform has an integral proportional to the integral of this power. This leads to the interpretation of the transform squared as some sort of power density in frequency. $|g(\omega)|^2 d\omega$ is then a power received in this frequency interval. When this energy interpretation isn't appropriate, $|g(\omega)|^2$ is called the "spectral density." A useful result appears by looking at the Fourier transform of this function.

$$\begin{aligned}
\int \frac{d\omega}{2\pi}|g(\omega)|^2 e^{-i\omega t} &= \int \frac{d\omega}{2\pi} g^*(\omega)e^{-i\omega t} \int dt'\, f(t')e^{i\omega t'} \\
&= \int dt'\, f(t') \int \frac{d\omega}{2\pi} g^*(\omega)e^{i\omega t'} e^{-i\omega t} \\
&= \int dt'\, f(t') \left[\int \frac{d\omega}{2\pi} g(\omega)e^{-i\omega(t'-t)} \right]^* \\
&= \int dt'\, f(t')f(t'-t)^* \qquad (15.24)
\end{aligned}$$

When you're dealing with a real f, this last integral is called the autocorrelation function. It tells you in some average way how closely related a signal is to the same signal at some other time. If the signal that you are examining is just noise then what happens now will be unrelated to what happened a few milliseconds ago and this autocorrelation function will be close to zero. If there is structure in the signal then this function gives a lot of information about it.

The left side of this whole equation involves two Fourier transforms ($f \to g$, then $|g|^2$ to it's transform). The right side of this theorem seems to be easier and more direct to compute than the left, so why is this relation useful? It is because of the existence of the FFT, the "Fast Fourier Transform," an algorithm that makes the process of Fourier transforming a set of data far more efficient than doing it by straight-forward numerical integration methods — faster by factors that reach into the thousands for large data sets.

Problems

15.1 Invert the Fourier transform, g, in Eq. (15.7).

15.2 What is the Fourier transform of $e^{ik_0x-x^2/\sigma^2}$? Ans: A translation of the $k_0 = 0$ case

15.3 What is the Fourier transform of xe^{-x^2/σ^2}?

15.4 What is the square of the Fourier transform operator? That is, what is the Fourier transform of the Fourier transform?

15.5 A function is defined to be

$$f(x) = \begin{cases} 1 & (-a < x < a) \\ 0 & (\text{elsewhere}) \end{cases}$$

What is the convolution of f with itself? $(f * f)(x)$ And graph it of course. Start by graphing both $f(x')$ and the other factor that goes into the convolution integral.
Ans: $(2a - |x|)$ for $(-2a < x < +2a)$, and zero elsewhere.

15.6 Two functions are

$$f_1(x) = \begin{cases} 1 & (a < x < b) \\ 0 & (\text{elsewhere}) \end{cases} \qquad \text{and} \qquad f_2(x) = \begin{cases} 1 & (A < x < B) \\ 0 & (\text{elsewhere}) \end{cases}$$

What is the convolution of f_1 with f_2? And graph it.

15.7 Derive these properties of the convolution:
(**a**) $f * g = g * f$ (**b**) $f * (g * h) = (f * g) * h$ (**c**) $\delta(f * g) = f * \delta g + g * \delta f$ where $\delta f(t) = tf(t)$, $\delta g(t) = tg(t)$, *etc.* (**d**) What are $\delta^2(f * g)$ and $\delta^3(f * g)$?

15.8 Show that you can rewrite Eq. (15.9) as

$$\mathcal{F}(f * g) = \mathcal{F}(f) \cdot \mathcal{F}(g)$$

where the shorthand notation $\mathcal{F}(f)$ is the Fourier transform of f.

15.9 Derive Eq. (15.10) from Eq. (15.9).

15.10 What is the analog of Eq. (15.10) for two different functions? That is, relate the scalar product of two functions,

$$\langle f_1, f_2 \rangle = \int_{-\infty}^{\infty} f_1^*(x) f_2(x)$$

to their Fourier transforms. Ans: $\int g_1^*(k) g_2(k)\, dk/2\pi$

15.11 In the derivation of the harmonic oscillator Green's function, and starting with Eq. (15.15), I assumed that the oscillator is underdamped: that $b^2 < 4km$. Now assume the reverse, the overdamped case, and repeat the calculation.

15.12 Repeat the preceding problem, but now do the critically damped case, for which $b^2 = 4km$. Compare your result to the result that you get by taking the limit of critical damping in the preceding problem and in Eq. (15.17).

15.13 Show that if $f(t)$ is real then the Fourier transform satisfies $g(-\omega) = g^*(\omega)$. What are the properties of g if f is respectively even or odd?

15.14 Evaluate the Fourier transform of

$$f(x) = \begin{cases} A\big(a - |x|\big) & (-a < x < a) \\ 0 & \text{(otherwise)} \end{cases}$$

How do the properties of the transform vary as the parameter a varies?
Ans: $2A(1 - \cos ka)/k^2$

15.15 Evaluate the Fourier transform of $Ae^{-\alpha|x|}$. Invert the transform to verify that it takes you back to the original function. Ans: $2\alpha/(\alpha^2 + k^2)$

15.16 Given that the Fourier transform of $f(x)$ is $g(k)$, what is the Fourier transform of the the function translated a distance a to the right, $f_1(x) = f(x - a)$?

15.17 Schroedinger's equation is

$$-i\hbar\frac{\partial\psi}{\partial t} = -\frac{\hbar^2}{2m}\frac{\partial^2\psi}{\partial x^2} + V(x)\psi$$

Fourier transform the whole equation with respect to x, and find the equation for$\Phi(k, t)$, the Fourier transform of $\psi(x, t)$. The result will *not* be a differential equation.
Ans: $-i\hbar\partial\Phi(k,t)/\partial t = (\hbar^2k^2/2m)\Phi + (v * \Phi)/2\pi$

15.18 Take the Green's function solution to Eq. (15.13) as found in Eq. (15.17) and take the limit as both k and b go to zero. Verify that the resulting single integral satisfies the original second order differential equation.

15.19 (a) In problem 15.18 you have the result that a double integral (undoing two derivatives) can be written as a single integral. Now solve the equation

$$\frac{d^3x}{dt^3} = F(t)$$

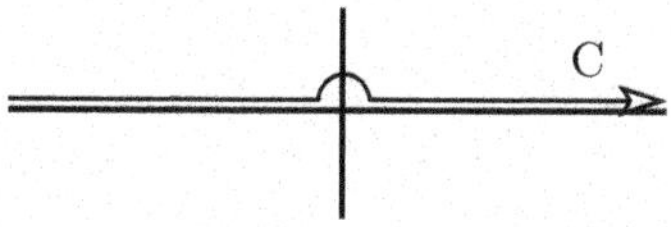

C

directly, using the same method as for Eq. (15.13). You will get a pole at the origin and how do you handle this, where the contour of integration goes straight through the origin? Answer: Push the contour up as in the figure. Why? This is what's called the "retarded solution" for which the value of $x(t)$ depends on only those values of $F(t')$ in the past. If you try any other contour to define the integral you will not get this property. (And sometimes there's a reason to make another choice.)

(**b**) Pick a fairly simple F and verify that this gives the right answer.
Ans: $\frac{1}{2}\int_{-\infty}^{t} dt'\, F(t')(t-t')^2$

15.20 Repeat the preceding problem for the fourth derivative. Would you care to conjecture what 3½ integrals might be? Then perhaps an arbitrary non-integer order?
Ans: $\frac{1}{6}\int_{-\infty}^{t} dt'\, F(t')(t-t')^3$

15.21 What is the Fourier transform of $xf(x)$? Ans: $ig'(k)$

15.22 Repeat the calculations leading to Eq. (15.21), but for the boundary conditions $u'(0) = 0 = u'(L)$, leading to the Fourier cosine transform.

15.23 For both the sine and cosine transforms, the original function $f(x)$ was defined for positive x only. Each of these transforms define an extension of f to negative x. This happens because you compute $g(k)$ and from it get an inverse transform. Nothing stops you from putting a negative value of x into the answer. What are the results?

15.24 What are the sine and cosine transforms of $e^{-\alpha x}$. In each case evaluate the inverse transform.

15.25 What is the sine transform of $f(x) = 1$ for $0 < x < L$ and $f(x) = 0$ otherwise. Evaluate the inverse transform.

15.26 Repeat the preceding calculation for the cosine transform. Graph the two transforms and compare them, including their dependence on L.

15.27 Choose any different way around the pole in problem 15.19, and compute the difference between the result with your new contour and the result with the old one. Note: Plan ahead before you start computing.

Calculus of Variations

The biggest step from derivatives with one variable to derivatives with many variables is from one to two. After that, going from two to three was just more algebra and more complicated pictures. Now the step will be from a finite number of variables to an infinite number. That will require a new set of tools, yet in many ways the techniques are not very different from those you know.

If you've never read chapter 19 of volume II of the Feynman Lectures in Physics, now would be a good time. It's a classic introduction to the area. For a deeper look at the subject, pick up MacCluer's book refered to in the Bibliography at the beginning of this book.

16.1 Examples

What line provides the shortest distance between two points? A straight line of course, no surprise there. But not so fast, with a few twists on the question the result won't be nearly as obvious. How do I measure the length of a curved (or even straight) line? Typically with a ruler. For the curved line I have to do successive approximations, breaking the curve into small pieces and adding the finite number of lengths, eventually taking a limit to express the answer as an integral. Even with a straight line I will do the same thing if my ruler isn't long enough.

Put this in terms of how you do the measurement: Go to a local store and purchase a ruler. It's made out of some real material, say brass. The curve you're measuring has been laid out on the ground, and you move along it, counting the number of times that you use the ruler to go from one point on the curve to another. If the ruler measures in decimeters and you lay it down 100 times along the curve, you have your first estimate for the length, 10.0 meters. Do it again, but use a centimeter length and you need 1008 such lengths: 10.08 meters.

That's tedious, but simple. Now do it again for another curve and compare their lengths. Here comes the twist: The ground is not at a uniform temperature. Perhaps you're making these measurements over a not-fully-cooled lava flow in Hawaii. Brass will expand when you heat it, so if the curve whose length you're measuring passes over a hot spot, then the ruler will expand when you place it down, and you will need to place it down fewer times to get to the end of the curve. You will measure the curve as shorter. Now it is not so clear which curve will have the shortest (measured) length. If you take the straight line and push it over so that it passes through a hotter region, then you may get a smaller result.

Let the coefficient of expansion of the ruler be α, assumed constant. For modest temperature changes, the length of the ruler is $\ell' = (1 + \alpha\Delta T)\ell$. The length of a curve as measured with this ruler is

$$\int \frac{d\ell}{1 + \alpha T} \tag{16.1}$$

Here I'm taking $T = 0$ as the base temperature for the ruler and $d\ell$ is the length you would use if everything stayed at this temperature. With this measure for length, it becomes an interesting problem to discover which path has the shortest "length." The formal term for the path of shortest length is geodesic.

In section 13.1 you saw integrals that looked very much like this, though applied to a different problem. There I looked at the time it takes a particle to slide down a curve under gravity. That time is the integral of $dt = d\ell/v$, where v is the particle's speed, a function of position along the path. Using conservation of energy, the expression for the time to slide down a curve was Eq. (13.6).

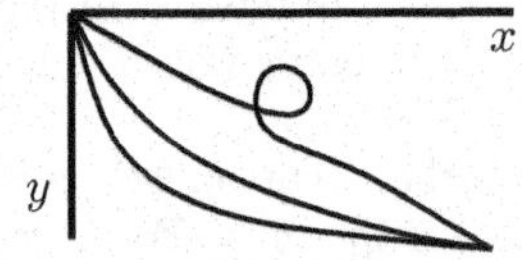

$$\int dt = \int \frac{d\ell}{\sqrt{(2E/m) + 2gy}} \tag{16.2}$$

In that chapter I didn't attempt to answer the question about which curve provides the quickest route to the end, but in this chapter I will. Even qualitatively you can see a parallel between these two problems. You get a shorter length by pushing the curve into a region of higher temperature. You get a shorter time by pushing the curve lower, (larger y). In the latter case, this means that you drop fast to pick up speed quickly. In both cases the denominator in the integral is larger. You can overdo it of course. Push the curve too far and the value of $\int d\ell$ itself can become too big. It's a balance.

In problems 2.35 and 2.39 you looked at the amount of time it takes light to travel from one point to another along various paths. Is the time a minimum, a maximum, or neither? In these special cases, you saw that this is related to the focus of the lens or of the mirror. This is a very general property of optical systems, and is an extension of some of the ideas in the first two examples above.

These questions are sometimes pretty and elegant, but are they related to anything else? Yes. Newton's classical mechanics can be reformulated in this language and it leads to powerful methods to set up the equations of motion in complicated problems. The same ideas lead to useful approximation techniques in electromagnetism, allowing you to obtain high-accuracy solutions to problems for which there is no solution by other means.

16.2 Functional Derivatives

It is time to get specific and to implement* these concepts. All the preceding examples can be expressed in the same general form. In a standard x-y rectangular coordinate system,

$$d\ell = \sqrt{dx^2 + dy^2} = dx\sqrt{1 + \left(\frac{dy}{dx}\right)^2} = dx\sqrt{1 + y'^2}$$

Then Eq. (16.1) is

$$\int_a^b dx\, \frac{\sqrt{1 + y'^2}}{1 + \alpha T(x, y)} \tag{16.3}$$

This measured length depends on the path, and I've assumed that I can express the path with y as a function of x. No loops. You can allow more general paths by using another parametrization: $x(t)$ and $y(t)$. Then the same integral becomes

$$\int_{t_1}^{t_2} dt\, \frac{\sqrt{\dot{x}^2 + \dot{y}^2}}{1 + \alpha T\big(x(t), y(t)\big)} \tag{16.4}$$

---

* If you find the methods used in this section confusing, you may prefer to look at an alternate approach to the subject as described in section 16.6. Then return here.

The equation (16.2) has the same form

$$\int_a^b dx\,\frac{\sqrt{1+y'^2}}{\sqrt{(2E/m)+2gy}}$$

And the travel time for light through an optical system is

$$\int dt = \int \frac{d\ell}{v} = \int_a^b dx\,\frac{\sqrt{1+y'^2}}{v(x,y)}$$

where the speed of light is some known function of the position.

In all of these cases the output of the integral depends on the path taken. It is a *functional* of the path, a scalar-valued function of a function variable. Denote the argument by square brackets.

$$I[y] = \int_a^b dx\, F\big(x, y(x), y'(x)\big) \tag{16.5}$$

The specific F varies from problem to problem, but the preceding examples all have this general form, even when expressed in the parametrized variables of Eq. (16.4).

The idea of differential calculus is that you can get information about a function if you try changing the independent variable by a small amount. Do the same thing here. Now however the independent variable is the whole path, so I'll change that path by some small amount and see what happens to the value of the integral I. This approach to the subject is due to Lagrange. The development in section 16.6 comes from Euler.

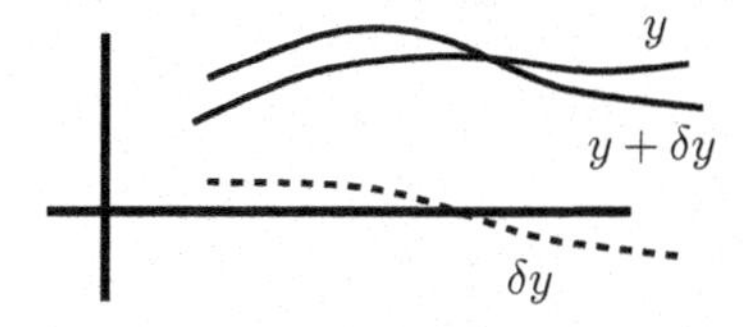

$$\begin{aligned}\Delta I &= I[y+\delta y] - I[y] \\ &= \int_{a+\Delta a}^{b+\Delta b} dx\, F\big(x, y(x)+\delta y(x), y'(x)+\delta y'(x)\big) - \int_a^b dx\, F\big(x, y(x), y'(x)\big)\end{aligned} \tag{16.6}$$

The (small) function $\delta y(x)$ is the vertical displacement of the path in this coordinate system. To keep life simple for the first attack on this problem, I'll take the special case for which the endpoints of the path are fixed. That is,

$$\Delta a = 0, \qquad \Delta b = 0, \qquad \delta y(a) = 0, \qquad \delta y(b) = 0$$

To compute the value of Eq. (16.6) use the power series expansion of F, as in section 2.5.

$$F(x+\Delta x, y+\Delta y, z+\Delta z) = F(x,y,z) + \frac{\partial F}{\partial x}\Delta x + \frac{\partial F}{\partial y}\Delta y + \frac{\partial F}{\partial z}\Delta z$$
$$+ \frac{\partial^2 F}{\partial x^2}\frac{(\Delta x)^2}{2} + \frac{\partial^2 F}{\partial x \partial y}\Delta x\, \Delta y + \cdots$$

For now look at just the lowest order terms, linear in the changes, so ignore the second order terms. In this application, there is no Δx.

$$F(x, y+\delta y, y'+\delta y') = F(x,y,y') + \frac{\partial F}{\partial y}\delta y + \frac{\partial F}{\partial y'}\delta y'$$

plus terms of higher order in δy and $\delta y'$.

Put this into Eq. (16.6), and

$$\delta I = \int_a^b dx \left[\frac{\partial F}{\partial y}\delta y + \frac{\partial F}{\partial y'}\delta y'\right] \tag{16.7}$$

For example, Let $F = x^2 + y^2 + y'^2$ on the interval $0 \le x \le 1$. Take a base path to be a straight line from $(0,0)$ to $(1,1)$. Choose for the change in the path $\delta y(x) = \epsilon x(1-x)$. This is simple and it satisfies the boundary conditions.

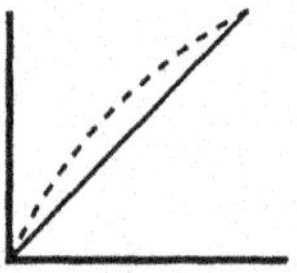

$$I[y] = \int_0^1 dx \left[x^2 + y^2 + y'^2\right] = \int_0^1 dx \left[x^2 + x^2 + 1^2\right] = \frac{5}{3}$$
$$I[y+\delta y] = \int_0^1 \left[x^2 + \big(x + \epsilon x(1-x)\big)^2 + \big(1 + \epsilon(1-2x)\big)^2\right]$$
$$= \frac{5}{3} + \frac{1}{6}\epsilon + \frac{11}{30}\epsilon^2 \tag{16.8}$$

The value of Eq. (16.7) is

$$\delta I = \int_0^1 dx \left[2y\delta y + 2y'\delta y'\right] = \int_0^1 dx \left[2x\epsilon x(1-x) + 2\epsilon(1-2x)\right] = \frac{1}{6}\epsilon$$

Return to the general case of Eq. (16.7) and you will see that I've explicitly used only *part* of the assumption that the endpoint of the path hasn't moved, $\Delta a = \Delta b = 0$. There's nothing in the body of the integral itself that constrains the change in the y-direction, and I had to choose the function δy by hand so that this constraint held. In order to use the equations $\delta y(a) = \delta y(b) = 0$ more generally, there is a standard trick: integrate by parts. You'll *always* integrate by parts in these calculations.

$$\int_a^b dx \frac{\partial F}{\partial y'}\delta y' = \int_a^b dx \frac{\partial F}{\partial y'}\frac{d\,\delta y}{dx} = \frac{\partial F}{\partial y'}\delta y\bigg|_a^b - \int_a^b dx \frac{d}{dx}\left(\frac{\partial F}{\partial y'}\right)\delta y(x)$$

This expression allows you to use the information that the path hasn't moved at its endpoints in the y direction either. The boundary term from this partial integration is

$$\frac{\partial F}{\partial y'}\delta y\Big|_a^b = \frac{\partial F}{\partial y'}\big(b, y(b)\big)\delta y(b) - \frac{\partial F}{\partial y'}\big(a, y(a)\big)\delta y(a) = 0$$

Put the last two equations back into the expression for δI, Eq. (16.7) and the result is

$$\delta I = \int_a^b dx\,\left[\frac{\partial F}{\partial y} - \frac{d}{dx}\left(\frac{\partial F}{\partial y'}\right)\right]\delta y \tag{16.9}$$

Use this expression for the same example $F = x^2 + y^2 + y'^2$ with $y(x) = x$ and you have

$$\delta I = \int_0^1 dx\,\left[2y - \frac{d}{dx}2y'\right]\delta y = \int_0^1 dx\,[2x - 0]\,\epsilon x(1-x) = \frac{1}{6}\epsilon$$

This is sort of like Eq. (8.16),

$$df = \vec G\cdot d\vec r = \operatorname{grad} f\cdot d\vec r = \nabla f\cdot d\vec r = \frac{\partial f}{\partial x_k}dx_k = \frac{\partial f}{\partial x_1}dx_1 + \frac{\partial f}{\partial x_2}dx_2 + \cdots$$

The differential change in the function depends linearly on the change $d\vec r$ in the coordinates. It is a sum over the terms with $dx_1, dx_2, \ldots$. This is a precise parallel to Eq. (16.9), except that the sum over discrete index k is now an integral over the continuous index x. The change in I is a linear functional of the change δy in the independent variable y; this δy corresponds to the change $d\vec r$ in the independent variable $\vec r$ in the other case. The coefficient of the change, instead of being called the gradient, is called the "functional derivative" though it's essentially the same thing.

$$\frac{\delta I}{\delta y} = \frac{\partial F}{\partial y} - \frac{d}{dx}\left(\frac{\partial F}{\partial y'}\right), \qquad \delta I[y,\delta y] = \int dx\,\frac{\delta I}{\delta y}\big(x, y(x), y'(x)\big)\,\delta y(x) \tag{16.10}$$

and for a change, I've indicated explicitly the dependence of δI on the two functions y and δy. This parallels the equation (8.13). The statement that this functional derivative vanishes is called the Euler-Lagrange equation.

Return to the example $F = x^2 + y^2 + y'^2$, then

$$\frac{\delta I}{\delta y} = \frac{\delta}{\delta y}\int_0^1 dx\,\left[x^2 + y^2 + y'^2\right] = 2y - \frac{d}{dx}2y' = 2y - 2y''$$

What is the minimum value of I? Set this derivative to zero.

$$y'' - y = 0 \Longrightarrow y(x) = A\cosh x + B\sinh x$$

The boundary conditions $y(0) = 0$ and $y(1) = 1$ imply $y = B\sinh x$ where $B = 1/\sinh 1$. The value of I at this point is

$$I[B\sinh x] = \int_0^1 dx\,\left[x^2 + B^2\sinh^2 x + B^2\cosh^2 x\right] = \frac{1}{3} + \coth 1 \tag{16.11}$$

Is it a minimum? Yes, but just as with the ordinary derivative, you have to look at the next order terms to determine that. Compare this value of $I[y] = 1.64637$ to the value 5/3 found for the nearby function $y(x) = x$, evaluated in Eq. (16.8).

Return to one of the examples in the introduction. What is the shortest distance between two points, but for now assume that there's no temperature variation. Write the length of a path for a function y between fixed endpoints, take the derivative, and set that equal to zero.

$$L[y] = \int_a^b dx\,\sqrt{1+y'^2}, \qquad \text{so}$$

$$\frac{\delta L}{\delta y} = -\frac{d}{dx}\frac{y'}{\sqrt{1+y'^2}} = -\frac{y''}{\sqrt{1+y'^2}} + \frac{y'^2\,y''}{\left(1+y'^2\right)^{3/2}} = \frac{-y''}{\left(1+y'^2\right)^{3/2}} = 0$$

For a minimum length then, $y'' = 0$, and that's a straight line. Surprise!

Do you really have to work through this mildly messy manipulation? Sometimes, but not here. Just notice that the derivative is in the form

$$\frac{\delta L}{\delta y} = -\frac{d}{dx} f(y') = 0 \tag{16.12}$$

so it doesn't matter what the particular f is and you get a straight line. $f(y')$ is a constant so y' must be constant too. Not so fast! See section 16.9 for another opinion.

16.3 Brachistochrone

Now for a tougher example, again from the introduction. In Eq. (16.2), which of all the paths between fixed initial and final points provides the path of least time for a particle sliding along it under gravity. Such a path is called a brachistochrone. This problem was first proposed by Bernoulli (one of them), and was solved by several people including Newton, though it's unlikely that he used the methods developed here, as the general structure we now call the calculus of variations was decades in the future.

Assume that the particle starts from rest so that $E = 0$, then

$$T[y] = \int_0^{x_0} dx\,\frac{\sqrt{1+y'^2}}{\sqrt{2gy}} \tag{16.13}$$

For the minimum time, compute the derivative and set it to zero.

$$\sqrt{2g}\,\frac{\delta T}{\delta y} = -\frac{\sqrt{1+y'^2}}{2y^{3/2}} - \frac{d}{dx}\frac{y'}{2\sqrt{y}\sqrt{1+y'^2}} = 0$$

This is starting to look intimidating, leading to an impossibly* messy differential equation. Is there another way? Yes. Why must x be the independent variable? What about using y? In the general setup leading to Eq. (16.10) nothing changes except the symbols, and you have

$$I[x] = \int dy\,F(y,x,x') \longrightarrow \frac{\delta I}{\delta x} = \frac{\partial F}{\partial x} - \frac{d}{dy}\left(\frac{\partial F}{\partial x'}\right) \tag{16.14}$$

* Only improbably. See problem 16.12.

Equation (16.13) becomes

$$T[x] = \int_0^{y_0} dy \, \frac{\sqrt{1 + x'^2}}{\sqrt{2gy}} \tag{16.15}$$

The function x does not appear explicitly in this integral, just its derivative $x' = dx/dy$. This simplifies the functional derivative, and the minimum time now comes from the equation

$$\frac{\delta I}{\delta x} = 0 - \frac{d}{dy}\left(\frac{\partial F}{\partial x'}\right) = 0 \tag{16.16}$$

This is much easier. $d()/dy = 0$ means that the object in parentheses is a constant.

$$\frac{\partial F}{\partial x'} = C = \frac{1}{\sqrt{2gy}} \frac{x'}{\sqrt{1 + x'^2}}$$

Solve this for x' and you get (let $K = C\sqrt{2g}$)

$$x' = \frac{dx}{dy} = \sqrt{\frac{K^2 y}{1 - K^2 y}}, \qquad \text{so} \qquad x(y) = \int dy \sqrt{\frac{K^2 y}{1 - K^2 y}}$$

This is an elementary integral. Let $2a = 1/K^2$

$$x(y) = \int dy \, \frac{y}{\sqrt{2ay - y^2}} = \int dy \, \frac{y}{\sqrt{a^2 - a^2 + 2ay - y^2}} = \int dy \, \frac{(y - a) + a}{\sqrt{a^2 - (y - a)^2}}$$

Make the substitution $(y - a)^2 = z$ in the first half of the integral and $(y - a) = a \sin\theta$ in the second half.

$$\begin{aligned} x(y) &= \frac{1}{2}\int dz \, \frac{1}{\sqrt{a^2 - z}} + \int \frac{a^2 \cos\theta d\theta}{\sqrt{a^2 - a^2 \sin^2\theta}} \\ &= -\sqrt{a^2 - z} + a\theta = -\sqrt{a^2 - (y - a)^2} + a \sin^{-1}\left(\frac{y - a}{a}\right) + C' \end{aligned}$$

The boundary condition that $x(0) = 0$ determines $C' = a\pi/2$, and the other end of the curve determines a: $x(y_0) = x_0$. You can rewrite this as

$$x(y) = -\sqrt{2ay - y^2} + a \cos^{-1}\left(\frac{a - y}{a}\right) \tag{16.17}$$

This is a cycloid. What's a cycloid and why does this equation describe one? See problem 16.2.

x-independent

In Eqs. (16.15) and (16.16) there was a special case for which the dependent variable was missing from F. That made the equation much simpler. What if the independent variable is missing? Does that provide a comparable simplification? Yes, but it's trickier to find.

$$I[y] = \int dx \, F(x, y, y') \longrightarrow \frac{\delta I}{\delta y} = \frac{\partial F}{\partial y} - \frac{d}{dx}\left(\frac{\partial F}{\partial y'}\right) = 0 \tag{16.18}$$

Use the chain rule to differentiate F with respect to x.

$$\frac{dF}{dx} = \frac{\partial F}{\partial x} + \frac{dy}{dx}\frac{\partial F}{\partial y} + \frac{dy'}{dx}\frac{\partial F}{\partial y'} \tag{16.19}$$

Multiply the Lagrange equation (16.18) by y' to get

$$y'\frac{\partial F}{\partial y} - y'\frac{d}{dx}\frac{\partial F}{\partial y'} = 0$$

Now substitute the term $y'(\partial F/\partial y)$ from the preceding equation (16.19).

$$\frac{dF}{dx} - \frac{\partial F}{\partial x} - \frac{dy'}{dx}\frac{\partial F}{\partial y'} - y'\frac{d}{dx}\frac{\partial F}{\partial y'} = 0$$

The last two terms are the derivative of a product.

$$\frac{dF}{dx} - \frac{\partial F}{\partial x} - \frac{d}{dx}\left[y'\frac{\partial F}{\partial y'}\right] = 0 \tag{16.20}$$

If the function F has no explicit x in it, the second term is zero, and the equation is now a derivative

$$\frac{d}{dx}\left[F - y'\frac{\partial F}{\partial y'}\right] = 0 \qquad \text{and} \qquad y'\frac{\partial F}{\partial y'} - F = C \tag{16.21}$$

This is already down to a first order differential equation. The combination $y'F_{y'} - F$ that appears on the left of the second equation is important. It's called the Hamiltonian.

16.4 Fermat's Principle

Fermat's principle of least time provides a formulation of geometrical optics. When you don't know about the wave nature of light, or if you ignore that aspect of the subject, it seems that light travels in straight lines — at least until it hits something. Of course this isn't fully correct, because when light hits a piece of glass it refracts and its path bends and follows Snell's equation. All of these properties of light can be described by Fermat's statement that the path light follows will be that path that takes the least* time.

$$T = \int dt = \int \frac{d\ell}{v} = \frac{1}{c}\int n\,d\ell \tag{16.22}$$

The total travel time is the integral of the distance $d\ell$ over the speed (itself a function of position). The index of refraction is $n = c/v$, where c is the speed of light in vacuum, so I can rewrite the travel time in the above form using n. The integral $\int n\,d\ell$ is called the optical path.

---

* Not always least. This just requires the first derivative to be zero; the second derivative is addressed in section 16.10. "Fermat's principle of stationary time" may be more accurate, but "Fermat's principle of least time" is entrenched in the literature.

From this idea it is very easy to derive the rule for reflection at a surface: angle of incidence equals angle of reflection. It is equally easy to derive Snell's law. (See problem 16.5.) I'll look at an example that would be difficult to do by any means *other* than Fermat's principle: Do you remember what an asphalt road looks like on a very hot day? If you are driving a car on a black colored road you may see the road in the distance appear to be covered by what looks like water. It has a sort of mirror-like sheen that is always receding from you — the "hot road mirage". You can never catch up to it. This happens because the road is very hot and it heats the air next to it, causing a strong temperature gradient near the surface. The density of the air decreases with rising temperature because the pressure is constant. That in turn means that the index of refraction will decrease with the rising temperature near the surface. The index will then be an increasing function of the distance above ground level, $n = f(y)$, and the travel time of light will depend on the path taken.

$$\int n\, d\ell = \int f(y)\, d\ell = \int f(y)\sqrt{1+x'^2}\, dy = \int f(y)\sqrt{1+y'^2}\, dx \tag{16.23}$$

What is $f(y)$? I'll leave that for a moment and then after carrying the calculation through for a while I can pick an f that is both plausible and easy to manipulate.

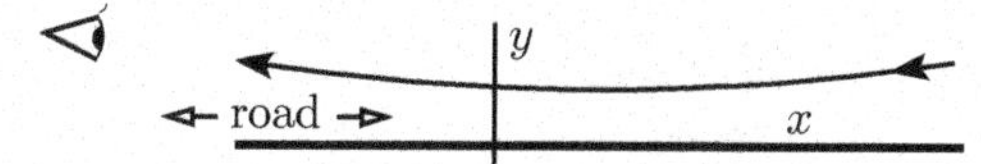

Should x be the independent variable, or y? Either should work, and I chose y because it seemed likely to be easier. (See problem 16.6 however.) The integrand then does not contain the dependent variable x.

$$\text{minimize} \int n\, d\ell = \int f(y)\sqrt{1+x'^2}\, dy \implies \frac{d}{dy}\frac{\partial}{\partial x'}\left[f(y)\sqrt{1+x'^2}\right] = 0$$

$$f(y)\frac{x'}{\sqrt{1+x'^2}} = C$$

Solve for x' to get

$$f(y)^2 x'^2 = C^2\left(1+x'^2\right) \qquad \text{so} \qquad x' = \frac{dx}{dy} = \frac{C}{\sqrt{f(y)^2 - C^2}} \tag{16.24}$$

At this point pick a form for the index of refraction that will make the integral easy and will still plausible represent reality. The index increases gradually above the road surface, and the simplest function works: $f(y) = n_0(1+\alpha y)$. The index increases linearly above the surface.

$$x(y) = \int \frac{C}{\sqrt{n_0^2(1+\alpha y)^2 - C^2}}\,dy = \frac{C}{\alpha n_0}\int dy \frac{1}{\sqrt{(y+1/\alpha)^2 - C^2/\alpha^2 n_0^2}}$$

This is an elementary integral. Let $u = y + 1/\alpha$, then $u = (C/\alpha n_0)\cosh\theta$.

$$x = \frac{C}{\alpha n_0}\int d\theta \implies \theta = \frac{\alpha n_0}{C}(x - x_0) \implies y = -\frac{1}{\alpha} + \frac{C}{\alpha n_0}\cosh\left((\alpha n_0/C)(x - x_0)\right)$$

C and x_0 are arbitrary constants, and x_0 is obviously the center of symmetry of the path. You can relate the other constant to the y-coordinate at that same point: $C = n_0(\alpha y_0+1)$.

Because the value of α is small for any real roadway, look at the series expansion of this hyperbolic function to the lowest order in α.

$$y \approx y_0 + \alpha(x - x_0)^2/2 \tag{16.25}$$

When you look down at the road you can be looking at an image of the sky. The light comes from the sky near the horizon down toward the road at an angle of only a degree or two. It then curves up so that it can enter your eye as you look along the road. The shimmering surface is a reflection of the distant sky or in this case an automobile — a mirage.

16.5 Electric Fields

The energy density in an electric field is $\epsilon_0 E^2/2$. For the static case, this electric field is the gradient of a potential, $\vec{E} = -\nabla\phi$. Its total energy in a volume is then

$$W = \frac{\epsilon_0}{2}\int dV\,(\nabla\phi)^2 \tag{16.26}$$

What is the minimum value of this energy? Zero of course, if ϕ is a constant. That question is too loosely stated to be much use, but keep with it for a while and it will be possible to turn it into something more precise and more useful. As with any other derivative taken to find a minimum, change the independent variable by a small amount. This time the variable is the function ϕ, so really this quantity W can more fully be written as a functional $W[\phi]$ to indicate its dependence on the potential function.

$$\begin{aligned} W[\phi+\delta\phi] - W[\phi] &= \frac{\epsilon_0}{2}\int dV\,\big(\nabla(\phi+\delta\phi)\big)^2 - \frac{\epsilon_0}{2}\int dV\,(\nabla\phi)^2 \\ &= \frac{\epsilon_0}{2}\int dV\,\big(2\nabla\phi\cdot\nabla\delta\phi + (\nabla\delta\phi)^2\big) \end{aligned}$$

Now pull out a vector identity from problem 9.36,

$$\nabla\cdot(f\vec{g}) = \nabla f\cdot\vec{g} + f\nabla\cdot\vec{g}$$

and apply it to the previous line with $f = \delta\phi$ and $\vec{g} = \nabla\phi$.

$$W[\phi+\delta\phi] - W[\phi] = \epsilon_0\int dV\,\big[\nabla\cdot(\delta\phi\nabla\phi) - \delta\phi\nabla^2\phi\big] + \frac{\epsilon_0}{2}\int dV\,(\nabla\delta\phi)^2$$

The divergence term is set up to use Gauss's theorem; this is the vector version of integration by parts.

$$W[\phi+\delta\phi] - W[\phi] = \epsilon_0\oint d\vec{A}\cdot(\nabla\phi)\delta\phi - \epsilon_0\int dV\,\delta\phi\nabla^2\phi + \frac{\epsilon_0}{2}\int dV\,(\nabla\delta\phi)^2 \tag{16.27}$$

If the value of the potential ϕ is specified everywhere on the boundary, then I'm not allowed to change it in the process of finding the change in W. That means that $\delta\phi$

vanishes on the boundary. That makes the boundary term, the $d\vec{A}$ integral, vanish. Its integrand is zero everywhere on the surface of integration.

In looking for a minimum energy I want to set the first derivative to zero, and that's the coefficient of the term linear in $\delta\phi$.

$$\frac{\delta W}{\delta\phi} = -\epsilon_0 \nabla^2\phi = 0$$

The function that produces the minimum value of the total energy (with these fixed boundary conditions) is the one that satisfies Laplace's equation. Is it really a minimum? Yes. In this instance it's very easy to see that. The extra term in the change of W is $\int dV\,(\nabla\delta\phi)^2$. That is positive no matter what $\delta\phi$ is.

That the correct potential function is the one having the minimum energy allows for an efficient approximate method to solve electrostatic problems. I'll illustrate this using an example borrowed from the Feynman Lectures in Physics and that you can also solve exactly: What is the capacitance of a length of coaxial cable? (Neglect the edge effects at the ends of the cable of course.) Let the inner and outer radii be a and b, and the length L. A charge density λ is on the inner conductor (and therefore $-\lambda$ on the inside of the outer conductor). It creates a radial electric field of size $\lambda/2\pi\epsilon_0 r$. The potential difference between the conductors is

$$\Delta V = \int_a^b dr\,\frac{\lambda}{2\pi\epsilon_0 r} = \frac{\lambda\ln(b/a)}{2\pi\epsilon_0} \tag{16.28}$$

The charge on the inner conductor is λL, so $C = Q/\Delta V = 2\pi\epsilon_0 L/\ln(b/a)$, where$\Delta\ V = V_b - V_a$.

The total energy satisfies $W = C\Delta V^2/2$, so for the given potential difference, knowing the energy is the same as knowing the capacitance.

This exact solution provides a laboratory to test the efficacy of a variational approximation for the same problem. The idea behind this method is that you assume a form for the solution $\phi(r)$. This assumed form must satisfy the boundary conditions, but it need not satisfy Laplace's equation. It should also have one or more free parameters, the more the better up to the limits of your patience. Now compute the total energy that this trial function implies and then use the free parameters to minimize it. This function with minimum energy is the best approximation to the correct potential among all those with your assumed trial function.

Let the potential at $r = a$ be V_a and at $r = b$ it is V_b. An example function that satisfies these conditions is

$$\phi(r) = V_a + (V_b - V_a)\frac{r-a}{b-a} \tag{16.29}$$

The electric field implied by this is $\vec{E} = -\nabla\phi = \hat{r}\,(V_a - V_b)/(b-a)$, a constant radial component. From (16.26), the energy is

$$\frac{\epsilon_0}{2}\int_a^b L\,2\pi r\,dr\left(\frac{d\phi}{dr}\right)^2 = \frac{\epsilon_0}{2}\int_a^b L\,2\pi r\,dr\left(\frac{V_b - V_a}{b-a}\right)^2 = \frac{1}{2}\pi L\epsilon_0\frac{b+a}{b-a}\Delta V^2$$

Set this to $C\Delta V^2/2$ to get C and you have

$$C_{\text{approx}} = \pi L\epsilon_0 \frac{b+a}{b-a}$$

How does this compare to the exact answer, $2\pi\epsilon_0 L/\ln(b/a)$? Let $x = b/a$.

$$\frac{C_{\text{approx}}}{C} = \frac{1}{2}\frac{b+a}{b-a}\ln(b/a) = \frac{1}{2}\frac{x+1}{x-1}\ln x$$

x:	1.1	1.2	1.5	2.0	3.0	10.0
ratio:	1.0007	1.003	1.014	1.04	1.10	1.41

Assuming a constant magnitude electric field in the region between the two cylinders is clearly not correct, but this estimate of the capacitance gives a remarkable good result even when the ratio of the radii is two or three. This is true even though I didn't even put in a parameter with which to minimize the energy. How much better will the result be if I do?

Instead of a linear approximation for the potential, use a quadratic.

$$\phi(r) = V_a + \alpha(r-a) + \beta(r-a)^2, \qquad \text{with} \qquad \phi(b) = V_b$$

Solve for α in terms of β and you have, after a little manipulation,

$$\phi(r) = V_a + \Delta V\frac{r-a}{b-a} + \beta(r-a)(r-b) \tag{16.30}$$

Compute the energy from this.

$$W = \frac{\epsilon_0}{2}\int_a^b L\,2\pi r\,dr\left[\frac{\Delta V}{b-a} + \beta(2r-a-b)\right]^2$$

Rearrange this for easier manipulation by defining $2\beta = \gamma\Delta V/(b-a)$ and $c = (a+b)/2$ then

$$\begin{aligned} W &= \frac{\epsilon_0}{2}2L\pi\left(\frac{\Delta V}{b-a}\right)^2\int\big((r-c)+c\big)dr\left[1+\gamma(r-c)\right]^2 \\ &= \frac{\epsilon_0}{2}2L\pi\left(\frac{\Delta V}{b-a}\right)^2\left[c(b-a)+\gamma(b-a)^3/6+c\gamma^2(b-a)^3/12\right] \end{aligned}$$

γ is a free parameter in this calculation, replacing the original β. To minimize this energy, set the derivative $dW/d\gamma = 0$, resulting in the value $\gamma = -1/c$. At this value of γ the energy is

$$W = \frac{\epsilon_0}{2}2L\pi\left(\frac{\Delta V}{b-a}\right)^2\left[\frac{1}{2}(b^2-a^2) - \frac{(b-a)^3}{6(b+a)}\right] \tag{16.31}$$

Except for the factor of $\Delta V^2/2$ this is the new estimate of the capacitance, and to see how good it is, again take the ratio of this estimate to the exact value and let $x = b/a$.

$$\frac{C'_{\text{approx}}}{C} = \ln x\,\frac{1}{2}\frac{x+1}{x-1}\left[1 - \frac{(x-1)^2}{3(x+1)^2}\right] \tag{16.32}$$

x:	1.1	1.2	1.5	2.0	3.0	10.0
ratio:	1.00000046	1.000006	1.00015	1.0012	1.0071	1.093

For only a one parameter adjustment, this provides very high accuracy. This sort of technique is the basis for many similar procedures in this and other contexts, especially in quantum mechanics.

16.6 Discrete Version

There is another way to find the functional derivative, one that more closely parallels the ordinary partial derivative. It is due to Euler, and he found it first, before Lagrange's discovery of the treatment that I've spent all this time on. Euler's method is perhaps more intuitive than Lagrange's, but it is not as easy to extend it to more than one dimension and it doesn't easily lend itself to the powerful manipulative tools that Lagrange's method does. This alternate method starts by noting that an integral is the limit of a sum. Go back to the sum and perform the derivative on *it*, finally taking a limit to get an integral. This turns the problem into a finite-dimensional one with ordinary partial derivatives. You have to decide which form of numerical integration to use, and I'll pick the trapezoidal rule, Eq. (11.15), with a constant interval. Other choices work too, but I think this entails less fuss. You don't see this approach as often as Lagrange's because it is harder to manipulate the equations with Euler's method, and the notation can become quite cumbersome. The trapezoidal rule for an integral is just the following picture, and all that you have to handle with any care are the endpoints of the integral.

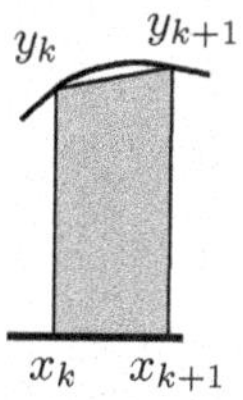

$y_k \quad y_{k+1}$

$x_k \quad x_{k+1}$

$$x_k = a + k\Delta, \quad 0 \le k \le N \qquad \text{where} \qquad \frac{b-a}{N} = \Delta$$

$$\int_a^b dx\, f(x) = \lim \left[\frac{1}{2}f(a) + \sum_1^{N-1} f(x_k) + \frac{1}{2}f(b)\right]\Delta$$

The integral Eq. (16.5) involves y', so in the same spirit, approximate this by the centered difference.

$$y'_k = y'(x_k) \approx \big(y(x_{k+1}) - y(x_{k-1})\big)/2\Delta$$

This evaluates the derivative *at* each of the coordinates $\{x_k\}$ instead of between them.

The discrete form of (16.5) is now

$$\begin{aligned} I_{\text{discrete}} &= \frac{\Delta}{2}F\big(a, y(a), y'(a)\big) \\ &+ \sum_1^{N-1} F\big(x_k, y(x_k), (y(x_{k+1}) - y(x_{k-1}))/2\Delta\big)\Delta + \frac{\Delta}{2}F\big(b, y(b), y'(b)\big) \end{aligned}$$

Not quite yet. What about $y'(a)$ and $y'(b)$? The endpoints $y(a)$ and $y(b)$ aren't changing, but that doesn't mean that the slope there is fixed. At these two points, I can't use the centered difference scheme for the derivative, I'll have to use an asymmetric form to give

$$\begin{aligned} I_{\text{discrete}} &= \frac{\Delta}{2}F\big(a, y(a), (y(x_1) - y(x_0))/\Delta\big) + \frac{\Delta}{2}F\big(b, y(b), (y(x_N) - y(x_{N-1}))/\Delta\big) \\ &+ \sum_1^{N-1} F\big(x_k, y(x_k), (y(x_{k+1}) - y(x_{k-1}))/2\Delta\big)\Delta \end{aligned} \tag{16.33}$$

When you keep the endpoints fixed, this is a function of $N-1$ variables, $\{y_k = y(x_k)\}$ for $1 \le k \le N-1$, and to find the minimum or maximum you simply take the partial derivative with respect to each of them. It is *not* a function of any of the $\{x_k\}$ because those are defined and fixed by the partition $x_k = a + k\Delta$. The clumsy part is keeping track of the notation. When you differentiate with respect to a particular y_ℓ, most of the terms in the sum (16.33) don't contain it. There are only three terms in the sum that contribute: ℓ and $\ell \pm 1$. In the figure $N = 5$, and the $\ell = 2$ coordinate (y_2) is being changed. For all the indices ℓ except the two next to the endpoints (1 and $N-1$), this is

$$\begin{aligned}\frac{\partial}{\partial y_\ell} I_{\text{discrete}} = \frac{\partial}{\partial y_\ell}\Big[&F\big(x_{\ell-1}, y_{\ell-1}, (y_\ell - y_{\ell-2})/2\Delta\big) + \\ &F\big(x_\ell, y_\ell, (y_{\ell+1} - y_{\ell-1})/2\Delta\big) + \\ &F\big(x_{\ell+1}, y_{\ell+1}, (y_{\ell+2} - y_\ell)/2\Delta\big)\Big]\Delta\end{aligned}$$

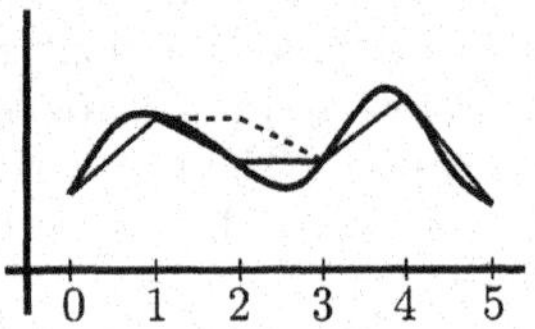

An alternate standard notation for partial derivatives will help to keep track of the manipulations:

D_1F is the derivative with respect to the first *argument*

The above derivative is then

$$\begin{aligned}&\Big[D_2F\big(x_\ell, y_\ell, (y_{\ell+1} - y_{\ell-1})/2\Delta\big) \\ &+\frac{1}{2\Delta}\big[D_3F\big(x_{\ell-1}, y_{\ell-1}, (y_\ell - y_{\ell-2})/2\Delta\big) - D_3F\big(x_{\ell+1}, y_{\ell+1}, (y_{\ell+2} - y_\ell)/2\Delta\big)\big]\Big]\Delta\end{aligned} \tag{16.34}$$

There is no D_1F because the x_ℓ is essentially an index.

If you now take the limit $\Delta \to 0$, the third argument in each function returns to the derivative y' evaluated at various x_ks:

$$\Big[D_2F(x_\ell, y_\ell, y'_\ell) + \frac{1}{2\Delta}\big[D_3F(x_{\ell-1}, y_{\ell-1}, y'_{\ell-1}) - D_3F(x_{\ell+1}, y_{\ell+1}, y'_{\ell+1})\big]\Big]\Delta$$

$$\begin{aligned}= &\Big[D_2F\big(x_\ell, y(x_\ell), y'(x_\ell)\big) \\ &+\frac{1}{2\Delta}\big[D_3F\big(x_{\ell-1}, y(x_{\ell-1}), y'(x_{\ell-1})\big) - D_3F\big(x_{\ell+1}, y(x_{\ell+1}), y'(x_{\ell+1})\big)\big]\Big]\Delta\end{aligned} \tag{16.35}$$

Now take the limit that $\Delta \to 0$, and the last part is precisely the definition of (minus) the derivative with respect to x. This then becomes

$$\frac{1}{\Delta}\frac{\partial}{\partial y_\ell} I_{\text{disc}} \to D_2F(x_\ell, y_\ell, y'_\ell) - \frac{d}{dx} D_3F(x_\ell, y_\ell, y'_\ell) \tag{16.36}$$

Translate this into the notation that I've been using and you have Eq. (16.10). Why did I divide by Δ in the final step? That's the equivalent of looking for the coefficient of both the dx and the δy in Eq. (16.10). It can be useful to retain the discrete approximation of Eq. (16.34) or (16.35) to the end of the calculation. This allows you to do numerical calculations in cases where the analytic equations are too hard to manipulate.

Again, not quite yet. The two cases $\ell = 1$ and $\ell = N - 1$ have to be handled separately. You need to go back to Eq. (16.33) to see how they work out. The factors show up in different places, but the final answer is the same. See problem 16.15.

It is curious that when formulating the problem this way, you don't seem to need a partial integration. The result came out automatically. Would that be true with some integration method other other than the trapezoidal rule? See problem 16.16.

16.7 Classical Mechanics

The calculus of variations provides a way to reformulate Newton's equations of mechanics. The results produce efficient techniques to set up complex problems and they give insights into the symmetries of the systems. They also provide alternate directions in which to generalize the original equations.

Start with one particle subject to a force, and the force has a potential energy function U. Following the traditions of the subject, denote the particle's kinetic energy by T. Picture this first in rectangular coordinates, where $T = m\vec{v}^{\,2}/2$ and the potential energy is $U(x_1, x_2, x_3)$. The functional S depends on the path $[x_1(t), x_2(t), x_3(t)]$ from the initial to the final point. The integrand is the Lagrangian, $L = T - U$.

$$S[\vec{r}\,] = \int_{t_1}^{t_2} L(\vec{r}, \dot{\vec{r}}\,)\, dt, \qquad \text{where} \qquad L = T - U = \frac{m}{2}\left(\dot{x}_1^2 + \dot{x}_2^2 + \dot{x}_3^2\right) - U(x_1, x_2, x_3) \tag{16.37}$$

The statement that the functional derivative is zero is

$$\frac{\delta S}{\delta x_k} = \frac{\partial L}{\partial x_k} - \frac{d}{dt}\left(\frac{\partial L}{\partial \dot{x}_k}\right) = -\frac{\partial U}{\partial x_k} - \frac{d}{dt}(m\dot{x}_k)$$

Set this to zero and you have

$$m\ddot{x}_k = -\frac{\partial U}{\partial x_k}, \qquad \text{or} \qquad m\frac{d^2\vec{r}}{dt^2} = \vec{F} \tag{16.38}$$

That this integral of $L\,dt$ has a zero derivative is $\vec{F} = m\vec{a}$. Now what? This may be elegant, but does it accomplish anything? The first observation is that when you state the problem in terms of this integral it is independent of the coordinate system. If you specify a path in space, giving the velocity at each point along the path, the kinetic energy and the potential energy are well-defined at each point on the path and the integral S is too. You can now pick whatever bizarre coordinate system that you want in order to do the computation of the functional derivative. Can't you do this with $\vec{F} = m\vec{a}$? Yes, but computing an acceleration in an odd coordinate system is a lot more work than computing a velocity. A second advantage will be that it's easier to handle constraints by this method. The technique of Lagrange multipliers from section 8.12 will apply here too.

Do the simplest example: plane polar coordinates. The kinetic energy is

$$T = \frac{m}{2}\left(\dot{r}^2 + r^2\dot{\phi}^2\right)$$

The potential energy is a function U of r and ϕ. With the of the Lagrangian defined as $T - U$, the variational derivative determines the equations of motion to be

$$\begin{aligned}
S[r,\phi] &= \int_{t_1}^{t_2} L\big(r(t),\phi(t)\big)\,dt \longrightarrow \\
\frac{\delta S}{\delta r} &= \frac{\partial L}{\partial r} - \frac{d}{dt}\frac{\partial L}{\partial \dot{r}} = mr\dot{\phi}^2 - \frac{\partial U}{\partial r} - m\ddot{r} = 0 \\
\frac{\delta S}{\delta \phi} &= \frac{\partial L}{\partial \phi} - \frac{d}{dt}\frac{\partial L}{\partial \dot{\phi}} = -\frac{\partial U}{\partial \phi} - m\frac{d}{dt}\left(r^2\dot{\phi}\right) = 0
\end{aligned}$$

These are the components of $\vec{F} = m\vec{a}$ in polar coordinates. If the potential energy is independent of ϕ, the second equation says that angular momentum is conserved: $mr^2\dot{\phi}$.

What do the discrete approximate equations (16.34) or (16.35) look like in this context? Look at the case of one-dimensional motion to understand an example. The Lagrangian is

$$L = \frac{m}{2}\dot{x}^2 - U(x)$$

Take the expression in Eq. (16.34) and set it to zero.

$$-\frac{dU}{dx}(x_\ell) + \frac{1}{2\Delta}\left[m(x_\ell - x_{\ell-2})/2\Delta - m(x_{\ell+2} - x_\ell)/2\Delta\right] = 0$$

or

$$m\frac{x_{\ell+2} - 2x_\ell + x_{\ell-2}}{(2\Delta)^2} = -\frac{dU}{dx}(x_\ell) \tag{16.39}$$

This is the discrete approximation to the second derivative, as in Eq. (11.12).

16.8 Endpoint Variation

Following Eq. (16.6) I restricted the variation of the path so that the endpoints are kept fixed. What if you don't? As before, keep terms to the first order, so that for example $\Delta t_b\,\delta y$ is out. Because the most common application of this method involves integrals with respect to time, I'll use that integration variable

$$\begin{aligned}
\Delta S &= \int_{t_a+\Delta t_a}^{t_b+\Delta t_b} dt\, L\big(t, y(t)+\delta y(t), \dot{y}(t)+\delta\dot{y}(t)\big) - \int_{t_a}^{t_b} dt\, L\big(t, y(t), \dot{y}(t)\big) \\
&= \int_{t_a+\Delta t_a}^{t_b+\Delta t_b} dt\left[L(t,y,\dot{y}) + \frac{\partial L}{\partial y}\delta y + \frac{\partial L}{\partial \dot{y}}\delta\dot{y}\right] - \int_{t_a}^{t_b} dt\, L\big(t, y(t), \dot{y}(x)\big) \\
&= \left[\int_{t_b}^{t_b+\Delta t_b} - \int_{t_a}^{t_a+\Delta t_a}\right] dt\, L(t,y,\dot{y}) + \int_{t_a}^{t_b} dt\left[\frac{\partial L}{\partial y}\delta y + \frac{\partial L}{\partial \dot{y}}\delta\dot{y}\right]
\end{aligned}$$

Drop quadratic terms in the second line: anything involving $(\delta y)^2$ or $\delta y\delta\dot{y}$ or $(\delta\dot{y})^2$. Similarly, drop terms such as $\Delta t_a\,\delta y$ in going to the third line. Do a partial integration on the last term

$$\int_{t_a}^{t_b} dt\,\frac{\partial L}{\partial \dot{y}}\frac{d\,\delta y}{dt} = \frac{\partial L}{\partial \dot{y}}\delta y\bigg|_{t_a}^{t_b} - \int_{t_a}^{t_b} dt\,\frac{d}{dt}\left(\frac{\partial L}{\partial \dot{y}}\right)\delta y(t) \tag{16.40}$$

The first two terms, with the Δt_a and Δt_b, are to first order

$$\left[\int_{t_b}^{t_b+\Delta t_b} - \int_{t_a}^{t_a+\Delta t_a}\right] dt\, L(t,y,\dot y) = L\big((t_b, y(t_b), \dot y(t_b)\big)\Delta t_b - L\big((t_a, y(t_a), \dot y(t_a)\big)\Delta t_a$$

This produces an expression for ΔS

$$\begin{aligned}\Delta S =& L\big((t_b, y(t_b), \dot y(t_b)\big)\Delta t_b - L\big((t_a, y(t_a), \dot y(t_a)\big)\Delta t_a \\ &+ \frac{\partial L}{\partial \dot y}(t_b)\delta y(t_b) - \frac{\partial L}{\partial \dot y}(t_a)\delta y(t_a) + \int_{t_a}^{t_b} dt\,\left[\frac{\partial L}{\partial y} - \frac{d}{dt}\left(\frac{\partial L}{\partial \dot y}\right)\right]\delta y\end{aligned} \tag{16.41}$$

Up to this point the manipulation was straight-forward, though you had to keep all the algebra in good order. Now there are some rearrangements needed that are not all that easy to anticipate, adding and subtracting some terms.

Start by looking at the two terms involving Δt_b and $\delta y(t_b)$ — the first and third terms. The change in the position of this endpoint is not simply $\delta y(t_b)$. Even if δy is identically zero the endpoint will change in both the t-direction and in the y-direction because of the slope of the curve ($\dot y$) and the change in the value of t at the endpoint (Δt_b).

The total movement of the endpoint at t_b is horizontal by the amount Δt_b, and it is vertical by the amount $\big(\delta y + \dot y\Delta t_b\big)$. To incorporate this, add and subtract this second term, $\dot y\Delta t$, in order to produce this combination as a coefficient.

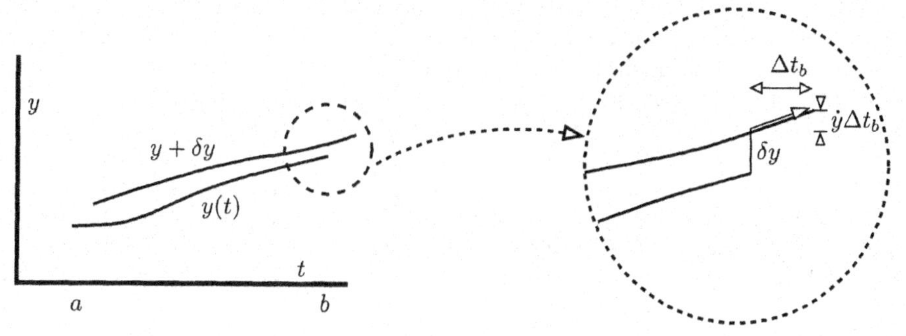

$$\begin{aligned}L\big((t_b, y(t_b), \dot y(t_b)\big)&\Delta t_b + \frac{\partial L}{\partial \dot y}(t_b)\delta y(t_b) \\ &= \left[L(t_b)\Delta t_b - \frac{\partial L}{\partial \dot y}(t_b)\dot y(t_b)\Delta t_b\right] + \left[\frac{\partial L}{\partial \dot y}(t_b)\dot y(t_b)\Delta t_b + \frac{\partial L}{\partial \dot y}(t_b)\delta y(t_b)\right] \\ &= \left[L - \frac{\partial L}{\partial \dot y}\dot y\right]\Delta t_b + \frac{\partial L}{\partial \dot y}\left[\dot y\Delta t_b + \delta y\right]\end{aligned} \tag{16.42}$$

Do the same thing at t_a, keeping the appropriate signs. Then denote

$$p = \frac{\partial L}{\partial \dot y}, \qquad H = p\dot y - L, \qquad \Delta y = \delta y + \dot y\Delta t$$

H is the Hamiltonian and Eq. (16.41) becomes Noether's theorem.

$$\Delta S = \left[p\Delta y - H\Delta t\right]_{t_a}^{t_b} + \int_{t_a}^{t_b} dt\, \left[\frac{\partial L}{\partial y} - \frac{d}{dt}\left(\frac{\partial L}{\partial \dot{y}}\right)\right]\delta y \tag{16.43}$$

If the equations of motion are satisfied, the argument of the last integral is zero. The change in S then comes only from the translation of the endpoint in either the time or space direction. If Δt is zero, and Δy is the same at the two ends, you are translating the curve in space — vertically in the graph. Then

$$\Delta S = p\Delta y\Big|_{t_a}^{t_b} = \big[p(t_b) - p(t_a)\big]\Delta y = 0$$

If the physical phenomenon described by this equation is invariant under spacial translation, then momentum is conserved.

If you do a translation in time instead of space and S is invariant, then Δt is the same at the start and finish, and

$$\Delta S = \big[-H(t_b) + H(t_a)\big]\Delta t = 0$$

This is conservation of energy. Write out what H is for the case of Eq. (16.37).

If you write this theorem in three dimensions and require that the system is invariant under rotations, you get conservation of angular momentum. In more complicated system, especially in field theory, there are other symmetries, and they in turn lead to conservation laws. For example conservation of charge is associated with a symmetry called "gauge symmetry" in quantum mechanical systems.

The equation (16.10), in which the variation δy had the endpoints fixed, is much like a directional derivative in multivariable calculus. For a directional derivative you find how a function changes as the independent variable moves along some specified direction, and in the variational case the direction was specified to be with functions that were tied down at the endpoints. The development of the present section is in the spirit of finding the derivative in all possible directions, not just a special set.

16.9 Kinks

In all the preceding analysis of minimizing solutions to variational problems, I assumed that everything is differentiable and that all the derivatives are continuous. That's not always so, and it is quite possible for a solution to one of these problems to have a discontinuous derivative somewhere in the middle. These are more complicated to handle, but just because of some extra algebra. An integral such as Eq. (16.5) is perfectly well defined if the integrand has a few discontinuities, but the partial integrations leading to the Euler-Lagrange equations are *not*. You can apply the Euler-Lagrange equations only in the intervals between any kinks.

If you're dealing with a problem of the standard form $I[x] = \int_a^b dt\, L(t, x, \dot{x})$ and you want to determine whether there is a kink along the path, there are some internal boundary conditions that have to hold. Roughly speaking they are conservation of momentum and conservation of energy, Eq. (16.44), and you can show this using the results of the preceding section on endpoint variations.

$$S = \int_{t_a}^{t_b} dt\, L(t, x, \dot{x}) = \int_{t_a}^{t_m} dt\, L + \int_{t_m}^{t_b} dt\, L$$

t_a t_m t_b

Assume there is a discontinuity in the derivative $\dot{x}$ at a point in the middle, t_m. The equation to solve is still $\delta S/\delta x = 0$, and for variations of the path that leave the endpoints and the middle point alone you have to satisfy the standard Euler-Lagrange differential equations on the two segments. Now however you also have to set the variation to zero for paths that leave the endpoints alone but move the middle point.

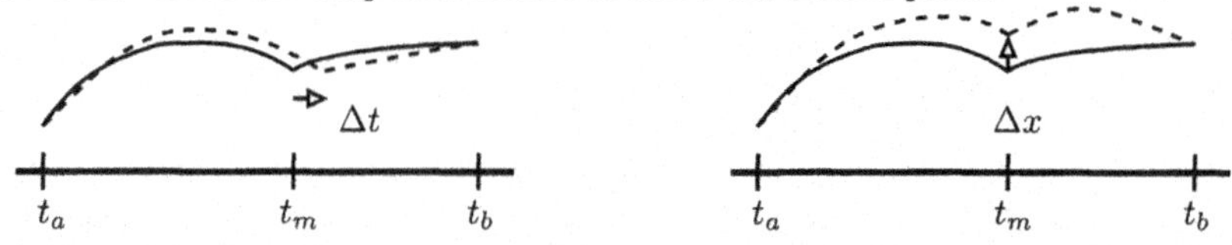

Apply Eq. (16.43) to each of the two segments, and assume that the differential equations are already satisfied in the two halves. For the sort of variation described in the last two figures, look at the endpoint variations of the two segments. They produce

$$\delta S = \Big[p\Delta x - H\Delta t\Big]_{t_a}^{t_m} + \Big[p\Delta x - H\Delta t\Big]_{t_m}^{t_b} = \big[p\Delta x - H\Delta t\big](t_m^-) - \big[p\Delta x - H\Delta t\big](t_m^+) = 0$$

These represent the contributions to the variation just above t_m and just below it. This has to vanish for arbitrary Δt and Δx, so it says

$$p(t_m^-) = p(t_m^+) \qquad \text{and} \qquad H(t_m^-) = H(t_m^+) \tag{16.44}$$

These equations, called the Weierstrass-Erdmann conditions, are two equations for the values of the derivative $\dot{x}$ on the two side of t_m. The two equations for the two unknowns may tell you that there is no discontinuity in the derivative, or if there is then it will dictate the algebraic equations that the two values of $\dot{x}$ must satisfy. More dimensions means more equations of course.

There is a class of problems in geometry coming under the general heading of Plateau's Problem. What is the minimal surface that spans a given curve? Here the functional is $\int dA$, giving the area as a function of the function describing the surface. If the curve is a circle in the plane, then the minimum surface is the spanning disk. What if you twist the circle so that it does not quite lie in a plane? Now it's a tough problem. What if you have two parallel circles? Is the appropriate surface a cylinder? (No.) This subject is nothing more than the mathematical question of trying to describe soap bubbles. They're not all spheres.

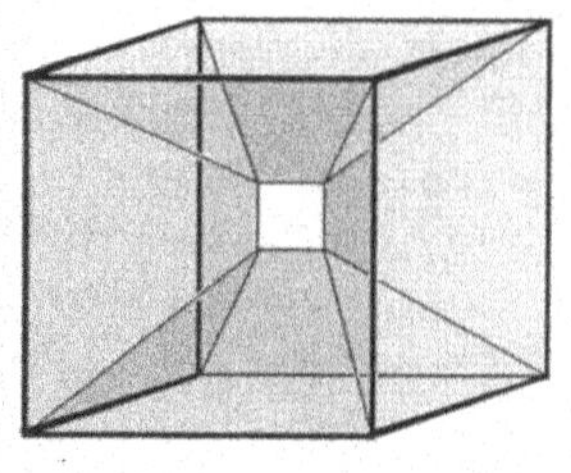

Do kinks happen often? They are rare in problems that usually come up in physics, and it seems to be of more concern to those who apply these techniques in engineering. For an example that you can verify for yourself however, construct a wire frame in the shape of a cube. You can bend wire or you can make it out of solder, which is much easier to manipulate. Attach a handle to one corner so that you can hold it. Now make a soap solution that you can use to blow bubbles. (A trick to make it work better is to add some glycerine.) Now dip the wire cube into the soap and see what sort of soap film will result, filling in the surfaces among the wires. It is not what you expect, and has several faces that meet at surprising angles. There is a square in the middle. This surface has minimum area among surfaces that span the cube.

Example

In Eq. (16.12), looking at the shortest distance between two points in a plane, I jumped to a conclusion. To minimize the integral $\int_a^b f(y')dx$, use the Euler-Lagrange differential equation:

$$\frac{\partial f}{\partial y} - \frac{d}{dx}\frac{\partial f}{\partial y'} = f'(y')y'' = 0$$

This seems to say that $f(y')$ is a constant or that $y'' = 0$, implying either way that $y = Ax + B$, a straight line. Now that you know that solutions can have kinks, you have to look more closely. Take the particular example

$$f(y') = \alpha y'^4 - \beta y'^2, \qquad \text{with} \qquad y(0) = 0, \qquad \text{and} \qquad y(a) = b \tag{16.45}$$

One solution corresponds to $y'' = 0$ and $y(x) = bx/a$. Can there be others?

Apply the conditions of Eq. (16.44) at some point between 0 and a. Call it x_m, and assume that the derivative is not continuous. Call the derivatives on the left and right (y'^-) and (y'^+). The first equation is

$$p = \frac{\partial L}{\partial y'} = 4\alpha y'^3 - 2\beta y', \qquad \text{and} \qquad p(x_m^-) = p(x_m^+)$$
$$4\alpha(y'^-)^3 - 2\beta(y'^-) = 4\alpha(y'^+)^3 - 2\beta(y'^+)$$
$$\left[(y'^-) - (y'^+)\right]\left[(y'^+)^2 + (y'^+)(y'^-) + (y'^-)^2 - \beta/2\alpha\right] = 0$$

If the slope is not continuous, the second factor must vanish.

$$(y'^+)^2 + (y'^+)(y'^-) + (y'^-)^2 - \beta/2\alpha = 0$$

This is one equation for the two unknown slopes. For the second equation, use the second condition, the one on H.

$$H = y'\frac{\partial f}{\partial y'} - f, \qquad \text{and} \qquad H(x_m^-) = H(x_m^+)$$
$$H = y'\left[4\alpha y'^3 - 2\beta y'\right] - \left[\alpha y'^4 - \beta y'^2\right] = 3\alpha y'^4 - \beta y'^2$$
$$\left[(y'^-) - (y'^+)\right]\left[(y'^+)^3 + (y'^+)^2(y'^-) + (y'^+)(y'^-)^2 + (y'^-)^3 - \beta\big((y'^+) + (y'^-)\big)/3\alpha\right] = 0$$

Again, if the slope is not continuous this is

$$(y'^+)^3 + (y'^+)^2(y'^-) + (y'^+)(y'^-)^2 + (y'^-)^3 - \beta\big((y'^+) + (y'^-)\big)/3\alpha = 0$$

These are two equations in the two unknown slopes. It looks messy, but look back at H itself first. It's even. That means that its continuity will always hold if the slope simply changes sign.

$$(y'^{+}) = -(y'^{-})$$

Can this work in the other (momentum) equation?

$$(y'^{+})^2 + (y'^{+})(y'^{-}) + (y'^{-})^2 - \beta/2\alpha = 0 \qquad \text{is now} \qquad (y'^{+})^2 = \beta/2\alpha$$

As long as α and β have the same sign, this has the solution

$$(y'^{+}) = \pm\sqrt{\beta/2\alpha}, \qquad (y'^{-}) = \mp\sqrt{\beta/2\alpha} \tag{16.46}$$

The boundary conditions on the original problem were $y(0) = 0$ and $y(a) = b$. Denote $\gamma = \pm\sqrt{\beta/2\alpha}$, and $x_1 = a/2 + b/2\gamma$, then

$$y = \begin{cases} \gamma x & (0 < x < x_1) \\ b - \gamma(x - a) & (x_1 < x < b) \end{cases} \tag{16.47}$$

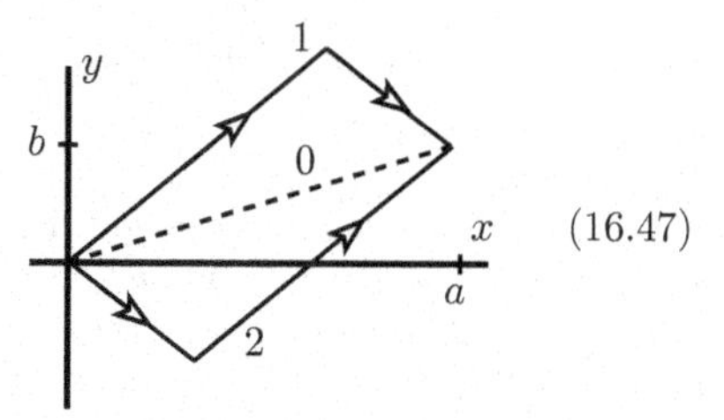

The paths labeled 0, 1, and 2 are three solutions that make the variational functional derivative vanish. Which is smallest? Does that answer depend on the choice of the parameters? See problem 16.19.

Are there any other solutions? After all, once you've found three, you should wonder if it stops there. Yes, there are many — infinitely many in this example. They are characterized by the same slopes, $\pm\gamma$, but they switch back and forth several times before coming to the endpoint. The same internal boundary conditions (p and H) apply at each corner, and there's nothing in their solutions saying that there is only one such kink.

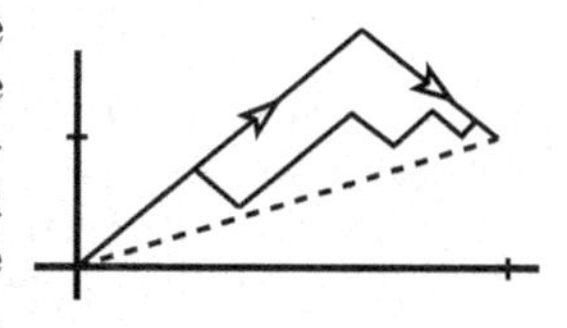

Do you encounter such weird behavior often, with an infinite number of solutions? No, but you see from this example that it doesn't take a very complicated integrand to produce such a pathology.

16.10 Second Order

Except for a couple of problems in optics in chapter two, 2.35 and 2.39, I've mostly ignored the question about minimum versus maximum.

- Does it matter in classical mechanics whether the integral, $\int L\,dt$ is minimum or not in determining the equations of motion? No.
- In geometric optics, does it matter whether Fermat's principle of least time for the path of the light ray is *really* minimum? Yes, in this case it does, because it provides information about the focus.

• In the calculation of minimum energy electric potentials in a capacitor does it matter? No, but only because it's *always* a minimum.
• In problems in quantum mechanics similar to the electric potential problem, the fact that you're dealing sometimes with a minimum and sometimes not leads to some serious technical difficulties.

How do you address this question? The same way you do in ordinary calculus: See what happens to the second order terms in your expansions. Take the same general form as before and keep terms through second order. Assume that the endpoints are fixed.

$$\begin{aligned} I[y] &= \int_a^b dx\, F\big(x, y(x), y'(x)\big) \\ \Delta I &= I[y+\delta y] - I[y] \\ &= \int_a^b dx\, F\big(x, y(x)+\delta y(x), y'(x)+\delta y'(x)\big) - \int_a^b dx\, F\big(x, y(x), y'(x)\big) \\ &= \int_a^b dx \left[\frac{\partial F}{\partial y}\delta y + \frac{\partial F}{\partial y'}\delta y' + \frac{\partial^2 F}{\partial y^2}\big(\delta y\big)^2 + 2\frac{\partial^2 F}{\partial y \partial y'}\delta y \delta y' + \frac{\partial^2 F}{\partial y'^2}\big(\delta y'\big)^2\right] \end{aligned} \tag{16.48}$$

If the first two terms combine to zero, this says the first derivative is zero. Now for the next terms.

Recall the similar question that arose in section 8.11. How can you tell if a function of two variables has a minimum, a maximum, or neither? The answer required looking at the matrix of all the second derivatives of the function — the Hessian. Now, instead of a 2×2 matrix as in Eq. (8.31) you have an integral.

$$\begin{aligned} \big\langle d\vec r, H\, d\vec r\big\rangle &= \begin{pmatrix} dx & dy \end{pmatrix} \begin{pmatrix} f_{xx} & f_{xy} \\ f_{yx} & f_{yy} \end{pmatrix} \begin{pmatrix} dx \\ dy \end{pmatrix} \\ &\longrightarrow \int_a^b dx \begin{pmatrix} \delta y & \delta y' \end{pmatrix} \begin{pmatrix} F_{yy} & F_{yy'} \\ F_{y'y} & F_{y'y'} \end{pmatrix} \begin{pmatrix} \delta y \\ \delta y' \end{pmatrix} \end{aligned}$$

In the two dimensional case $\nabla f = 0$ defines a minimum if the product $\big\langle d\vec r, H\, d\vec r\big\rangle$ is positive for all possible directions $d\vec r$. For this new case the "directions" are the possible functions δy and its derivative $\delta y'$.

The direction to look first is where $\delta y'$ is big. The reason is that I can have a very small δy that has a very big $\delta y'$: $10^{-3} \sin\big(10^6 x\big)$. If ΔI is to be positive in every direction, it has to be positive in this one. That requires $F_{y'y'} > 0$.

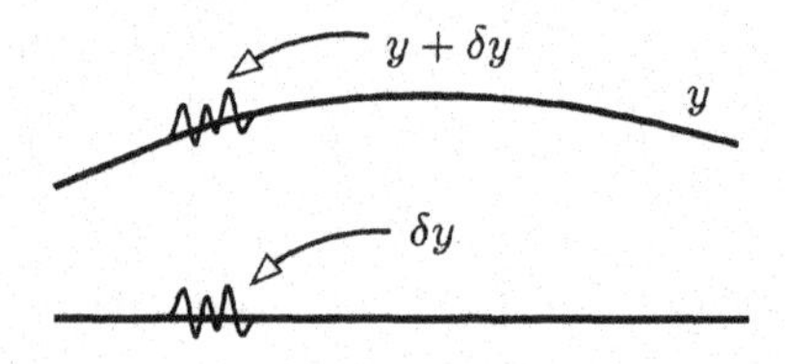

Is it really that simple? No. First the δy terms can be important too, and second y can itself have several components. Look at the latter first. The final term in Eq. (16.48) should be

$$\int_a^b dx\, \frac{\partial^2 F}{\partial y'_m \partial y'_n}\delta y'_m \delta y'_n$$

This set of partial derivatives of F is *at each point along the curve* a Hessian. At each point it has a set of eigenvalues and eigenvectors, and if all along the path all the eigenvalues are always positive, it meets the first, necessary conditions for the original functional to be a minimum. If you look at an example from mechanics, for which the independent variable is time, these y_n' terms are then $\dot{x}_n$ instead. Terms such as these typically represent kinetic energy and you expect that to be positive.

An example:

$$S[\vec{r}\,] = \int_0^T dt\, L(x, y, \dot{x}, \dot{y}, t) = \int_0^T dt\, \frac{1}{2}\left[\dot{x}^2 + \dot{y}^2 + 2\gamma t\dot{x}\dot{y} - x^2 - y^2\right]$$

This is the action for a particle moving in two dimensions (x, y) with the specified Lagrangian. The equation of motion are

$$\frac{\delta S}{\delta x} = -x - \ddot{x} - \gamma(t\ddot{y} + \dot{y}) = 0$$
$$\frac{\delta S}{\delta y} = -y - \ddot{y} - \gamma(t\ddot{x} + \dot{x}) = 0$$

If $\gamma = 0$ you have two independent harmonic oscillators.

The matrix of derivatives of L with respect to $\dot{x} = \dot{y}_1$ and $\dot{y} = \dot{y}_2$ is

$$\frac{\partial^2 L}{\partial \dot{y}_m \partial \dot{y}_n} = \begin{pmatrix} 1 & \gamma t \\ \gamma t & 1 \end{pmatrix}$$

The eigenvalues of this matrix are $1 \pm \gamma t$, with corresponding eigenvectors $\begin{pmatrix} 1 \\ 1 \end{pmatrix}$ and $\begin{pmatrix} 1 \\ -1 \end{pmatrix}$. This Hessian then says that S should be a minimum up to the time $t = 1/\gamma$, but not after that. This is also a singular point of the differential equations for x and y.

Focus

When the Hessian made from the $\delta y'^2$ terms has only positive eigenvalues everywhere, the preceding analysis might lead you to believe that the functional is always a minimum. Not so. That condition is necessary; it is not sufficient. It says that the functional is a minimum with respect to rapidly oscillating δy. It does not say what happens if δy changes gradually over the course of the integral. If this happens, and if the length of the interval of integration is long enough, the $\delta y'$ terms may be the small ones and the $(\delta y)^2$ may then dominate over the whole length of the integral. This is exactly what happens in the problems 2.35, 2.39, and 16.17.

When this happens in an optical system, where the functional $T = \int d\ell/v$ is the travel time along the path, it signals something important. You have a focus. An ordinary light ray obeys Fermat's principle that T is stationary with respect to small changes in the path. It is a minimum if the path is short enough. A focus occurs when light can go from one point to another by many different paths, and for still longer paths the path is neither minimum nor maximum.

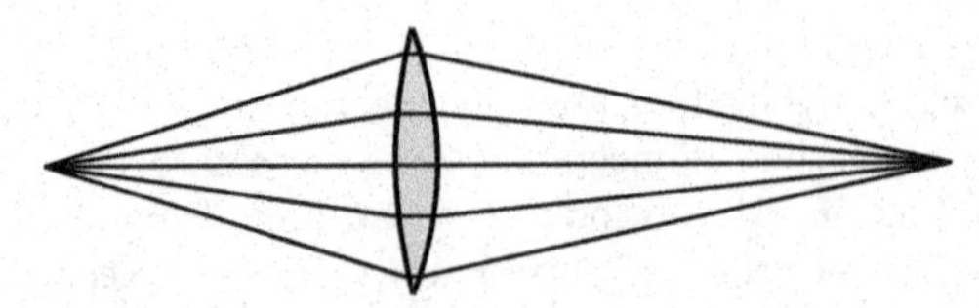

In the integral for T, where the starting point and the ending point are the source and image points, the second order variation will be zero for these long, gradual changes in the path. The straight-line path through the center of the lens takes *least* time if its starting point and ending point are closer than this source and image. The same path will be a saddle (neither maximum nor minimum) if the points are farther apart than this. This sort of observation led to the development of the mathematical subject called "Morse Theory," a topic that has had applications in studying such diverse subjects as nuclear fission and the gravitational lensing of light from quasars.

Thin Lens

This provides a simple way to understand the basic equation for a thin lens. Let its thickness be t and its radius r.

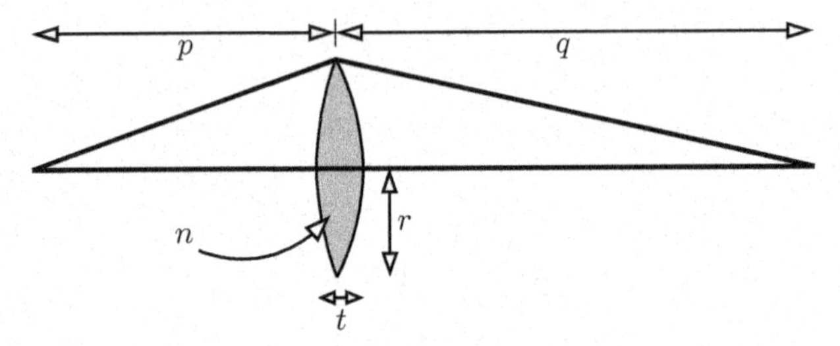

Light that passes through this lens along the straight line through the center moves more slowly as it passes through the thickness of the lens, and takes a time

$$T_1 = \frac{1}{c}(p+q-t) + \frac{n}{c}t$$

Light that take a longer path through the edge of the lens encounters no glass along the way, and it takes a time

$$T_2 = \frac{1}{c}\left[\sqrt{p^2+r^2} + \sqrt{q^2+r^2}\right]$$

If p and q represent the positions of a source and the position of its image at a focus, these two times should be equal. At least they should be equal in the approximation that the lens is thin and when you keep terms only to the second order in the variation of the path.

$$T_2 = \frac{1}{c}\left[p\sqrt{1+r^2/p^2} + q\sqrt{1+r^2/q^2}\right] = \frac{1}{c}\left[p\big(1+r^2/2p^2\big) + q\big(1+r^2/2q^2\big)\right]$$

Equate T_1 and T_2.

$$(p+q-t) + nt = \left[p\big(1+r^2/2p^2\big) + q\big(1+r^2/2q^2\big)\right]$$

$$\begin{aligned}(n-1)t &= \frac{r^2}{2p} + \frac{r^2}{2q} \\ \frac{1}{p} + \frac{1}{q} &= \frac{2(n-1)t}{r^2} = \frac{1}{f}\end{aligned} \tag{16.49}$$

This is the standard equation describing the focusing properties of thin lenses as described in every elementary physics text that even mentions lenses. The focal length of the lens is then $f = r^2/2(n-1)t$. That is *not* the expression you usually see, but it is the same. See problem 16.21. Notice that this equation for the focus applies whether the lens is double convex or plano-convex or meniscus:). If you allow the thickness t to be negative (equivalent to saying that there's an extra time delay at the edges instead of in the center), then this result still works for a diverging lens, though the analysis leading up to it requires more thought.

Exercises

1 For the functional $F[x] = x(0) + \int_0^\pi dt\, \big(x(t)^2 + \dot x(t)^2\big)$ and the function $x(t) = 1 + t^2$, evaluate $F[x]$.

2 For the functional $F[x] = \int_0^1 dt\, x(t)^2$ with the boundary conditions $x(0) = 0$ and $x(1) = 1$, what is the minimum value of F and what function x give it? Start by drawing graphs of various x that satisfy these boundary conditions. Is there any reason to require that x be a continuous function of t?

3 With the functional F of the preceding exercise, what is the functional derivative $\delta F/\delta x$?

Problems

16.1 You are near the edge of a lake and see someone in the water needing help. What path do you take to get there in the shortest time? You can run at a speed v_1 on the shore and swim at a probably slower speed v_2 in the water. Assume that the edge of the water forms a straight line, and express your result in a way that's easy to interpret, not as the solution to some quartic equation. Ans: Snell's Law.

16.2 The cycloid is the locus of a point that is on the edge of a circle that is itself rolling along a straight line — a pebble caught in the tread of a tire. Use the angle of rotation as a parameter and find the parametric equations for $x(\theta)$ and $y(\theta)$ describing this curve. Show that it is Eq. (16.17).

16.3 In Eq. (16.17), describing the shortest-time slide of a particle, what is the behavior of the function for $y \ll a$? In figuring out the series expansion of $w = \cos^{-1}(1-t)$, you may find it useful to take the cosine of both sides. Then you should be able to find that the two lowest order terms in this expansion are $w = \sqrt{2t} - t^{3/2}/12\sqrt{2}$. You will need both terms. Ans: $x = \sqrt{2y^3/a}\,/3$

16.4 The dimensions of an ordinary derivative such as dx/dt is the quotient of the dimensions of the numerator and the denominator (here L/T). Does the same statement apply to the functional derivative?

16.5 Use Fermat's principle to derive both Snell's law and the law of reflection at a plane surface. Assume two straight line segments from the starting point to the ending point and minimize the total travel time of the light. The drawing applies to Snell's law, and you can compute the travel time of the light as a function of the coordinate x at which the light hits the surface and enters the higher index medium.

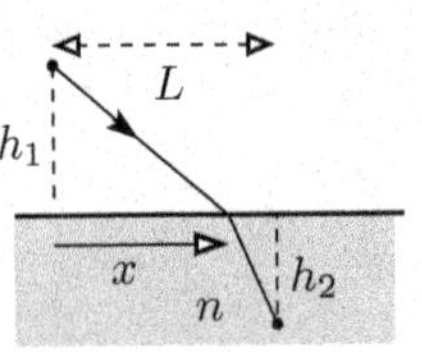

16.6 Analyze the path of light over a roadway starting from Eq. (16.23) but using x as the independent variable instead of y.

16.7 (**a**) Fill in the steps leading to Eq. (16.31). And do you understand the point of the rearrangements that I did just preceding it? Also, can you explain why the form of the function Eq. (16.30) should have been obvious without solving any extra boundary conditions? (**b**) When you can explain that in a few words, then what general cubic polynomial can you use to get a still better result?

16.8 For the function $F(x, y, y') = x^2 + y^2 + y'^2$, explicitly carry through the manipulations leading to Eq. (16.41).

16.9 Use the explicit variation in Eq. (16.8) and find the minimum of that function of ϵ. Compare that minimum to the value found in Eq. (16.11). Ans: 1.64773

16.10 Do either of the functions, Eqs. (16.29) or (16.30), satisfy Laplace's equation?

16.11 For the function $F(x,y,y') = x^2 + y^2 + y'^2$, repeat the calculation of δI only now keep all the higher order terms in δy and $\delta y'$. Show that the solution Eq. (16.11) is a minimum.

16.12 Use the techniques leading to Eq. (16.21) in order to solve the brachistochrone problem Eq. (16.13) again. This time use x as the independent variable instead of y.

16.13 On a right circular cylinder, find the path that represents the shortest distance between two points. $d\ell^2 = dz^2 + R^2 d\phi^2$.

16.14 Put two circular loops of wire in a soap solution and draw them out, keeping their planes parallel. If they are fairly close you will have a soap film that goes from one ring to the other, and the minimum energy solution is the one with the smallest area. What is the shape of this surface? Use cylindrical coordinates to describe the surface. It is called a catenoid, and its equation involves a hyperbolic cosine.

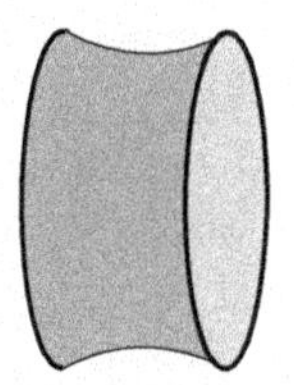

16.15 There is one part of the derivation going from Eq. (16.33) to (16.36) that I omitted: the special cases of $\ell = 1$ and $\ell = N - 1$. Go back and finish that detail, showing that you get the same result even in this case.

16.16 Section 16.6 used the trapezoidal rule for numerical integration and the two-point centered difference for differentiation. What happens to the derivation if (**a**) you use Simpson's rule for integration or if (**b**) you use an asymmetric differentiation formula such as $y'(0) \approx [y(h) - y(0)]/h$?

16.17 For the simple harmonic oscillator, $L = m\dot{x}^2/2 - m\omega^2 x^2/2$. Use the time interval $0 < t < T$ so that $S = \int_0^T L\,dt$, and find the equation of motion from $\delta S/\delta x = 0$. When the independent variable x is changed to $x + \delta x$, keep the second order terms in computing δS this time and also make an explicit choice of

$$\delta x(t) = \epsilon \sin(n\pi t/T)$$

For integer $n = 1, 2\ldots$ this satisfies the boundary conditions that $\delta x(0) = \delta x(T) = 0$. Evaluate the change is S through second order in ϵ (that is, do it exactly). Find the conditions on the interval T so that the solution to $\delta S/\delta x = 0$ is in fact a minimum. Then find the conditions when it isn't, and what is special about the T for which $S[x]$ changes its structure? Note: This T is defined independently from ω. It's specifies an arbitrary time interval for the integral.

16.18 Eq. (16.37) describes a particle with a specified potential energy. For a charge in an electromagnetic field let $U = qV(x_1, x_2, x_3, t)$ where V is the electric potential. Now how do you include magnetic effects? Add another term to L of the form $C\dot{\vec{r}}\cdot\vec{A}(x_1, x_2, x_3, t)$. Figure out what the Lagrange equations are, making $\delta S/\delta x_k = 0$. What value must C have in order that this matches $\vec{F} = q(\vec{E} + \vec{v} \times \vec{B}) = m\vec{a}$ with $\vec{B} = \nabla \times \vec{A}$? What is $\vec{E}$ in terms of V and $\vec{A}$? Don't forget the chain rule. Ans: $C = q$ and then $\vec{E} = -\nabla V - \partial\vec{A}/\partial t$

16.19 (**a**) For the solutions that spring from Eq. (16.46), which of the three results shown have the largest and smallest values of $\int f\,dx$? Draw a graph of $f(y')$ and see where the

characteristic slope of Eq. (16.46) is with respect to the graph.
(**b**) There are circumstances for which these kinked solutions, Eq. (16.47) do and do not occur; find them and explain them.

16.20 What are the Euler-Lagrange equations for $I[y] = \int_a^b dx\, F(x, y, y', y'')$?

16.21 The equation for the focal length of a thin lens, Eq. (16.49), is not the traditional one found in most texts. That is usually expressed in terms of the radii of curvature of the lens surfaces. Show that this is the same. Also note that Eq. (16.49) is independent of the optics sign conventions for curvature.

16.22 The equation (16.25) is an approximate solution to the path for light above a hot road. Is there a function $n = f(y)$ representing the index of refraction above the road surface such that this equation would be its exact solution?

16.23 On the first page of this chapter, you see the temperature dependence of length measurements. (**a**) Take a metal disk of radius a and place it centered on a block of ice. Assume that the metal reaches an equilibrium temperature distribution $T(r) = T_0(r^2/a^2 - 1)$. The temperature at the edge is $T = 0$, and the ruler is calibrated there. The disk itself remains flat. Measure the distance from the origin straight out to the radial coordinate r. Call this measured radius s. Measure the circumference of the circle at this radius and then express this circumference in terms of the measured radius s.
(**b**) On a sphere of radius R (constant temperature) start at the pole ($\theta = 0$) and write the distance along the arc at constant ϕ down to the angle θ. Now go around the circle at this constant θ and write its circumference. Express this circumference in terms of the distance you just wrote for the "radius" of this circle.
(**c**) Show that the geometry you found in (**a**) is the same as that in (**b**) and find the radius of the sphere that this "heat metric" expresses. Ans: $R = a/2\sqrt{\alpha T_0(1 - \alpha T_0)}$

16.24 Using the same techniques as in section 16.5, apply these methods to two concentric spheres. Again, use a linear and then a quadratic approximation. Before you do this, go back to Eq. (16.30) and and see if you can arrive at that form directly, *without* going through all the manipulations of solving for α and β. That is, determine how you could have gotten to (16.30) easily. Check some numbers against the exact answer.

16.25 For the variational problem Eq. (16.45) one solution is $y = bx/a$. Assume that α, $\beta > 0$ and determine if this is a minimum or maximum or neither. Do this also for the other solution, Eq. (16.47). Ans: The first is min if if $b/a > \sqrt{\beta/6\alpha}$. The kinked solution is always a minimum.

16.26 If you can construct glass with a variable index of refraction, you can make a flat lens with an index that varies with distance from the axis. What function of distance must the index $n(r)$ be in order that this flat cylinder of glass of thickness t has a focal length f? All small angles and thin lenses of course. Ans: $n(r) = n(0) - r^2/2ft$.

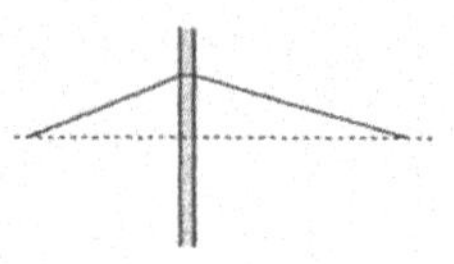

Densities and Distributions

Back in section 12.1 I presented a careful and full definition of the word "function." This is useful even though you should already have a pretty good idea of what the word means. If you haven't read that section, now would be a good time. The reason to review it is that this definition doesn't handle all the cases that naturally occur. This will lead to the idea of a "generalized function."

There are (at least) two approaches to this subject. One that relates it to the ideas of functionals as you saw them in the calculus of variations, and one that is more intuitive and is good enough for most purposes. The latter appears in section 17.5, and if you want to jump there first, I can't stop you.

17.1 Density

What *is* density? If the answer is "mass per unit volume" then what does that mean? It clearly doesn't mean what it says, because you aren't required* to use a cubic meter.

It's a derivative. Pick a volume ΔV and find the mass in that volume to be Δm. The average volume-mass-density in that volume is $\Delta m/\Delta V$. If the volume is the room that you're sitting in, the mass includes you and the air and everything else in the room. Just as in defining the concept of velocity (instantaneous velocity), you have to take a limit. Here the limit is

$$\lim_{\Delta V\to 0}\frac{\Delta m}{\Delta V}=\frac{dm}{dV} \tag{17.1}$$

Even this isn't quite right, because the volume could as easily shrink to zero by approaching a line, and that's not what you want. It has to shrink to a point, but the standard notation doesn't let me say that without introducing more symbols than I want.

Of course there are other densities. If you want to talk about paper or sheet metal you may find area-mass-density to be more useful, replacing the volume ΔV by an area ΔA. Maybe even linear mass density if you are describing wires, so that the denominator is $\Delta\ell$. And why is the numerator a mass? Maybe you are describing volume-charge-density or even population density (people per area). This last would appear in mathematical notation as dN/dA.

This last example manifests a subtlety in all of these definitions. In the real world, you can't take the limit as $\Delta A\to 0$. When you count the number of people in an area you can't very well let the area shrink to zero. When you describe mass, remember that the world is made of atoms. If you let the volume shrink too much you'll either be between or inside the atoms. Maybe you will hit a nucleus; maybe not. This sort of problem means that you have to stop short of the mathematical limit and let the volume shrink to some size that still contains many atoms, but that is small enough so the quotient $\Delta m/\Delta V$ isn't significantly affected by further changing ΔV. Fortunately, this fine point seldom gets in the way, and if it does, you'll know it fast. I'll ignore it. If you're bothered by it remember that you are accustomed to doing the same thing when you approximate a sum by an integral. The world is made of atoms, and any common computation about a physical system will really involve a sum over all the atoms in the system (*e.g.* find the

* even by the panjandrums of the Système International d'Unités

center of mass). You never do this, preferring to do an integral instead even though this is an approximation to the sum over atoms.

If you know the density — when the word is used unqualified it commonly means volume-mass-density — you find mass by integration over the volume you have specified.

$$m = \int_V \rho\, dV \tag{17.2}$$

You can even think of this as a new kind of function $m(V)$: input a specification for a volume of space; output a mass. That's really what density provides, a prescription to go from a volume specification to the amount of mass within that volume.

For the moment, I'll restrict the subject to linear mass density, and so that you simply need the coordinate along a straight line,

$$\lambda(x) = \frac{dm}{dx}(x), \qquad \text{and} \qquad m = \int_a^b \lambda(x)\, dx \tag{17.3}$$

If λ represents a function such as Ax^2 $(0 < x < L)$, a bullwhip perhaps, then this is elementary, and $m_{\text{total}} = AL^3/3$. I want to look at the reverse specification. Given an interval, I will specify the amount of mass in that interval and work backwards. The first example will be simple. The interval $x_1 \le x \le x_2$ is denoted $[x_1, x_2]$. The function m has this interval for its argument.*

$$m\big([x_1,x_2]\big) = \begin{cases} 0 & (x_1 \le x_2 \le 0) \\ Ax_2^3/3 & (x_1 \le 0 \le x_2 \le L) \\ AL^3/3 & (x_1 \le 0 \le L \le x_2) \\ A\big(x_2^3 - x_1^3\big)/3 & (0 \le x_1 \le x_2 \le L) \\ A\big(L^3 - x_1^3\big)/3 & (0 \le x_1 \le L \le x_2) \\ 0 & (L \le x_1 \le x_2) \end{cases} \tag{17.4}$$

The density Ax^2 $(0 < x < L)$ is of course a much easier way to describe the same distribution of mass. This distribution function, $m\big([x_1,x_2]\big)$, comes from integrating the density function $\lambda(x) = Ax^2$ on the interval $[x_1, x_2]$.

Another example is a variation on the same theme. It is slightly more involved, but still not too bad.

$$m\big([x_1,x_2]\big) = \begin{cases} 0 & (x_1 \le x_2 \le 0) \\ Ax_2^3/3 & (x_1 \le 0 \le x_2 < L/2) \\ Ax_2^3/3 + m_0 & (x_1 \le 0 < L/2 \le x_2 \le L) \\ AL^3/3 + m_0 & (x_1 \le 0 < L \le x_2) \\ A\big(x_2^3 - x_1^3\big)/3 & (0 \le x_1 \le x_2 < L/2) \\ A\big(x_2^3 - x_1^3\big)/3 + m_0 & (0 \le x_1 < L/2 \le x_2 \le L) \\ A\big(L^3 - x_1^3\big)/3 + m_0 & (0 \le x_1 \le L/2 < Ll2) \\ A\big(x_2^3 - x_1^3\big)/3 & (L/2 < x_1 \le x_2 \le L) \\ A\big(L^3 - x_1^3\big)/3 & (L/2 < x_1 \le L \le x_2) \\ 0 & (L \le x_1 \le x_2) \end{cases} \tag{17.5}$$

* I'm abusing the notation here. In (17.2) m is a number. In (17.4) m is a function. You're used to this, and physicists do it all the time despite reproving glances from mathematicians.

If you read through all these cases, you will see that the sole thing that I've added to the first example is a point mass m_0 at the point $L/2$. What density function λ will produce this distribution? Answer: *No function will do this.* That's why the concept of a "generalized function" appeared. I could state this distribution function in words by saying

"Take Eq. (17.4) and if $[x_1, x_2]$ contains the point $L/2$ then add m_0."

That there's no density function λ that will do this is inconvenient but not disastrous. When the very idea of a density was defined in Eq. (17.1), it started with the distribution function, the mass within the volume, and only arrived at the definition of a density by some manipulations. The density is a type of derivative and not all functions are differentiable. The function $m\big([x_1, x_2]\big)$ or $m\big(V\big)$ is more fundamental (if less convenient) than is the density function.

17.2 Functionals

$$F[\phi] = \int_{-\infty}^{\infty} dx\, f(x)\phi(x)$$

defines a scalar-valued function of a function variable. Given any (reasonable) function ϕ as input, it returns a scalar. That is a functional. This one is a linear functional because it satisfies the equations

$$F[a\phi] = aF[\phi] \qquad \text{and} \qquad F[\phi_1 + \phi_2] = F[\phi_1] + F[\phi_2] \tag{17.6}$$

This isn't a new idea, it's just a restatement of a familiar idea in another language. The mass density can define a useful functional (*linear* density for now). Given $dm/dx = \lambda(x)$ what is the total mass?

$$\int_{-\infty}^{\infty} dx \lambda\,(x) 1 = M_{\text{total}}$$

Where is the center of mass?

$$\frac{1}{M_{\text{total}}} \int_{-\infty}^{\infty} dx \lambda\,(x) x = x_{\text{cm}}$$

Restated in the language of functionals,

$$F[\phi] = \int_{-\infty}^{\infty} dx \lambda\,(x)\phi(x) \qquad \text{then} \qquad M_{\text{total}} = F[1], \qquad x_{\text{cm}} = \frac{1}{M_{\text{total}}} F[x]$$

If, instead of mass density, you are describing the distribution of grades in a large class or the distribution of the speed of molecules in a gas, there are still other ways to use this sort of functional. If $dN/dg = f(g)$ is the grade density in a class (number of students per grade interval), then with $F[\phi] = \int dg\, f(g)\phi(g)$

$$N_{\text{students}} = F[1], \qquad \text{mean grade} = \bar{g} = \frac{1}{N_{\text{students}}} F[g], \tag{17.7}$$

$$\text{variance} = \sigma^2 = \frac{1}{N_{\text{students}}} F[(g - \bar{g})^2], \qquad \text{skewness} = \frac{1}{N_{\text{students}}\sigma^3} F[(g - \bar{g})^3]$$

$$\text{kurtosis excess} = \frac{1}{N_{\text{students}}\sigma^4} F[(g - \bar{g})^4] - 3$$

Unless you've studied some statistics, you will probably never have heard of skewness and kurtosis excess. They are ways to describe the shape of the density function, and for a Gaussian both these numbers are zero. If it's skewed to one side the skewness is non-zero. [Did I really say that?] The kurtosis excess compares the flatness of the density function to that of a Gaussian.

The Maxwell-Boltzmann function describing the speeds of molecules in an ideal gas is at temperature T

$$f_{\mathsf{MB}}(v) = \left(\frac{m}{2\pi kT}\right)^{3/2} 4\pi v^2 e^{-mv^2/2kT} \tag{17.8}$$

dN/dv is the number of molecules per speed interval, but this function F_{MB} is normalized differently. It is instead $(dN/dv)/N_{\mathsf{total}}$. It is the fraction of the molecules per speed interval. That saves carrying along a factor specifying the total number of molecules in the gas. I could have done the same thing in the preceding example of student grades, defining the functional $F_1 = F/N_{\mathsf{students}}$. Then the equations (17.7) would have a simpler appearance, such as $\overline{g} = F_1[g]$. For the present case of molecular speeds, with $F[\phi] = \int_0^\infty dv\, f_{\mathsf{MB}}(v)\phi(v)$,

$$F[1] = 1, \qquad F[v] = \overline{v} = \sqrt{\frac{8kT}{\pi m}}, \qquad F[mv^2/2] = \overline{\text{K.E.}} = \frac{3}{2}kT \tag{17.9}$$

Notice that the mean kinetic energy is not the kinetic energy that corresponds to the mean speed.

Look back again to section 12.1 and you'll see not only a definition of "function" but a definition of "functional." It looks different from what I've been using here, but look again and you will see that when you view it in the proper light, that of chapter six, they are the same. Equations (12.3)–(12.5) involved vectors, but remember that when you look at them as elements of a vector space, functions are vectors too.

Functional Derivatives

In section 16.2, equations (16.6) through (16.10), you saw a development of the functional derivative. What does that do in this case?

$$F[\phi] = \int dx\, f(x)\phi(x), \qquad \text{so} \qquad F[\phi + \delta\phi] - F[\phi] = \int dx\, f(x)\delta\phi(x)$$

The functional derivative is the coefficient of $\delta\phi$ and dx, so it is

$$\frac{\delta F}{\delta \phi} = f \tag{17.10}$$

That means that the functional derivative of $m([x_1, x_2])$ in Eq. (17.4) is the linear mass density, $\lambda(x) = Ax^2$, $(0 < x < L)$. Is there such a thing as a functional integral? Yes, but not here, as it goes well beyond the scope of this chapter. Its development is central in quantum field theory.

17.3 Generalization

Given a function f, I can create a linear functional F using it as part of an integral. What sort of linear functional arises from f'? Integrate by parts to find out. Here I'm going to have to assume that f or ϕ or both vanish at infinity, or the development here won't work.

$$F[\phi] = \int_{-\infty}^{\infty} dx\, f(x)\phi(x), \qquad \text{then}$$

$$\int_{-\infty}^{\infty} dx f'(x)\phi(x) = f(x)\phi(x)\Big|_{-\infty}^{\infty} - \int_{-\infty}^{\infty} dx\, f(x)\phi'(x) = -F[\phi'] \tag{17.11}$$

In the same way, you can relate higher derivatives of f to the functional F. There's another restriction you need to make: For this functional $-F[\phi']$ to make sense, the function ϕ has to be differentiable. If you want higher derivatives of f, then ϕ needs to have still higher derivatives.

If you know everything about F, what can you determine about f? If you assume that all the functions you're dealing with are smooth, having as many derivatives as you need, then the answer is simple: *everything.* If I have a rule by which to get a number $F[\phi]$ for every (smooth) ϕ, then I can take a special case for ϕ and use it to find f. Use a ϕ that drops to zero very rapidly away from some given point; for example if n is large this function drops off rapidly away from the point x_0.

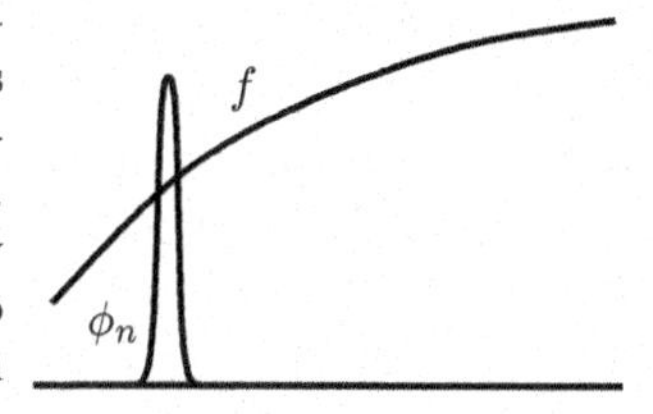

$$\phi_n(x) = \sqrt{\frac{n}{\pi}} e^{-n(x-x_0)^2}$$

I've also arranged so that its integral of ϕ_n over all x is one. If I want the value of f at x_0 I can do an integral and take a limit.

$$F[\phi_n] = \int_{-\infty}^{\infty} dx\, f(x)\phi_n(x) = \int_{-\infty}^{\infty} dx\, f(x)\sqrt{\frac{n}{\pi}} e^{-n(x-x_0)^2}$$

As n increases to infinity, all that matters for f is its value at x_0. The integral becomes

$$\lim_{n\to\infty} \int_{-\infty}^{\infty} dx\, f(x_0)\sqrt{\frac{n}{\pi}} e^{-n(x-x_0)^2} = f(x_0) \tag{17.12}$$

This means that I can reconstruct the function f if I know everything about the functional F. To get the value of the derivative $f'(x_0)$ instead, simply use the function $-\phi_n'$ and take the same limit. This construction is another way to look at the functional derivative. The equations (17.10) and (17.12) produce the same answer.

You say that this doesn't sound very practical? That it is an awfully difficult and roundabout way to do something? Not really. It goes back to the ideas of section 17.1. To define a density you have to know how much mass is contained in an arbitrarily specified volume. That's a functional. It didn't look like a functional there, but it is. You just have to rephrase it to see that it's the same thing.

As before, do the case of linear mass density. $\lambda(x)$ is the density, and the functional $F[\phi] = \int_{-\infty}^{\infty} dx\lambda(x)\phi(x)$. Then as in Eq. (17.4), $m([x_1, x_2])$ is that mass contained in the interval from x_1 to x_2 and you can in turn write it as a functional.

$$\text{Let } \chi \text{ be the step function } \chi(x) = \begin{cases} 1 & (x_1 \le x \le x_2) \\ 0 & (\text{otherwise}) \end{cases}$$

$$\text{then} \qquad \int_{x_1}^{x_2} dx\lambda(x) = F[\chi] = m([x_1, x_2])$$

What happens if the function f is itself not differentiable? I can still define the original functional. Then I'll see what implications I can draw from the functional itself. The first and most important example of this is for f a step function.

$$f(x) = \theta(x) = \begin{cases} 1 & (x \ge 0) \\ 0 & (x < 0) \end{cases} \qquad F[\phi] = \int_{-\infty}^{\infty} dx\theta(x)\phi(x) \tag{17.13}$$

θ has a step at the origin, so of course it's not differentiable there, but if it were possible in some way to define its derivative, then when you look at its related functional, it should give the answer $-F[\phi']$ as in Eq. (17.11), just as for any other function. What then is $-F[\phi']$?

$$-F[\phi'] = -\int_{-\infty}^{\infty} dx\theta(x)\phi'(x) = -\int_{0}^{\infty} dx\phi'(x) = -\phi(x)\Big|_0^{\infty} = \phi(0) \tag{17.14}$$

This defines a perfectly respectable linear functional. Input the function ϕ and output the value of the function at zero. It easily satisfies Eq. (17.6), but there is no function f that when integrated against ϕ will yield this result. Still, it is so useful to be able to do these manipulations as if such a function exists that the notation of a "delta function" was invented. This is where the idea of a generalized function enters.

Green's functions

In the discussion of the Green's function solution to a differential equation in section 4.6, I started with the differential equation

$$m\ddot{x} + kx = F(t)$$

and found a general solution in Eq. (4.34). This approach pictured the external force as a series of small impulses and added the results from each.

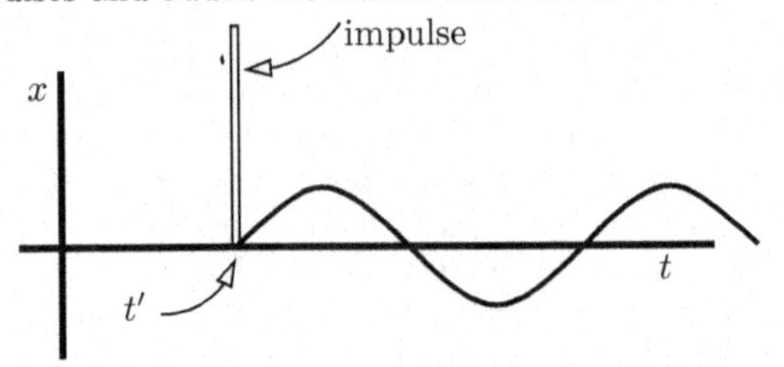

$$x(t) = \int_{-\infty}^{\infty} dt'\, G(t-t')F(t') \qquad \text{where} \qquad G(t) = \begin{cases} \frac{1}{m\omega_0}\sin\omega_0 t & (t \ge 0) \\ 0 & (t < 0) \end{cases} \tag{17.15}$$

I wrote it a little differently there, but it's the same. Can you verify that this is a solution to the stated differential equation? Simply plug in, do a couple of derivatives and see what happens. One derivative at a time:

$$\frac{dx}{dt} = \int_{-\infty}^{\infty} dt'\, \dot G(t-t')F(t') \qquad \text{where} \qquad \dot G(t) = \begin{cases} \frac{1}{m}\cos\omega_0 t & (t \ge 0) \\ 0 & (t < 0) \end{cases} \tag{17.16}$$

Now for the second derivative. *Oops.* I can't differentiate $\dot G$. It has a step at $t = 0$.

This looks like something you saw in a few paragraphs back, where there was a step in the function θ. I'm going to handle this difficulty now by something of a kludge. In the next section you'll see the notation that makes this manipulation easy and transparent. For now I will subtract and add the discontinuity from $\dot G$ by using the same step function θ.

$$\begin{aligned} \dot G(t) &= \begin{cases} \frac{1}{m}\big[\cos\omega_0 t - 1 + 1\big] & (t \ge 0) \\ 0 & (t < 0) \end{cases} \\ &= \left\{\begin{matrix} \frac{1}{m}\big[\cos\omega_0 t - 1\big] & (t \ge 0) \\ 0 & (t < 0) \end{matrix}\right\} + \frac{1}{m}\theta(t) = \dot G_0(t) + \frac{1}{m}\theta(t) \end{aligned} \tag{17.17}$$

The (temporary) notation here is that $\dot G_0$ is the part of $\dot G$ that doesn't have the discontinuity at $t = 0$. That part is differentiable. The expression for dx/dt now has two terms, one from the $\dot G_0$ and one from the θ. Do the first one:

$$\frac{d}{dt}\int_{-\infty}^{\infty} dt'\, \dot G_0(t-t')F(t') = \int_{-\infty}^{\infty} dt'\, \frac{d}{dt}\dot G_0(t-t')F(t')$$

$$\text{and} \qquad \frac{d}{dt}\dot G_0(t) = \begin{cases} \frac{1}{m}[-\omega_0\sin\omega_0 t] & (t \ge 0) \\ 0 & (t < 0) \end{cases}$$

The original differential equation involved $m\ddot x + kx$. The $\dot G_0$ part of this is

$$m\int_{-\infty}^{\infty} dt' \left\{\begin{matrix} \frac{1}{m}[-\omega_0\sin\omega_0(t-t')] & (t \ge t') \\ 0 & (t < t') \end{matrix}\right\} F(t') \\ + k\int_{-\infty}^{\infty} dt' \left\{\begin{matrix} \frac{1}{m\omega_0}\sin\omega_0(t-t') & (t \ge t') \\ 0 & (t < t') \end{matrix}\right\} F(t')$$

Use $k = m\omega_0^2$, and this is zero.

Now go back to the extra term in θ. The kx terms doesn't have it, so all that's needed is

$$m\ddot x + kx = m\frac{d}{dt}\int_{-\infty}^{\infty} dt'\, \frac{1}{m}\theta(t-t')F(t') = \frac{d}{dt}\int_{-\infty}^{t} dt'\, F(t') = F(t)$$

This verifies yet again that this Green's function solution works.

17.4 Delta-function Notation

Recognizing that it would be convenient to be able to differentiate non-differentiable functions, that it would make manipulations easier if you could talk about the density of a point mass ($m/0 =?$), and that the impulsive force that appears in setting up Green's functions for the harmonic oscillator isn't a function, what do you do? If you're Dirac, you invent a notation that works.

Two functionals are equal to each other if they give the same result for all arguments. Does that apply to the functions that go into defining them? If

$$\int_{-\infty}^{\infty} dx\, f_1(x)\phi(x) = \int_{-\infty}^{\infty} dx\, f_2(x)\phi(x)$$

for all test functions ϕ (smooth, infinitely differentiable, going to zero at infinity, whatever constraints we find expedient), does it mean that $f_1 = f_2$? Well, no. Suppose that I change the value of f_1 at exactly one point, adding 1 there and calling the result f_2. These functions aren't equal, but underneath an integral sign you can't tell the difference. In terms of the functionals they define, they are essentially equal: "equal in the sense of distributions."

Extend this to the generalized functions

The equation (17.14) leads to specifying the functional

$$\delta[\phi] = \phi(0) \tag{17.18}$$

This delta-functional isn't a help in doing manipulations, so define the notation

$$\int_{-\infty}^{\infty} dx \delta\,(x)\phi(x) = \delta[\phi] = \phi(0) \tag{17.19}$$

This notation isn't an integral in the sense of something like section 1.6, and $\delta(x)$ isn't a function, but the notation allows you effect manipulations just as if they were. Note: the symbol δ here is not the same as the δ in the functional derivative. We're just stuck with using the same symbol for two different things. Blame history and look at problem 17.10.

You can treat the step function as differentiable, with $\theta' = \delta$, and this notation leads you smoothly to the right answer.

$$\text{Let}\quad \theta_{x_0}(x) = \begin{cases} 1 & (x \geq x_0) \\ 0 & (x < x_0) \end{cases} = \theta(x - x_0), \quad \text{then} \quad \int_{-\infty}^{\infty} dx \theta_{\,x_0}(x)\phi(x) = \int_{x_0}^{\infty} dx \phi\,(x)$$

The derivative of this function is

$$\frac{d}{dx}\theta_{x_0}(x) = \delta(x - x_0)$$

You show this by

$$\int_{-\infty}^{\infty} dx\, \frac{d\theta_{x_0}(x)}{dx}\phi(x) = -\int_{-\infty}^{\infty} dx\theta_{\,x_0}(x)\phi'(x) = -\int_{x_0}^{\infty} dx\phi'(x) = \phi(x_0)$$

The idea of a generalized function is that you can manipulate it *as if* it were an ordinary function provided that you put the end results of your manipulations under an integral.

The above manipulations for the harmonic oscillator, translated to this language become

$$m\ddot{G} + kG = \delta(t) \qquad \text{for} \qquad G(t) = \begin{cases} \frac{1}{m\omega_0} \sin \omega_0 t & (t \geq 0) \\ 0 & (t < 0) \end{cases}$$

Then the solution for a forcing function $F(t)$ is

$$x(t) = \int_{-\infty}^{\infty} G(t - t')F(t')\,dt'$$

because

$$m\ddot{x} + kx = \int_{-\infty}^{\infty} (m\ddot{G} + kG)F(t')\,dt' = \int_{-\infty}^{\infty} \delta(t - t')F(t')\,dt' = F(t)$$

This is a lot simpler. Is it legal? Yes, though it took some serious mathematicians (Schwartz, Sobolev) some serious effort to develop the logical underpinnings for this subject. The result of their work is: It's o.k.

17.5 Alternate Approach

This delta-function method is so valuable that it's useful to examine it from more than one vantage. Here is a very different way to understand delta functions, one that avoids an explicit discussion of functionals. Picture a sequence of smooth functions that get narrower and taller as the parameter n gets bigger. Examples are

$$\sqrt{\frac{n}{\pi}}e^{-nx^2}, \qquad \frac{n}{\pi}\frac{1}{1+n^2x^2}, \qquad \frac{1}{\pi}\frac{\sin nx}{x}, \qquad \frac{n}{\pi}\operatorname{sech} nx \tag{17.20}$$

Pick any one such sequence and call it $\delta_n(x)$. (A "delta sequence") The factors in each case are arranged so that

$$\int_{-\infty}^{\infty} dx\delta_n(x) = 1$$

As n grows, each function closes in around $x = 0$ and becomes very large there. Because these are perfectly smooth functions there's no question about integrating them.

$$\int_{-\infty}^{\infty} dx\delta_n(x)\phi(x) \tag{17.21}$$

makes sense as long as ϕ doesn't cause trouble. You will typically have to assume that the ϕ behave nicely at infinity, going to zero fast enough, and this is satisfied in the physics applications that we need. For large n any of these functions looks like a very narrow spike. If you multiply one of these δ_ns by a mass m, you have a linear mass density that is (for large n) concentrated near to a point: $\lambda(x) = m\delta_n(x)$. Of course you can't take the limit as $n \to \infty$ because this doesn't have a limit. If you could, then that would be the density for a point mass: $m\delta(x)$.

What happens to (17.21) as $n \to \infty$? For large n any of these delta-sequences approaches zero everywhere except at the origin. Near the origin $\phi(x)$ is very close to $\phi(0)$, and the function δ_n is non-zero in only the tiny region around zero. If the function ϕ is simply continuous at the origin you have

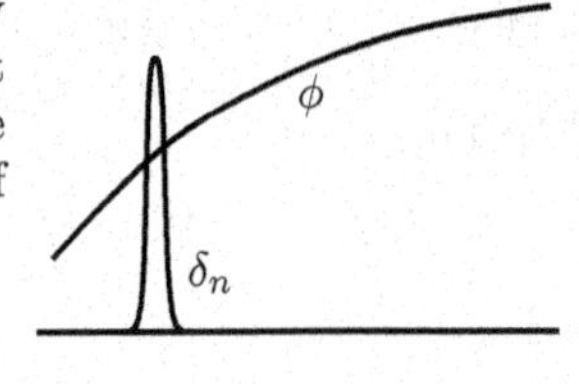

$$\lim_{n\to\infty} \int_{-\infty}^{\infty} dx\,\delta_n(x)\phi(x) = \phi(0) \cdot \lim_{n\to\infty} \int_{-\infty}^{\infty} dx\,\delta_n(x) = \phi(0) \tag{17.22}$$

At this point I can introduce a notation:

$$\text{“}\int_{-\infty}^{\infty} dx\,\delta(x)\phi(x)\text{”} \qquad MEANS \qquad \lim_{n\to\infty} \int_{-\infty}^{\infty} dx\,\delta_n(x)\phi(x) \tag{17.23}$$

In this approach to distributions the collection of symbols on the left has for its definition the collection of symbols on the right. Those in turn, such as $\int$, have definitions that go back to the fundamentals of calculus. You *cannot* move the limit in this last integral under the integral sign. You can't interchange these limits because the limit of δ_n is not a function.

In this development you say that the delta function is a notation, not for a function, but for a process (but then, so is the integral sign). That means that the underlying idea always goes back to a familiar, standard manipulation of ordinary calculus. If you have an equation involving such a function, say

$$\theta'(x) = \delta(x), \qquad \text{then this } \textit{means} \qquad \theta_n'(x) = \delta_n(x) \qquad \text{and that}$$

$$\lim_{n\to\infty} \int_{-\infty}^{\infty} dx\,\theta_n'(x)\phi(x) = \lim_{n\to\infty} \int_{-\infty}^{\infty} dx\,\delta_n(x)\phi(x) = \phi(0) \tag{17.24}$$

How can you tell if this equation is true? Remember now that in this interpretation these are sequences of ordinary, well-behaved functions, so you can do ordinary manipulations such as partial integration. The functions θ_n are smooth and they rise from zero to one as x goes from $-\infty$ to $+\infty$. As n becomes large, the interval over which this rise occurs will become narrower and narrower. In the end these θ_n will approach the step function $\theta(x)$ of Eq. (17.13).

$$\int_{-\infty}^{\infty} dx\,\theta_n'(x)\phi(x) = \theta_n(x)\phi(x)\Big|_{-\infty}^{\infty} - \int_{-\infty}^{\infty} dx\,\theta_n(x)\phi'(x)$$

The functions ϕ go to zero at infinity — they're "test functions" — and that kills the boundary terms, leaving the last integral standing by itself. Take the limit as $n \to \infty$ on it. You can take the limit inside the integral now, because the limit of θ_n is a perfectly good function, even if it is discontinuous.

$$-\lim_{n} \int_{-\infty}^{\infty} dx\,\theta_n(x)\phi'(x) = -\int_{-\infty}^{\infty} dx\,\theta(x)\phi'(x) = -\int_{0}^{\infty} dx\,\phi'(x) = -\phi(x)\Big|_{0}^{\infty} = \phi(0)$$

This is precisely what the second integral in Eq. (17.24) is. This is the proof that $\theta' = \delta$. Any proof of an equation involving such generalized functions requires you to integrate the equation against a test function and to determine if the resulting integral becomes an identity as $n \to \infty$. This implies that it now makes sense to differentiate a discontinuous function — as long as you mean differentiation "in the sense of distributions." That's the jargon you encounter here. An equation such as $\theta' = \delta$ makes sense only when it is under an integral sign and is interpreted in the way that you just saw.

In these manipulations, where did I use the particular form of the delta sequence? Never. A particular combination such as

$$\theta_n(x) = \frac{1}{2}\bigl[1 + \frac{2}{\pi}\tan^{-1} nx\bigr], \qquad \text{and} \qquad \delta_n(x) = \frac{n}{\pi}\frac{1}{1 + n^2x^2} \tag{17.25}$$

never appeared. Any of the other sequences would have done just as well, and all that I needed was the *properties* of the sequence, not its particular representation. You can even use a delta sequence that doesn't look like any of the functions in Eq. (17.20).

$$\delta_n(x) = 2\sqrt{\frac{2n}{\pi}}e^{inx^2} \tag{17.26}$$

This turns out to have all the properties that you need, though again, you don't have to invoke its explicit form.

What is $\delta(ax)$? Integrate $\delta_n(ax)$ with a test function.

$$\lim_n \int_{-\infty}^{\infty} dx\,\delta_n(ax)\phi(x) = \lim_n \int_{-\infty}^{\infty} \frac{dy}{a}\,\delta_n(y)\phi(y/a)$$

where $y = ax$. Actually, this isn't quite right. If $a > 0$ it is fine, but if a is negative, then when $x \to -\infty$ you have $y \to +\infty$. You have to change the limits to put it in the standard form. You can carry out that case for yourself and verify that the expression covering both cases is

$$\lim_n \int_{-\infty}^{\infty} dx\,\delta_n(ax)\phi(x) = \lim_n \int_{-\infty}^{\infty} \frac{dy}{|a|}\,\delta_n(y)\phi(y/a) = \frac{1}{|a|}\phi(0)$$

Translate this into the language of delta functions and it is

$$\delta(ax) = \frac{1}{|a|}\delta(x) \tag{17.27}$$

You can prove other relations in the same way. For example

$$\delta(x^2 - a^2) = \frac{1}{2|a|}\bigl[\delta(x-a) + \delta(x+a)\bigr] \qquad \text{or} \qquad \delta\bigl(f(x)\bigr) = \sum_k \frac{1}{|f'(x_k)|}\delta(x - x_k) \tag{17.28}$$

In the latter equation, x_k is a root of f, and you sum over all roots. Notice that it doesn't make any sense if you have a double root. Just try to see what $\delta(x^2)$ would mean. The last of these identities contains the others as special cases. Eq. (17.27) implies that δ is even.

17.6 Differential Equations

Where do you use these delta functions? Start with differential equations. I'll pick one that has the smallest number of technical questions associated with it. I want to solve the equation

$$\frac{d^2 f}{dx^2} - k^2 f = F(x) \tag{17.29}$$

subject to conditions that $f(x)$ should approach zero for large magnitude x. I'll assume that the given function F has this property too.

But first: Let k and y be constants, then solve

$$\frac{d^2 g}{dx^2} - k^2 g = \delta(x - y) \tag{17.30}$$

I want a solution that is well-behaved at infinity, staying finite there. This equality is "in the sense of distributions" recall, and I'll derive it a couple of different ways, here and in the next section.

First way: Treat the δ as simply a spike at $x = y$. Everywhere else on the x-axis it is zero. Solve the equation for two cases then, $x < y$ and $x > y$. In both cases the form is the same.

$$g'' - k^2 g = 0, \qquad \text{so} \qquad g(x) = Ae^{kx} + Be^{-kx}$$

For $x < y$, I want $g(x)$ to go to zero far away, so that requires the coefficient of e^{-kx} to be zero. For $x > 0$, the reverse is true and only the e^{-kx} can be present.

$$g(x) = \begin{cases} Ae^{kx} & (x < y) \\ Be^{-kx} & (x > y) \end{cases}$$

Now I have to make g satisfy Eq. (17.30) *at* $x = y$.

Compute dg/dx. But wait. This is impossible unless the function is at least continuous. If it isn't then I'd be differentiating a step function and I don't want to do that (at least not yet). That is

$$g(y-) = Ae^{ky} = g(y+) = Be^{-ky} \tag{17.31}$$

This is one equation in the two unknowns A and B. Now differentiate.

$$\frac{dg}{dx} = \begin{cases} Ake^{kx} & (x < y) \\ -Bke^{-kx} & (x > y) \end{cases} \tag{17.32}$$

This is in turn differentiable everywhere except at $x = y$. There it has a step

$$\text{discontinuity in } g' = g'(y+) - g'(y-) = -Bke^{-ky} - Ake^{ky}$$

This means that (17.32) is the sum of two things, one differentiable, and the other a step, a multiple of θ.

$$\frac{dg}{dx} = \text{differentiable stuff} + \big(-Bke^{-ky} - Ake^{ky}\big)\theta(x - y)$$

The differentiable stuff satisfies the differential equation $g'' - k^2 g = 0$. For the rest, Compute d^2g/dx^2

$$\bigl(-Bke^{-ky} - Ake^{ky}\bigr)\frac{d}{dx}\theta(x-y) = \bigl(-Bke^{-ky} - Ake^{ky}\bigr)\delta(x-y)$$

Put this together and remember the equation you're solving, Eq. (17.30).

$$\begin{aligned} &g'' - k^2 g = 0 \text{ (from the differentiable stuff)} \\ &\qquad + \bigl(-Bke^{-ky} - Ake^{ky}\bigr)\delta(x-y) \quad = \delta(x-y) \end{aligned}$$

Now there are two equations for A and B, this one and Eq. (17.31).

$$\begin{aligned} Ae^{ky} &= Be^{-ky} \\ -Bke^{-ky} - Ake^{ky} &= 1 \end{aligned} \qquad \text{solve these to get} \qquad \begin{aligned} A &= -e^{-ky}/2k \\ B &= -e^{ky}/2k \end{aligned}$$

Finally, back to g.

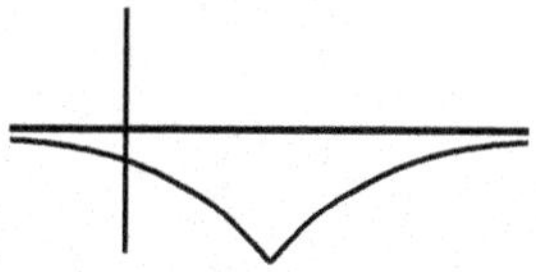

$$g(x) = \begin{cases} -e^{k(x-y)}/2k & (x < y) \\ -e^{-k(x-y)}/2k & (x > y) \end{cases} \tag{17.33}$$

When you get a fairly simple form of solution such as this, you have to see if you could have saved some work, perhaps replacing brute labor with insight? Of course. The original differential equation (17.30) is symmetric around the point y. It's plausible to look for a solution that behaves the same way, using $(x - y)$ as the variable instead of x. Either that or you could do the special case $y = 0$ and then change variables at the end to move the delta function over to y. See problem 17.11.

There is standard notation for this function; it *is* a Green's function.

$$G(x,y) = \left\{ \begin{matrix} -e^{k(x-y)}/2k & (x < y) \\ -e^{-k(x-y)}/2k & (x > y) \end{matrix} \right\} = -e^{-k|x-y|}/2k \tag{17.34}$$

Now to solve the original differential equation, Eq. (17.29). Substitute into this equation the function

$$\int_{-\infty}^{\infty} dy\, G(x,y)F(y)$$

$$\begin{aligned} \left[\frac{d^2}{dx^2} - k^2\right]\int_{-\infty}^{\infty} dy\, G(x,y)F(y) &= \int_{-\infty}^{\infty} dy \left[\frac{d^2G(x,y)}{dx^2} - k^2 G(x,y)\right] F(y) \\ &= \int_{-\infty}^{\infty} dy \delta\,(x-y)F(y) = F(x) \end{aligned}$$

This is the whole point of delta functions. They make this sort of manipulation as easy as dealing with an ordinary function. Easier, once you're used to them.

For example, if $F(x) = F_0$ between $-x_0$ and $+x_0$ and zero elsewhere, then the solution for f is

$$\int dy\, G(x,y)F(y) = -\int_{-x_0}^{x_0} dy\, F_0 e^{-k|x-y|}/2k$$

$$= -\frac{F_0}{2k}\begin{cases} \int_{-x_0}^{x_0} dy\, e^{-k(x-y)} & (x > x_0) \\ \int_{-x_0}^{x} dy\, e^{-k(x-y)} + \int_{x}^{x_0} dy\, e^{-k(y-x)} & (-x_0 < x < x_0) \\ \int_{-x_0}^{x_0} dy\, e^{-k(y-x)} & (x < -x_0) \end{cases}$$

$$= -\frac{F_0}{k^2}\begin{cases} e^{-kx}\sinh kx_0 & (x > x_0) \\ [1 - e^{-kx_0}\cosh kx] & (-x_0 < x < x_0) \\ e^{kx}\sinh kx_0 & (x < -x_0) \end{cases} \tag{17.35}$$

You can see that this resulting solution is an even function of x, necessarily so because the original differential equation is even in x, the function $F(x)$ is even in x, and the boundary conditions are even in x.

Other Differential Equations

Can you apply this method to other equations? Yes, many. Try the simplest first order equation:

$$\frac{dG}{dx} = \delta(x) \longrightarrow G(x) = \theta(x)$$

$$\frac{df}{dx} = g(x) \longrightarrow f(x) = \int_{-\infty}^{\infty} dx'\, G(x-x')g(x') = \int_{-\infty}^{x} dx'\, g(x')$$

which clearly satisfies $df/dx = g$.

If you try $d^2G/dx^2 = \delta(x)$ you explain the origin of problem 1.48.

Take the same equation $d^2G/dx^2 = \delta(x-x')$ but in the domain $0 < x < L$ and with the boundary conditions $G(0) = 0 = G(L)$. The result is

$$G(x) = \begin{cases} x(x'-L)/L & (0 < x < x') \\ x'(x-L)/L & (x' < x < L) \end{cases} \tag{17.36}$$

0 x' L

17.7 Using Fourier Transforms

Solve Eq. (17.30) another way. Fourier transform everything in sight.

$$\frac{d^2g}{dx^2} - k^2 g = \delta(x-y) \rightarrow \int_{-\infty}^{\infty} dx \left[\frac{d^2g}{dx^2} - k^2 g\right] e^{-iqx} = \int_{-\infty}^{\infty} dx\,\delta(x-y)e^{-iqx} \tag{17.37}$$

The right side is designed to be easy. For the left side, do some partial integrations. I'm looking for a solution that goes to zero at infinity, so the boundary terms will vanish. See Eq. (15.12).

$$\int_{-\infty}^{\infty} dx\,\big[-q^2 - k^2\big]g(x)e^{-iqx} = e^{-iqy} \tag{17.38}$$

The left side involves only the Fourier transform of g. Call it $\tilde{g}$.

$$\left[-q^2-k^2\right]\tilde{g}(q)=e^{-iqy}, \qquad \text{so} \qquad \tilde{g}(q)=-\frac{e^{-iqy}}{q^2+k^2}$$

Now invert the transform.

$$g(x)=\int_{-\infty}^{\infty}\frac{dq}{2\pi}\,\tilde{g}(q)e^{iqx}=-\int_{-\infty}^{\infty}\frac{dq}{2\pi}\,\frac{e^{iq(x-y)}}{q^2+k^2}$$

Do this by contour integration, where the integrand has singularities at $q=\pm ik$.

$$-\frac{1}{2\pi}\int_{C_1}dq\,\frac{e^{iq(x-y)}}{k^2+q^2}$$

The poles are at $\pm ik$, and the exponential dominates the behavior of the integrand at large $|q|$, so there are two cases: $x>y$ and $x<y$. Pick the first of these, then the integrand vanishes rapidly as $q\to+i\infty$.

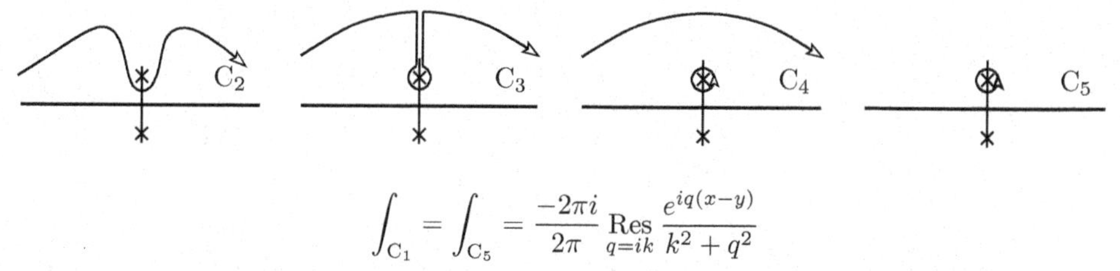

$$\int_{C_1}=\int_{C_5}=\frac{-2\pi i}{2\pi}\operatorname*{Res}_{q=ik}\frac{e^{iq(x-y)}}{k^2+q^2}$$

Compute the residue.

$$\frac{e^{iq(x-y)}}{k^2+q^2}=\frac{e^{iq(x-y)}}{(q-ik)(q+ik)}\approx\frac{e^{iq(x-y)}}{(q-ik)(2ik)}$$

The coefficient of $1/(q-ik)$ is the residue, so

$$g(x)=-\frac{e^{-k(x-y)}}{2k} \qquad (x>y) \tag{17.39}$$

in agreement with Eq. (17.33). The $x<y$ case is yours.

17.8 More Dimensions

How do you handle problems in three dimensions? $\delta(\vec{r})=\delta(x)\delta(y)\delta(z)$. For example I can describe the charge density of a point charge, dq/dV as $q\delta(\vec{r}-\vec{r}_0)$. The integral of this is

$$\int q\delta(\vec{r}-\vec{r}_0)\,d^3r=q$$

as long as the position $\vec{r}_0$ is inside the volume of integration. For an example that uses this, look at the potential of a point charge, satisfying Poisson's equation.

$$\nabla^2 V = -\rho/\epsilon_0$$

What is the potential for a specified charge density? Start by finding the Green's function.

$$\nabla^2 G = \delta(\vec{r}), \qquad \text{or instead do:} \qquad \nabla^2 G - k^2 G = \delta(\vec{r}) \tag{17.40}$$

The reason for starting with the latter equation is that you run into some problems with the first form. It's too singular. I'll solve the second one and then take the limit as $k \to 0$. The Fourier transform method is simpler, so use that.

$$\int d^3r \big[\nabla^2 G - k^2 G\big] e^{-i\vec{q}\cdot\vec{r}} = 1$$

When you integrate by parts (twice) along each of the three integration directions dx, dy, and dz, you pull down a factor of $-q^2 = -q_x^2 - q_y^2 - q_z^2$ just as in one dimension.

$$\int d^3r \big[-q^2 G - k^2 G\big] e^{-i\vec{q}\cdot\vec{r}} = 1 \qquad \text{or} \qquad \tilde{G}(\vec{q}) = \frac{-1}{q^2 + k^2}$$

where $\tilde{G}$ is, as before, the Fourier transform of G.

Now invert the transform. Each dimension picks up a factor of $1/2\pi$.

$$G(\vec{r}) = -\int \frac{d^3q}{(2\pi)^3} \frac{e^{i\vec{q}\cdot\vec{r}}}{q^2 + k^2}$$

This is a three dimensional integral, and the coordinate system to choose is spherical. q^2 doesn't depend on the direction of $\vec{q}$, and the single place that this direction matters is in the dot product in the exponent. The coordinates are q, θ, ϕ for the vector $\vec{q}$, and since I can pick my coordinate system any way I want, I will pick the coordinate axis along the direction of the vector $\vec{r}$. Remember that in this integral $\vec{r}$ is just some constant.

$$G = -\frac{1}{(2\pi)^3} \int q^2\, dq \sin\theta\, d\theta\, d\phi\, \frac{e^{iqr\cos\theta}}{q^2 + k^2}$$

The integral $d\phi$ becomes a factor 2π. Let $u = \cos\theta$, and that integral becomes easy. All that is left is the dq integral.

$$G = -\frac{1}{(2\pi)^2} \int_0^\infty \frac{q^2\, dq}{q^2 + k^2} \frac{1}{iqr} \big[e^{iqr} - e^{-iqr}\big]$$

More contour integrals: There are poles in the q-plane at $q = \pm ik$. The q^2 factors are even. The q in the denominator would make the integrand odd except that the combination of exponentials in brackets are also odd. The integrand as a whole is even. I will then extend the limits to $\pm\infty$ and divide by 2.

$$G = -\frac{1}{8\pi^2 ir}\int_{-\infty}^{\infty}\frac{q\,dq}{q^2+k^2}\left[e^{iqr}-e^{-iqr}\right]$$

There are two terms; start with the first one. $r > 0$, so $e^{iqr} \to 0$ as $q \to +i\infty$. The contour is along the real axis, so push it toward $i\infty$ and pick up the residue at $+ik$.

$$-\frac{1}{8\pi^2 ir}\int = -\frac{1}{8\pi^2 ir}2\pi i \mathop{\mathrm{Res}}_{q=ik}\frac{q}{(q-ik)(q+ik)}e^{iqr} = -\frac{1}{8\pi^2 ir}2\pi i\frac{ik}{2ik}e^{-kr} \tag{17.41}$$

For the second term, see problem 17.15. Combine it with the preceding result to get

$$G = -\frac{1}{4\pi r}e^{-kr} \tag{17.42}$$

This is the solution to Eq. (17.40), and if I now let k go to zero, it is $G = -1/4\pi r$.

Just as in one dimension, once you have the Green's function, you can write down the solution to the original equation.

$$\nabla^2 V = -\rho/\epsilon_0 \implies V = -\frac{1}{\epsilon_0}\int d^3r'\,G(\vec r,\vec r\,')\rho(\vec r\,') = \frac{1}{4\pi\epsilon_0}\int d^3r'\,\frac{\rho(\vec r\,')}{|\vec r-\vec r\,'|} \tag{17.43}$$

State this equation in English, and it says that the potential of a single point charge is $q/4\pi\epsilon_0 r$ and that the total potential is the sum over the contributions of all the charges. Of course this development also provides the Green's function for the more complicated equation (17.42).

Applications to Potentials

Let a charge density be $q\delta(\vec r)$. This satisfies $\int d^3r\,\rho = q$. The potential it generates is that of a point charge at the origin. This just reproduces the Green's function.

$$\phi = \frac{1}{4\pi\epsilon_0}\int d^3r'\,q\,\delta(\vec r\,')\frac{1}{|\vec r-\vec r\,'|} = \frac{q}{4\pi\epsilon_0 r} \tag{17.44}$$

What if $\rho(\vec r) = -p\,\partial\delta(\vec r)/\partial z$? [Why $-p$? Patience.] At least you can say that the dimensions of the constant p are charge times length. Now use the Green's function to compute the potential it generates.

$$\begin{aligned}\phi = \frac{1}{4\pi\epsilon_0}\int d^3r'\,(-p)\frac{\partial\delta(\vec r\,')}{\partial z'}\frac{1}{|\vec r-\vec r\,'|} &= \frac{p}{4\pi\epsilon_0}\int d^3r'\,\delta(\vec r\,')\frac{\partial}{\partial z'}\frac{1}{|\vec r-\vec r\,'|}\\ &= \frac{p}{4\pi\epsilon_0}\frac{\partial}{\partial z'}\frac{1}{|\vec r-\vec r\,'|}\bigg|_{\vec r\,'=0}\end{aligned} \tag{17.45}$$

This is awkward, so I'll use a little trick. Make a change of variables in (17.45) $\vec u = \vec r - \vec r\,'$, then

$$\frac{\partial}{\partial z'} \to -\frac{\partial}{\partial u_z}, \qquad \frac{p}{4\pi\epsilon_0}\frac{\partial}{\partial z'}\frac{1}{|\vec r-\vec r\,'|}\bigg|_{\vec r\,'=0} = -\frac{p}{4\pi\epsilon_0}\frac{\partial}{\partial u_z}\frac{1}{u} \tag{17.46}$$

(See also problem 17.19.) Cutting through the notation, this last expression is just

$$\phi = \frac{-p}{4\pi\epsilon_0}\frac{\partial}{\partial z}\frac{1}{r} = \frac{-p}{4\pi\epsilon_0}\frac{\partial}{\partial z}\frac{1}{\sqrt{x^2+y^2+z^2}} = \frac{p}{4\pi\epsilon_0}\frac{z}{\left(x^2+y^2+z^2\right)^{3/2}}$$
$$= \frac{p}{4\pi\epsilon_0}\frac{z}{r^3} = \frac{p}{4\pi\epsilon_0}\frac{\cos\theta}{r^2} \tag{17.47}$$

The expression $-\partial(1/r)/\partial z$ is such a simple way to compute this, the potential of an electric dipole, that it is worth trying to understand why it works. And in the process, *why* is this an electric dipole? The charge density, the source of the potential is a derivative, and a derivative is (the limit of) a difference quotient. This density is just that of two charges, and they produce potentials just as in Eq. (17.44). There are two point charges, with delta-function densities, one at the origin the other at $-\hat{z}\,\Delta z$.

$$\rho = \frac{-p}{\Delta z}\left[\delta(\vec{r}+\hat{z}\,\Delta z) - \delta(\vec{r})\right] \qquad \text{gives potential} \qquad \frac{-p}{4\pi\epsilon_0\Delta z}\left[\frac{1}{|\vec{r}+\hat{z}\,\Delta z|} - \frac{1}{r}\right] \tag{17.48}$$

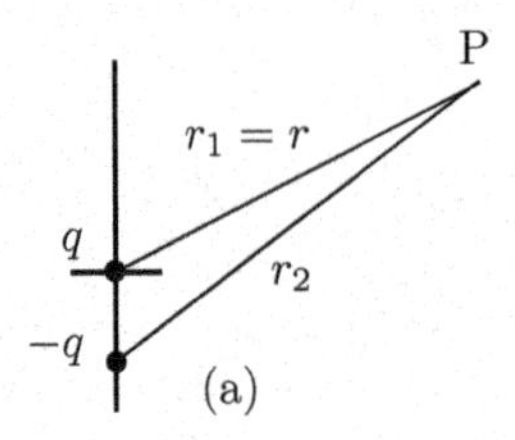

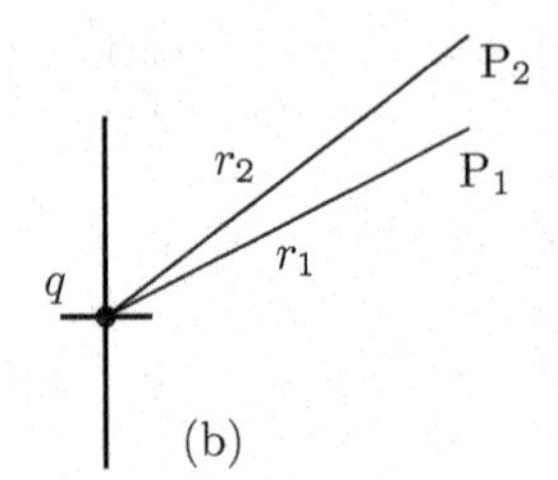

The picture of the potential that arises from this pair of charges is (a). A negative charge ($-q = -p/\Delta z$) at $-\hat{z}\,\Delta z$ and a corresponding positive charge $+q$ at the origin. This picture explains why this charge density, $\rho(\vec{r}) = -p\,\partial\delta(\vec{r})/\partial z$, represent an electric dipole. Specifically it represents a dipole whose vector representation points toward $+\hat{z}$, from the negative charge to the positive one. It even explains why the result (17.47) has the sign that it does: The potential ϕ is positive along the positive z-axis and negative below. That's because a point on the positive z-axis is closer to the positive charge than it is to the negative charge.

Now the problem is to understand *why* the potential of this dipole ends up with a result as simple as the derivative $(-p/4\pi\epsilon_0)\partial(1/r)/\partial z$: The potential at the point P in the figure (a) comes from the two charges at the two distances r_1 and r_2. Construct figure (b) by moving the line r_2 upward so that it starts at the origin. Imagine a new charge configuration consisting of only *one* charge, $q = p/\Delta z$ at the origin. Now evaluate the potentials that this single charge produces at two different points P_1 and P_2 that are a distanceΔ z apart. Then subtract them.

$$\phi(P_2) - \phi(P_1) = \frac{q}{4\pi\epsilon_0}\left[\frac{1}{r_2} - \frac{1}{r_1}\right]$$

In the notation of the preceding equations this is

$$\phi(P_2) - \phi(P_1) = \frac{q}{4\pi\epsilon_0}\left[\frac{1}{|\vec{r}+\hat{z}\,\Delta z|} - \frac{1}{r}\right] = \frac{p}{4\pi\epsilon_0\Delta z}\left[\frac{1}{|\vec{r}+\hat{z}\,\Delta z|} - \frac{1}{r}\right]$$

Except for a factor of (-1), this is Eq. (17.48), and it explains why the potential caused by an ideal electric dipole at the origin can be found by taking the potential of a point charge and differentiating it.

$$\phi_{\text{point charge}} = \frac{q}{4\pi\epsilon_0 r} \qquad\qquad \phi_{\text{dipole}} = -a\frac{\partial}{\partial z}\phi_{\text{point charge}}$$

Here, the electric dipole strength is qa.

Can you repeat this process? Yes. Instead of two opposite point charges near each other, you can place two opposite point dipoles near each other.

$$\phi_{\text{linear quadrupole}} = -a\frac{\partial}{\partial z}\phi_{\text{dipole}} \tag{17.49}$$

This is the potential from the charge density $\rho = +Q\,\partial^2\delta(\vec{r})/\partial z^2$, where $Q = qa^2$. [What about $\partial^2/\partial x\partial y$?]

Exercises

1 What is the analog of Eq. (17.4) for the linear mass density $\lambda(x) = C$ (a constant) for $0 < x < L$ and zero otherwise?

2 Take the preceding mass density and add a point mass m_0 at $x = L/2$. What is the distribution $m([x_1, x_2])$ now?

3 Use the λ from the first exercise and define the functional $F[\phi] = \int_{-\infty}^{\infty} dx\lambda(x)\phi(x)$. What is the total mass, $F[1] = M$? What is the mean position of the mass, $F[x]/M$?

4 As in the preceding exercise, what are the variance, the skewness, and the kurtosis excess?

5 What is $\int_0^1 dx\delta(x - x_0)$?

6 Pick any two of Eq. (17.20) and show that they are valid delta sequences.

7 What is $\int_{-\infty}^{x} dt\delta(t)$?

Problems

17.1 Calculate the mean, the variance, the skewness, and the kurtosis excess for a Gaussian: $f(g) = Ae^{-B(g-g_0)^2}$ $(-\infty < g < \infty)$. Assume that this function is normalized the same way that Eq. (17.8) is, so that its integral is one.

17.2 Calculate the mean, variance, skewness, and the kurtosis excess for a flat distribution, $f(g) = \text{constant}$, $(0 < g < g_{\mathsf{max}})$. Ans: Var $= g_{\mathsf{m}}^2/12$ kurt. exc. $= -6/5$

17.3 Derive the results stated in Eq. (17.9). Compare $m\bar{v}^2/2$ to $\overline{\text{K.E.}}$ Compare this to the results of problem 2.48.

17.4 Show that you can rewrite Eq. (17.16) as an integral $\int_{-\infty}^{t} dt' \, \frac{1}{m} \cos \omega_0(t-t')F(t')$ and differentiate this directly, showing yet again that (17.15) satisfies the differential equation.

17.5 What are the units of a delta function?

17.6 Show that

$$\delta\big(f(x)\big) = \delta(x - x_0)/|f'(x_0)|$$

where x_0 is the root of f. Assume just one root for now, and the extension to many roots will turn this into a sum as in Eq. (17.28).

17.7 Show that

(**a**) $x\delta'(x) = -\delta(x)$ (**b**) $x\delta(x) = 0$

(**c**) $\delta'(-x) = -\delta'(x)$ (**d**) $f(x)\delta(x - a) = f(a)\delta(x - a)$

17.8 Verify that the functions in Eq. (17.20) satisfy the requirements for a delta sequence. Are they normalized to have an integral of one? Sketch each. Sketch Eq. (17.26). It is complex, so sketch both parts. How can a delta sequence be complex? Verify that the imaginary part of this function doesn't contribute.

17.9 What is the analog of Eq. (17.25) if δ_n is a sequence of Gaussians: $\sqrt{n/\pi}e^{-nx^2}$? Ans: $\theta_n(x) = \frac{1}{2}\left[1 + \text{erf}\left(x\sqrt{n}\right)\right]$

17.10 Interpret the functional derivative of the functional in Eq. (17.18): $\delta\,\delta[\phi]/\delta\phi$. Despite appearances, this actually makes sense. Ans: $\delta(x)$

17.11 Repeat the derivation of Eq. (17.33) but with less labor, selecting the form of the function g to simplify the work. In the discussion following this equation, reread the comments on this subject.

17.12 Verify the derivation of Eq. (17.35). Also examine this solution for the cases that x_0 is very large and that it is very small.

17.13 Fill in the steps in section 17.7 leading to the Green's function for $g'' - k^2 g = \delta$.

17.14 Derive the analog of Eq. (17.39) for the case $x < y$.

17.15 Calculate the contribution of the second exponential factor leading to Eq. (17.41).

17.16 Starting with the formulation in Eq. (17.23), what is the result of δ' and of δ'' on a test function? Draw sketches of a typical δ_n, δ'_n, and δ''_n.

17.17 If $\rho(\vec{r}) = qa^2\,\partial^2\delta(\vec{r})/\partial z^2$, compute the potential *and* sketch the charge density. You should express your answer in spherical coordinates as well as rectangular, perhaps commenting on the nature of the results and relating it to functions you have encountered before. You can do this calculation in either rectangular or spherical coordinates.
Ans: $(2qa^2/4\pi\epsilon_0)P_2(\cos\theta)/r^3$

17.18 What is a picture of the charge density $\rho(\vec{r}) = qa^2\,\partial^2\delta(\vec{r})/\partial x\,\partial y$?
(Planar quadrupole) What is the potential for this case?

17.19 In Eq. (17.46) I was not at all explicit about which variables are kept constant in each partial derivative. Sort this out for both $\partial/\partial z'$ and for $\partial/\partial u_z$.

17.20 Use the results of the problem 17.16, showing graphs of δ_n and its derivatives. Look again at the statements leading up to Eq. (17.31), that g is continuous, and ask what would happen if it is not. Think of the right hand side of Eq. (17.30) as a δ_n too in this case, and draw a graph of the left side of the same equation if g_n is assumed to change very fast, approaching a discontinuous function as $n \to \infty$. Demonstrate by looking at the graphs of the left and right side of the equation that this *can't* be a solution and so that g must be continuous as claimed.

17.21 Calculate the mean, variance, skewness, and the kurtosis excess for the density $f(g) = A[\delta(g) + \delta(g-g_0) + \delta(g-xg_0)]$. See how these results vary with the parameter x.
Ans: skewness $= 2^{-3/2}(1+x)(x-2)(2x-1)/(1-x+x^2)$
kurt. excess $= -3 + \frac{3}{4}(1+x^4+(1-x)^4)/(1-x+x^2)^2$

17.22 Calculate the potential of a linear quadrupole as in Eq. (17.49). Also, what is the potential of the planar array mentioned there? You should be able to express the first of these in terms of familiar objects.

17.23 (If this seems out of place, it's used in the next problems.) The unit square, $0 < x < 1$ and $0 < y < 1$, has area $\int dx\,dy = 1$ over the limits of x and y. Now change the variables to

$$u = \tfrac{1}{2}(x+y) \qquad \text{and} \qquad v = x - y$$

and evaluate the integral, $\int du\,dv$ over the square, showing that you get the same answer. You have only to work out all the limits. Draw a picture. This is a special example of how to change multiple variables of integration. The single variable integral generalizes

$$\text{from} \qquad \int f(x)\,dx = \int f(x)\,\frac{dx}{du}\,du \qquad \text{to} \qquad \int f(x,y)\,dx\,dy = \int f(x,y)\frac{\partial(x,\,y)}{\partial(u,\,v)}\,du\,dv$$

where

$$\frac{\partial(x,\,y)}{\partial(u,\,v)} = \det\begin{pmatrix} \partial x/\partial u & \partial x/\partial v \\ \partial y/\partial u & \partial y/\partial v \end{pmatrix}$$

For the given change from x, y to u, v show that this Jacobian determinant is one. A discussion of the Jacobian appears in many advanced calculus texts.

17.24 Problem 17.1 asked for the mean and variance for a Gaussian, $f(g) = Ae^{-B(g-g_0)^2}$. Interpreting this as a distribution of grades in a class, what is the resulting distribution of the average of any two students? That is, given this function for all students, what is the resulting distribution of $(g_1 + g_2)/2$? What is the mean of this and what is the root-mean-square deviation from the mean? How do these compare to the original distribution? To do this, note that $f(g)dg$ is the fraction of students in the interval g to $g + dg$, so $f(g_1)f(g_2)dg_1\,dg_2$ is the fraction for both. Now make the change of variables

$$x = \frac{1}{2}(g_1 + g_2) \qquad \text{and} \qquad y = g_1 - g_2$$

then the fraction of these coordinates between x and $x + dx$ and y and $y + dy$ is

$$f(g_1)f(g_2)dx\,dy = f(x + y/2)f(x - y/2)dx\,dy$$

Note where the result of the preceding problem is used here. For fixed x, integrate over all y in order to give you the fraction between x and $x + dx$. That is the distribution function for $(g_1 + g_2)/2$. [Complete the square.] Ans: Another Gaussian with the same mean and with rms deviation from the mean decreased by a factor $\sqrt{2}$.

17.25 Same problem as the preceding one, but the initial function is

$$f(g) = \frac{a/\pi}{a^2 + g^2} \qquad (-\infty < g < \infty)$$

In this case however, you don't have to evaluate the mean and the rms deviation. Show why not. [Residue Theorem.] Ans: The result reproduces the original $f(g)$ exactly, with no change in the spread. These two problems illustrate examples of "stable distributions," for which the distribution of the average of two variables has the same form as the original distribution, changing at most the widths. There are an infinite number of other stable distributions, but there are precisely three that have simple and explicit forms. These examples show two of them.

17.26 Same problem as the preceding two, but the initial function is

(a) $f(g) = 1/g_{\text{max}}$ for $0 < g < g_{\text{max}}$ (b) $f(g) = \frac{1}{2}\big[\delta(g - g_1) + \delta(g - g_2)\big]$

17.27 In the same way as defined in Eq. (17.10), what is the functional derivative of Eq. (17.5)?

17.28 Rederive Eq. (17.27) by choosing an explicit delta sequence, $\delta_n(x)$.

17.29 Verify the result in Eq. (17.36).

Index